Functional Carbon Materials

Online at: https://doi.org/10.1088/978-0-7503-4972-7

Functional Carbon Materials

Edited by
Jianmin Ma
School of Chemistry, Tiangong University, Tianjin, China

Jiantie Xu
School of Environment and Energy, South China University of Technology, Guangzhou 510640, China

IOP Publishing, Bristol, UK

ISBN   978-0-7503-4972-7 (ebook)
ISBN   978-0-7503-4970-3 (print)
ISBN   978-0-7503-4973-4 (myPrint)
ISBN   978-0-7503-4971-0 (mobi)

DOI   10.1088/978-0-7503-4972-7

Version: 20221201

IOP ebooks

British Library Cataloguing-in-Publication Data: A catalogue record for this book is available from the British Library.

Published by IOP Publishing, wholly owned by The Institute of Physics, London

IOP Publishing, No.2 The Distillery, Glassfields, Avon Street, Bristol, BS2 0GR, UK

US Office: IOP Publishing, Inc., 190 North Independence Mall West, Suite 601, Philadelphia, PA 19106, USA

Cover image: Computer illustration showing the hexagonal carbon structure of a nanotube, or buckytube. Image credit: Kateryna Kon / Science Photo Library.

Contents

6 Graphite and its main applications   **6-1**

Haiying Lu, Xianghong Chen, Jiakui Zhang, Yu Lei, Wenlu Min and Jiantie Xu

7 The structures, synthesis, properties, and applications of diamond   7-1

Sheng-Yi Xie and Fuyang Liu

8 Activated carbon   8-1

Daxiong Wu, Wen Ma, Qinghe Yu and Jianmin Ma

Preface

Carbon is widely used in many fields, such as electronics, photonics, electro-chemistry, and semiconductors; it is one of the most important materials, since it can exhibit many interesting properties. It can used in electronic devices, energy storage, catalysts, medicine, and environmental applications. In particular, it has many different structures and can thus be found in the form of different carbon materials such as fullerene, carbon fibers and nanofibers, carbon nanotubes, graphene, graphite, graphdiyne, diamonds, activated carbon, and carbon aerogels. Here, it is worth pointing out that the discovery of graphdiyne by Professor Yuliang Li in 2010 has also accelerated the development of carbon.

To help more researchers to learn about carbon materials, Professor Jiantie Xu and I decided to edit this book on functional carbon materials together with some carbon scientists. The main authors are Professor Xiangyue Meng, Yurui Xue, and Yuliang Li (University of the Chinese Academy of Sciences), Professor Weidong Zhou and Chuangang Hu (Beijing University of Chemical Technology), Professor Jianmin Ma (Tiangong University), Professor Xinhua Liu (Beihang University), Professor Chenxin Ran (Northwestern Polytechnical University), Professor Yonghua Chen (Nanjing Tech University), Professor Jiantie Xu and Xinwen Peng (South China University of Technology), Professor Sheng-Yi Xie (Hunan University), Fuyang Liu (Center for High Pressure Science and Technology Advanced Research). This book includes the wide range of contents, i.e. fullerene, carbon fibers and nanofibers, carbon nanotubes, graphene, graphite, graphdiyne, diamonds, activated carbon, carbon aerogels. Although there are many topics to be discussed, as mentioned above, we were not able to include all this content in one book. We hope that we can give you more about carbon materials in the future.

This book will be a useful handbook on carbon for students and researchers who are interested in materials. Finally, we hope that this book will help them gain useful knowledge about the various kinds of carbon and their applications.

Editor biography

Jianmin Ma

 **Jianmin Ma** is Professor at Tiangong University. He received his BS degree in Chemistry from the Shanxi Normal University in 2003 and PhD degree in Materials Physics and Chemistry from Nankai University in 2011. During 2011–2015, he also conducted the research in several overseas universities as a postdoctoral research associate. He serves as the Academic Editor for *Rare Metals*, the Associate Editor for *Chinese Chemical Letters*, and editorial board member for *Journal of Energy Chemistry, Nano-Micro Letters, Journal of Physics: Condensed Matter, JPhys Energy*, and others. His research interest focuses on the energy storage devices and components including metal anodes and electrolytes, and theoretical calculations from Density Functional Theory and Molecular Dynamics to Finite Element Analysis.

Jiantie Xu

 Prof. Xu received his PhD degree in Engineering from the University of Wollongong, Australia. During his PhD study, he moved to Prof. Liming Dai's group at Case Western Reserve University for a half-year's visiting. He completed postdoctoral studies in Prof. Liming Dai's group and his PhD supervisors Profs. Shixue Dou & Huakun Liu's group. Since September 2017, he started as a group leader and professor at South China University of Technology. His research interests mainly focus on advanced materials for various types of energy storage/conversion devices and recycling of spent lithium ion batteries. Prof. Xu has published 70+ papers with a total citation of >7000 times and a H-index of 39.

List of contributors

Xianghong Chen
National Engineering Laboratory for VOCs Pollution Control Technology and Equipment, Guangdong Provincial Key Laboratory of Solid Wastes Pollution Control and Recycling, School of Environment and Energy, South China University of Technology, Guangzhou, China

Yonghua Chen
Key Laboratory of Flexible Electronics (KLOFE) and Institution of Advanced Materials (IAM), Nanjing Tech University (NanjingTech), Nanjing, Jiangsu, China

Yuling Chen
State Key Laboratory of Pulp and Paper Engineering, South China University of Technology, Guangzhou, China

Weiyin Gao
Frontiers Science Center for Flexible Electronics, Xi'an Institute of Flexible Electronics (IFE) and Xi'an Institute of Biomedical Materials and Engineering, Northwestern Polytechnical University, Xi'an, China

Chuangang Hu
State Key Laboratory of Organic–Inorganic Composites, Center for Soft Matter Science and Engineering, College of Chemical Engineering, Beijing University of Chemical Technology, Beijing, China

Yu Lei
National Engineering Laboratory for VOCs Pollution Control Technology and Equipment, Guangdong Provincial Key Laboratory of Solid Wastes Pollution Control and Recycling, School of Environment and Energy, South China University of Technology, Guangzhou, China

Yuliang Li
Institute of Chemistry, Chinese Academy of Sciences, Beijing, China

Fuyang Liu
Center for High Pressure Science and Technology Advanced Research, Beijing, China

Xinhua Liu
School of Transportation Science and Engineering, Beihang University, Beijing, China

Yongde Long
State Key Laboratory of Organic–Inorganic Composites, Center for Soft Matter Science and Engineering, College of Chemical Engineering, Beijing University of Chemical Technology, Beijing, China

Haiying Lu
National Engineering Laboratory for VOCs Pollution Control Technology and Equipment, Guangdong Provincial Key Laboratory of Solid Wastes Pollution Control and Recycling, School of Environment and Energy, South China University of Technology, Guangzhou, China

Mingsheng Lv
School of Optoelectronics, University of Chinese Academy of Sciences, Beijing, China

Jianmin Ma
School of Chemistry, Tiangong University, Tianjin, China

Wen Ma
Company: Youke Publishing Co., Ltd., Beijing, China

Xiangyue Meng
School of Optoelectronics, University of Chinese Academy of Sciences, Beijing, China

Wenlu Min
National Engineering Laboratory for VOCs Pollution Control Technology and Equipment, Guangdong Provincial Key Laboratory of Solid Wastes Pollution Control and Recycling, School of Environment and Energy, South China University of Technology, Guangzhou, China

Xinwen Peng
State Key Laboratory of Pulp and Paper Engineering, South China University of Technology, Guangzhou, China

Chenxin Ran
Frontiers Science Center for Flexible Electronics, Xi'an Institute of Flexible Electronics (IFE) and Xi'an Institute of Biomedical Materials and Engineering, Northwestern Polytechnical University, Xi'an, China

Daxiong Wu
School of Physics and Electronics, Hunan University, Changsha, China

Sheng-Yi Xie
School of Physics and Electronics, Hunan University, Changsha, China

Jiantie Xu
National Engineering Laboratory for VOCs Pollution Control Technology and Equipment, Guangdong Provincial Key Laboratory of Solid Wastes Pollution Control and Recycling, School of Environment and Energy, South China University of Technology, Guangzhou, China

Yurui Xue
Institute of Chemistry, Chinese Academy of Sciences, Beijing, China

Shichun Yang
School of Transportation Science and Engineering, Beihang University, Beijing, China

Wang Yang
State Key Laboratory of Pulp and Paper Engineering, South China University of Technology, Guangzhou, China

Wu Yang
State Key Laboratory of Pulp and Paper Engineering, South China University of Technology, Guangzhou, China

Fenghui Ye
State Key Laboratory of Organic–Inorganic Composites, Center for Soft Matter Science and Engineering, College of Chemical Engineering, Beijing University of Chemical Technology, Beijing, China

Qinghe Yu
GRIMAT Engineering Institute Co., Ltd., Beijing, China

Jiakui Zhang
National Engineering Laboratory for VOCs Pollution Control Technology and Equipment, Guangdong Provincial Key Laboratory of Solid Wastes Pollution Control and Recycling, School of Environment and Energy, South China University of Technology, Guangzhou, China

Weidong Zhou
Beijing University of Chemical Technology, State Key Laboratory of Organic–Inorganic Composites, Beijing, China

Zhajun Zhang
School of Optoelectronics, University of Chinese Academy of Sciences, Beijing, China

IOP Publishing

Functional Carbon Materials

Jianmin Ma and Jiantie Xu

Chapter 1

Fullerenes

Mingsheng Lv, Zhajun Zhang and Xiangyue Meng

A fullerene is a kind of all-carbon molecule that has a cage structure composed of pentagons and hexagons. Due to their unique physical and chemical properties, fullerenes have opened a new era of carbon research since 1985, when they were discovered, directly promoting the beginning of nanoscience research. This chapter provides a general introduction to fullerenes and their derivatives. First, the history of the discovery and the structural characteristics of fullerenes are reviewed. The synthesis and isolation of fullerenes are then summarized. Finally, the physical and chemical properties of fullerenes and their derivatives are discussed, together with their applications.

1.1 Discovery of fullerenes

Carbon is a non-metallic element denoted by the symbol C; it is in the IVA group of the second period in the periodic table. Carbon is the fifteenth most abundant element in the Earth's crust, and the fourth most abundant element in the Universe by mass after hydrogen, helium, and oxygen. Carbon's abundance, its unique diversity of organic compounds, and its unusual ability to form polymers at the temperatures commonly encountered on Earth enable this element to serve as a common element in all known life. A series of carbon compounds forms the foundation of life. Carbon accounts for about 18.5% of the mass in the human body, and is the second most abundant element, being second only to oxygen. Carbon was one of the first elements to be discovered by humans and is one of the most widely used elements, not only as old but also new materials. Currently, scientists are working to develop a series of new carbon materials, such as fullerenes, graphene and carbon nanotubes, etc. Progress in the development of new carbon materials will continue to be recorded as a chapter in the overall story of materials science [1].

Fullerenes are new types of carbon material. They are all-carbon molecules that have cage structures composed of pentagons and hexagons, among which [60]fullerene is a molecule with a perfectly symmetrical structure like that of a football (figures 1.1(a) and (b)). The discovery of fullerenes greatly expanded the number of allotropes of carbon; previously, the only allotropes of carbon were diamond, graphite, and amorphous carbon

doi:10.1088/978-0-7503-4972-7ch1

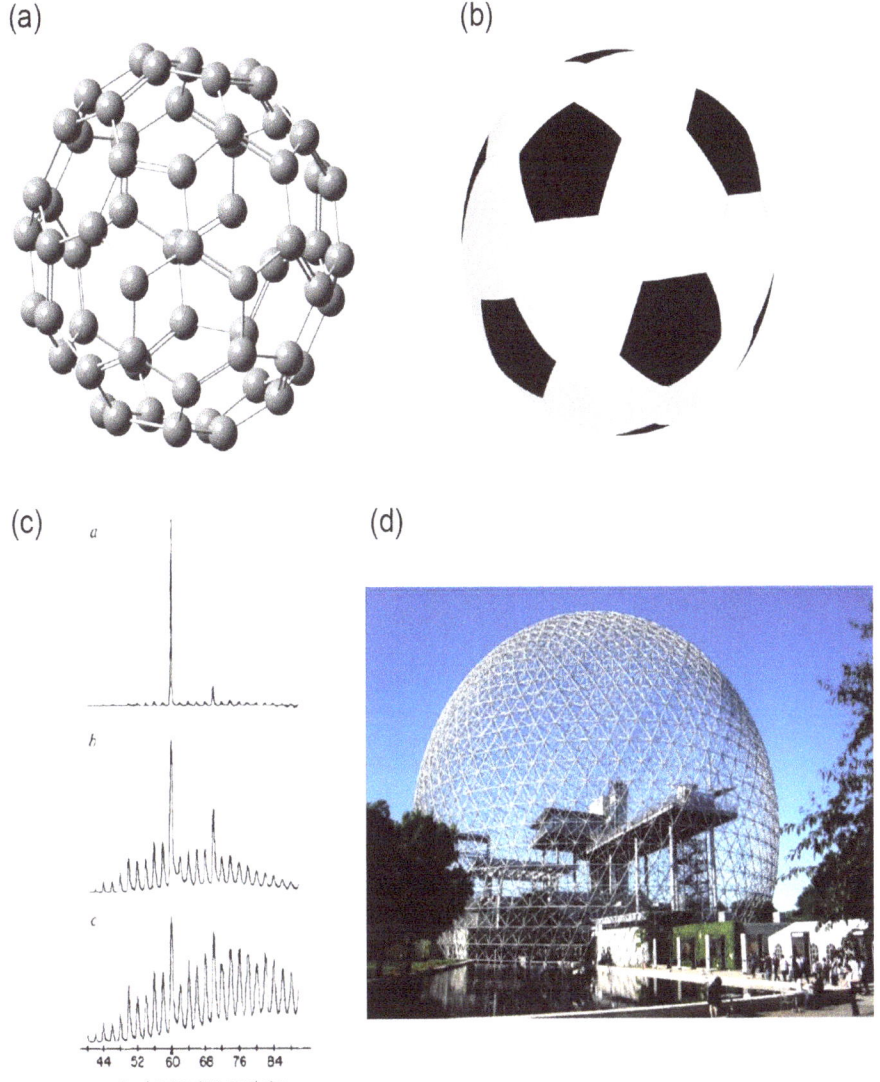

Figure 1.1. (a) Structure of [60]fullerene. (b) A football. The structure of [60]fullerene is similar to that of a football. (c) Time-of-flight mass spectra of fullerenes prepared by laser vaporization of graphite and cooled in a supersonic beam. The spectra in parts *c* to *a* differ in the extent of helium collisions occurring in the supersonic nozzle (from small to large). The spectrum in *c* proved the existence of C_{60} [4], reproduced with permission © Nature. (d) The US pavilion at the Montreal Expo.

(such as carbon black and coal). Fullerenes have aroused strong interest in researchers as soon as they were discovered due to their unique physical and chemical properties, which have great potential in materials science, electronics, and nanotechnology.

As early as the 1970s, the Japanese theoretical chemist Eiji Osawa predicted the existence of the football-shaped C_{60} molecule [2, 3]. In 1985, Harold Kroto, Richard E. Smalley, Robert Curl, *et al* reported an new all-carbon molecule in the journal

Figure 1.2. Three winners of the 1996 Nobel Prize in Chemistry: Harold Kroto, Robert Curl, and Richard E Smalley (from left to right).

Figure 1.3. Discovery of [60]fullerene in interstellar dust [6].

Nature [4], which marked the unveiling of the mystery of fullerene. They were studying the formation of long-chain carbon in interstellar carbon dust, and found the signal of [60]fullerene in the mass spectrum produced by the laser vaporization of graphite (figure 1.1(c)); they were the first to propose its closed-cage molecular structure. Because this molecular structure is very similar to the architectural work of the architect Buckminster Fuller, it was named buckminsterfullerene in his honor. The discovery of fullerene ushered in a new era of elemental carbon research and directly promoted the beginning of nanoscience research because of the unique electronic and geometric structures and special physical and chemical properties of fullerene. In 1990, Huffman, Krätschmer, and Fostiropoulos first reported a method for the large-scale synthesis of C_{60} [5], which led to an explosion of research into fullerene. In 1991, the journal *Science* selected it as the 'star molecule' of the year. On December 10, 1996, in recognition of the discovery of buckminsterfullerene and other fullerenes, Harold Kroto, Richard E. Smalley, and Robert Curl were awarded the Nobel Prize in Chemistry (figure 1.2).

As mentioned above, fullerene was originally accidentally explored as part of research into, and exploration of, interstellar dust (figure 1.3). In fact, geologists and astronomers also have found traces of fullerenes in nature. This has not only promoted the further understanding of the formation of fullerenes, but also helped us to understand the origin of life, the cosmos, geology, etc. In 1992, American scientist R P Buseck *et al* first reported the existence of [60]fullerene and [70]fullerene in a Cambrian rock near St. Petersburg, Russia [7]. Since then, there have been some reports of the existence of fullerenes in nature [8, 9]. In 2010, scientists discovered the infrared signal of fullerenes in interstellar dust clouds around 6500 light-years away with the help of the Spitzer Space Telescope [10]. The astronomer Letizia Stanghellini of the National Optical Astronomy Observatory in Tucson, Arizona said 'It is possible that buckyballs from outer space provided seeds for life on Earth.' [6]. Fullerenes have existed for billions of years, much longer than human history, and may be ubiquitous in the vast Universe. Maybe they are treasures given to us by nature, but we have not yet fully explored their charms.

1.2 Structural characteristics of fullerenes

Fullerenes are closed-cage-like molecules made of sp^2 hybrid carbon atoms. Each carbon atom in the fullerene carbon cage forms three carbon–carbon covalent bonds with the other three adjacent carbon atoms. This unique bonding method makes the carbon skeleton of fullerene form a series of polyhedrons (carbocyclic rings); each carbon atom is an vertex of the polyhedron, and all the carbon–carbon covalent bonds are edges of the polyhedron. Mathematically, a simple polyhedron satisfies Euler's polyhedron formula, which stipulates that the number of vertexes (V), the number of edges (E), and the number of faces (F) satisfy $V+F-E=2$. For fullerenes, V is equal to the number of carbon atoms (n) that make up the carbon cage. Each carbon atom forms three carbon–carbon bonds with three adjacent carbon atoms, E is equal to $3n/2$, and since the number of edges (E) of a polyhedron must be an integer, it can be determined that n must be an even number. Therefore, fullerenes must be carbon cluster molecules that have an even number of carbon atoms. According to Euler's polyhedron formula, the number of faces (the number of carbocyclic rings of the fullerene) is $F = n/2 + 2$. This equation determines the relationship between the number of carbon atoms in the carbon cage of fullerene and the carbocyclic ring, which is satisfied by the entire fullerene family.

1.2.1 Isolated pentagon rule

Generally, the carbocyclic ring of the fullerene carbon cage is usually pentagonal or hexagonal, but following further research, fullerenes containing tetragons and heptagons have also been found. Thus, the fullerenes containing only pentagons and hexagons are known as classic fullerenes, and the fullerenes containing tetragons or heptagons are known as non-classical fullerenes. For a classic fullerene, assuming that it contains p pentagons and h hexagons, the number of carbocyclic rings is given by $F = h + P$; similarly, the number of carbon atoms is given by $n=(5p + 6h)/3$. All classic fullerenes are composed of 12 pentagons, that is, $p = 12$ and $h = n/2 - 10$. Because h is an integer greater than or equal to zero, which means that the number of carbon atoms in fullerenes must not be less than 20, the smallest fullerene is C_{20}. However, when the number of carbon atoms is 22, it is impossible to construct a fullerene from 12 pentagons and one hexagon. Therefore, we can conclude that the number of carbon atoms in a fullerene needs to satisfy $n \geqslant 20$, $n \neq 22$.

Due to the different arrangements of pentagons and hexagons, fullerenes may have a large number of isomers. In particular, the arrangement of pentagons has a huge impact on the physical and chemical properties of fullerenes. In 1987, Kroto proposed the 'isolated pentagon rule (IPR)': he considered that in a stable fullerene structure, all pentagons are separated by hexagons; a structure containing adjacent pentagons would cause a big bending tension. If a fullerene's spherical surface contains two or more adjacent pentagons, its carbon cage is unstable due to a large amount of tension [11]. This rule is widely accepted in the field of fullerene research because it is consistent with experimental facts; almost all unmodified fullerene molecules that have been synthesized have strictly followed this rule. Therefore, whether the pentagons are adjacent can be used as the basis for classifying fullerenes, and fullerenes can be divided into IPR fullerenes and non-IPR fullerenes. As the number of atoms increases, the number of fullerene isomers also increases exponentially. For fullerenes with a specific carbon number, the number of non-IPR isomers is much greater than the number of IPR isomers. When $n < 60$ or $60 < n < 70$, all fullerenes inevitably contain adjacent pentagons and therefore do not meet the IPR rule. For [60]fullerene and [70]fullerene, only I_h-C_{60} and D_{5h}-C_{70} meet the IPR rule, which theoretically have 1811 and 8148 isomers containing adjacent pentagons, respectively. When $n > 70$, the corresponding number of isomers that meet the IPR rule greatly increases, but theoretically, the number of isomers that do not meet the IPR rule still has an absolute advantage. Regarding the change in the number of carbon atoms, fullerenes with $n < 60$ are called small fullerenes, and those with $n > 70$ are called higher fullerenes. Fullerenes have a cage-like structure, so the cavity inside the carbon cage can embed specific atoms or clusters of atoms. When fullerenes are embedded with atoms or clusters of atoms, such fullerenes are called endohedral fullerenes. Correspondingly, fullerenes that do not embed endohedral atoms or clusters are called hollow fullerenes. In addition, if one or more carbon atoms in a fullerene are replaced by heteroatoms (such as B, N, P, O, S, Se, Te, Si, or Ge), the resulting material is called a heterofullerene.

1.2.2 Hollow fullerenes

The different bonding modes of pentagons and hexagons in fullerenes with the same number of carbon atoms lead to isomers, and the number of isomers increases sharply with an increase in the number of carbon atoms. In order to distinguish these isomers, a set of effective naming methods is particularly helpful. Fowler and Manolopoulos developed a widely accepted naming method, the spiral algorithm, in 1995, which provides a unique way to number of fullerene isomers [12]. Regarding the spiral algorithm, we refer the reader to An Atlas of Fullerenes written by Fowler and Manolopoulos for details [12]; we will only briefly introduce it here. As shown in figure 1.4, the connection relationship of fullerenes can be shown by a Schlegel diagram [12]; each straight line represents a carbon–carbon bond, which can clearly show pentagons and hexagons. The face of fullerene can be spirally unwound into pentagons and hexagons, and satisfies the condition that there is still an exit when the spiral passes through the last face. Specifically, the first face in the spiral can be any face of fullerene, and the spiral can pass through any edge of the first face; the second face can be any face that shares a common edge with the first face; similarly, the third face can be any face that shares a common edge with the first and second

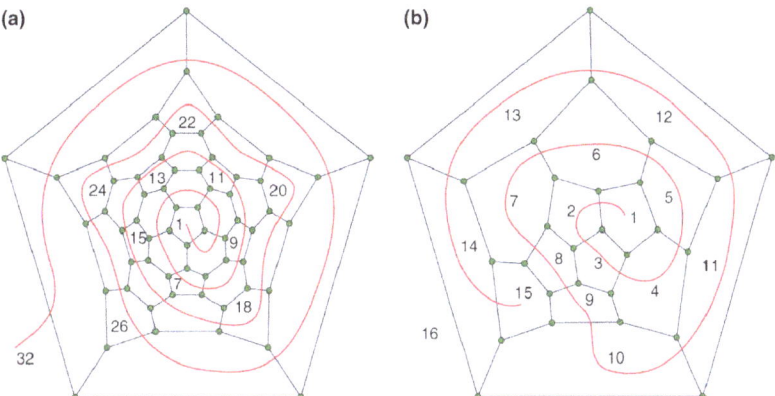

Figure 1.4. (a) Canonical spiral of Ih-C_{60}; (b) $D_2(1)$-C_{28} with a failing face spiral [14], reproduced with permission © Wiley.

faces. Once the first three faces have been selected, the order of appearance of all subsequent faces is determined according to the direction of the spiral. Therefore, there are $6N$ ($12\times5\times2 + (n/2 - 10) \times 6\times2=6n$) ways to untie the fullerene spirally. However, a fullerene has a certain symmetry, so a lot of ways to untie it are equivalent. It should be noted that spirals sometimes cannot be completed (figure 1.4(b)). When the spiral passes through each face round by round, a spiral sequence is obtained. As shown in figure 1.4(a), when the spiral passes through a pentagon, '5' is recorded in the sequence, similarly, when the spiral passes through a hexagon, '6' is recorded in the sequence. Thus, I_h-C_{60} in figure 1.4(a) can be represented by the sequence: 566666565656565666565656565666665. (In figure 1.4(a), the position of the pentagon is marked with a serial number.) There are multiple sequences for the same isomer, so it is necessary to choose a sequence to uniquely represent this isomer. Usually, the sequence with the smallest value corresponding to the sequence is selected, which is also called the canonical spiral sequence. The sequence 566666565656565666565656565666665 is the canonical sequence for I_h-C_{60}; interested readers can try it for themselves. Reversing the above process, we can construct isomers. All fullerenes have $n/2 + 2$ faces, of which 12 are pentagons and the remaining $n/2 - 10$ are hexagons. According to the available permutations and combinations, there are $(n/2 + 2)!/12!(n/2 - 10)!$ methods. We can try these methods in turn to see whether they conform to the spiral algorithm and keep the ones that are consistent with the rule. A large number of the sequences which meet the requirements of the spiral algorithm are equivalent. Therefore, we need to check them in all 6N ways to find the equivalent sequences and retain the canonical spiral sequences. This process is very complicated, but it can be easily solved with a computer program [12].

When naming the isomers of fullerenes C_n, we need to add the highest symmetry of the carbon cage and its isomer number in the spiral algorithm before C_n, and the isomer number should be marked with brackets after the symmetry, for example: C_s (39663)-C_{82}. In the spiral algorithm, the number value of the isomer meeting the IPR rule is generally the largest. Therefore, when numbering IPR fullerenes, we use the simplified version of the numbering system, which is that is only the IPR fullerenes are numbered in the order of the spiral algorithm, the number of the first IPR isomer is one, and if there is only one IPR fullerene, the number one is omitted, for example I_h-C_{60}, $I_h(7)$-C_{80} (which corresponds to $I_h(39712)$-C_{80} in normal spiral algorithm). For non-IPR fullerenes, the number of isomers must be completely numbered using the spiral algorithm. The advantage of this is that the IPR and non-IPR fullerenes can be directly distinguished from their names, because the spiral algorithm number now reported for non-IPR fullerene isomers is generally a relatively large number [13]. Therefore, for consistency, we also adopt the above rules, but the numbering is omitted in places where it does not affect the discussion.

Figure 1.5 shows some hollow fullerenes with less than 100 carbon atoms. Recently, more and more hollow fullerenes have been separated and characterized,

Figure 1.5. Some fullerene structures with n <100 carbon atoms [16], reproduced with permission © Institute of Physics.

(a)   (b)

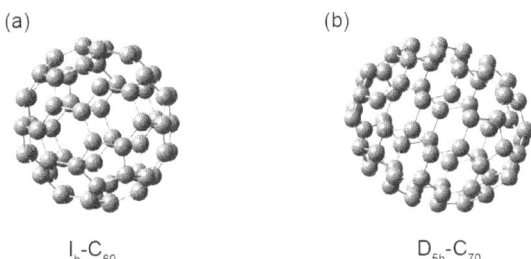

I_h-C_{60} D_{5h}-C_{70}

Figure 1.6. Structure of (a) I_h-C_{60} and (b) D_{5h}-C_{70}.

so that we cannot introduce all of them in this book. The two most abundant and stable fullerenes are [60]fullerene and [70]fullerene, so in the following, we introduce the structural characteristics of these two fullerenes.

The first fullerene discovered, [60]fullerene, is the most special and significant of all fullerenes and is the smallest fullerene that obeys the IPR rule. It is also the fullerene with the highest yield. Theoretically, it has 1812 possible structural isomers, of which I_h-C_{60} satisfies the IPR rule [11]. I_h-C_{60} was first proposed by Kroto, Curl, and Smalley [4], which marked the discovery of fullerenes and the beginning of carbon nanomaterial research. I_h-C_{60} consists of 12 pentagons and 20 hexagons; each pentagon is surrounded by five hexagons, and each hexagon is surrounded by three pentagons and three hexagons. As shown in figure 1.6(a), its structure is just same as the structure of a football in daily life, so it is also informally called a buckyball. I_h-C_{60} has I_h symmetry and is currently the most symmetrical fullerene synthesized; each carbon atom in its cage has the same chemical environment, so that there is only one signal peak on ^{13}C NMR, a chemical shift of 143 ppm. However, there are two types of C–C bond in I_h-C_{60}: the bond between two hexagons is called a [6,6] bond, and the bond between hexagons and pentagons is called a [5,6] bond. Although I_h-C_{60} has a large delocalized π_{60}^{60} bond, the [6,6] and [5,6] bonds exhibit different chemical properties. X-ray single-crystal diffraction of I_h-C_{60} shows that the length of the [6,6] bond is 1.355 Å and that of the [5,6] bond is 1.467 Å. Therefore, the [6,6] bond shows more C=C bond properties (the double bond length value of ethylene is 1.34 Å) [15], which is the reason why the addition reaction occurs easily on the [6,6] bond.

With an increase in the number of carbon atoms, the size of the fullerene carbon cage continues to increase. The abundance of [70]fullerene is second only to that of [60]fullerene, and it has high stability because of its unique carbon cage structure. As in the case of [60]fullerene, [70]fullerene also has an isomer that meets the requirements of the IPR rule, namely D_{5h}-C_{70}, which contains 12 pentagons and 25 hexagons (figure 1.6(b)). As is the case for I_h-C_{60}, all of its pentagons are surrounded by hexagons, which is also the reason for the relative stability of D_{5h}-C_{70}. From a structural point of view, D_{5h}-C_{70} can be regarded as an elongated form of I_h-C_{60} that has a belt of five hexagons inserted at its equator. As shown in figure 1.7, if all ten carbon atoms on the equator of D_{5h}-C_{70} are removed, two hemispherical fragments are obtained, which can be rotated by 36° and spliced together to obtain I_h-C_{60}. Due

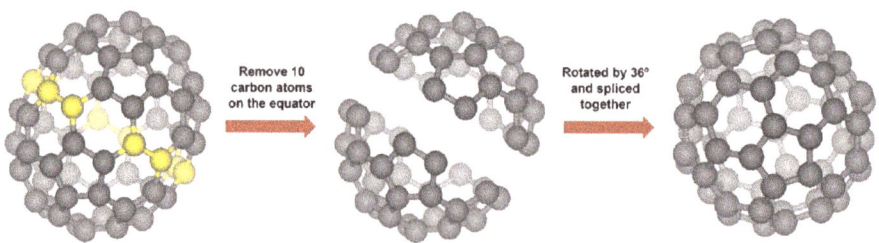

Figure 1.7. Removal of the carbon on the equator of D_{5h}-C_{70}, which can then be spliced into I_h-C_{60}.

to the extra ten carbon atoms, there are fused hexagonal structures on the equator of D_{5h}-C_{70} equator that have high aromaticity and low chemical reactivity. Therefore, the main chemical reactions of D_{5h}-C_{70} are mostly concentrated at both ends.

Other hollow fullerenes have been reported, such as C_{76}, C_{78}, C_{80}, C_{82}, C_{84}, C_{86}, C_{88}, C_{90}, C_{96}, etc [17–24]. Although mass spectrometry results show that there are hollow fullerenes with large carbon cages, it is difficult to separate pure fullerenes because the more carbon atoms are involved, the more isomers are possible. Therefore, only a few studies of large-cage hollow fullerenes have been performed.

1.2.3 Endohedral fullerenes

As mentioned above, fullerenes have a cage-like structure; therefore, it is possible to use them as a robust container for atoms or clusters of atoms. In 1985, shortly after the discovery of [60]fullerene, Hetah *et al* noticed the its hollow structure, which could be used as a nano container to hold atoms, molecules, or clusters of atoms. They performed a laser vaporization experiment with a low-density graphite disk which had been soaked in a boiling saturated aqueous solution of $LaCl_3$ and detected the signal of $C_{60}La$ by mass spectrometry [25]. In 1991, this group made further progress: they used a laser to vaporize the La_2O_3/graphite mixture, and obtained the macro-level endohedral fullerene $La@C_{82}$ for the first time [26]. Such fullerenes were known as 'endohedral,' and the same group was also the first to propose the use of the '@' symbol to represent this endohedral fullerene: the endohedral species are written first, followed by the symbol '@,' and then by the carbon cage [26]. This naming method has been used ever since. In 1993, Saunders and co-workers discovered the presence of noble gas (He and Ne) endohedral fullerenes generated by the arc-discharge method for the first time; they then used a high-temperature pressurization method to embed a series of noble gases (He, Ne, Ar, Kr, Xe) into fullerene carbon cages [27, 28]. The field has undergone dramatic development over the last 30 years, and a wide variety of endohedral fullerenes have been discovered and separated. According to whether the endohedral element is metallic, endohedral fullerenes can be roughly divided into two categories: non-metal endohedral fullerenes (nMEFs) and endohedral metallofullerenes (EMFs) [29]. Figure 1.8 shows the periodic table of the endohedral fullerenes [30].

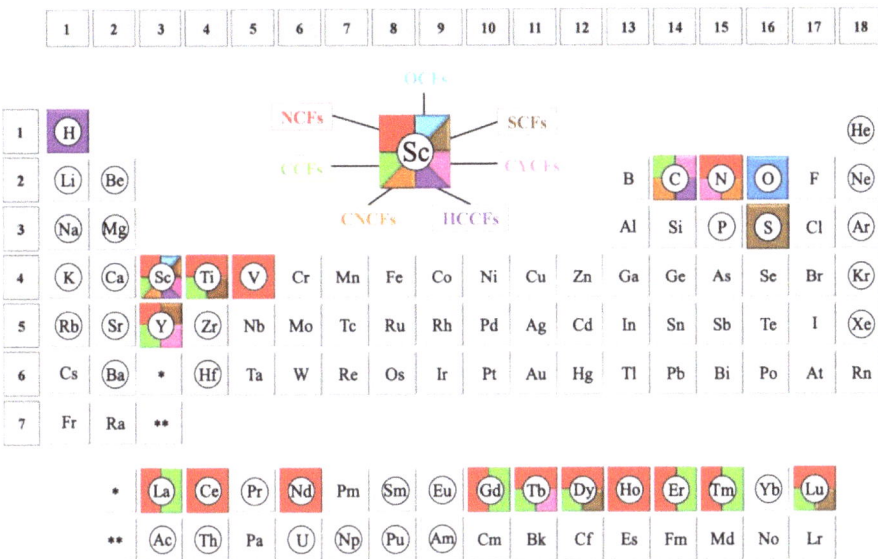

Figure 1.8. A modified periodic table in which all endohedral clusterfullerenes are highlighted and classified using different color codes. The circles indicate elements which are able to form endohedral fullerenes, including both conventional metallofullerenes and endohedral clusterfullerenes. See [30] for further details. Source: [30], reproduced with permission © Royal Society of Chemistry.

1.2.4 Heterofullerenes

In addition to hollow fullerenes and endohedral fullerenes, heterofullerenes represent the third type of fullerene, in which at least one carbon atom is replaced by non-carbon atoms such as B, N, P, O, S, Se, Te, Si, or Ge [31]. Doping with heteroatoms changes the electronic, optical, and chemical properties of fullerenes. Because the electronic configuration of heteroatoms is different from that of the carbon atom, heterofullerenes have broad application prospects in superconductors, optoelectronic devices, organic ferromagnets, etc. Nitrogen (N) and boron (B) atoms are similar in atomic radius to carbon atoms and are considered to be the most promising candidates for doping into the carbon cage of fullerenes. Consequently, the replacement of an odd number of C atoms by trivalent atoms such as nitrogen or boron leads to radicals (such as $C_{59}N\cdot$), which can be stabilized by dimerization (e.g. $(C_{59}N)_2$), whereas the replacement of an even number of C atoms directly results in closed-shell system [31]. In addition, azafullerenes can effectively reduce the influence of tension caused by adjacent pentagons and improve the stability of the fullerene carbon cage [32], which is of great significance for obtaining non-IPR fullerenes and for further exploring the structure of fullerenes. In fact, the research into heterofullerenes is mainly focused on azafullerenes. To date, azafullerenes containing one nitrogen atom have been prepared on a large scale, while research is still at the theoretical stage for B, Si, P, and other non-carbon atoms.

1.3 Synthesis of fullerenes

Since [60]fullerene was first reported in 1985 [4], fullerenes have received extensive attention from researchers in different fields due to their special structure and unique physical and chemical properties. However, before 1990, research into fullerenes was mainly focused on theory. To study the properties and reactivity of fullerenes, sizable quantities of fullerenes are required. Therefore, the synthesis of fullerenes is very significant for the study of fullerenes. In 1990, Huffman, Krätschmer, and Fostiropoulos first reported a method for the large-scale synthesis of [60]fullerene [5], which led to an explosion of research into [60]fullerene. Since then, after continuous exploration and research, scientists have discovered dozens of ways to synthesize fullerenes, such as the laser vaporization method, arc discharge vaporization of graphite, the chemical vapor deposition method, etc. In this section, we will introduce some of the synthesis methods used for fullerenes.

1.3.1 Laser vaporization method

In 1985, Kroto, Smalley, Curl, *et al* used pulsed laser beams to vaporize a graphite disk. The carbon vapor produced was rapidly cooled in a high-density helium atmosphere to form a series of carbon cluster products, and they observed the C_{60} signal in the mass spectrum for the first time [4] (figure 1.9). This method was also used for the first synthesis of metal fullerenes. A few days after C_{60} was first reported, Heath *et al* used a laser to vaporize a $LaCl_3$/graphite mixture and detected the signal of $La@C_{60}$ in its mass spectrum [25]. They made a simple revision to the laser vaporization method originally used in 1985: they performed the synthesis in an oven at 1200 °C, which improved the yield of C_{60} or endohedral metallofullerenes ($La@C_{82}$) [26]. However, it is difficult to synthesise a large amount of product using this method, so it is unsuitable for the large-scale production of fullerenes. In addition, expensive equipment and complex operation also limit the application of laser vaporization. However, this method is one of the effective ways to synthesize single-walled carbon nanotubes (SWCNTs) [33, 34].

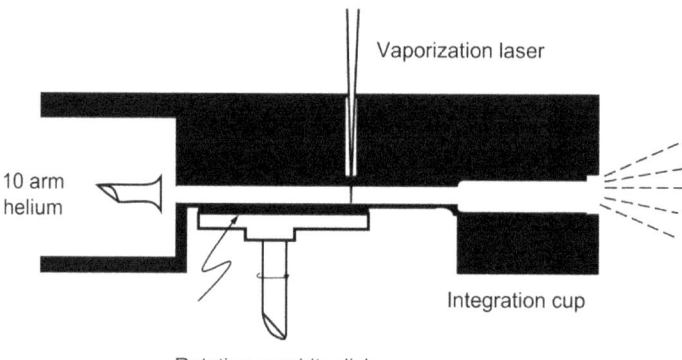

Figure 1.9. Schematic of the device used to prepare fullerenes by laser vaporization [4], reproduced with permission © Wiley.

1.3.2 Arc evaporation method (the Krätschmer–Huffman method)

Arc evaporation (AE), also called arc discharge, is a technique used to synthesize fullerenes as well as SWCNTs and multi-walled carbon nanotubes [16]; it is the most commonly used synthesis method for fullerenes and endohedral metallofullerenes due to its advantages of simple equipment, high output, low cost, safety, and reliability. As mentioned above, Huffman *et al* first used this method for the bulk production of [60]fullerene in 1990 [5]. Specifically, a requisite for fullerene formation is carbon atoms in the gaseous state; in this process, pure graphite is the source of carbon atoms [5, 35, 36]. In this process, graphite electrodes are evaporated at ~100 Torr (13.3322 kPa) of helium to obtain pure graphitic carbon soot with a few percent of fullerenes (figure 1.10(a)). The resulting black soot is gently scraped from the collecting surface inside the evaporation chamber and used for the next stage of processing (to be introduced in the next section) [5]. The Krätschmer–Huffman method is capable of producing gram quantities of C_{60} per day [37].

Although the arc evaporation is the earliest method used for the mass production of fullerenes, because the graphite electrodes are consumed in the process of fullerene synthesis, the distance between the electrodes changes. This not only affects the electrical characteristics of the plasma but also the radiation level and the subsequent exchange between the plasma and its surroundings, causing the arc between the graphites to become unstable and affecting the synthesis of fullerenes [38, 39]. In 1990, Smalley and co-workers improved this method; their intent was to provide a cookbook-level recipe for the production of sufficient quantities for general chemical experiments that would be suitable for most laboratories [37]. A 6 mm outside diameter such that the bulk of the power was dissipated in the arc and not in ohmic heating of the rod. They called this a 'contact arc' and thought that the best operation occurred when the electrodes were just barely touching. Using this generator, it is possible to obtain ~10 g of black soot within a few hours, which greatly improves the yield of fullerenes [37]. Subsequently, researchers conducted a series of studies on the arc discharge method and found that the efficiency of

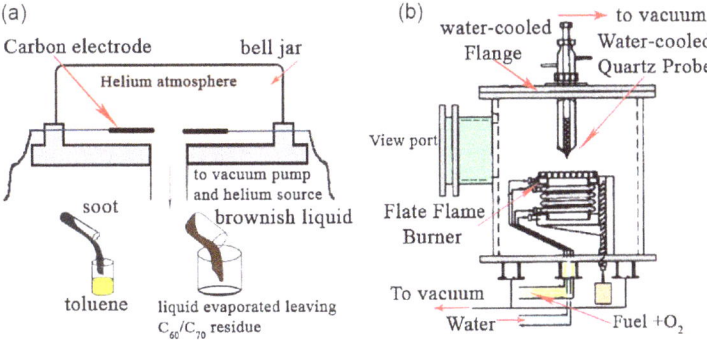

Figure 1.10. Schematic of fullerene synthesis methods: (a) the arc evaporation method and (b) the combustion method [35], reproduced with permission © Elsevier.

fullerene preparation depends on many parameters, such as inert gas pressure and purity, current intensity, catalysts, etc [40–43].

Arc evaporation is one of the methods most commonly used for the synthesis of fullerenes, and most of the new fullerene structures are synthesized by this method. However, the product obtained by this method contains impurities such as amorphous carbon and graphite, which are difficult to separate [42]. In addition, this method consumes a lot of energy and is limited by the length of the graphite rod, so it is not suitable for large-scale industrial production. However, this method is the most popular method for small-scale production in the laboratory [42].

1.3.3 Combustion method

The production of a sooty flame by a hydrocarbon feedstock is the most efficient method for the preparation of empty-cage fullerenes [39]. In 1987, Homann *et al* found positive and negative ions of the types C_{2n} ($30 \leqslant 2n \leqslant 210$) and $C_{2n+1}H_x$ ($x \leqslant 2$) in hydrocarbon flames, and they detected their mass spectrum signals; some of the products had the same molecular weights as the C_{60} and C_{70} obtained by arc evaporation [44]. However, this was not enough to confirm that the all-carbon molecules they obtained were fullerenes. Howard and co-workers used premixed benzene and oxygen with argon as a diluent for incomplete combustion in a burner chamber. They extracted the soot from flames and separated it into its components using high-performance liquid chromatography (HPLC) to obtain fullerenes [45, 46]. This was the first time that flame synthesis was found to be an effective alternative method for the production of fullerenes. After further research, they found that in this method, flame conditions such as the C/O ration of the premixed gas, the pressure and temperature of the burner chamber, the types and concentration of the dilution gas, the residence time and gas velocity used in the combustion of benzene, and the flame temperature influence the fullerene amount and composition [35, 47–49]. Figure 1.10(b) shows the schematic of a combustion method for synthesizing fullerenes.

The combustion method has the advantages of being usable for long-term continuous preparation, simple operation, and low energy consumption, which provides new possibilities for the large-scale production of fullerenes. Thus, the combustion method has now become the mainstream method of industrialized fullerene production around the world. In 2001, Mitsubishi Corporation established the world's first fullerene production line at the ton level; they then declared that the annual output of fullerenes produced by the combustion method could reach thousands of tons [50]. The industrial production of fullerenes has laid a solid foundation for their practical application.

1.3.4 Radio-frequency plasma method

The preparation of fullerenes by the radio-frequency (RF) plasma method uses a rapidly alternating electromagnetic field to cause the rapid vibration or rotation of particles inside the carbon raw material. This causes the thermal effect of friction,

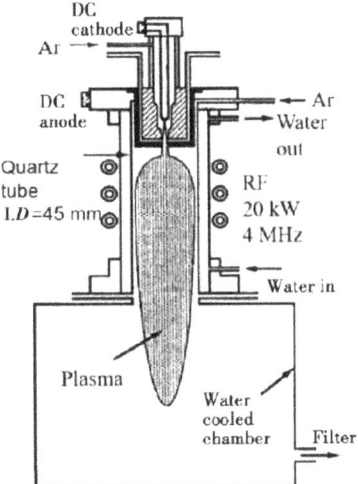

Figure 1.11. Schematic diagram of fullerene production based on a combination of RF plasma and DC [51], reproduced with permission © Academic International Press.

which decomposes the carbon raw material, producing small carbon molecules. These small carbon molecules then interact and combine to form fullerenes during the cooling process.

In 1992, Yoshie *et al* reported a novel method to fabricate fullerenes with a 7% yield by carbon particle evaporation in a hybrid plasma, which is characterized by the superposition of an RF plasma and a direct-current (DC) arc jet operated at atmospheric pressure (figure 1.11) [51]. Following further research and improvement, many instances of the use of this method to synthesize hollow fullerenes have been reported [52–54]. In addition, this method can also be used to synthesize endohedral metallofullerenes [55, 56].

As mentioned above, the RF plasma method can effectively synthesize hollow fullerenes and endohedral metallofullerenes and is a potential alternative to the arc discharge method widely used in laboratories. However, compared with the arc discharge method, the RF plasma method consumes an average of 20% more energy to product the same amount of fullerenes [57], and the device is more complicated. So, this method needs to be further improved.

1.3.5 Chemical synthesis

Chemical synthesis is a method for the directional synthesis of specific fullerenes through various chemical reactions. Its principle is to synthesize the precursors of fullerene using aromatic compounds as raw materials. The precursors undergo chemical bond breakage and reconstruction under certain conditions to form a specific fullerene molecule. Although the various methods introduced above can easily synthesize fullerenes, it is possible to synthesize fullerenes with a defined structure by the chemical synthesis of fullerenes, and this method has broad significance for studying the formation mechanism and modification of fullerenes.

In 2001, Scoot *et al* reported that a polycyclic aromatic hydrocarbon (PAH) $C_{60}H_{30}$ had been synthesized in nine steps by a conventional laboratory method. This $C_{60}H_{30}$ incorporates all 60 carbon atoms and 75 of the 90 carbon–carbon bonds required to form [60]fullerene; subjecting it to laser irradiation at 337 nm induces hydrogen loss and the formation of C_{60}, as detected by mass spectrometry. They found that a specifically labeled $[^{13}C_3]C_{60}H_{30}$ retains all three of the labeled atoms during the cage formation process and confirmed that the C_{60} is directly formed by molecular transformation from the $C_{60}H_{30}$ PAH—i.e. not by fragmentation and recombination in the gas phase [58]. However, the yield of fullerene synthesized by this method is very low, and the fullerene product cannot be separated. Therefore, they undertook further research: through molecular design, they changed three key positions of $C_{60}H_{30}$ PAH to chlorine substituents. The precursor molecule $C_{60}H_{27}Cl_3$ of C_{60} was synthesized through 11 steps of chemical reactions. Through flash vacuum pyrolysis (FVP) at 1100 °C, the $C_{60}H_{27}Cl_3$ PAH can form [60]fullerene [59]. The result shows that this method can synthesize isolable quantities of C_{60}, and no other fullerenes are formed as by-products, which is the first time that C_{60} has been obtained through total synthesis in the true sense [59]. This method should make the directed laboratory preparation of other fullerenes possible as well, including those not accessible by graphite vaporization (figure 1.12) [59]. In 2021, Nakamura *et al* used single-molecular atomic-resolution real-time transmission electron microscopy (SMART-TEM) technology to take a high-resolution transmission electron microscopic video of the electron-beam-induced bottom-up synthesis of C_{60} through the cyclodehydrogenation of a $C_{60}H_{30}$ PAH, which was deposited on graphene as a substrate [60]. During the reaction, $C_{60}H_{30}$ transformed into C_{60} via a total of 15 cyclodehydrogenation steps. Understanding these processes is not only of high value to the comprehension of fundamental chemistry but also is directly related to technological application, utilizing electron beam lithography for the construction of nanoarchitectures [60, 61].

The mainstream fullerene preparation methods have uncontrolled reactions and unpredictable products, while chemical synthesis methods can prepare fullerene molecules in a targeted manner and hopefully obtain the larger fullerenes (C_n, $n>100$) that are difficult to synthesize by other methods. This is of great significance for studying the formation mechanism of fullerenes and expanding the application

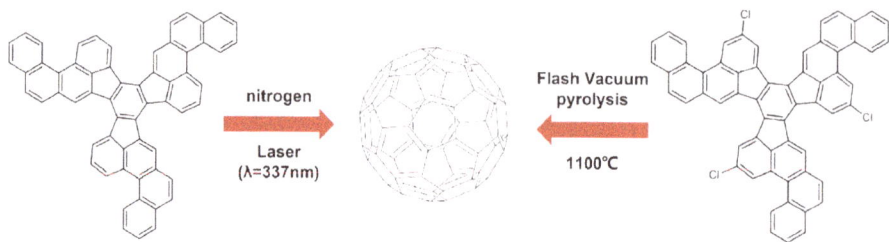

Figure 1.12. Synthesis of C_{60} through the polycyclic aromatic hydrocarbons $C_{60}H_{30}$ or $C_{60}H_{27}Cl_3$ [58, 59].

range of fullerenes. However, there are still many problems in the preparation of fullerenes by chemical synthesis, such as low yield and complicated steps, which need to be further studied.

In addition to the methods introduced above, the synthesis methods of fullerenes include solar graphite evaporation [62], chemical vapor deposition (CVD) [63], etc. However, these methods have problems such as low yield or high cost. The mass production of fullerenes provides new possibilities for subsequent research and applications, and also provides a new path for the commercialization of fullerenes. In summary, it is still very meaningful to continue to develop new synthesis methods for the low-cost, high-yield, and large-scale preparation of fullerenes.

1.4 Isolation of fullerenes

In the previous section, we introduced several commonly used methods of fullerene synthesis. Since the discovery of fullerenes, researchers have developed many preparation methods to increase the yield of fullerenes. No matter which method is used to obtain carbon soot, the main product is basically C_{60}, C_{70}, or other mixed products consisting of fullerenes, small molecular carbon clusters, and PAHs. The physical and chemical properties of these products are very similar, so the efficient separation and extraction of fullerenes are very critical problems in fullerene research. Therefore, in this section, we introduce the extraction and separation of fullerenes.

The purification of fullerenes involves a series of complex physical and chemical processes, which are mainly divided into two parts: extraction and separation. Extraction refers to the extraction of fullerenes from the carbon soot produced during the synthesis process. Separation refers to the separation of different fullerenes from each other, such as C_{60} and C_{70}. Many applications need to increase the purity of C_{60} or C_{70} fullerenes to more than 98%, so this requires multiple separation steps, which greatly increase the cost of fullerenes [42]. Therefore, when evaluating the extraction and separation of fullerenes, we need to take into account many factors, such as cost, efficiency, effects, and environmental friendliness. In the following, we will specifically introduce these extraction and separation techniques.

1.4.1 Extraction technology

At present, the technology used to extract fullerenes is relatively mature. The existing technology can already realize the low-cost and high-efficiency operation of the extraction process. The fullerenes produced in the extraction stage can be purchased in kilogram or ton amounts on a commercial basis. In simple terms, extraction consists of the selection of an appropriate solvent based on the difference in solubility of a fullerene and its by-products (such as amorphous carbon or graphite), which only dissolves the fullerene but not its by-products. Today, the methods used to extract fullerenes from carbon soot mainly include Soxhlet extraction (SE) and ultrasonic-assisted extraction (UAE) [64]. Thus, the choice of different solvents and methods is very significant for the high-efficiency extraction of fullerenes.

Fullerenes are different from the other components of carbon soot, which are slightly soluble in alkanes and extremely difficult to dissolve in common polar solvents, such as water, alcohol, and acetone. However, fullerenes have a certain aromaticity due to their unique electronic structure, so they have good solubility in benzene, toluene, carbon disulfide, etc. In 1993, Ruoff *et al* systematically studied the room-temperature solubility of pure C_{60} in 47 common organic solvents; the solubilities covered a wide range from 0.01 mg mL^{-1} in methanol to 50 mg mL^{-1} in 1-chloronaphthalene [65], which provided a very crucial basis for the subsequent selection of fullerene extraction solvents (also see section 1.5.1).

Soxhlet extraction (SE) is a method of extracting compounds from solid substances. In 1879, Franz von Soxhlet invented the Soxhlet extractor [66]. Generally, if the compound to be extracted has limited solubility in the solvent and the impurities are not soluble in the solvent, the SE method can be used. The SE method uses the principle of solvent reflux and siphoning, so that carbon soot can be extracted by hot pure solvent every time. At the same time, the fullerenes and their derivatives in the carbon soot are gradually dissolved and enriched in the flask by the solvent. The enriched extract is then concentrated by a rotary evaporator to prepare for the next step of separation. In 1991, Parker *et al* first used the SE method for the extraction of a fullerene [67]. They compared the extraction efficiency before and after using this method with benzene as the solvent; the results showed that the amount of fullerene obtained by the SE method was twice the amount of fullerene obtained by the previous method of high-temperature reflux extraction [67]. This shows that the SE method can effectively extract fullerenes.

However, due to the small size of the Soxhlet extractor, it is only suitable for the laboratory and not for large-scale extraction. In addition, the SE method requires a high temperature, which may cause active fullerenes to react or decompose during the extraction process; thus, this method is only suitable for the extraction of relatively stable fullerenes. Therefore, UAE is often used for the large-scale extraction of unstable fullerenes. The principle of UAE is to use high-speed, strong vibration, the aerobic effect, and the stirring effect generated by an ultrasonic wave to release fullerenes from a carbon soot mixture and then diffuse them into an organic extractant, which is a physical extraction method. This method does not change the structure of the material, and the extraction process is carried out under relatively stable conditions, which reduces the damage to the sample. In addition, the UAE method is suitable for large-scale extraction and has high efficiency. Therefore, we often use this method to extract the crude product of fullerenes in the laboratory [68]. The UAE method can extract a large amount of carbon soot in a short time, saving time and effort, and greatly improving the efficiency of fullerene extraction.

In addition to the above two methods, there are some other methods such as pressurized liquid extraction (PLE) [69, 70], mechanical extraction (ME) [71], etc. However, those methods have various problems.

1.4.2 Separation technology

Although fullerenes can be effectively extracted from carbon soot by different extraction technologies, these extracted fullerenes are still mixed liquids which contain fullerenes of different carbon cage sizes and different fullerene isomers. The physical and chemical properties of these fullerenes are very similar, so suitable separation techniques are needed to separate them. The separation of fullerenes requires a series of complex physical and chemical process, which makes it difficult to obtain high-purity C_{60} and C_{70}. Commonly used separation methods include chromatography, recrystallization, sublimation, selective complexation, etc. However, there is no particularly efficient, economical, and environmentally friendly method [72], which leads to the relatively high price of high-purity fullerenes. Generally speaking, high-purity fullerenes are usually supplied in quantities of grams or milligrams; the price of 99.9% pure C_{60} is generally around US$105 per gram, while the price of C_{70} is even higher: 99% pure C_{70} generally costs US$310 per gram or even more. Therefore, high-efficiency and low-cost separation technology is critical to the development and application of fullerenes. We introduce several separation methods in this section.

1.4.2.1 Chromatography

Chromatography is a common separation and analysis method that has a very wide range of applications in analytical chemistry, organic chemistry, and biochemistry and is one of the most effective ways to separate the fullerene mixture extracted from carbon soot. Chromatography can be divided into column chromatography and high-performance liquid chromatography (HPLC), among which HPLC can effectively separate fullerenes without being restricted by sample volatilization and sample thermal stability when separating fullerenes. The principle of chromatography is that the different fullerenes in the mixture have different affinities for the stationary phase, i.e. the different fullerenes stay longer or shorter in the stationary phase depending on their interactions with surface sites. Thus, they travel at different apparent velocities in a mobile fluid, causing them to separate.

For column chromatography, the most important factor affecting separation efficiency is the stationary phase. Depending on the stationary phase, column chromatography is categorised into silica gel column, graphite column, neutral alumina column, activated charcoal flash chromatography column, etc. After the macro-preparation of fullerenes became possible, Taylor *et al* first used column chromatography to separate fullerenes in 1990. They used neutral alumina as the stationary phase and n-hexane as the eluent to successfully separate milligrams of C_{60} and C_{70} [73]. However, due to the low solubility of fullerenes in n-hexane, this process was extremely solvent-consuming, and the operation process was complicated, which limited its further application. In 1992, Chatterjee *et al* reported the fast one-step separation and purification of fullerenes; their idea was to adapt the solvent recycling mechanism of a Soxhlet extraction apparatus to a chromatographic column. Although the amount of solvent used was reduced, the amount of purified fullerenes obtained was still low [74]. Silica gel was also used as a stationary phase, but no matter what mobile phase was used, its retention of fullerenes was very weak, and its selectivity was poor [75]. Scrivens *et al* proposed the use of activated carbon as the stationary phase in 1992. They used the most common Norit A activated carbon, however, they encountered irreversible adsorption of fullerenes in activated carbon. Even if the activated carbon and silica gel were packed at a certain ratio, a large amount of the fullerene still remained in the stationary phase and could not be eluted [76]. With further research, more and more simple and inexpensive chromatographic methods have been developed [77–79].

HPLC is one of the most effective methods for the separation of fullerenes. Currently commonly used columns are the C_{18} column, the Buckyprep column, the Buckyprep-M column, etc. The C_{18} column is a commonly used reverse-phase chromatographic column, which uses polymeric octadecylsilica bonded phases as the packing. It is widely used in the separation of fullerenes [80]. The Buckyprep column is a charge-transfer column that uses pyrenylpropyl-bonded silica as the stationary phase, which is specially designed for HPLC fullerene separation. It uses toluene as the mobile phase and its injection volume is 35 times that of a standard C_{18} column, which has become a standard column for the separation of fullerenes. Buckyprep-M is specially designed for the separation of metallofullerenes; it uses phenothiazinyl-bonded silica as the stationary phase. Compared to its retention of fullerenes, it has strong retention of metallofullerenes. In addition to the above-mentioned chromatographic columns, other options are the pentabromobenzyl (PBB) column, the 2-(1-pyrenyl)ethyl (PYE) column, the nitrophenylethyl (NPE) column, etc (figure 1.13). Interested readers can find the relevant information by themselves.

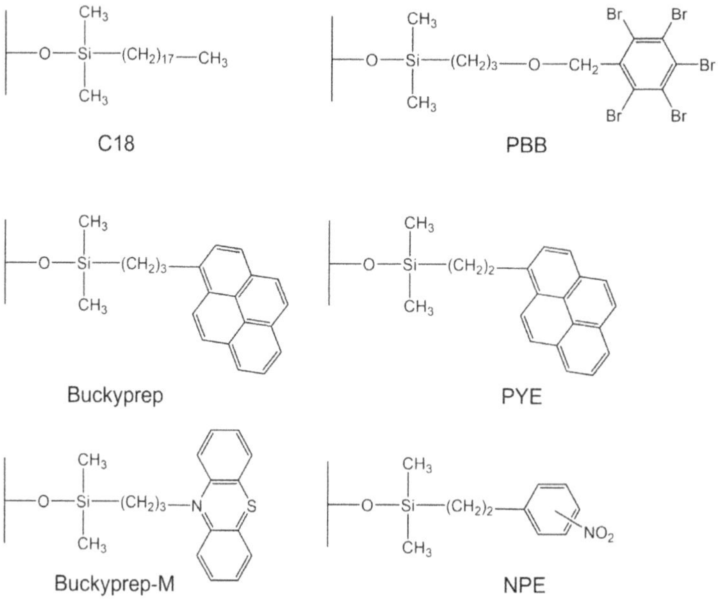

Figure 1.13. Molecular structure of the stationary phases of commonly used chromatographic columns.

At present, HPLC is the most effective separation method for the separation of fullerenes, and it is currently the only method that can effectively separate high-purity fullerenes. However, chromatography also has some basic disadvantages, such as the limited column loadings, the time-consuming process, the irreversible adsorption, and the requirement for large quantities of the stationary and mobile phases [81]. Therefore, it very important to develop new chromatography methods that can purify fullerenes faster, with simpler operation and cheaper prices.

1.4.2.2 Recrystallization

The separation of fullerenes by recrystallization is based on the different solubilities of fullerenes in organic solvents at different temperatures. The selection of an appropriate solvent and crystallization conditions enables the purification and separation of fullerenes. In the carbon soot extract, the main components of fullerenes are C_{60} and C_{70}. The solvent capacities of the two kinds of fullerenes are different for different organic solvents and temperatures. For example, in the temperature range of 20 °C–80 °C, the solubility of C_{60} in o-xylene reaches a maximum at about 30 °C, and its solubility above 30 °C decreases with increasing temperature, while in the same temperature range, the solubility of C_{70} in o-xylene increases monotonously with the increase of temperature, which provides conditions that can be used to separate fullerenes [82].

The main step of the recrystallization method is generally to dissolve the initially separated components in a suitable solvent to make a saturated solution; the solvent is allowed to evaporate slowly, separating the different types of fullerene. Let us take C_{60} and C_{70} as example to briefly introduce the industrial process of recrystallization (figure 1.14). When recrystallizing C_{60} fullerene, we first use common solvents such as petroleum ether and acetone to ultrasonically wash the crude fullerenes extracted from carbon soot, in order to remove aromatics and other small-molecule products. We then carry out suction filtration and put the mixture in a vacuum drying oven to dry, using a mortar to grind it into fine particles for later use. We next weigh a certain amount of the milled fullerene sample, select an appropriate organic solvent to dissolve it, concentrate it to a saturated solution after full ultrasonication, heat it to 80 °C and stir it for a period of time; we then filter it while hot and collect the filter residue and filtrate, respectively. The C_{60} fullerene is enriched and crystallized in the filter residue. The above steps are repeated to get higher-purity C_{60}. The process of recrystallizing C_{70} is similar; in this case, C_{70} stays in the filtrate. The filtrate is stored at a temperature lower than −20 °C for a period of time and then quickly filtered. The C_{70} is enriched and crystallized in the filter residue obtained this time. The above steps are repeated several times to obtain a C_{70} sample with higher purity.

The characteristics of the recrystallization method are that the solvent and remaining fullerene can be recovered, the equipment is simple, and continuous batch separation and purification operations can be carried out. However, it is worth pointing out that it is currently difficult to obtain fullerenes with a purity of more than 99.9% by recrystallization. The recrystallization method is a promising method, but at present, the solubility and the parameters of various fullerenes in different solvents are not perfect and still need further research.

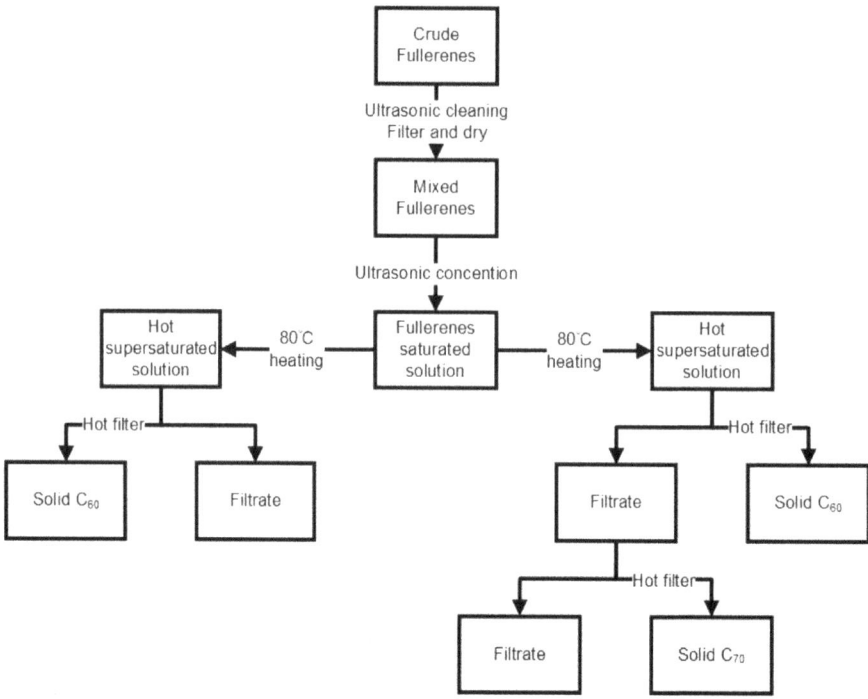

Figure 1.14. Process flowchart showing the separation of C_{60} and C_{70} by recrystallization.

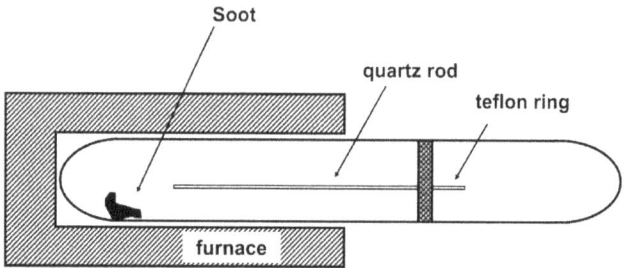

Figure 1.15. Diagram of a device used for the separation of fullerenes by the sublimation method [83], reproduced with permission © American Chemical Society.

1.4.2.3 Sublimation

The sublimation method is a method of separating fullerenes based on the principle that different fullerene molecules have different intermolecular forces; thus, the degree of difficulty of volatilization is different. The sublimation method was one of the important separation methods in early fullerene research. The advantage of the sublimation method is that it avoids the use of extraction solvents and simplifies the extraction process of fullerenes. In addition, the sublimation method can be used as a supplement to extract some fullerenes that are insoluble in conventional organic solvents. For example, in 1993, Yeretzian *et al* designed a sublimation device (figure 1.15). The quartz tube was evacuated to vacuum and then carbon soot was placed at the end of the quartz tube and heated. Different fullerenes adhere to different positions on the axis of the quartz rod because of their different melting points [83]. They used this method to successfully extract C_{74}, which is insoluble in common organic solvents [83].

However, the sublimation method also has many shortcomings; for example, the experimental conditions are difficult to control, the separation amount is small each time, the sublimation temperature difference between the different fullerene isomers is small and they are difficult to separate, etc. Therefore, the separation of fullerenes by sublimation is rarely used now, and only played an important role in the early stages of fullerene research.

1.4.2.4 Selective complexation (supramolecular chemistry)

At present, one of the most active areas of research into fullerene separation is the search for suitable receptors to bind fullerenes and then release them through host–guest interactions based on supramolecular chemistry to obtain a pure product [81]. The method that uses designed molecular receptors for fullerene separation is called selective complexation technology. Complementarity in size, shape, structure, and electronic donor-acceptor behaviors that rely on the interaction between molecular receptors and fullerenes are very important for the design of molecular receptors [81]. In fact, as early as 1990, after scientists used aromatic hydrocarbon reagents to extract fullerenes from carbon soot, the interaction between fullerene molecules and π-conjugated planar aromatic hydrocarbons attracted great attention. In 1994, Suzuki et al considered that host–guest chemistry using calix[n]arenes might provide a breakthrough in the facile purification of C_{60}. They synthesized 5,11,17,29,35,41,47-octa-tert-butylcalix[8]arene-49,50,51,52,53,54,55,56-octol, whose diameter is about 8.6 Å; it can selectively form a 1:1 complex with C_{60} by wrapping it [84]. The results showed that gram quantities of C_{60} with 99.8 wt% purity could be isolated from carbon soot by fractional precipitation with this calix[n]arenes, which is a very convenient and efficient C60 purification method that does not require any expensive apparatus [84]. Subsequently, there were a series of studies of the possibility of using supramolecular chemistry to separate fullerenes [81, 85–87]. For example, Yoshizawa et al reported that a novel M_2L_2 molecular tube capable of binding fullerene C_{60} was synthesized from bispyridine ligands with embedded anthracene panels and Ag(I) hinges (figure 1.16) [87]. This molecular tube could not only encapsulate C_{60} but also

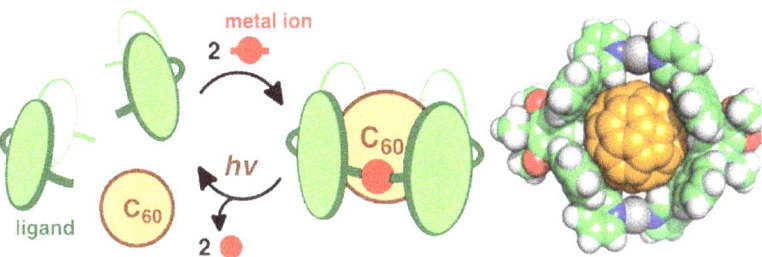

Figure 1.16. Illumination realizes the selective encapsulation and release of fullerenes by an M_2L_2 molecular tube [87], reproduced with permission © American Chemical Society.

C_{60} derivatives with large functional groups, and the fullerene guest could then be released by using the ideal, noninvasive external stimulus, light [87].

The selective complexation method provides a new possibility for overcoming the difficulties of fullerene separation and realizing the sustainable development of the fullerene discipline. However, this method still has many problems, such as relatively high cost, which need further research to solve.

In the previous section, we mentioned that the application of the combustion method makes the large-scale production of fullerenes possible, which in turn reduces the price of mixed fullerenes significantly. However, the price of high-purity fullerenes is still high, which seriously hinders the application prospects of fullerenes. This mainly because the fullerene separation and purification processes still have major problems, such as high energy consumption and low efficiency. With the application of fullerenes and their derivatives in other fields, there is a confirmed demand for a high-volume extraction and separation process. Thus, it is critical to develop new low-cost, high-efficiency separation and purification processes.

1.5 Physical properties of fullerenes

1.5.1 Solubility

The solubility of fullerenes is a very significant physical property. As mentioned in the previous section, the choice of solvent is very important for the purification process of fullerenes. According to the principle of similar compatibility [60], fullerenes with I_h symmetry are a non-polar molecules, which are extremely difficult to dissolve in common polar solvents such as water, acetone, and alcohol; they is slightly soluble in alkanes, and soluble in non-polar solvents such as toluene, xylenes, and carbon disulfide. In 1993, Ruoff *et al* systematically studied the room-temperature solubility of pure C_{60} in 47 common organic solvents; the results are shown in table 1.1 [65].

The solubility of [70]fullerene follows qualitatively similar trends. Table 1.2 shows its solubility in some common solvents.

1.5.2 Spectral properties

1.5.2.1 UV–visible spectroscopy

UV–visible spectroscopy is the most direct method used to study the light absorption properties of fullerenes. The absorption spectra of [60]fullerene and [70]fullerene dissolved in 1,2-Dichlorobenzene solvent are shown in figure 1.17(a) [89]. The absorption spectrum of fullerene is mainly caused by the transition of $\pi \rightarrow \pi^*$. It can be seen that the absorption of fullerenes is mainly concentrated between 200 and 700 nm [90]. Based on their light absorption characteristics, the colors of [60] fullerene and [70]fullerene solutions are purple red and wine red, respectively (figure 1.17(b)) [89]. Many applications have been developed because the absorption spectra of fullerene and their derivatives are mainly concentrated in the UV region, as shown in section 1.7.

Table 1.1. The solubility of [60]fullerene in different solvents [65]. Source: [65], reproduced with permission © American Chemical Society.

Solvent	Solubility (mg mL^{-1})
Alkanes	
n-Pentane	0.005
Cyclopentane	0.002
n-Hexane	0.043
Cyclohexane	0.036
n-Decane	0.071
Decalins	4.6
cis-Decalin	2.2
trans-Decalin	1.3
Haloalkanes	
Dichloromethane	0.26
Chloroform	0.16
Carbon tetrachloride	0.32
1,2-Dibromethane	0.50
Trichloroethylene	1.4
Tetrachloroethylene	1.2
Dichlorodifluoroethane	0.020
1,1,2-Trichlorotrifluoroethane	0.014
1,1,2,2-Tetrachloroethane	5.3
Polar solvents	
Methanol	0.000
Ethanol	0.001
Nitromethane	0.000
Nitroethane	0.002
Acetone	0.001
Acetonitrile	0.000
N-methyl-2-pyrrolidone	0.89
Benzenes	
Benzene	1.7
Toluene	2.8
Xylenes	5.2
Mesitylene	1.5
Tetralin	16
o-Cresol	0.014
Benzonitrile	0.41
Fluorobenzene	0.59
Nitrobenzene	0.80
Bromobenzene	3.3
Anisole	5.6
Chlorobenzene	7.0
1,2-Dichlorobenzene	27
1,2,4-Trichlorobenzene	8.5

	Naphthalenes	
1-Methylnaphthalene		33
Dimethylnaphthalenes		36
1-Phenylnaphthalene		50
1-Chloronaphthalene		51
	Miscellaneous	
Carbon disulfide		7.9
Tetrahydrofuran		0.000
Tetrahydrothiophene		0.030
2-Methylthiophene		6.8
Pyridine		0.89

Table 1.2. The solubility of [70]fullerene in different solvents [88], reproduced with permission © Institute of Physics.

Solvents	Solubility (mg mL^{-1})
1,2-Dichlorobenzene	36.2
Carbon disulfide	9.875
Xylene	3.985
Toluene	1.406
Benzene	1.3
Carbon tetrachloride	0.121
n-Hexane	0.013
Cyclohexane	0.08
Pentane	0.002
Octane	0.042
Decane	0.053
Dodecane	0.098
Heptane	0.047
Isopropanol	0.0021
Mesitylene	1.472
Dichloromethane	0.080

1.5.2.2 Infrared and Raman spectra

The vibrational spectrum (Fourier-transform infrared spectroscopy (FTIR), Raman) has high structural sensitivity, which can be combined with theoretical calculation to determine the structure of fullerene molecules. Figures 1.18(a) and (b) show the FTIR spectra of [60]fullerene and [70]fullerene. Due to the symmetry of fullerenes, their FTIR spectra are relatively simple. For [60]fullerene, four characteristic vibration signals

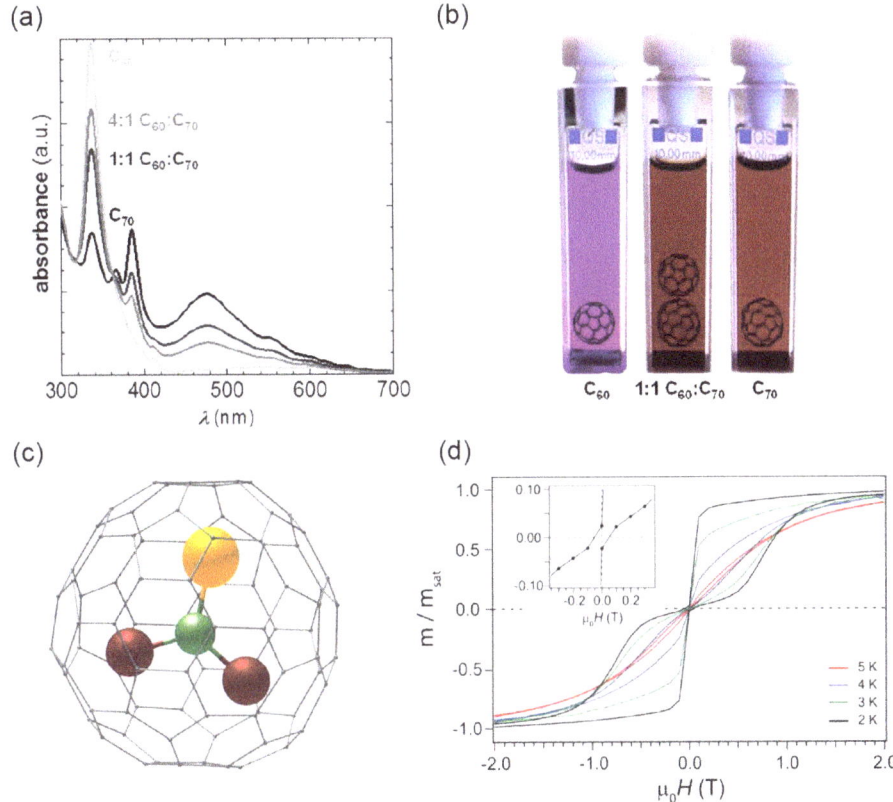

Figure 1.17. (a) UV–Vis absorbance spectra of 0.01 mg mL^{-1} fullerene solutions in 1,2-Dichlorobenzene with varying C_{70} fractions. (b) Solutions of ~2 mg mL^{-1} C_{60}, C_{70} and a 1:1 C_{60}:C_{70} mixture in 1,2-Dichlorobenzene. Source for (a), (b): [89], reproduced with permission © Royal Society of Chemistry. (c) Chemical structure of $DySc_2N@I_h\text{-}C_{80}$. (d) Temperature-dependent magnetization curves for $DySc_2N@Ih\text{-}C_{80}$. Source for (c), (d): [91], reproduced with permission © American Chemical Society.

appear at 1429 cm^{-1}, 1183 cm^{-1}, 577 cm^{-1}, and 528 cm^{-1} [92, 93]. For [70]fullerene, as a result of its lower symmetry, the FTIR spectrum is more complex [92, 93].

Raman spectroscopy is also a useful research method for fullerenes. According to theoretical calculations, [60]fullerene belongs to the I_h point group and has ten Raman vibration modes, whose characteristic peaks at 271 cm^{-1} $H_g(1)$, 431 cm^{-1} $H_g(2)$, 495 cm^{-1} $A_g(1)$, 568 cm^{-1} $H_g(3)$, 771 cm^{-1} $H_g(4)$, 1099 cm^{-1} $H_g(5)$, 1249 cm^{-1} $H_g(6)$, 1424 cm^{-1} $H_g(7)$, 1468 cm^{-1} $A_g(2)$, 1573cm^{-1} $H_g(8)$ have all been observed experimentally (figure 1.18(c)) [92]. Due to the lower symmetry of [70] fullerene, there are five unequal carbon atoms in [70]fullerene. This structure leads to a more complicated intramolecular vibration mode in [70]fullerene, but as in the case of [60]fullerene, the Raman characteristic peaks are mainly concentrated in the range of 1600–200 cm^{-1} (figure 1.18(d)) [92].

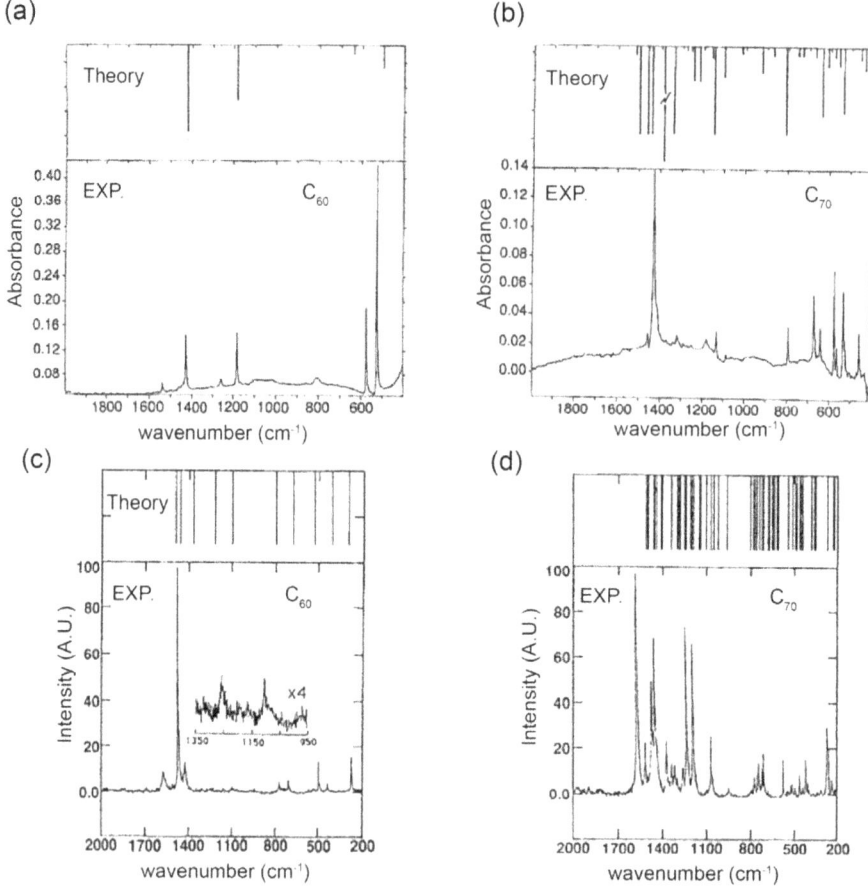

Figure 1.18. FTIR spectra of purified films of (a) C_{60} and (b) C_{70} on KBr substrates. Unpolarized Raman spectra of purified films of (c) C_{60} and (d) C_{70} on a Si (100) substrate. Source: [92], reproduced with permission © Academic International Press.

1.5.3 Magnetic properties

Magnetic materials are among the most widely used materials in our daily lives, from information storage to communication technology, medical diagnosis, and aerospace, which are all related to magnetic materials [94]. Faced with the needs of social development, traditional magnetic materials are increasingly unable to meet the requirements. Thus, it is crucial to develop novel magnetic materials. Stable hollow fullerenes are non-magnetic, but EMFs are a type of magnetic material. EMFs have variable electronic structures due to their different endohedral metals. After the endohedral transfer of electrons to the fullerene carbon cage and reorganization, some EMFs have an unpaired electron on the molecular orbital, which gives these molecules paramagnetism. These unpaired electrons can be distributed on the carbon cage or in clusters inside the carbon cage. When the fullerene is endohedral and has a lanthanide metal containing multiple unpaired 4f electrons, the fullerene thus formed also exhibits paramagnetism.

In addition, EMFs have emerged in the field of single-molecule magnets (SMMs) in recent years, due to their advantages of high stability and easy control of the molecular structure. An SMM is a monomolecular compound that has super-paramagnetic behavior below a certain blocking temperature at the molecular scale; such compounds have huge potential applications in high-density information storage and quantum computers, because they represent the smallest scale of magnetic storage devices [94–96]. Since Greber *et al* reported the first metal nitride clusterfullerene SMM $DySc_2N@I_h$-C_{80} in 2012 (figures 1.17(c) and (d)) [91], a large number of endohedral fullerene SMMs have been reported [97–100]. We believe that the magnetic properties of fullerenes will have more extensive applications in the near future.

1.5.4 Superconductivity

Although superconductivity was discovered more than a century ago, it still has an important position in modern science and has broad application prospects in the fields of superconducting power generation, power transmission, and energy storage. Because of the special structure of fullerenes, the π electrons are delocalized on the entire carbon cage, so the π orbital can be approximated as an atomic orbital, and the movements of all π electrons on the carbon cage are without resistance, which gives fullerenes special charge transport performance. However, fullerene molecules are strongly bound to the π electrons, and it is difficult for the π electrons to move freely between molecules, so fullerenes themselves are not good conductors. Fullerenes are electron deficient and have a strong electron affinity (see section 1.6). Therefore, fullerenes can interact with alkali metals to form stable ionic compounds that are superconductors. This change in charge conductivity behavior can be easily compared to adding an excess of water to several adjacent pools of the same height, so that the pools are connected to each other, and the excess water can move freely between several pools.

As early as 1991, Stephens *et al* prepared K_3C_{60}, which had superconducting properties [101]. Subsequently, researchers set off an upsurge in the study of fullerene superconductors [102–106]. Although the performance of fullerene super-conductors to date still cannot be compared to those of other types of super-conductor, their special structure provides a special model that can be used to study the mechanism of superconductivity. Therefore, the study of fullerene superconduc-tivity is still very meaningful.

In addition to some of the physical properties introduced above, fullerenes and their derivatives also have some special physical properties, such as nonlinear optics and photoconductivity [107, 108]. These unique physical properties of fullerenes are key to their wide range of applications. However, the study of their physical properties has lagged behind the extensive in-depth study of their chemical proper-ties. We expect that more of the physical properties of fullerenes and their derivatives will be explored in the future.

1.6 Chemical properties of fullerenes

As the only carbon allotropes with defined molecular structures, fullerenes have attracted great attention due to their unique structure and special physical and chemical properties. Since Huffman, Krätschmer, and Fostiropoulos first discovered a procedure for the preparation of bulk quantities of C_{60} in the early 1990s [5], scientists have conducted a lot of research into their chemical properties. Fullerenes can be functionalized through chemical reactions in order to regulate their properties. For example, the solubility of fullerenes in most solvents is generally low, although they are the only known allotropes of carbon that can be dissolved in common solvents at room temperature. We can modify the appropriate functional groups to improve their solubility while maintaining the electronic properties of fullerenes. The chemical modification of fullerenes is usually divided into three categories: the first is chemical modification outside the fullerene cage, the second is chemical modification within the cage to form endohedral fullerenes, and the last is the replacement of carbon atoms on the fullerene cage with heteroatoms to form heterofullerenes. In this section, we mainly focus on chemical modification outside the fullerene cage.

1.6.1 Bonding and electronic properties of fullerenes

Fullerenes form a family of carbon cages which have high stability due to their unique structures, but this does not mean that they are completely unreactive. A typical case is [60]fullerene, which is a hollow symmetric molecule like a football composed of 60 carbons connected by 20 hexagons and 12 pentagons. Although there is only one signal in the ^{13}C NMR spectrum, there are two types of carbon–carbon bond: the bond between two hexagons is called the [6,6] bond, and the bond between hexagons and pentagons is called the [5,6] bond. The x-ray single-crystal diffraction of [60] fullerene shows that the length value of the [6,6] bond is 1.355 Å and that of the [5,6] bond is 1.467 Å (the double bond length value of ethylene is 1.34 Å) [15]. Thus the [6,6] bond is more like a double bond, which is more prone to addition reactions. Another reason that the [6,6] bond is prone to chemical reactions is because of its large angular strain. Sp^2-hybridized carbon atoms are at an energy minimum in planar graphite. However, in order to form a cage or tube structure made of fullerenes, the surface must be bent, which produces a large angular strain. When certain double bonds of [60]fullerene are saturated by some chemical reactions, the angular strain is released. For example, the [6,6] bond of fullerenes is electrophilic; when chemical reactions change the bond from an sp^2-hybridized orbital (about 120°) to an sp^3-hybridized orbital (about 109.5°), the reduction in the Gibbs free energy enhances the stability of [60]fullerene. Thus, fullerenes have the potential to form addition products by opening the [6,6] bond.

Fullerenes' spherical shape has far-reaching consequences for their reactivity. The interconnection of the P orbital on the outer sphere is greater than that on the inner sphere (each two carbon atoms are connected by a sp^2-hybridized orbital to form a σ bond; each carbon atom is left with a P orbital—such orbitals are perpendicular to the plane of the σ bond and parallel to each other—forming a π bond; and all π electrons form a complex π-π conjugate system, which is approximately spherical), which is why fullerenes can sometimes act as electron donors. Although the carbon atoms in fullerenes are all conjugated [60], fullerene is not a very aromatic compound [60]. Fullerenes have 60 π electrons, but a closed-shell configuration requires 72 electrons [109]. In general, fullerenes are electron deficient and display obvious electrophilic properties. Theoretical computation shows that the lowest unoccupied molecular orbital (LUMO) (t_{1u}-symmetry) and LUMO+1 (t_{1g}-symmetry) of [60]fullerene exhibit comparatively low energies and are triply degenerate (figure 1.19(a)) [110–112]; thus, they can accept up to six electrons. Conventional cyclic voltammetry (CV) and differential pulse voltammetry (DPV) can only detect four reduction potentials; however, the use of a 1:1.5 mixed solvent made of acetonitrile and toluene under vacuum conditions can detect six reduction potentials (figure 1.19(b)) [113]. Therefore, [60]fullerene is prone to nucleophilic addition reactions, cycloaddition reactions, reduction reactions, free radical addition reactions, etc. In the next part, we will briefly introduce the chemical reactions of fullerenes.

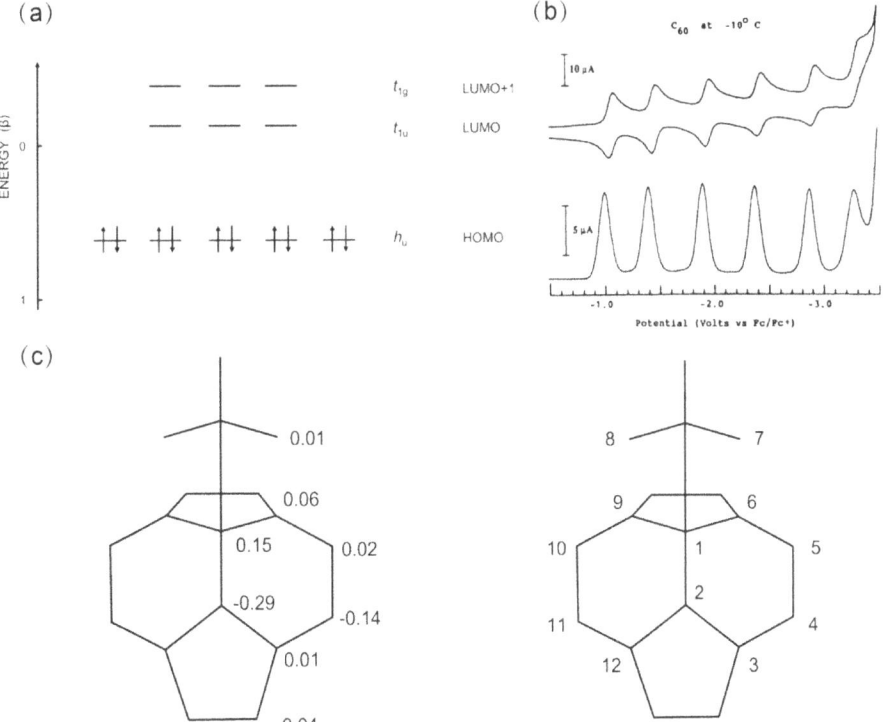

Figure 1.19. (a) Schematic representation of the Hückel molecular orbital (HMO) of C_{60} (highest occupied molecular orbital (HOMO), LUMO, LUMO+1) [112], reproduced with permission © Elsevier. (b) Cyclic voltammogram of C_{60} in acetonitrile–toluene with TBAPF$_6$ as the supporting electrolyte at −10 °C using CV at a 100 mV s^{-1} scan rate (upper) and DPV at a 25 mV s^{-1} scan rate with a 50 mV pulse, a 50 ms pulse width, and a 300 ms period (lower) [113], reproduced with permission © Elsevier. (c) Charge density calculation of RC$_{60}^-$ (R = -H, -Me, -^{t}Bu) [114], reproduced with permission © John Wiley and Sons.

1.6.2 Chemical reactions of fullerenes

1.6.2.1 Nucleophilic addition reaction

Fullerenes are electron-deficient conjugated systems; their chemical properties are similar to those of conjugated polyenes, and they have have considerable electronegativity. Therefore, fullerenes are prone to nucleophilic reactions. When nucleophiles (Nu^-) attack C_{60}, electrons are transferred to the C_{60}, forming the active intermediate $Nu_nC_{60}^{n-}$, which is unstable and undergoes further reactions: (a) when it is in an environment that contains an electrophile (E^+), such as a proton (H^+) or a carbocation, $C_{60}E_nNu_n$ is formed; (b) when it is in an environment that contains a neutral electrophile (E-X), such as an alkyl halide (RX), the active intermediate $Nu_nC_{60}^{n-}$ further attacks E-X, producing $C_{60}E_nNu_n$; (c) when a leaving group is included in the Nu^-, the final product can be formed by an intramolecular nucleophilic substitution (S_Ni), such as the Bingel reaction. Although the different types of addition can form many isomers, [1,2] addition is the main product. When the steric hindrance is relatively large, [1,4] addition or even [1,6] addition products may also be formed [115]. As shown in figure 1.19(c), the theoretical calculation of RC_{60}^- (R=H, Me, t-Bu) charge density shows that the negative charge is not delocalized; the highest electron density is that of C-2 next to the sp^3 carbon atom across the 6-6 bond, followed by that of C-4 (C-11), which explains the regioselectivity of the nucleophilic addition reactions [114]. The nucleophilic addition reaction is one way to achieve fullerene chemical modification; in this section, we will briefly introduce some nucleophilic addition reactions of fullerenes.

(1) The hydroalkylation and hydroarylation of fullerenes

As mentioned above, [60]fullerene can easily react with carbanion-containing nucleophiles such as Grignard reagents or organolithium reagents to form $R_nCn_{60}^-$ [116–120]. If the reaction is carried out in toluene, one can immediately see the formation of a precipitate after the metal compound is added, which is the corresponding $C_{60}M_nR_n$. Their protonation yields the hydrofullerene derivatives

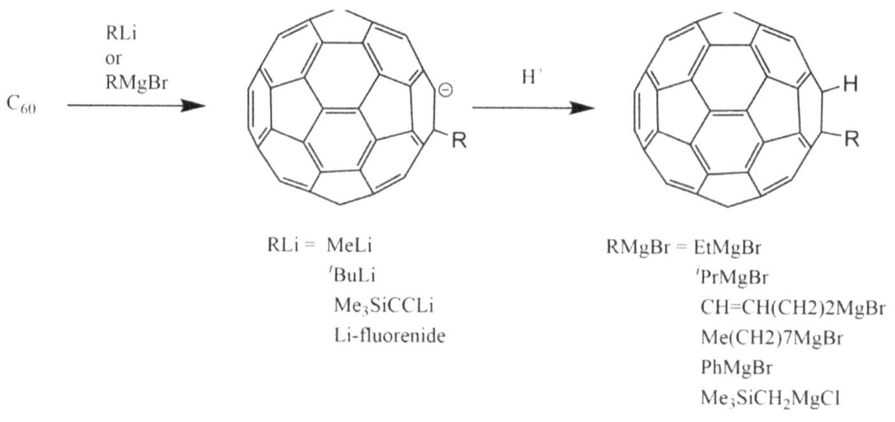

RLi = MeLi	RMgBr = EtMgBr
'BuLi	'PrMgBr
Me₃SiCCLi	CH=CH(CH2)2MgBr
Li-fluorenide	Me(CH2)7MgBr
	PhMgBr
	Me₃SiCH₂MgCl

Scheme 1.1. Hydroalkylation of C_{60}.

$C_{60}H_nR_n$. Highly alkylated $R_nCn_{60}^-$ can be soluble in tetrahydrofuran (THF), which can be quenched with MeI to form $C_{60}Me_nR_n$. In order to obtain highest yield of pure mono-adduct $C_{60}HR$, which is [1,2] addition, the nucleophile can be added dropwise by titration. When the optimal equivalent of the added nucleophile is reached, 0.01 mol L^{-1} hydrochloric acid in methanol solution is used to quench the mixture (scheme 1.1) [121, 122]. Obviously, the optimal equivalents of different nucleophiles are different: for organolithium reagents with strong nucleophilic ability, 1.2 eq is the best equivalent; for milder Grignard reagents, 5–27 eq are required [114, 122].

Metal acetylides are widely used in alkynylation in organic chemistry; they are nucleophiles that add to a variety of electrophilic and unsaturated substrates. They have relatively good stability and low nucleophilicity, which makes their reactivity with C_{60} lower than those of Grignard reagents. Although their reaction rates are slow, a variety of C_{60} ethynyl derivatives can still be obtained. For example, Komatsu *et al* reported the first fullerene–acetylene hybrids [123]; they used ((trimethylsilyl)ethynyl)lithium to react with C_{60} and then quenched the reactants with acid to obtain the final product at a 45% yield [124].

(2) Cyclopropanation: the Bingel reaction

As mentioned above, when the nucleophile (Nu) contains a leaving group, the reactive intermediates $Nu_nC_{60}^{n-}$ or $Nu_nC_{70}^{n-}$ can form fullerene derivatives through an S_{Ni} reaction. For example, an α-halogenated ester or an α-halogenated ketone loses a proton under the action of a strong base; it can be used as a nucleophile to attack C_{60} and then undergoes the cyclopropanation reaction (Bingel reaction), resulting in a product [125]. Cyclopropanation is one of the most significant ways in which fullerenes can be chemically modified. It can synthesize a significant class of fullerene derivatives, namely methanofullerenes, which have the advantages of high stability and little interference with the carbon cage structure caused by substituents; they are the most widely studied fullerene derivatives and have huge application prospects. In addition, methanofullerenes can also be prepared by the cycloaddition reaction, which is introduced later; in this section, we mainly introduce the Bingel reaction.

In 1993, Bingel reported that C_{60} can react smoothly with diethyl bromomalonate and NaH in a toluene solution at room temperature, where the NaH is used as an auxiliary base to pull out the α-H of diethyl bromomalonate (scheme 1.2) [125]. This nucleophilic cyclopropanation reaction can also carried out by reacting C_{60} with malonic acid derivatives or 1,3-diketone derivatives in the presence of iodine or CBr_4 and a base (usually 1,8-diazabicyclo[5.4.0]undec7-ene (DBU)) [126–128]. The principle is to generate an α-halogenated ester or an α-halogenated ketone *in situ* and then react it with C_{60}. This method can omit the purification process of intermediate halogenated compounds and can obtain good yields under optimized reaction conditions [126–128]. The advantage of this method is that the cyclo-propanation is achieved under near-neutral conditions and eliminates the alkali necessary for the Bingel reaction [129]. In summary, some functional groups or molecules can be easily modified to C_{60} by the Bingel reaction, which greatly expands the application range of fullerenes.

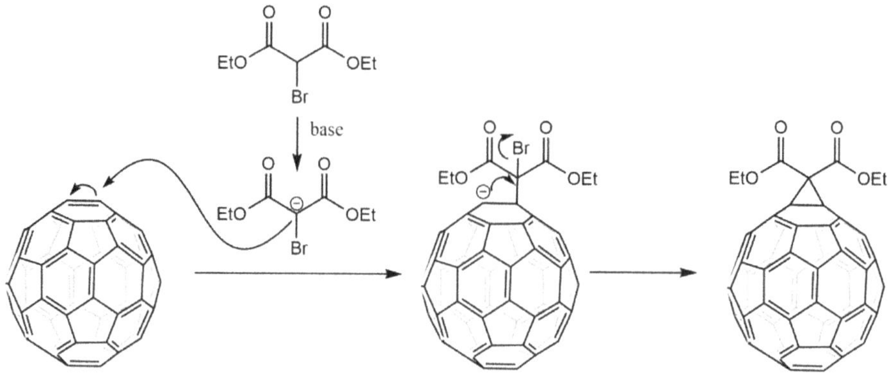

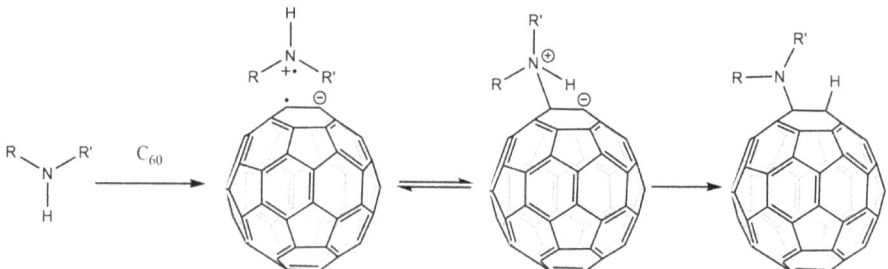

Scheme 1.2. The mechanism of the Bingel reaction.

(3) Other nucleophilic addition reactions

Scheme 1.3. Reaction mechanism of an amine and a fullerene

In addition to the abovementioned examples of carbanions as nucleophiles that attack C_{60}, many other compounds can also react with fullerenes as nucleophiles, such as aliphatic primary and secondary amines [130, 131], hydroxides and alkoxides [132, 133], silyllithium [134], etc. It is worth mentioning that in the reaction of amines and fullerenes, the first step is a single-electron transfer from the amine to C_{60} to give the C_{60} radical anion, and then through free radical coupling to form zwitterions (scheme 1.3) [130]. We will not introduce these reactions in detail here; the interested reader can read related references.

1.6.2.2 Cycloaddition reaction

We already know that the bond length of the [6,6] bond of C_{60} is closer to the bond length of the ethylene double bond than it is to the length of the [5,6] bond [15]. Thus, the [6,6] bond is more like a double bond and can be used as a dienophile for cycloaddition reactions [120, 135, 136]. Cycloaddition is the most important way to chemically modify fullerenes because of its high yield and easy control; almost all functional groups can be connected to C_{60} through cycloaddition, which provides a powerful tool for the functionalization of fullerenes. Some cycloaddition products exhibit excellent chemical and thermal stability and have a wide range of application prospects; for example, [6,6]-Phenyl-C_{61}-butyric acid methyl ester (PCBM) is a significant organic optoelectronic material (we will introduce its application in section 1.7), which can be synthesized by the cycloaddition reaction. In this section, we will briefly introduce the cycloaddition reaction of fullerenes.

[1 + 2] Cycloaddition

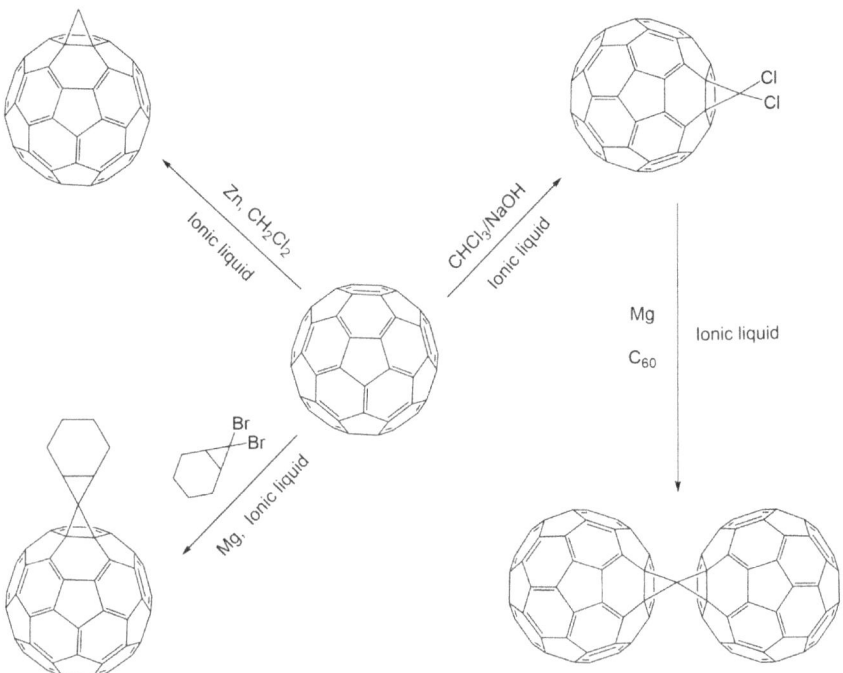

Scheme 1.4. The [1 + 2] cycloaddition reaction of C_{60} with carbene.

[1 + 2] Cycloaddition is one of the methods for preparing methanofullerenes; we have introduced another method of preparation, the Bingel reaction, above. In [1 + 2] cycloaddition, carbene and diazo compounds are often used as reactants to react with fullerene.

Singlet carbene can be selectively added to the [6,6] bond of fullerene to obtain regioselective methanofullerene [137, 138]. For example, Zhu *et al* reported that the dehalogenation of the polyhalides $CHCl_3$ and CH_2I_2, 7,7-dibromobicyclo(4,1,0) heptane by treatment with Zn, Mg, or NaOH powder, followed by an *in situ* reaction with a fullerene in the presence of ultrasonic irradiation and in the ionic liquid solvent 1-methyl-3-octylimidazolium tetrafluoroborate [bmim][PF_4] or 1-butyl-3-methylimidazolium hexafluorophosphate [omim][BF_4] produced [6,6]-junction cycloaddition products at yields of 53%–79% (scheme 1.4) [137].

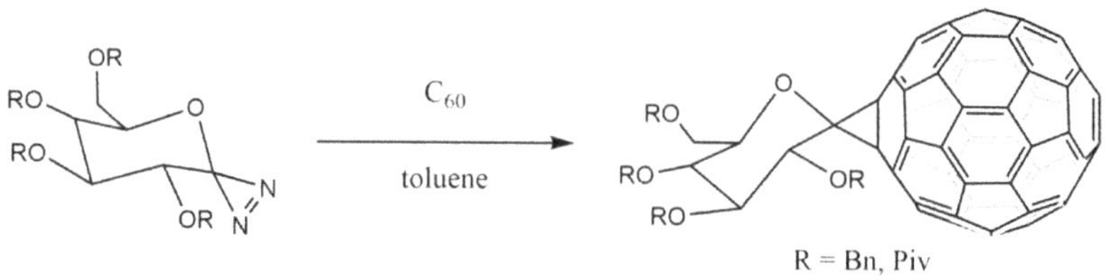

Scheme 1.5. The [1 + 2] cycloaddition reaction of C_{60} with carbene.

The target product can also be obtained by reacting a diazonium compound with a fullerene (diazonium compounds are used as precursors for the synthesis of carbene). For example, Vasella *et al* reported that C_{60} reacts with the corresponding carbene produced by the thermal extrusion of N_2 from O-benzyl- and O-pivaloyl-protected diazirine to obtain 1,2-methano-bridged sugar monoadducts (scheme 1.5) [139]. It is worth noting that this method can also be used to obtain [6,6] addition products.

(2) [2 + 2] Cycloaddition

Compared with [1 + 2]cycloaddition, [2 + 2]cycloaddition is less reported. The reaction of fullerenes with benzyne [140], electron-rich alkynes or alkenes [141, 142], and α-cyclohexenone [143] is due to [2 + 2] cycloaddition. A typical example of [2 + 2] cycloaddition is the reaction of C_{60} and α-cyclohexenone. Under illumination, α-cyclohexenone is excited and produces triplet intermediates, which then react with C_{60} (scheme 1.6) [143].

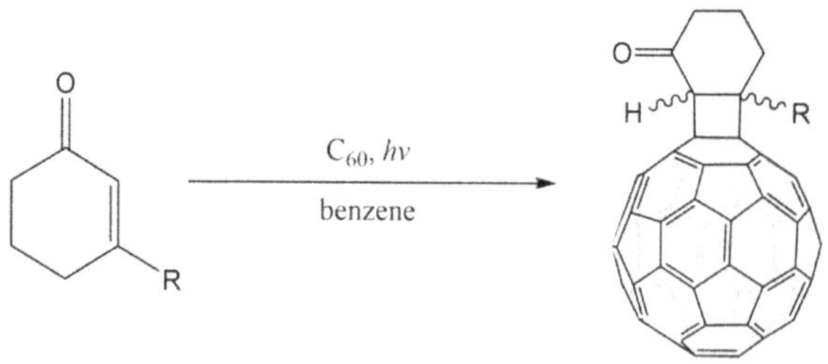

Scheme 1.6. The [2 + 2] cycloaddition reaction of C_{60} with α-cyclohexenone.

(3) [3 + 2] Cycloaddition

Diazomethane [144, 145], diazoacetate [146, 147], azide [148], etc. can react with fullerenes via [3 + 2] cycloaddition (the Prato reaction). Indeed, this is one of the most frequently used tools to functionalize fullerenes due to the fact that cyclo-addition always occurs on the [6,6] bonds of fullerenes. For the [3 + 2] cycloaddition reaction of diazomethane with C_{60}, diazomethane first undergoes 1,3-dipolar cycloaddition with the [6,6] bond of C_{60}, and then N_2 is removed by photochemistry or a heat treatment to form two different isomers (scheme 1.7), a [5,6]-open isomer and a [6,6]-closed isomer [144, 145]. For other diazo compounds, such as diphenyldiazomethane, a mixture of different isomers is also formed [145]. As is the case for the diazomethane reaction, when the cycloaddition reaction of C_{60} and diazoacetate is carried out in refluxing toluene, ^{1}H NMR has shown that three kinds of isomer, a–c, are produced at a ratio of 1:1:3 (scheme 1.8) [146]. The formation of the [5,6]-open isomer c is kinetically favorable, while the [6,6]-closed isomer a is the most stable thermodynamically [146]. Wudl *et al* creatively used these reactions to obtain a series of fullerene derivatives, which opened up new possibilities for the application of fullerenes [149].

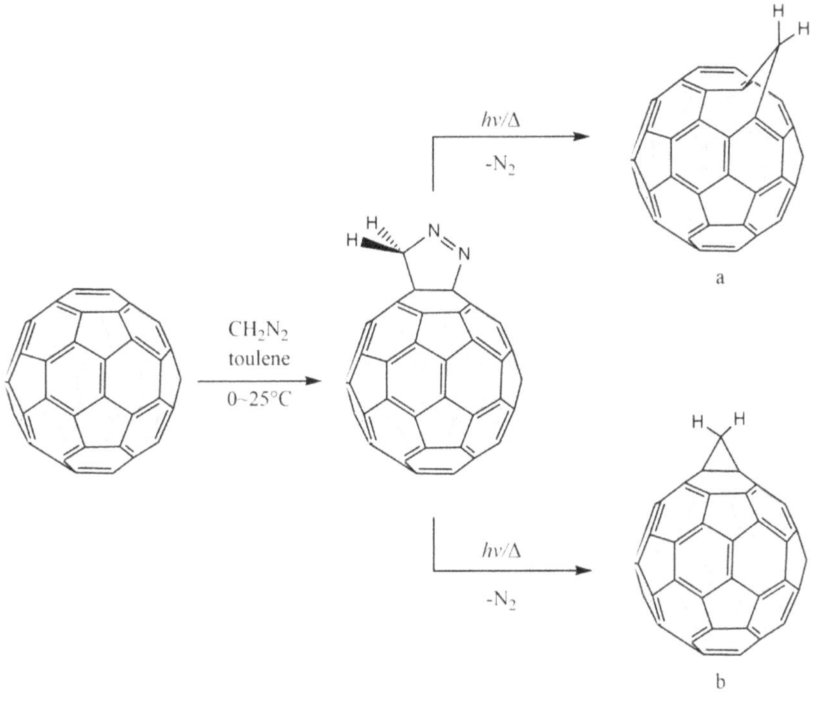

Scheme 1.7. The [3 + 2] cycloaddition reaction of C_{60} with diazomethane.

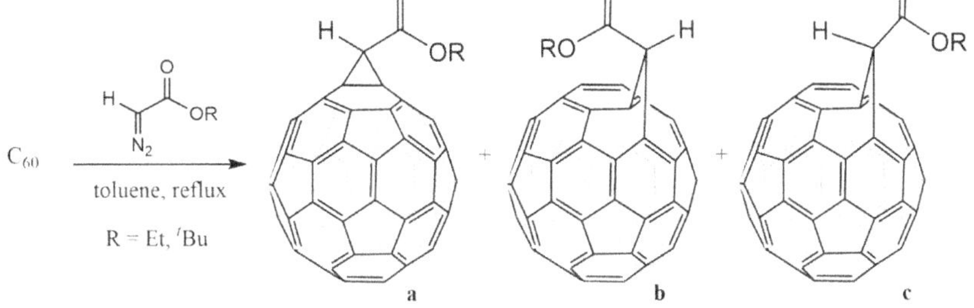

Scheme 1.8. The [3 + 2] cycloaddition reaction of C_{60} with diazoacetate.

(4) [4 + 2] Cycloaddition

Fullerenes can react with different dienes to obtain [6,6] bond cycloaddition products because the [6,6] bond of fullerene is dienophilic and its reactivity is equivalent to that of maleic anhydride [120, 135, 136]. Cyclopentadiene and anthracene [150], butadiene, and tetrathiafulvalene (TTF) derivatives [151, 152] can undergo the Diels–Alder (D–A) reaction with fullerenes. Depending on the reactivity of the dienes, different reaction conditions are required, such as heating, light, microwave radiation, or reflux in a high-boiling-point solvent [153, 154]. For example, to react C_{60} with cyclopentadiene, 1 eq of cyclopentadiene was added to a C_{60} toluene solution at room temperature, which obtained the D–A mono-addition product with a yield of 74% [150]. However for the anthracene reaction, more severe reaction conditions are required; the excess anthracene is refluxed in toluene for reaction (scheme 1.9) [120, 150, 155]. In short, the [4 + 2] cycloaddition reaction has become one of the indispensable methods with which chemists can construct fullerene derivatives.

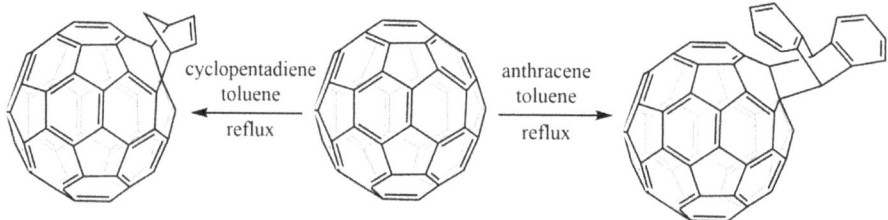

Scheme 1.9. The [4 + 2] cycloaddition reaction of C_{60} with cyclopentadiene or anthracene.

(5) Synthesis of PCBM

PCBM has been one of the most successful receptor materials for organic solar cells in the past three decades (we will introduce it in detail in section 1.7); it can be synthesized by a cycloaddition reaction. Wudl and co-workers reported the preparation of PCBM by a one-pot method in which C_{60} was added directly to a solution of tosylhydrazone anions in the presence of a base at 70 °C [149, 156]. This procedure makes it possible to generate diazo compounds in situ, without the need for purification prior to the addition of C_{60} [149]. As shown in scheme 1.10, the first step produced mainly [5,6]-open isomer, which may occur for a variety of reasons. First, two cycloaddition mechanisms are equally likely in the thermal reaction of diazo compounds with C_{60}: (1) carbene may form due to the thermal decomposition of diazo compounds, followed by synchronous addition to the [6,6] bond of C_{60}; (2) the diazo compound may react with C_{60} via 1,3-dipolar cycloaddition and form both isomers as a result; the [5,6]-open isomer is the product of kinetic advantage. It is then converted into thermodynamically stable PCBM by heat or light [156].

1.6.2.3 Radical addition

Fullerenes and their derivatives are effective free radical scavengers due to their unique electronic structures; as a result, they are called 'free radical sponges' [157].

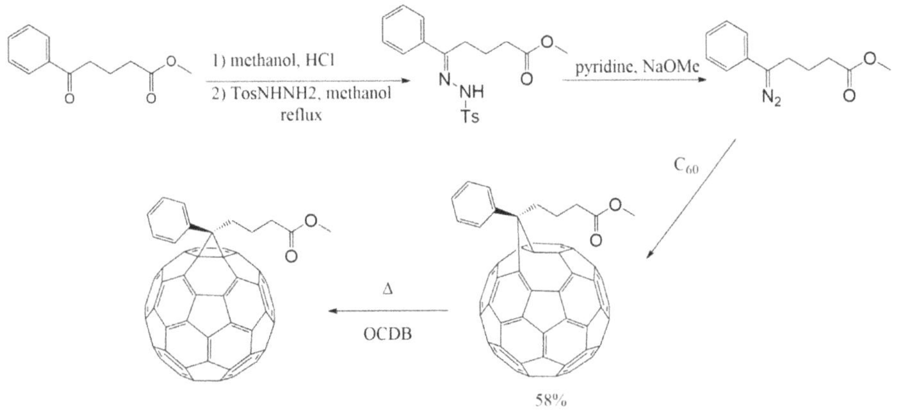

Scheme 1.10. Synthesis route of PCBM.

Therefore, they have huge application prospects in the fields of cosmetics, biomedicine, etc [35, 158]. Fullerenes can undergo radical addition with carbon radicals [159–161], silicon radicals [162, 163], oxygen radicals [164, 165], nitrogen radicals [166, 167], etc. The radicals R· can be obtained by the photolysis of suitable precursors. The photolysis of free radicals can be achieved by irradiating saturated C_{60} benzene or tert-butylbenzene solutions containing a slight excess of free radical precursors with UV light; the precursors include alkyl bromides, carbon tetrachloride, dialkymercury compounds, hydrocarbons (RH), di-tert-butylperoxide, etc [168]. It is also possible to generate radicals by directly illuminating halides (scheme 1.11) [168]. In addition, free radicals can be generated by pyrolysis [168, 169]. Radical addition can be studied by electron spin-resonance spectroscopy (ESR), for example, to study the reaction of tert-butyl radicals (tBu·) and C_{60}; the ESR spectrum of tBuC_{60} consists of ten narrow lines, which are appropriate for a hyperfine interaction between the unpaired electron and nine equivalent protons of the tert-butyl group (figure 1.20) [159]. In addition, the positional information of the unpaired electrons can also be obtained from the ^{13}C satellites in the ESR spectra; the results show that most of the unpaired electrons do not delocalize over the entire sphere but are located on the two fused hexagons on the surface of C_{60} [159, 170, 171]. Research into the radical addition reaction of fullerenes provided a basis for the subsequent work on fullerenes in the fields of biomedicine and cosmetics.

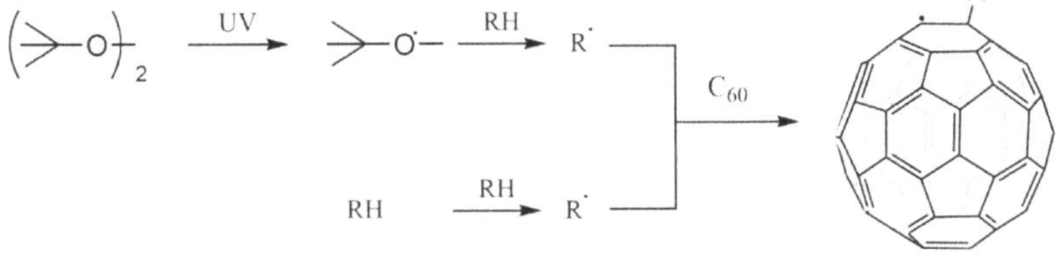

Scheme 1.11. Mechanism of the radical addition reaction.

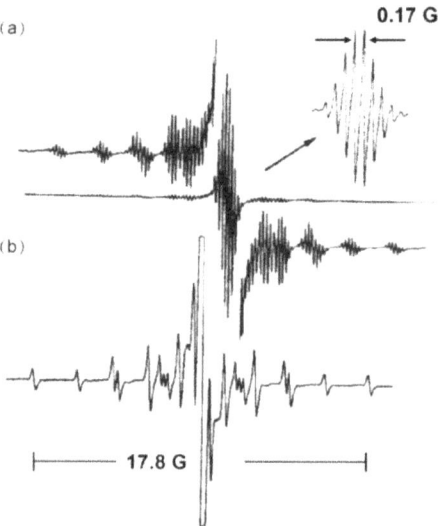

Figure 1.20. ESP spectra of $t\mathrm{BuC}_{60}^{\cdot}$ in benzene at 80 °C: (a) $(\mathrm{CH_3})_3\mathrm{CC}_{60}^{\cdot}$ and (b) $(\mathrm{CD_3})_3\mathrm{CC}_{60}^{\cdot}$ show 13C hyperfine satellites [159], reproduced with permission © American Chemical Society.

In this section, we briefly introduced some of the chemical properties and reactions of fullerenes. Thanks to these unique chemical properties and chemical reactions, we can chemically modify fullerenes and design a variety of functionalized fullerene derivatives, which has greatly expanded the application prospects of fullerenes. Therefore, in the next sections, we will specifically introduce the applications of fullerenes and their derivatives.

1.7 Applications of fullerenes

As mentioned in the previous sections, fullerenes and their derivatives have the characteristics of excellent electron transport, antioxidant activity, catalytic performance, superconductivity, and good biocompatibility because of their unique physical and chemical properties. Therefore, they have good application prospects in the fields of organic electronics, hydrogen storage, biomedicine, superconductors, cosmetics, etc. In this section, we will give a general introduction to the applications of fullerenes and their derivatives.

1.7.1 Organic electronics

Fullerenes and their derivatives have brilliant electron-acquiring and electron-transporting capabilities due to their special stability in the nanoscale range and their unique three-dimensional conjugated electronic structure. Therefore, fullerenes and their derivatives have significant application prospects in organic solar cells (OSCs), perovskite solar cells (PSCs), organic field-effect transistors (OFETs), and other organic electronics fields. In this section, we will give a detailed introduction to the applications of fullerenes and their derivatives in organic electronics.

1.7.1.1 Organic solar cells

OSCs are considered to be an emerging, promising, renewable source of electricity due to their advantages of simple device structure, low cost, light weight, flexibility, simple production process, and capability to be fabricated into flexible and semi-transparent devices [172–180]. During the past few decades, OSCs have experienced continuous advancement and their power conversion efficiency (PCE) has exceeded 18% with the fast evolution of new materials [181–185], making OSCs even more commercially relevant. Fullerenes and their derivatives have high electron affinity, low electron recombination energy, relatively high electron mobility, and good miscibility with electron donors; thus, they are ideal electron acceptor materials for OSCs [186–191].

OSCs are usually heterojunction devices composed of an active layer consisting of an electron donor and an acceptor, carrier transport layers, and two electrodes with different work functions. This kind of heterojunction device can be divided into planar heterojunction and bulk heterojunction (BHJ) devices (figures 1.21(a) and

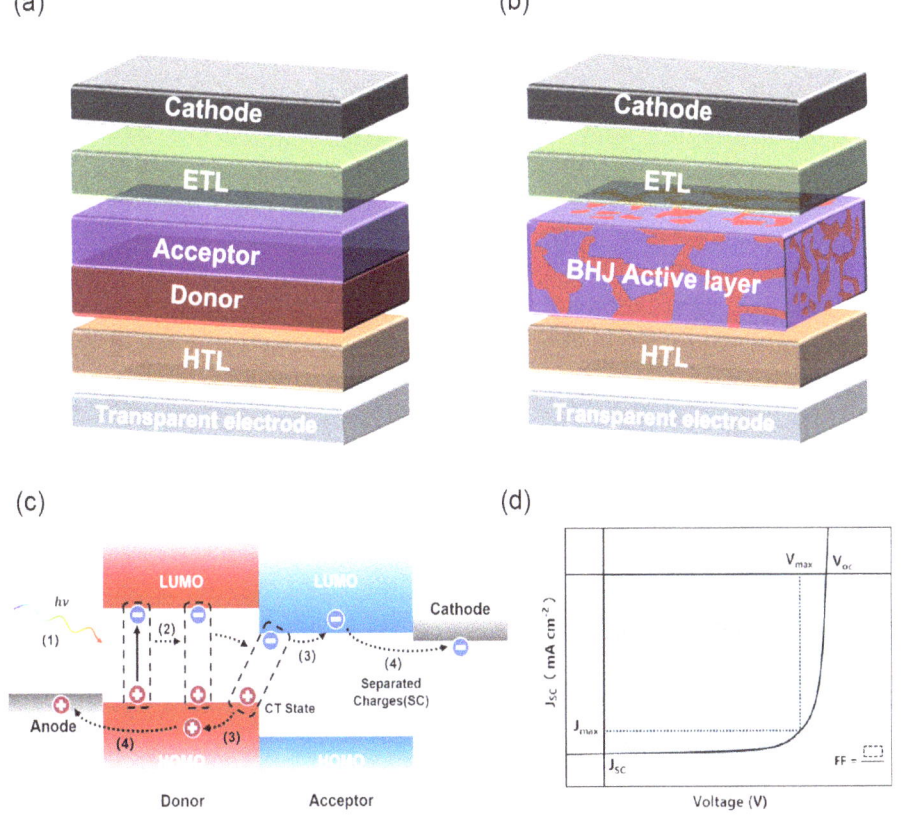

Figure 1.21. OSC device structures: (a) a planar heterojunction device and (b) a bulk heterojunction device. (c) Working principle of organic photovoltaic (OPV) devices. (d) The J–V curve of OSCs.

(b)) [192, 193]. As shown in figure 1.22, the electron donor material is usually composed of small organic molecule donor materials or conjugated polymer donor materials; acceptor materials usually consist of fullerenes and their derivatives, conjugated polymer acceptor materials, and non-fullerene organic small-molecule acceptor materials [178, 194–196]. In this section, we only introduce fullerenes and their derivatives, for other acceptor materials, we refer the reader to the related references [178, 194].

The first organic photovoltaic (OPV) device was made by Kearns and Calvin in 1958 [197]. In 1986, Ching W Tang of the Eastman Kodak Company first proposed the concept of the heterojunction and pioneered research into organic donor/

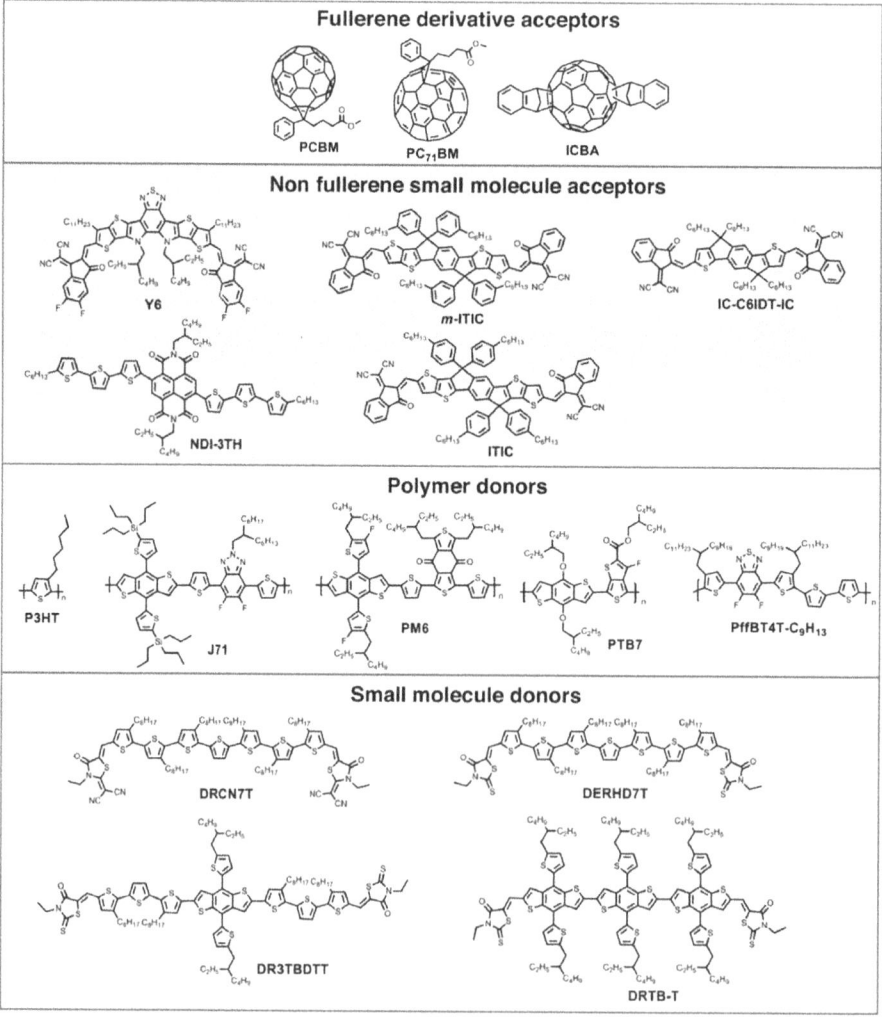

Figure 1.22. Chemical structures of the usual donor and acceptor materials [195], reproduced with permission © John Wiley and Sons.

acceptor heterojunction solar cells [192]. In 1992, Sariciftci and co-workers found that excited-state electrons could be injected from poly[2-methoxy-5-(2-ethylhexyloxy)-1,4-phenylenevinylene] (MEH-PPV) into [60]fullerene extremely quickly, which created an upsurge of polymer/fullerene solar-cell research [198]. Heeger *et al* blended [6,6]-Phenyl-C_{61}-butyric acid methyl ester ($PC_{61}BM$) with MEH-PPV to prepare the first BHJ OSC device with a nanoscale interpenetrating network structure in 1995, which solved the problem of exciton diffusion and transport and increased the contact area between the donor and the acceptor [193]. To date, the BHJ structure is still the most successful OSC structure [181–184, 199].

Unmodified fullerenes (such as [60]fullerene and [70]fullerene) are not suitable for spin coating due to their low solubility in organic solvents [200]. To overcome the disadvantage of the poor solubility of fullerenes and improve the material compatibility between fullerene acceptors and polymer donors, researchers chemically modified fullerenes to obtain derivatives with good solubility. For clarity, we will first introduce some basic parameters of OPVs. As shown in figure 1.21(c), the photoelectric conversion process of OSCs mainly consists of four processes: (1) absorption of incident light and Frenkel exciton generation; (2) diffusion of the excitons to a donor-acceptor interface; (3) dissociation of the excitons across donor-acceptor interfaces; (4) charge-carrier transport and collection. The four most significant photovoltaic performance metrics of solar cells are: the open-circuit voltage (V_{OC}), the short-circuit current (J_{SC}), the fill factor (FF), and the power conversion efficiency (PCE) (figure 1.21(d)). V_{OC} is the voltage between the anode and cathode of the battery when the external circuit is open; J_{SC} is the current density through the device when the external circuit is a short circuit; the FF is defined the ratio of the available power at the maximum power point (P_m) to the product of V_{OC} and J_{SC}; and the PCE is defined as the ratio of the product of J_{SC}, V_{OC}, and FF to the input sunlight power (P_{in}). Therefore, to increase the PCE of OSCs, we must start by increasing V_{OC}, J_{SC}, and FF.

Methanofullerene derivatives have the advantages of high stability and less disturbance of the carbon cage structure by the substituents; they are the most widely studied fullerene derivatives and are widely used in organic electronics. Currently, $PC_{61}BM$ and its derivative [6,6]-Phenyl-C_{71}-butyric acid methyl ester ($PC_{71}BM$) are the most commonly used fullerene electron acceptors; their molecular structure is shown in figure 1.22. $PC_{61}BM$ and $PC_{71}BM$ have high solubility in common organic solvents and are suitable for OSCs prepared by the spin-coating method. Another advantage they have is their energy level, which is matched to those of most electron donors. As mentioned above, the LUMO energy level of fullerenes is a very crucial parameter, as it directly affects the V_{OC} [178, 199, 201]. CV shows that the LUMO levels of $PC_{61}BM$ and $PC_{71}BM$ are both −3.91 eV, while the highest occupied molecular orbital (HOMO) levels are −5.93 eV and −5.87 eV, respectively (figure 1.23(a)) [202]. As shown in figure 1.23(b), $PC_{61}BM$ absorbs weakly in the visible region of the spectrum [202]. In contrast, $PC_{71}BM$ can absorb more visible light, which is why high-efficiency OSCs often use $PC_{71}BM$ instead of $PC_{61}BM$ [202]. Experiments have shown that when they are blended with different types of conjugated polymer donor, fullerene derivatives exhibit ultrafast (subpicosecond)

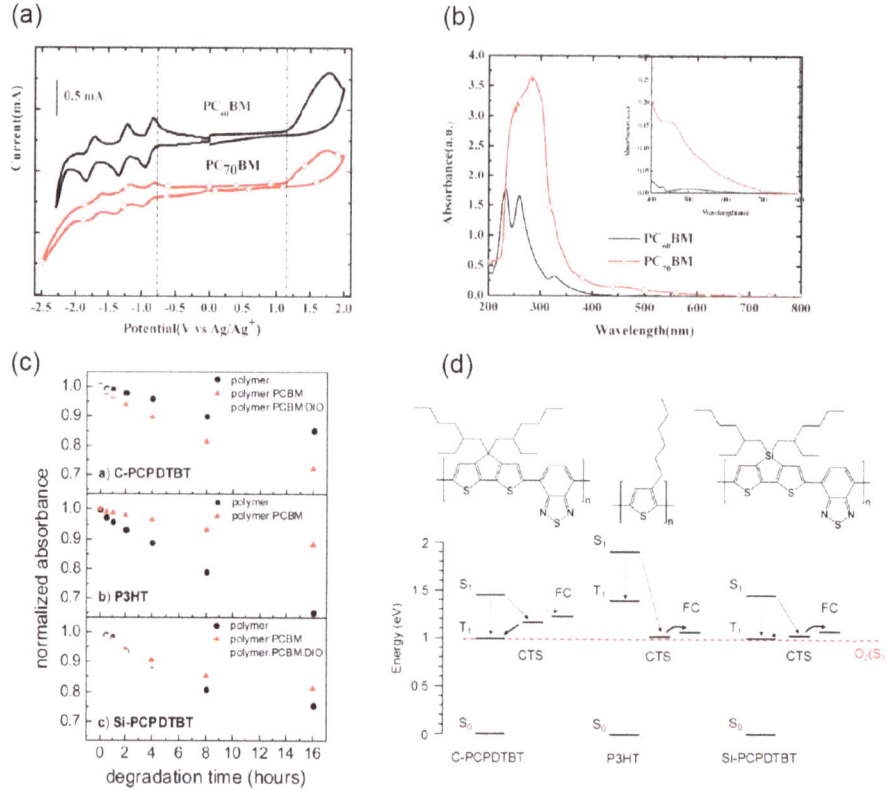

Figure 1.23. (a) Cyclic voltammograms of $PC_{61}BM$ and $PC_{71}BM$. (b) UV–Vis absorptions of $PC_{61}BM$ and $PC_{71}BM$. Source for (a), (b): [202], reproduced with permission © Royal Society of Chemistry. (c) Temporal development of absorbance (in the respective absorption maxima, normalized to the initial values) during the degradation of polymer films under a solar simulator in air. (d) State diagram of the three investigated polymer:$PC_{61}BM$ blends (here, CTS is the energetic level of the charge-transfer state between the polymer and $PC_{61}BM$; FC is the energetic level of the free charge carriers). Source for (c), (d): [205], reproduced with permission © American Chemical Society.

photogenerated electron transport characteristics [203]. Femtosecond laser spectro-scopy was used to study the electron transmission process of the poly[2-methoxy-5-(3,7-dimethyloctyloxy) phenylenevinylene-1,4-diyl] (MDMO-PPV):$PC_{61}BM$ system; the results showed that the forward transmission (30 ps) is nine orders of magnitude faster than the reverse transmission process [204]. Fullerene has a relatively high exciton diffusion orientation, because its derivatives have isotropic (for [60]fullerene derivatives) or relatively isotropic (for [70]fullerene derivatives) electron-accepting properties. Furthermore, fullerene derivatives have a mobility as high as 6 cm^2 V^{-1} s^{-1}, and the diffusion length of excitons in pure fullerene is about 40 nm [191]. The electron mobilities of $PC_{61}BM$ and $PC_{71}BM$ are relatively high and can effectively extract charge carriers; therefore, the polymer solar cells have a high FF.

In addition, fullerenes are also related to the stability of OSCs (figures 1.23(c) and (d)) [190, 205]. Distler *et al* reported that $PC_{61}BM$ has at least four different effects on the photodegradation of polymers, which are active simultaneously [205]. Three of them stabilize the polymer (light screening, excited-state quenching, and chemical effects (i.e. radical scavenging and hydroperoxide cleavage)); experiments show that $PC_{61}BM$ has a stabilizing effect on poly(3-hexylthiophene-2,5-diyl) (P3HT) and poly[2,6-(4,4'-bis(2-ethylhexyl)dithieno[3,2b:2',3'-d]silole)-alt-4,7-(2,1,3-ben-zothiadiazole)] (Si-PCPDTBT). The fourth, destabilization by $PC_{61}BM$, is related to the enhanced generation of triplets in these blends via the charge-transfer (CT) state; in the presence of oxygen, the triplet state may sensitize the formation of chemically active oxygen species, which destroy the stability of the device. For P3HT and Si-PCPDTBT, the triplet populations are reduced in the presence of $PC_{61}BM$, which means that $PC_{61}BM$ plays a stabilizing role. However, in poly[2,6-(4,4-bis-(2-ethylhexyl)-4H-cyclopenta[2,1-b;3,4-b']dithiophene)-alt-4,7(2,1,3-benzothiadia-zole)] (C-PCPDTBT):PCBM, the CT state level is energetically higher than the polymer triplet level, which is conducive to the formation of triplet polymer and is not conductive to the stability of the device [205].

In the past 30 years, fullerene derivatives have been the most widely used acceptor materials in OSCs. Therefore, scientists have done a lot of research into the modification of $PC_{61}BM$. In addition to $PC_{61}BM$ and $PC_{71}BM$, many electron acceptors of fullerene derivatives have been developed [202, 206, 207]. For instance, Li and co-workers designed a type of new fullerene derivative, 1',1',4',4'-tetrahydro-di[1,4]methanonaphthaleno[5,6]fullerene-C_{60} (ICBA) with an electron-rich indenyl group (figure 1.24), which further improved the LUMO energy level and the V_{OC} of

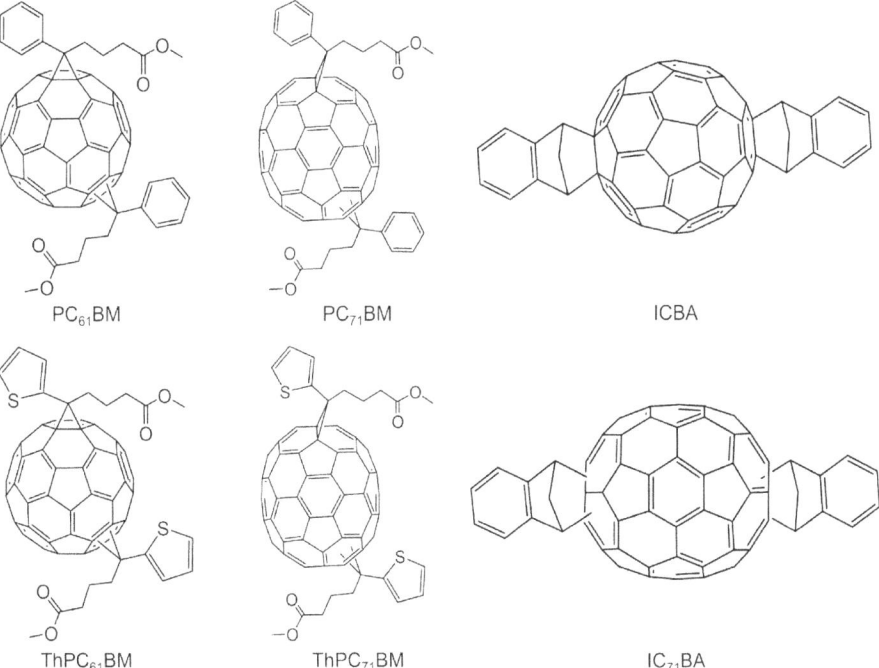

Figure 1.24. Some molecular structures of double-addition methanofullerene.

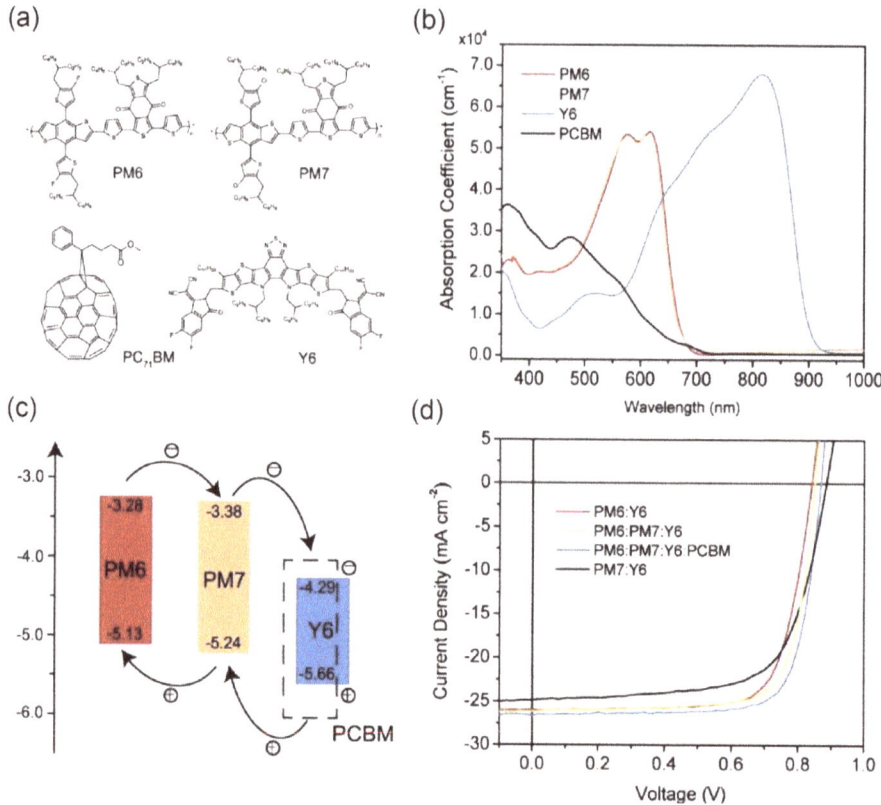

Figure 1.25. (a) Molecular structures of PM6, PM7, Y6, and PC$_{71}$BM. (b) Thin-film absorption coefficients. (c) Energy level alignment and double-cascading transport pathways for a quaternary system. (d) *J–V* curves of binary, ternary, and quaternary devices under constant incident light intensity (AM1.5G, 100 mW cm^{-2}). Source for (a–d): [184], reproduced with permission © Springer Nature.

the corresponding devices [208]. Blom *et al* synthesized adduct PCBM (bis-PCBM) for OSCs (figure 1.24), which effectively improved V_{OC} and the PCE in bis-PCBM: P3HT devices [209]. In addition, some other molecular structures of double-addition methanofullerene are shown in figure 1.24.

With the development of high-performance acceptor–donor–acceptor (A-D-A)-type non-fullerene (NF) electron acceptors, the PCEs of NF OSCs have increased dramatically since 2015 and exceed those of fullerene-based devices [178, 194, 199]. However, fullerene derivatives are still used as the third component in ternary blend OSCs [183, 184, 190]. Today, fullerenes and their derivatives are still seen in state-of-the-art OSCs due to their ability to tune morphology and improve stability [183, 184]. For example, Zhang *et al* used a quaternary blend (PM6:PM7:Y6:PC$_{71}$BM) strategy to produce OSCs with a PCE of more than 18% (figure 1.25) [184]. It should be pointed out that the fullerene derivatives not only play an irreplaceable role in the development of high-performance OSCs but also in understanding the degradation of OSCs [190]. Thus, it is necessary to continue to study the application of fullerene derivatives in OSCs.

1.7.1.2 Perovskite solar cells (PSCs)

In recent years, perovskite solar cells (PSCs) have been of great interest and have been intensively investigated by researchers from many different disciplines [210–212]. The active layer of a PSC is a perovskite-type organic–inorganic hybrid semiconductor material with the general formula ABX_3 (figure 1.26(a)), where A is a monovalent cation (such as MA^+, FA^+, or Cs^+), B is a divalent metal ion (such as Pb^{2+} or Sn^{2+}), and X is a halogen ion (such as I^-, Br^-, or F^-) [210, 213]. Currently, the most common perovskite material for high-efficiency PSCs is $MAPbI_3$; its bandgap is ~1.5 eV, approaching the ideal bandgap for single-junction solar cells [214]. Perovskite materials, as exemplified by $MAPbI_3$, have suitable bandgaps, high absorption coefficients, and excellent carrier transport properties, making them ideal active-layer materials for solar cells [214–217]. At the same time, the method used to prepare perovskite thin-film materials is simple; it can be realized by a low-temperature solution-processing method, which provides the basis for the preparation of high-performance, low-cost PSCs [211, 218, 219]. PSCs have progressed speedily in the past decade, and the PCE of single-junction PSCs has improved significantly from 3.8% [220] to the current certified value of 25.5% [185], which rivals those of commercial silicon solar cells. PSCs are developed from dye-sensitized solar cells [216], as shown in figure 1.26(b). Their device structures include mesoporous and planar heterojunctions, and the planar heterojunctions can be further subdivided into n–i–p and p–i–n planar heterojunctions [221, 222]. Fullerenes and their derivatives are used as the electron

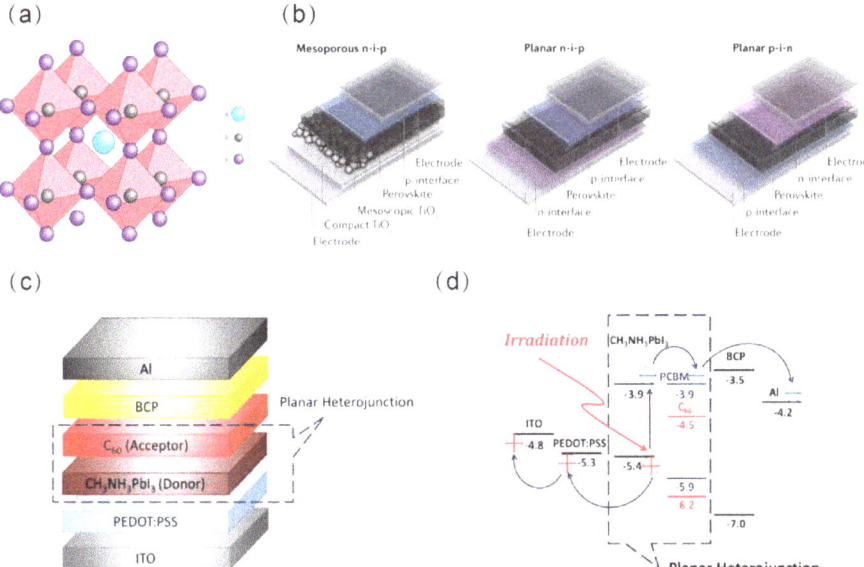

Figure 1.26. (a) Crystal structure of perovskite [211], reproduced with permission © Springer Nature. (b) Device structure of PSCs: mesoporous n-i-p, planar n-i-p, and p-i-n PSCs [222], reproduced with permission © Springer Nature. (c) Perovskite/fullerene planar heterojunction device structure and (d) energy levels. Source for (c), (d): [223], reproduced with permission © John Wiley and Sons.

transport layer or interface modification layer materials for PSCs because they have relatively high electron mobility, good film-forming properties, and suitable energy levels for perovskite materials [223, 224]. There are more and more applications of fullerenes in PSCs because of their increased efficiency, reduced hysteresis, and improved stability [225]. In this section, we will mainly introduce the applications of fullerenes in PSCs.

Fullerenes and their derivatives can be used as electron transport layers (ETLs). In 2013, Jeng *et al* introduced [60]fullerene and its derivatives $PC_{61}BM$ and ICBA into PSCs as ETLs for the first time [223]. They used the device structure of ITO/ poly(2,3-dihydrothieno-1,4-dioxin)-poly(styrenesulfonate) (PEDOT:PSS)/$MAPbI_3$/ C_{60} or C_{60} derivatives/bathocuproine (BCP)/Al, and only achieved a maximum PCE of 3.9% [223]. The device structure and energy levels of the corresponding materials are shown in figures 1.26(c) and (d) [223]. However, this work opened a precedent for the application of fullerenes and their derivatives in PSCs. Subsequently, Lam *et al* optimized the thickness of perovskite and the conditions of film formation, which improved the PCE by 7.4% based on $MAPbI_3$/$PC_{61}BM$ [226]. Huang *et al* spin coated a layer of $PC_{61}BM$ and ICBA onto a perovskite layer and then evaporated a layer of [60]fullerene to form a double fullerene electron transport layer, which increased the FF to more than 80% and the PCE to 12.2% [227]. Wu *et al* used $PC_{71}BM$ as the ETL in PSCs, achieving a V_{OC} as high as 1.05 V, a FF of 78%, and a PCE of 16.31% [228]. Huang and co-workers used solvent annealing of $PC_{61}BM$ and improved the PCE to more than 20% [229]. Tin perovskite is gaining popularity as promising candidate with which to address the toxicity and theoretical efficiency limitation of lead perovskite [214, 230, 231]. However, tin perovskite V_{OC} reported today is generally only 0.6 V, which much lower than the V_{OC} of lead perovskite [214, 230, 231]. In 2020, Jiang *et al* introduced the fullerene derivative ICBA, which has a higher energy level, into tin perovskite as an ETL, which increased the V_{OC} to 0.94 V, which is more than 50% higher than that of 0.6 V for the device based on PCBM [230]. In addition, the device has a PCE of 12.4% (figures 1.27(a) and (b)) [230].

Fullerenes can not only be used as ETLs, but can also play a significant role in reducing the hysteresis effect of PSCs by passivating the traps at the surfaces and grain boundaries of perovskite thin films, which can increase the PCE of PSCs [225, 232]. PSCs have a $J–V$ curve hysteresis effect, which limits their further development. The so-called hysteresis effect means that different results are obtained for forward sweeps than for reverse sweeps when testing the $J–V$ curve [233]. In 2014, Shao *et al* inserted a PCBM/C_{60} double fullerene layer between the perovskite and the cathode as the electron acceptor and collection layer [232]. The results showed that compared with control devices (without PCBM), the device with PCBM annealing at 100 °C for 45 min could effectively suppress the hysteresis effect and increase the PCE by 204%, from 7.3% to 14.9% (figure 1.27(c)) [232]. Photoluminescence (PL) spectroscopy and thermal admittance spectroscopy (TAS) proved that PCBM can diffuse to the perovskite grain boundaries in the top layer and effectively passivate defects (figure 1.27(d)), which is the reason that the current hysteresis and PCE of the device are improved [232].

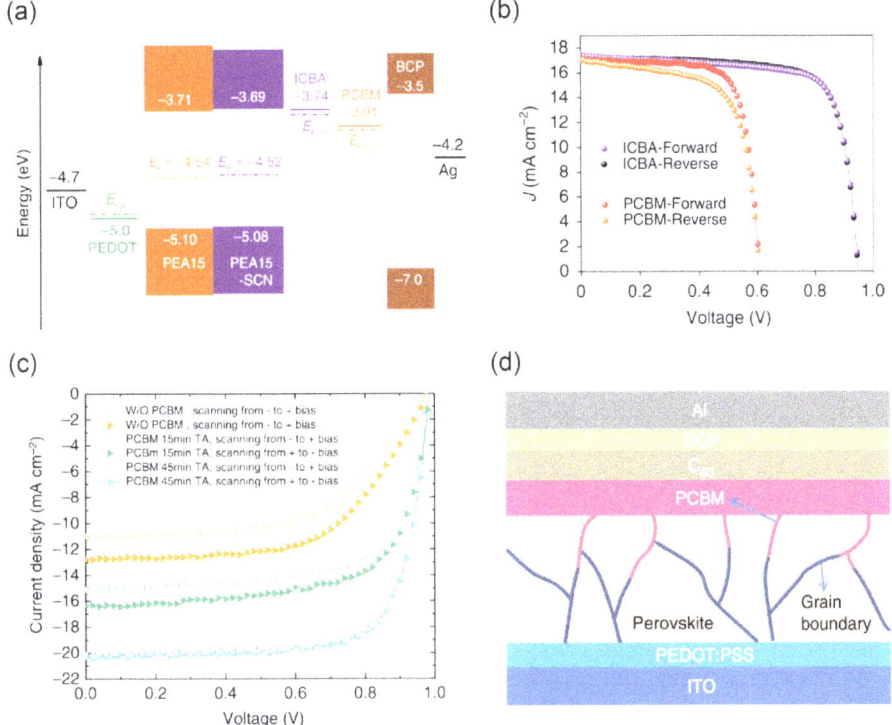

Figure 1.27. (a) Schematic illustration of energy levels. Dashed lines represent the quasi-Fermi levels of ICBA (E_{Fn-I}), PCBM (E_{Fn-p}), and PEDOT (E_{Fp}). (b) The J–V curves of a certified PEA15-SCN device with ICBA and the champion device of PEA15-SCN film with PCBM. Source for (a), (b): [230], reproduced with permission © Springer Nature. (c) The J–V curves for devices without a PCBM layer (orange), with a PCBM layer thermal annealing (TA) for 15 min (green) and 45 min (blue), respectively. (d) Schematic of PCBM diffusion to the perovskite grain boundary and the passivation of defects. Source for (c), (d): [232], reproduced with permission © Springer Nature.

In addition to the commonly used $PC_{61}BM$, $PC_{71}BM$, and ICBA, many functionalized fullerene derivatives have been applied to PSCs, such as Lewis base functionalized fullerene derivatives, carboxyl and hydroxyl functionalized fullerene derivatives, halogen functionalized fullerene derivatives, and cross-linked fullerene derivatives [234]. In short, fullerenes and their derivatives have played a crucial role in improving device performance and stability during the rapid evolution of PSCs. We believe that in the field of PSCs in the near future, fullerenes and their derivatives will have more and more significant applications and provide new possibilities for the commercialization of PSCs.

1.7.1.3 Organic field-effect transistors (OFETs)

The field-effect transistor (FET) is a semiconductor device that uses an electric field effect to control the flow of current; it is one of the most basic electronic components [235, 236]. The OFET is a FET that uses organic semiconductors as its active layers;

it is a basic building block of organic circuits, a vital organic semiconductor device, and one of the essential frontiers in the organic semiconductor materials and device research field [235, 237, 238]. The OFET has a wide range of application scenarios in the fields of organic sensors, organic storage devices, flexible displays, e-paper, smart wearable devices, and RF identification due to its light weight, low cost, ability to be folded, and its suitability for large area preparation [235–239].

Organic semiconductor materials can be divided into two categories: one type is the P-type organic semiconductor material, which is used to transport holes; the other is the N-type organic semiconductor material, which is used to transport electrons. Currently, in-depth research into P-type organic semiconductor materials is taking place; their mobility is relatively high and their device performance is good. Comparatively, N-type OFETs are as yet poorly developed. Therefore, the development of N-type organic semiconductor materials is one of the challenges of OFET materials. At present, most N-type OFET materials are fullerenes and their derivatives due to their high electron mobility. In 1995, Haddon and co-workers first used [60]fullerene as the active material with which to make an OFET; its field-effect mobility was 0.08 cm^2 (V·s)$^{-1}$ and its on–off ratio was 10^6 [240]. After that, they first used [70]fullerene to make an N-type OFET; its field-effect mobility was 0.002 cm^2 (V·s)$^{-1}$, which was lower than that of the device using [60]fullerene. They believed that the reason for the difference was the distinct electronic structures and symmetries of [60] and [70]fullerene [241]. In 2012, Jen *et al* systematically studied the relationship between the structure of ten fullerene derivatives and the performance of OFETs; the results showed that small structural alternations, functional patterns, and the number of addends on fullerene derivatives strongly affect their mobilities [242]. At present, the performance of fullerene-based OFETs is still lower than that of P-type OFETs, which needs further research.

In summary, fullerenes and their derivatives are widely used in OSCs, PSCs, and OFETs, although fullerenes and their derivatives are organic semiconductors and their properties cannot be compared with those of inorganic semiconductors such as Si or GaAs. We believe that organic electronics has great application prospects. With the rapid development of organic electronics, fullerenes and their derivatives will play an irreplaceable role in the near future.

1.7.2 Biomedicine

Fullerenes and their derivatives have multiple activities, such as antioxidant, cell-protective, antimicrobial, and photodynamic activities; they can be used for drug delivery and tumor treatment and they are a type of candidate material that has great potential in the field of biomedicine [243–248]. However, unmodified fullerenes are insoluble in water and need to be modified to make them water soluble in biological systems. Fullerenes, as one of the carbon nanomaterials used in the next generation of biomedicine, have made great progress in recent years. In this section, we will briefly introduce the application of fullerenes and their derivatives in biomedicine.

1.7.2.1 Antioxidative stress and radical scavenging

If we want to treat aging, then we must understand aging, and preferably at the molecular level. In 1956, Dr Denham Harman proposed the free radical theory of aging (FRRA) for the first time [249]. Free radicals are atoms or groups with unpaired electrons, which are produced by cellular metabolism or abnormal reactions. An imbalance between oxidation and antioxidation in the human body produces a large number of oxides, such as reactive oxygen species (ROS), reactive nitrogen species (RNS), etc. Free radicals have high reactivity and can damage biological molecules such as proteins, lipids, and DNA, which in turn can cause a variety of diseases [35, 250, 251]. With the gradual deepening of research into free radicals, scientists have become more and more aware that measures to remove excess free radicals are beneficial to the prevention and treatment of certain diseases. Therefore, the development and utilization of high-efficiency and non-toxic free radical scavengers has become a research hotspot in the life sciences.

Fullerenes have the reputation of being 'free radical sponges.' As mentioned in the previous section, fullerenes and their derivatives are large electron-deficient conjugated systems, which can quench ROS molecules such as superoxide (O_2^-), hydroxyl radicals (OH), and hydrogen peroxide (H_2O_2) and can scavenge free radicals and resist oxidative stress, thereby protecting cells. In 1999, Wang *et al* studied the antioxidant effects of C_{60}, vitamin E, and three C_{60} derivatives on the prevention of lipid peroxidation induced by superoxide and hydroxyl radicals. The results showed that both liposoluble and water-soluble fullerene derivatives could effectively protect lipids from radical-initiated peroxidation and breakdown of membrane integrity; a liposoluble fullerene derivative even showed stronger effects than vitamin E in the prevention of lipid peroxidation [252]. In 2006, H Takada *et al* reported that [60]fullerene could quickly capture free radical molecules, i.e. much faster than β-carotene [158]. In 2017, Zhou *et al* synthesized two novel materials, C_{60}–OH and C_{70}–OH, and studied their antioxidant properties. The results showed that the novel C_{60}–OH and C_{70}–OH have high stability in water. An experiment using a mouse model for single and reduplicative chemotherapy-induced liver injury demonstrated that they have protective effects in the chemo-therapeutic process (figure 1.28(a)) [158]. The two novel biocompatible [60]/[70] fullerenols may be promising protective agents that can satisfy the demand for future clinical chemotherapy [243]. All this shows that fullerenes and their derivatives are a kind of antioxidant with good performance and that they have great application potential. In addition, fullerene derivatives may also inhibit the internal physiological aging of organisms. Interestingly, Baati *et al* fed experimental rats poisoned by carbon tetrachloride (CCl_4) with fullerenes dissolved in olive oil for seven consecutive months. The results showed that the rats not only had no significant adverse reactions but also that their average lifespan was extended from two years to five years (figure 1.28(b)). The researchers thought that the effect on lifespan was mainly due to the attenuation of age-associated increases in oxidative stress [247].

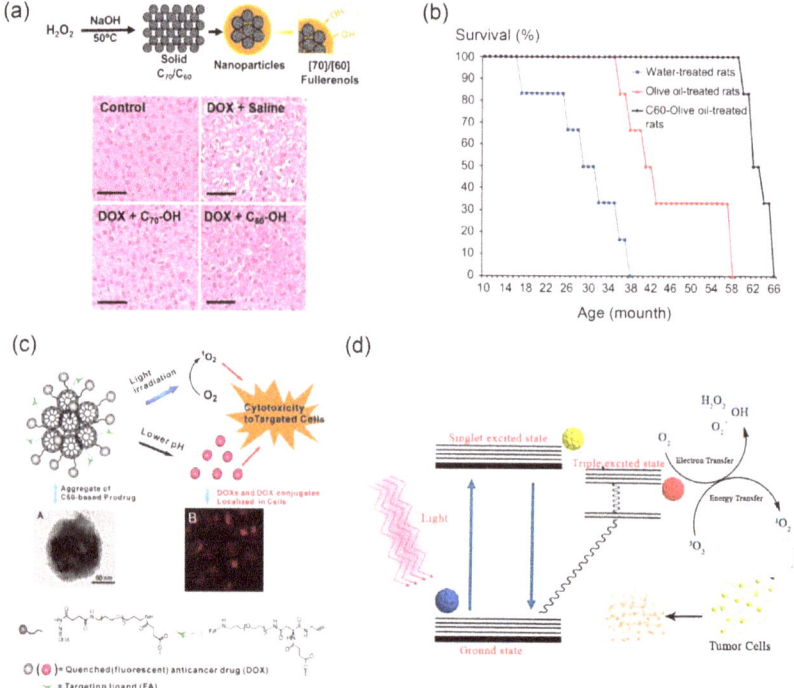

Figure 1.28. (a) Two novel biocompatible [60]/[70] fullerenols provide a potent defense against oxidative injury [243], reproduced with permission © American Chemical Society. (b) Survival rates of the surviving animals after treatment (oral gavages) at reiterated doses (1.7 mg kg^{-1} bw) of water, olive oil, and C$_{60}$-olive oil [247], reproduced with permission © Elsevier. (c) Schematic illustration of a fullerene-based aggregate with dual anticancer actions [248], reproduced with permission © John Wiley and Sons. (d) Schematic of photodynamic therapeutic (PDT) with a fullerene as a photosensitizer [35], reproduced with permission © Elsevier.

1.7.2.2 Drug delivery

Fullerenes have many advantages for drug delivery; for example, fullerenes can be modified by various groups according to the requirements, unique carbon cage structure, photosensitive activity, etc. These advantages enable fullerenes to be used for excellent drug delivery to achieve drug targeting and slow, controlled delivery. For example, Fan *et al* modified [60]fullerene using folic acid (as a targeting ligand) and doxorubicin (DOX, a type of anticancer drug). This derivative could agglomerate in water to form nanoscale aggregates, which demonstrates active targeting and pH-responsive chemotherapy. It was able to enter folate-receptor-positive cancer cells and kill the cells via intracellular release of the active drug form (figure 1.28(c)) [248]. In addition, this DOX–fullerene aggregate prodrug also had photodynamic therapeutic (PDT) properties and, as a photosensitizer, produced singlet oxygen under light to kill cancer cells (the principle of PDT is shown in figure 1.28(d)) [35, 248]. The combined effect of chemotherapy and PDT increased the therapeutic efficacy of the DOX–fullerene aggregate prodrug [248]. This drug design approach provides useful insights into designing fullerenes and their derivatives and improving their applicability in other prodrug systems for targeted

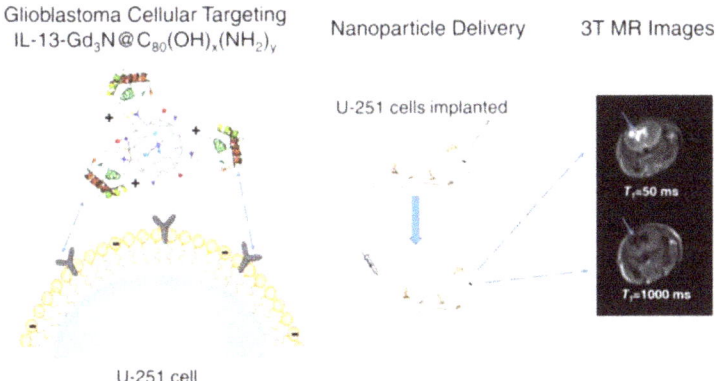

Glioblastoma Cellular Targeting
IL-13-Gd$_3$N@C$_{80}$(OH)$_x$(NH$_2$)$_y$

Nanoparticle Delivery

3T MR Images

U-251 cells implanted

T_1=50 ms

T_1=1000 ms

U-251 cell

Figure 1.29. The use of Gd$_3$N@C$_{80}$(OH)$_x$(NH$_2$)$_y$-targeted MRI to detect glioblastoma tumor cells [245], reproduced with permission © American Chemical Society.

cancer therapy [248]. In short, fullerenes and their derivatives have great application potential for drug and gene delivery.

1.7.2.3 MRI contrast agent for noninvasive imaging

Magnetic resonance imaging (MRI), an important nonradiative and high-spatial-resolution imaging technique, has wide clinical use. However, enhancing the imaging quality and the precision of diagnosis is still a challenge. The development of endohedral fullerenes (ENFs) as next-generation diagnostic and therapeutic drug platforms is an active area of both chemistry and cancer research. Many gadolinium (Gd)-containing metallofullerenes have been reported as diagnostic MRI contrast agents [245]. For example, Li and co-workers reported a new interleukin-13 amino-coated gadolinium metallofullerene, which was able to target MRI detection of glioblastoma tumors (figure 1.29) [245]. In short, fullerene contrast agents have the characteristics of low toxicity and high efficiency and the ability to perform targeted imaging of biological organs; as a result, they have huge application prospects.

In summary, fullerenes and their derivatives have many beneficial applications in biomedicine due to their unique physical and chemical properties. However, there are still some problems that need to be solved; for example, the issue of fullerene toxicity remains controversial. At present, the applications of fullerenes and derivatives in biomedicine are mostly limited to *in vitro* and *in vivo* research and have not yet reached the final stages of clinical trials; we still need to conduct more in-depth characterization and research on them [35]. We believe that in the near future, fullerenes and their derivatives will play a key role in the diagnosis, treatment, and pre-treatment of major human diseases and will bring breakthrough progress to modern medicine.

1.7.3 Cosmetics

As mentioned above, fullerenes and their derivatives are effective free radical scavengers and antioxidants due to their unique electronic structure. They can be used in sunscreen, skin whitening, and antiaging products [157]. Skin aging is related to cell oxidative stress damage; when fullerenes and their derivatives enter the human body, they can interact with active free radicals to prevent aging. However, fullerenes only have a very limited solubility in water because of their hydrophobic nature, and therefore their future application in cosmetics is hindered; they must be made hydrophilic to play their best roles in human tissues. Therefore, it is necessary to develop new water-soluble fullerenes and their derivatives to improve the biocompatibility and practicality of fullerenes in cosmetics [253–256].

There are already some fullerene-containing cosmetics on the market. In addition, researchers have conducted some clinical trials on fullerene-containing cosmetics [256, 257]. For example, Kato *et al* conducted a clinical evaluation of [60]fullerene dissolved in squalene for anti-wrinkle cosmetics in 2010. They conducted continuous tests on the skin condition of 23 young or middle-aged women who used the lipofullerene-containing skin cream LF-SQ twice a day in the morning and evening [257]. The result showed that, compared with the placebo group, the continuous use of LF-SQ for eight weeks could significantly reduce the area of wrinkles [257]. In addition, there are some reports that fullerene and their derivatives can be used as hair growth agents to induce the production of new hair follicles [258].

In summary, fullerenes and their derivatives have become one of the most popular research hotspots in the cosmetics industry due to their excellent antioxidant capacities. Many cosmetics containing fullerene ingredients are already on sale. However, their principles and mechanisms still need further research.

1.7.4 Catalysts

Fullerenes and their derivatives have a very attractive prospect in the field of catalysts because of their unique molecular structures, special electronic properties, and dispersion performance for metal catalysts. At present, there are two main approaches to fullerenes as catalysts: one is to mix fullerenes with other semi-conductor materials to enhance the photocatalytic performance of the resulting composite materials; this approach only uses a simple physical method to mix the two materials. The other is to directly use fullerenes and their derivatives as catalysts [105, 259]. Fullerene materials have strong electrophilicity, can stabilize free radicals, and promote the breaking and generation of strong chemical bonds. Therefore, fullerenes can be directly used as new catalysts to catalyze singlet oxygen reactions, hydrogenation reactions, hydrogen transfer reactions, nitrogen fixation reactions, diamond synthesis, etc [260–265]. In addition, fullerenes are a weak π electron donors and can be complexed with transition metals such as Ni, Pd, Ru, Rh, etc [266]. Fullerene–metal complexes have special catalytic properties which can catalyze various types of reaction, such as hydrogenation reactions, hydrosilation reactions, coupling reactions, etc.

1.7.5 Other applications of fullerenes

In addition to the abovementioned applications, fullerenes and their derivatives are also widely used in many other fields, such as superconductors [105, 106], nonlinear optics [267, 268], lubricants [269, 270], hydrogen storage [271, 272], etc. We cannot introduce all of them in detail here; however, the reader can learn about them from related references.

1.7.6 Summary and prospects

Fullerene were discovered more than 30 years ago, which ushered in a new era of carbon element research. The application of fullerenes and their derivatives has received continuous research and extensive attention from scientists around the world. These research activities have directly promoted an upsurge in nanoscience research. To date, researchers have achieved surprising results from basic and applied research into fullerenes. As a result of their unique molecular structures and special physical and chemical properties, fullerenes have had a profound impact on numerous disciplines, such as physics, chemistry, materials science, and biology. They also have wide and attractive application prospects in the fields of organic electronics, biomedicine, daily chemicals, energy, catalysts, superconductors, non-linear optics, lubricants, etc.

Although fullerenes and their derivatives have broad application prospects, they are still in the laboratory stage at present, and still have a long way to go before they will be ready for large-scale practical applications. On the one hand, we still need to conduct more in-depth research into fullerenes; on the other hand, the industrialization and marketization of fullerenes need to be further explored. We believe that fullerenes will have more applications soon.

References

[1] Walker P L 1972 *Carbon* **10** ii–382
[2] Osawa E 1970 Superaromaticity *Kagaku* **25** 854–63
[3] Osawa E, Kroto H W, Fowler P W and Wasserman E 1993 *Phil. Trans.* **343** 1–8
[4] Kroto H W, Heath J R, O'Brien S C, Curl R F and Smalley R E 1985 *Nature* **318** 162–3
[5] Krätschmer W, Lamb L D, Fostiropoulos K and Huffman D R 1990 *Nature* **347** 354–8
[6] Atkinson N 2010 Buckyballs Could Be Plentiful in the Universe https://universetoday.com/76732/buckyballs-could-be-plentiful-in-the-universe/
[7] Buseck P R, Tsipursky S J and Hettich R 1992 *Science* **257** 215
[8] Heymann D, Chibante L P F, Brooks R R, Wolbach W S and Smalley R E 1994 *Science* **265** 645
[9] Becker L, Bada J L, Winans R E, Hunt J E, Bunch T E and French B M 1994 *Science* **265** 642
[10] Cami J, Bernard-Salas J, Peeters E and Malek S E 2010 *Science* **329** 1180
[11] Kroto H W 1987 *Nature* **329** 529–31
[12] Fowler P W and Manolopoulos D E 1995 *An Atlas of Fullerenes* (Mineola, NY: Dover Publications)
[13] Aihara J, Nakagami Y and Sekine R 2015 *J. Phys. Chem.* A **119** 6542–50
[14] Schwerdtfeger P, Wirz L N and Avery J 2015 *Wiley Interdiscip Rev. Comput. Mol. Sci.* **5** 96–145
[15] Liu S, Lu Y-J, Kappes M M and Ibers J A 1991 *Science* **254** 408

[16] Tománek D 2014 *Guide Through the Nanocarbon Jungle* (San Rafael, CA: Morgan & Claypool)

[17] Wang C-R, Sugai T, Kai T, Tomiyama T and Shinohara H 2000 *Chem. Commun.* **2000** 557–8

[18] Ettl R, Chao I, Diederich F and Whetten R L 1991 *Nature* **353** 149–53

[19] Diederich F, Whetten R L, Thilgen C, Ettl R, Chao I T O and Alvarez M M 1991 *Science* **254** 1768

[20] Kikuchi K *et al* 1992 *Chem. Phys. Lett.* **188** 177–80

[21] Kikuchi K, Nakahara N, Wakabayashi T, Suzuki S, Shiromaru H, Miyake Y, Saito K, Ikemoto I, Kainosho M and Achiba Y 1992 *Nature* **357** 142–5

[22] Miyake Y, Minami T, Kikuchi K, Kainosho M and Achiba Y 2000 *Mol. Cryst. Liq. Cryst. A* **340** 553–8

[23] Yang H, Mercado B Q, Jin H, Wang Z, Jiang A, Liu Z, Beavers C M, Olmstead M M and Balch A L 2011 *Chem. Commun.* **47** 2068–70

[24] Yang H, Jin H, Che Y, Hong B, Liu Z, Gharamaleki J A, Olmstead M M and Balch A L 2012 *Chemistry* **18** 2792–6

[25] Heath J R, O'Brien S C, Zhang Q, Liu Y, Curl R F, Tittel F K and Smalley R E 1985 *J. Am. Chem. Soc.* **107** 7779–80

[26] Chai Y, Guo T, Jin C, Haufler R E, Chibante L P F, Fure J, Wang L, Alford J M and Smalley R E 1991 *J. Phys. Chem.* **95** 7564–8

[27] Saunders M, Cross R J, Jiménez-Vázquez H A, Shimshi R and Khong A 1996 *Science* **271** 1693

[28] Saunders M, Jiménez-Vázquez H A, Cross R J and Poreda R J 1993 *Science* **259** 1428

[29] Popov A A 2017 *Endohedral Fullerenes: Electron Transfer and Spin* (Cham: Springer)

[30] Yang S, Wei T and Jin F 2017 *Chem. Soc. Rev.* **46** 5005–58

[31] Vostrowsky O and Hirsch A 2006 *Chem. Rev.* **106** 5191–207

[32] Ewels C P 2006 *Nano Lett.* **6** 890–5

[33] Guo T, Nikolaev P, Thess A, Colbert D T and Smalley R E 1995 *Chem. Phys. Lett.* **243** 49–54

[34] Puretzky A A, Schittenhelm H, Fan X, Lance M J, Allard L F and Geohegan D B 2002 *Phys. Rev. B* **65** 245425

[35] Goodarzi S, D Ros T, Conde J, Sefat F and Mozafari M 2017 *Mater. Today* **20** 460–80

[36] Withers J C, Loutfy R O and Lowe T P 1997 *Fullerene Sci. Technol.* **5** 1–31

[37] Haufler R E, Conceicao J, Chibante L P F, Chai Y, Byrne N E, Flanagan S, Haley M M, O'Brien S C and Pan C *et al* 1990 *J. Phys. Chem.* **94** 8634–6

[38] Huczko A, Lange H, Byszewski P, Poplawska M and Starski A 1997 *J. Phys. Chem. A* **101** 1267–9

[39] Pinzón J R, Villalta-Cerdas A and Echegoyen L 2012 Fullerenes, carbon nanotubes, and graphene for molecular electronics *Unimolecular and Supramolecular Electronics I (Topics in Current Chemistry* vol 312) (Berlin: Springer) pp 127–74

[40] Dudnik A I, Osipova I V, Nikolaev N S and Churilov G N 2020 *Fullerenes, Nanotubes and Carbon Nanostructures* **28** 697–701

[41] Churilov G N, Krätschmer W, Osipova I V, Glushenko G A, Vnukova N G, Kolonenko A L and Dudnik A I 2013 *Carbon* **62** 389–92

[42] Wang Hao Z N, Cheng Z, Bo Y and Peng Q 2021 *Materials Reports* **35** 71–7

[43] Kareev I E, Nekrasov V M and Bubnov V P 2015 *Tech. Phys.* **60** 102–6

[44] Gerhardt P, Löffler S and Homann K H 1987 *Chem. Phys. Lett.* **137** 306–10

[45] Howard J B, McKinnon J T, Makarovsky Y, Lafleur A L and Johnson M E 1991 *Nature* **352** 139–41

[46] Baum R 1991 *Chemical & Engineering News Archive* **69** 6

[47] Howard J B, McKinnon J T, Johnson M E, Makarovsky Y and Lafleur A L 1992 *J. Phys. Chem.* **96** 6657–62

[48] Pope C J and Howard J B 1996 *Tetrahedron* **52** 5161–78

[49] Goel A, Hebgen P, Vander Sande J B and Howard J B 2002 *Carbon* **40** 177–82

[50] Murayama H, Tomonoh S, Alford J M and Karpuk M E 2005 *Fullerenes, Nanotubes and Carbon Nanostructures* **12** 1–9

[51] Yoshie K i, Kasuya S, Eguchi K and Yoshida T 1992 *Appl. Phys. Lett.* **61** 2782–3

[52] Todorovic-Marković B, Marković Z, Mohai I, Károly Z, Gál L, Föglein K, Szabó P T and Szépvölgyi J 2003 *Chem. Phys. Lett.* **378** 434–9

[53] Fulcheri L, Fabry F and Rohani V 2012 *Carbon* **50** 4524–33

[54] Szépvölgyi J, Marković Z, Todorović-Marković B, Nikolić Z, Mohai I, Farkas Z, Tóth M, Kováts É, Scheier P and Feil S 2006 *Plasma Chem. Plasma Process.* **26** 597–608

[55] Kaneko T, Abe S, Ishida H and Hatakeyama R 2007 *Phys. Plasmas* **14** 110705

[56] Krokos E 2010 *J. Phys. Chem.* C **114** 7626–30

[57] Anctil A, Babbitt C W, Raffaelle R P and Landi B J 2011 *Environ. Sci. Technol.* **45** 2353–9

[58] Boorum M M, Vasilev Y V, Drewello T and Scott L T 2001 *Science* **294** 828

[59] Scott L T, Boorum M M, McMahon B J, Hagen S, Mack J, Blank J, Wegner H and de Meijere A 2002 *Science* **295** 1500

[60] Lungerich D, Hoelzel H, Harano K, Jux N, Amsharov K Y and Nakamura E 2021 *ACS Nano* **15** 12804–14

[61] Winter A *et al* 2018 *Carbon* **128** 106–16

[62] Chibante L P F, Thess A, Alford J M, Diener M D and Smalley R E 1993 *J. Phys. Chem.* **97** 8696–700

[63] Manawi Y M, Ihsanullah, Samara A, Al-Ansari T and Atieh M A 2018 *Materials (Basel)* **11** 822

[64] Eacute R, Albero R A, Miguel B, Tadeo E, Eacute J, Aacute L and Nchez-Brunete C 2013 *Anal. Sci.* **29** 533–8

[65] Ruoff R S, Tse D S, Malhotra R and Lorents D C 1993 *J. Phys. Chem.* **97** 3379–83

[66] Jensen W B 2007 *J. Chem. Educ.* **84** 1913

[67] Parker D H, Wurz P, Chatterjee K, Lykke K R, Hunt J E, Pellin M J, Hemminger J C, Gruen D M and Stock L M 1991 *J. Am. Chem. Soc.* **113** 7499–503

[68] Vítek P, Jehlička J, Frank O, Hamplová V, Pokorná Z, Juha L and Boháček Z 2009 *Fullerenes, Nanotubes Carbon Nanostructures* **17** 109–22

[69] Shareef A, Li G and Kookana R S 2010 *Environ. Chem.* **7** 292–7

[70] Capp C, Wood T D, Marshall A G and Coe J V 1994 *J. Am. Chem. Soc.* **116** 4987–8

[71] Churilov G N, Elesina V I, Dudnik A I and Vnukova N G 2019 *Fullerenes, Nanotubes Carbon Nanostructures* **27** 225–32

[72] Kwok K S, Chan Y C, Ng K M and Wibowo C 2009 *AIChE J.* **56** 1801–12

[73] Taylor R, Hare J P, Abdul-Sada A a K and Kroto H W 1990 *J. Chem. Soc., Chem. Commun.* **20** 1423–5

[74] Chatterjee K, Parker D H, Wurz P, Lykke K R, Gruen D M and Stock L M 1992 *J. Org. Chem.* **57** 3253–4

[75] Diack M, Compton R N and Guiochon G 1993 *J. Chromatogr.* A **639** 129–40

[76] Scrivens W A, Bedworth P V and Tour J M 1992 *J. Am. Chem. Soc.* **114** 7917–9

[77] Scrivens W A, Cassell A M, North B L and Tour J M 1994 *J. Am. Chem. Soc.* **116** 6939–40

[78] Komatsu N, Ohe T and Matsushige K 2004 *Carbon* **42** 163–7

[79] Komatsu N, Kadota N, Kimura T, Kikuchi Y and Arikawa M 2007 *Fullerenes, Nanotubes Carbon Nanostructures* **15** 217–26

[80] Jinno K, Uemura T, Ohta H, Nagashima H and Itoh K 1993 *Anal. Chem.* **65** 2650–4

[81] Yi H *et al* 2017 *Chem. Eng. J.* **330** 134–45

[82] Zhou X, Liu J, Jin Z, Gu Z, Wu Y and Sun Y 1997 *Fullerene Sci. Technol.* **5** 285–90

[83] Yeretzian C, Wiley J B, Holczer K, Su T, Nguyen S, Kaner R B and Whetten R L 1993 *J. Phys. Chem.* **97** 10097–101

[84] Suzuki T, Nakashima K and Shinkai S 1994 *Chem. Lett.* **23** 699–702

[85] Araki K, Akao K, Ikeda A, Suzuki T and Shinkai S 1996 *Tetrahedron Lett.* **37** 73–6

[86] Garcia-Simon C, Garcia-Borras M, Gomez L, Parella T, Osuna S, Juanhuix J, Imaz I, Maspoch D, Costas M and Ribas X 2014 *Nat. Commun.* **5** 5557

[87] Kishi N, Akita M, Kamiya M, Hayashi S, Hsu H F and Yoshizawa M 2013 *J. Am. Chem. Soc.* **135** 12976–9

[88] Bezmel'nitsyn V N, Eletskii A V and Okun M V 1998 *Phys.-Usp.* **41** 1091–114

[89] Diaz de Zerio Mendaza A, Bergqvist J, Bäcke O, Lindqvist C, Kroon R, Gao F, Andersson M R, Olsson E, Inganäs O and Müller C 2014 *J. Mater. Chem.* A **2** 14354–9

[90] Hare J P, Kroto H W and Taylor R 2013 *Chem. Phys. Lett.* **589** 57–60

[91] Westerstrom R *et al* 2012 *J. Am. Chem. Soc.* **134** 9840–3

[92] Meilunas R, Chang R P H, Liu S, Jensen M and Kappes M M 1991 *J. Appl. Phys.* **70** 5128–30

[93] Hare J P, Dennis T J, Kroto H W, Taylor R, Allaf A W, Balm S and Walton D R M 1991 *J. Chem. Soc., Chem. Commun.* **6** 412–3

[94] Guan R, Chen M and Yang S 2020 *Chin. Sci. Bull.* **65** 2209

[95] Sorace L, Benelli C and Gatteschi D 2011 *Chem. Soc. Rev.* **40** 3092–104

[96] Cornia A, Mannini M, Sainctavit P and Sessoli R 2011 *Chem. Soc. Rev.* **40** 3076–91

[97] Westerström R *et al* 2014 *Phys. Rev.* B **89** 060406

[98] Nie M *et al* 2019 *Nano Res.* **12** 1727–31

[99] Brandenburg A, Krylov D S, Beger A, Wolter A U B, Büchner B and Popov A A 2018 *Chem. Commun.* **54** 10683–6

[100] Yang S *et al* 2013 *Sci. Rep.* **3** 1487

[101] Stephens P W, Mihaly L, Lee P L, Whetten R L, Huang S-M, Kaner R, Deiderich F and Holczer K 1991 *Nature* **351** 632–4

[102] Varma C M, Zaanen J and Raghavachari K 1991 *Science* **254** 989

[103] Margadonna S and Prassides K 2002 *J. Solid State Chem.* **168** 639–52

[104] Wang S-Z, Ren M-Q, Han S, Cheng F-J, Ma X-C, Xue Q-K and Song C-L 2021 *Commun. Phys.* **4** 114

[105] Vakros J, Panagiotou G, Kordulis C, Lycourghiotis A, Vougioukalakis G C, Angelis Y and Orfanopoulos M 2003 *Catal. Lett.* **89** 269–73

[106] Zhou O, Fleming R M, Murphy D W, Rosseinsky M J, Ramirez A P, van Dover R B and Haddon R C 1993 *Nature* **362** 433–5

[107] Wang Y and Cheng L T 1992 *J. Phys. Chem.* **96** 1530–2

[108] Gong Q, Sun Y, Xia Z, Zou Y H, Gu Z, Zhou X and Qiang D 1992 *J. Appl. Phys.* **71** 3025–6

[109] Hirsch A, Chen Z and Jiao H 2000 *Angew. Chem. Int. Ed.* **39** 3915–7

[110] Rosén A and Wästberg B 1989 *J. Chem. Phys.* **90** 2525–6

[111] Hirsch A, Brettreich M and Wudl F 2004 Reduction *Fullerenes: Chemistry and Reactions* (Hoboken, NJ: Wiley) ch 2 pp 49–72

[112] Haddon R C, Brus L E and Raghavachari K 1986 *Chem. Phys. Lett.* **125** 459–64

[113] Echegoyen L and Echegoyen L E 1998 *Acc. Chem. Res.* **31** 593–601

[114] Hirsch A 1994 Nucleophilic additions *The Chemistry of the Fullerenes* (Hoboken, NJ: Wiley) ch 3 pp 56–78

[115] Hirsch A 1999 Principles of fullerene reactivity *Fullerenes and Related Structures (Topics in Current Chemistry* vol 199) ed A Hirsch (Berlin: Springer)

[116] Fagan P J, Krusic P J, Evans D H, Lerke S A and Johnston E 1992 *J. Am. Chem. Soc.* **114** 9697–9

[117] Keshavarz-K M, Knight B, Srdanov G and Wudl F 1995 *J. Am. Chem. Soc.* **117** 11371–2

[118] Nagashima H, Saito M, Kato Y, Goto H, Osawa E, Haga M and Itoh K 1996 *Tetrahedron* **52** 5053–64

[119] Nagashima H, Terasaki H, Kimura E, Nakajima K and Itoh K 1994 *J. Organic Chem.* **59** 1246–8

[120] Wudl F, Hirsch A, Khemani K C, Suzuki T, Allemand P M, Koch A, Eckert H, Srdanov G and Webb H M 1992 *Fullerenes* (Washington, DC: American Chemical Society) pp 161–75

[121] Hirsch A, Soi A and Karfunhel H R 1992 *Angew. Chem. Int. Ed. Engl.* **31** 766–8

[122] Hirsch A, Grösser T, Skiebe A and Soi A 1993 *Chem. Ber.* **126** 1061–7

[123] Komatsu K, Murata Y, Takimoto N, Mori S, Sugita N and Wan T S M 1994 *J. Organic Chem.* **59** 6101–2

[124] Anderson H L, Faust R, Rubin Y and Diederich F 1994 *Angew. Chem. Int. Ed. Engl.* **33** 1366–8

[125] Bingel C 1993 *Chem. Ber.* **126** 1957–9

[126] Nierengarten J-F and Nicoud J-F 1997 *Tetrahedron Lett.* **38** 7737–40

[127] Nierengarten J-F, Gramlich V, Cardullo F and Diederich F 1996 *Angew. Chem. Int. Ed. Engl.* **35** 2101–3

[128] Camps X and Hirsch A 1997 *J. Chem. Soc., Perkin Trans.* **1** 1595–6

[129] Hino T, Kinbara K and Saigo K 2001 *Tetrahedron Lett.* **42** 5065–7

[130] Skiebe A, Hirsch A, Klos H and Gotschy B 1994 *Chem. Phys. Lett.* **220** 138–40

[131] Schick G, Kampe K-D and Hirsch A 1995 *J. Chem. Soc., Chem. Commun.* **1995** 2023–4

[132] Naim A and Shevlin P B 1992 *Tetrahedron Lett.* **33** 7097–100

[133] Wang G-W, Shu L-H, Wu S-H, Wu H-M and Lao X-F 1995 *J. Chem. Soc., Chem. Commun.* **10** 1071–2

[134] Kusukawa T and Ando W 1996 *Angew. Chem. Int. Ed. Engl.* **35** 1315–7

[135] Kräutler B and Maynollo J 1996 *Tetrahedron* **52** 5033–42

[136] Krätler B and Puchberger M 1993 *Helv. Chim. Acta* **76** 1626–31

[137] Yinghuai Z, Bahnmueller S, Chibun C, Carpenter K, Hosmane N S and Maguire J A 2003 *Tetrahedron Lett.* **44** 5473–6

[138] Diederich F, Isaacs L and Philp D 1994 *Chem. Soc. Rev.* **23** 243–55

[139] Vasella A, Uhlmann P, Waldraff C A A, Diederich F and Thilgen C 1992 *Angew. Chem. Int. Ed. Engl.* **31** 1388–90

[140] Tsuda M, Ishida T, Nogami T, Kurono S and Ohashi M 1992 *Chem. Lett.* **21** 2333–4

[141] Zhang X, Fan A and Foote C S 1996 *J. Org. Chem.* **61** 5456–61

[142] Zhang X, Romero A and Foote C S 1993 *J. Am. Chem. Soc.* **115** 11024–5

[143] Wilson S R, Kaprinidis N, Wu Y and Schuster D I 1993 *J. Am. Chem. Soc.* **115** 8495–6

[144] Smith A B, Strongin R M, Brard L, Furst G T, Romanow W J, Owens K G and King R C 1993 *J. Am. Chem. Soc.* **115** 5829–30

[145] SuzuKi T, Li Q, Khemani K C, Wudl F and Almarsson Ö 1991 *Science* **254** 1186

[146] Isaacs L, Wehrsig A and Diederich F 1993 *Helv. Chim. Acta* **76** 1231–50

[147] Wudl F 1992 *Acc. Chem. Res.* **25** 157–61

[148] Prato M, Li Q C, Wudl F and Lucchini V 1993 *J. Am. Chem. Soc.* **115** 1148–50

[149] Hummelen J C, Knight B W, LePeq F, Wudl F, Yao J and Wilkins C L 1995 *J. Organic Chem.* **60** 532–8

[150] Tsuda M, Ishida T, Nogami T, Kurono S and Ohashi M 1993 *J. Chem. Soc., Chem. Commun.* **16** 1296–8

[151] Hudhomme P 2006 *C.R. Chim.* **9** 881–91

[152] Llacay J, Mas M, Molins E, Veciana J, Powell D and Rovira C 1997 *Chem. Commun.* **7** 659–60

[153] Langa F, de la Cruz P, de la Hoz A, Díaz-Ortiz A and Díez-Barra E 1997 *Contemp. Org. Synth.* **4** 373–86

[154] Langa F, de la Cruz P, Espíldora E, García J J, Pérez M C and de la Hoz A 2000 *Carbon* **38** 1641–6

[155] Rubin Y, Khan S, Freedberg D I and Yeretzian C 1993 *J. Am. Chem. Soc.* **115** 344–5

[156] Biglova Y N 2021 *Beilstein J. Org. Chem.* **17** 630–70

[157] McEwen C N, McKay R G and Larsen B S 1992 *J. Am. Chem. Soc.* **114** 4412–4

[158] Takada H, Kokubo K, Matsubayashi K and Oshima T 2006 *Bioscience, Biotechnology, Biochemistry* **70** 3088–93

[159] Morton J R, Preston K F, Krusic P J, Hill S A and Wasserman E 1992 *J. Phys. Chem.* **96** 3576–8

[160] Lu S, Jin T, Bao M and Yamamoto Y 2011 *J. Am. Chem. Soc.* **133** 12842–8

[161] Tzirakis M D and Orfanopoulos M 2008 *Org. Lett.* **10** 873–6

[162] Kusukawa T and Ando W 1998 *J. Organomet. Chem.* **559** 11–22

[163] Akasaka T, Ando W, Kobayashi K and Nagase S 1993 *J. Am. Chem. Soc.* **115** 10366–7

[164] Gan L, Huang S, Zhang X, Zhang A, Cheng B, Cheng H, Li X and Shang G 2002 *J. Am. Chem. Soc.* **124** 13384–5

[165] You X, Li F-B and Wang G-W 2014 *J. Organic Chem.* **79** 11155–60

[166] Li F-B, Liu T-X and Wang G-W 2008 *J. Organic Chem.* **73** 6417–20

[167] Si W, Lu S, Bao M, Asao N, Yamamoto Y and Jin T 2014 *Org. Lett.* **16** 620–3

[168] Hirsch A and Brettreich M 2004 Radical Additions *Fullerenes: Chemistry and Reactions* (Hoboken, NJ: Wiley) ch 6 pp 213–30

[169] Okamura H, Terauchi T, Minoda M, Fukuda T and Komatsu K 1997 *Macromolecules* **30** 5279–84

[170] Morton J R, Negri F and Preston K F 1994 *Can. J. Chem.* **72** 776–82

[171] Morton J R, Negri F and Preston K F 1998 *Acc. Chem. Res.* **31** 63–9

[172] Sariciftci N S, Smilowitz L, Heeger A J and Wudl F 1992 *Science* **258** 1474

[173] Yu G, Gao J, Hummelen J C, Wudl F and Heeger A J 1995 *Science* **270** 1789

[174] Heeger A J 2014 *Adv. Mater.* **26** 10–27

[175] Li G, Zhu R and Yang Y 2012 *Nat. Photonics* **6** 153–61

[176] Li Y 2012 *Acc. Chem. Res.* **45** 723–33

[177] Dou L, Liu Y, Hong Z, Li G and Yang Y 2015 *Chem. Rev.* **115** 12633–65

[178] Hou J, Inganas O, Friend R H and Gao F 2018 *Nat. Mater.* **17** 119–28

[179] Inganas O 2018 *Adv. Mater.* **30** e1800388

[180] Yan T, Song W, Huang J, Peng R, Huang Ł and Ge Z 2019 *Adv. Mater.* **31** e1902210

[181] Cai Y *et al* 2021 *Adv. Mater.* **33** e2101733

[182] Li C *et al* 2021 *Nat. Energy* **6** 605–13

[183] Lin Y *et al* 2020 *ACS Energy Lett.* **5** 3663–71

[184] Zhang M *et al* 2021 *Nat. Commun.* **12** 309

[185] NREL 2021 Best Research-Cell Efficiency Chart https://nrel.gov/pv/assets/pdfs/best-research-cell-efficiencies.20200104.pdf

[186] Frankevich E, Maruyama Y and Ogata H 1993 *Chem. Phys. Lett.* **214** 39–44

[187] Gudaev O A, Malinovsky V K, Okotrub A V and Shevtsov Y V 1998 *Fullerene Sci. Technol.* **6** 433–43

[188] Reed C A and Bolskar R D 2000 *Chem. Rev.* **100** 1075–120

[189] Thompson B C and Frechet J M 2008 *Angew. Chem. Int. Ed. Engl.* **47** 58–77

[190] Yan L and Ma C-Q 2021 *Energy Technology* **9** 2000920

[191] Peumans P, Yakimov A and Forrest S R 2003 *J. Appl. Phys.* **93** 3693–723

[192] Tang C W and Vanslyke S A 1987 *Appl. Phys. Lett.* **51** 913

[193] Yu G, Gao J, Hummelen J C, Wudl F and Heeger A J 1995 *Science* **270** 1789–91

[194] Zhao F, Zhang H, Zhang R, Yuan J, He D, Zou Y and Gao F 2020 *Adv. Energy Mater.* **10** 2002746

[195] Cui C and Li Y 2021 *Aggregate* **2** e31

[196] Yang W *et al* 2021 *Joule* **5** 1209–30

[197] Kearns D and Calvin M 1958 *J. Chem. Phys.* **29** 950–1

[198] Sariciftci N S, Smilowitz L, Heeger A J and Wudl F 1992 *Science* **258** 1474–6

[199] Liu J *et al* 2016 *Nat. Energy* **1** 16089

[200] Peumans P and Forrest S R 2001 *Appl. Phys. Lett.* **79** 126–8

[201] Bredas J-L 2014 *Mater. Horiz.* **1** 17–9

[202] He Y and Li Y 2011 *Phys. Chem. Chem. Phys.* **13** 1970–83

[203] Kraabel B, McBranch D, Sariciftci N S, Moses D and Heeger A J 1994 *Phys. Rev. B: Condens. Matter* **50** 18543–52

[204] Brabec C J, Zerza G, Cerullo G, De Silvestri S, Luzzati S, Hummelen J C and Sariciftci S 2001 *Chem. Phys. Lett.* **340** 232–6

[205] Distler A, Kutka P, Sauermann T, Egelhaaf H-J, Guldi D M, Di Nuzzo D, Meskers S C J and Janssen R A J 2012 *Chem. Mater.* **24** 4397–405

[206] Li C-Z, Yip H-L and Jen A K Y 2012 *J. Mater. Chem.* **22** 4161–77

[207] Lai Y-Y, Cheng Y-J and Hsu C-S 2014 *Energy Environ. Sci.* **7** 1866–83

[208] He Y, Chen H-Y, Hou J and Li Y 2010 *J. Am. Chem. Soc.* **132** 1377–82

[209] Lenes M, Wetzelaer G-J A H, Kooistra F B, Veenstra S C, Hummelen J C and Blom P W M 2008 *Adv. Mater.* **20** 2116–9

[210] Correa-Baena J-P, Abate A, Saliba M, Tress W, Jesper Jacobsson T, Grätzel M and Hagfeldt A 2017 *Energy Environ. Sci.* **10** 710–27

[211] Green M A, Ho-Baillie A and Snaith H J 2014 *Nat. Photonics* **8** 506–14

[212] Zhao Y and Zhu K 2016 *Chem. Soc. Rev.* **45** 655–89

[213] Li W, Wang Z, Deschler F, Gao S, Friend R H and Cheetham A K 2017 *Nature Rev. Mater.* **2** 16099

[214] Jiang X, Zang Z, Zhou Y, Li H, Wei Q and Ning Z 2021 *Accounts Mater. Res.* **2** 210–9

[215] Lin Q, Armin A, Burn P L and Meredith P 2016 *Acc. Chem. Res.* **49** 545–53

[216] Snaith H J 2013 *J. Physical Chem. Lett.* **4** 3623–30

[217] Frost J M, Butler K T, Brivio F, Hendon C H, van Schilfgaarde M and Walsh A 2014 *Nano Lett.* **14** 2584–90

[218] Jeon N J, Noh J H, Kim Y C, Yang W S, Ryu S and Seok S I 2014 *Nat. Mater.* **13** 897–903

[219] Di Giacomo F, Fakharuddin A, Jose R and Brown T M 2016 *Energy Environ. Sci.* **9** 3007–35

[220] Kojima A, Teshima K, Shirai Y and Miyasaka T 2009 *J. Am. Chem. Soc.* **131** 6050–1

[221] Bai Y, Meng X and Yang S 2018 *Adv. Energy Mater.* **8** 1701883

[222] Luo D, Su R, Zhang W, Gong Q and Zhu R 2019 *Nature Rev. Mater.* **5** 44–60

[223] Jeng J Y, Chiang Y F, Lee M H, Peng S R, Guo T F, Chen P and Wen T C 2013 *Adv. Mater.* **25** 3727–32

[224] Castro E, Murillo J, Fernandez-Delgado O and Echegoyen L 2018 *J. Mater. Chem.* C **6** 2635–51

[225] Fang Y, Bi C, Wang D and Huang J 2017 *ACS Energy Lett.* **2** 782–94

[226] Sun S, Salim T, Mathews N, Duchamp M, Boothroyd C, Xing G, Sum T C and Lam Y M 2014 *Energy Environ. Sci.* **7** 399–407

[227] Wang Q, Shao Y, Dong Q, Xiao Z, Yuan Y and Huang J 2014 *Energy Environ. Sci.* **7** 2359–65

[228] Chiang C-H, Tseng Z-L and Wu C-G 2014 *J. Mater. Chem.* A **2** 15897–903

[229] Shao Y, Yuan Y and Huang J 2016 *Nat. Energy* **1** 15001

[230] Jiang X *et al* 2020 *Nat. Commun.* **11** 1245

[231] Wu T, Liu X, Luo X, Lin X, Cui D, Wang Y, Segawa H, Zhang Y and Han L 2021 *Joule* **5** 863–86

[232] Shao Y, Xiao Z, Bi C, Yuan Y and Huang J 2014 *Nat. Commun.* **5** 5784

[233] Snaith H J, Abate A, Ball J M, Eperon G E, Leijtens T, Noel N K, Stranks S D, Wang J T, Wojciechowski K and Zhang W 2014 *J. Phys. Chem. Lett.* **5** 1511–15

[234] Jia L, Chen M and Yang S 2020 *Mater. Chem. Front.* **4** 2256–82

[235] Di C A, Zhang F and Zhu D 2013 *Adv. Mater.* **25** 313–30

[236] Torsi L, Magliulo M, Manoli K and Palazzo G 2013 *Chem. Soc. Rev.* **42** 8612–28

[237] Sokolov A N, Tee B C K, Bettinger C J, Tok J B H and Bao Z 2012 *Acc. Chem. Res.* **45** 361–71

[238] Lipomi D J, Vosgueritchian M, Tee B C K, Hellstrom S L, Lee J A, Fox C H and Bao Z 2011 *Nat. Nanotechnol.* **6** 788–92

[239] Baude P F, Ender D A, Haase M A, Kelley T W, Muyres D V and Theiss S D 2003 *Appl. Phys. Lett.* **82** 3964–6

[240] Haddon R C, Perel A S, Morris R C, Palstra T T M, Hebard A F and Fleming R M 1995 *Appl. Phys. Lett.* **67** 121–3

[241] Haddon R C 1996 *J. Am. Chem. Soc.* **118** 3041–2

[242] Li C-Z, Chueh C-C, Yip H-L, Zou J, Chen W-C and Jen A K Y 2012 *J. Mater. Chem.* **22** 14976–81

[243] Zhou Y, Li J, Ma H, Zhen M, Guo J, Wang L, Jiang L, Shu C and Wang C 2017 *ACS Appl. Mater. Interfaces* **9** 35539–47

[244] Bosi S, D Ros T, Spalluto G and Prato M 2003 *Eur. J. Med. Chem.* **38** 913–23

[245] Li T *et al* 2015 *J. Am. Chem. Soc.* **137** 7881–8

[246] Shi J *et al* 2013 *Biomaterials* **34** 251–61

[247] Baati T, Bourasset F, Gharbi N, Njim L, Abderrabba M, Kerkeni A, Szwarc H and Moussa F 2012 *Biomaterials* **33** 4936–46

[248] Fan J, Fang G, Zeng F, Wang X and Wu S 2013 *Small* **9** 613–21

[249] Harman D 1956 *J. Gerontology* **11** 298–300

[250] Roberts R A, Smith R A, Safe S, Szabo C, Tjalkens R B and Robertson F M 2010 *Toxicology* **276** 85–94

[251] Lobo V, Patil A, Phatak A and Chandra N 2010 *Pharmacogn. Rev.* **4** 118–26

[252] Wang I C, Tai L A, Lee D D, Kanakamma P P, Shen C K F, Luh T-Y, Cheng C H and Hwang K C 1999 *J. Med. Chem.* **42** 4614–20

[253] Xiao L, Takada H, Maeda K, Haramoto M and Miwa N 2005 *Biomed. Pharmacother.* **59** 351–8

[254] Takada H and Matsubayashi K 2006 Process for producing PVP-fulleren complex and aqueous solution thereof *Patent* WO117877

[255] Williams R M, Verhoeven J W, Crielaard W and Hellingwerf K J 1996 *Recl. Trav. Chim. Pays-Bas* **115** 72–6

[256] Kato S, Aoshima H, Saitoh Y and Miwa N 2010 *J. Photochem. Photobiol.,* B **98** 99–105

[257] Kato S, Taira H, Aoshima H, Saitoh Y and Miwa N 2010 *J. Nanosci. Nanotechnol.* **10** 6769–74

[258] Zhou Z, Lenk R, Dellinger A, MacFarland D, Kumar K, Wilson S R and Kepley C L 2009 *Nanomedicine* **5** 202–7

[259] Fu H, Xu T, Zhu S and Zhu Y 2008 *Environ. Sci. Tech.* **42** 8064–9

[260] Orfanopoulos M and Kambourakis S 1994 *Tetrahedron Lett.* **35** 1945–8

[261] Li B and Xu Z 2009 *J. Am. Chem. Soc.* **131** 16380–2

[262] Malhotra R, McMillen D F, Tse D S, Lorents D C, Ruoff R S and Keegan D M 1993 *Energy Fuels* **7** 685–6

[263] Nishibayashi Y, Saito M, Uemura S, Takekuma S-i, Takekuma H and Yoshida Z-i 2004 *Nature* **428** 279–80

[264] Vul' A Y, Davidenko V M, Kidalov S V, Ordan'yan S S and Yashin V A 2001 *Tech. Phys. Lett.* **27** 384–6

[265] Berseth P A, Harter A G, Zidan R, Blomqvist A, Araújo C M, Scheicher R H, Ahuja R and Jena P 2009 *Nano Lett.* **9** 1501–5

[266] Sokolov V I 2007 *Russian J. Coordination Chem.* **33** 711–24

[267] Liu Z-B, Xu Y-F, Zhang X-Y, Zhang X-L, Chen Y-S and Tian J-G 2009 *J. Phys. Chem.* B **113** 9681–6

[268] Shurpo N A, Serov S V, Shmidt A V, Margaryan H L and Kamanina N V 2009 *Diam. Relat. Mater.* **18** 931–4

[269] Lee K, Hwang Y, Cheong S, Kwon L, Kim S and Lee J 2009 *Curr. Appl. Phys.* **9** e128-31

[270] Bhushan B, Gupta B K, Van Cleef G W, Capp C and Coe J V 1993 *Tribol. Trans.* **36** 573–80

[271] Pupysheva O V, Farajian A A and Yakobson B I 2008 *Nano Lett.* **8** 767–74

[272] Yoon M, Yang S, Hicke C, Wang E, Geohegan D and Zhang Z 2008 *Phys. Rev. Lett.* **100** 206806

Chapter 2

Graphdiyne

Weidong Zhou, Yurui Xue and Yuliang Li

Graphdiyne has been an important kind of carbon material since it was synthesized by Professor Yuliang Li in 2010. The discovery of graphdiyne has created opportunities for both fundamental and applied research into carbon materials in the fields of energy and catalysis. After describing its rapid development, we will introduce the advances in the fundamental and applied research results of the graphdiyne field. More importantly, we offer an overview of the applications of graphdiyne in catalysis, energy conversion, and energy storage. Finally, we also highlight the future perspective on graphdiyne.

2.1 Introduction

Carbon materials have experienced a long history of development. The application of carbon materials is considered to have promoted the progress of human society and the development of other materials. Carbon allotropes can be constructed using different types of hybridization (sp, sp^2, and sp^3) and have different structures and properties. In the last two to three decades, the synthesis and separation of new carbon allotropes with different dimensions have been a focus of research. Scientists have discovered new types of carbon allotrope, such as zero-dimensional fullerenes, one-dimensional carbon nanotubes, and two-dimensional (2D) graphene, which have attracted much attention due to their special structures and chemical and physical properties and have become the hotspot and frontier of international academic research [1–4]. These developments have ushered in a new stage of carbon material research. Inspired by these developments, scientists are committed to discovering and developing new carbon allotropes. The sp-hybridized acetylenic bond ($-C\equiv C-$) presents many unique advantages, such as a linear structure, the absence of cis–trans isomers, and strong conjugation. The hybridization of carbon allotropes via sp hybridization has therefore become a very interesting research topic. Researchers have long been eager to produce sp-hybridized carbon allotropes, which are considered to have excellent electrical and optical performances, and are key materials for the next generation of new electronic and optoelectronic devices. However, the lack of breakthroughs in the synthesis methods for traditional carbon materials has seriously impeded advances in the synthesis of graphdiyne.

In 2010, Professor Yuliang Li was the first to report the successful synthesis of the sp- and sp^2-cohybridized carbon material, graphdiyne (GDY, named 'shimoque' in Chinese), on the surface of copper foil by a facile chemical method [5]. This brought a new member into the family of carbon materials and opened up a new field for research into carbon materials. GDY is the first chemically synthesized all-carbon material and creates a precedent for the artificial synthesis of new carbon allotropes. In the GDY structure, each sp^2-hybridized benzene ring is linked by six sp-hybridized diacetylenic bonds, which is fundamentally different from the sp^3 and sp^2 hybridization of traditional carbon materials.

The successful synthesis of GDY has promoted research into the properties and applications of GDY. The special chemical structure of GDY endows it with many unique and fascinating properties, including rich chemical bonds, natural pores, highly conjugated structures, uneven dispersion of surface charge, a natural adjustable bandgap, excellent conductivity, etc. The high distribution of acetylenic bonds in GDY makes the surface charge distribution of GDY extremely uneven, which endows it with more active sites and high intrinsic activity and thus effectively promotes the catalytic reaction process [6]. In addition, GDY is the only carbon material that can be grown in a controlled manner on the surfaces of arbitrary substrates at ambient temperatures and pressures. Moreover, GDY presents high chemical activity and supports 'chemistry,' which are great advantages for chemical modification. As it benefits from these exclusive properties, GDY is expected to be a perfect and unique new carbon allotrope, and has exhibited transformative properties and performances in diverse fields such as energy, catalysis, optical devices, electrochemical intelligence and information devices, and photoelectric conversion.

The discovery of GDY creates opportunities for fundamental and applied research into carbon materials in the fields of energy, catalysis, photoelectricity, environmental science, the life sciences, electronic information, etc. and represents the development trend of carbon materials in the future. GDY-related research has entered a period of rapid development. Recently, Clarivate and the Chinese Academy of Sciences released their joint report 'Research Fronts 2020,' which identifies the hottest and emerging areas of scientific research. 'Graphdiyne research' was highlighted and placed in the top ten research fronts in both chemistry and materials science. The accelerating world of GDY has received extensive attention from researchers. GDY research has been carried out in more than 50 countries and regions over the world.

In this chapter, we will introduce advances in the fundamental and applied research results of the GDY field [7–10]. More importantly, we will focus on the application of GDY in catalysis [11, 12], energy conversion, and energy storage [13, 14].

2.2 Preparation of GDY and its derivatives

2.2.1 Classical Cu-surface synthesis

Using copper catalysis in the presence of pyridine and at a relatively low temperature (<100 °C), a typical GDY was prepared through the cross-coupling reaction of hexaethylbenzene (HEB) [15]. In detail, the monomer of hexaethynylbenzene was prepared through the de-trimethylsilylation of hexakis[(trimethylsilyl)ethynyl]benzene.

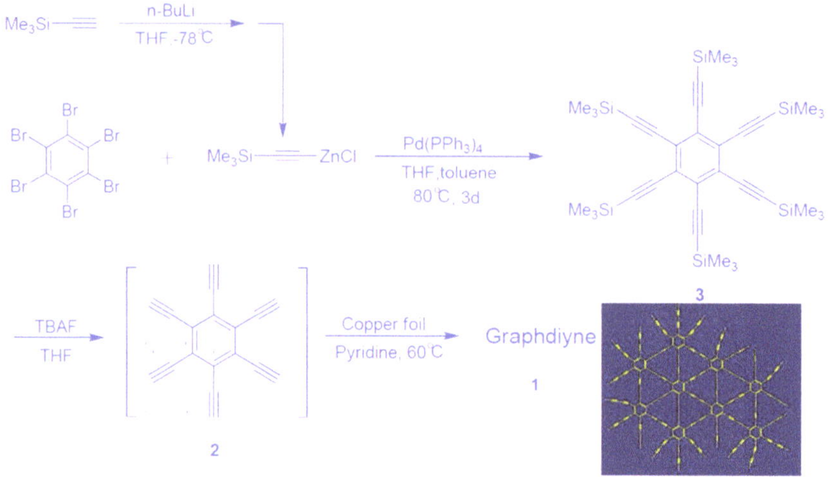

Figure 2.1. The synthesis route of GDY films on the Cu surface [15], reproduced with permission © Royal Society of Chemistry.

GDY was produced by the subsequent cross-coupling reaction of the monomer of hexaethynylbenzene on the surface of copper foil in the presence of pyridine. In this process, the copper foil not only functioned as the catalyst for the cross-coupling reaction but was also the substrate for the growing GDY film. In fact, trace amounts of Cu(I) and Cu(II) ions on copper are the real catalysts in this cross-coupling reaction (figure 2.1).

Using this preparation technology (which has good repeatability), different GDY derivatives have successfully been prepared, including boron-GDY (BGDY) [16], halogen-GDY (Cl-GDY in figure 2.2(a) and F-GDY in figure 2.2(b)) [17, 18], Ben-GDY (figure 2.2(c)) [19], hydrogen-GDY (HsGDY in figure 2.2(d)) [20], methyl substituted graphdiyne (MsGDY in (figure 2.2(e)) [21], cyano-graphdiyne (CN-GDY) [22], GDY containing a tetraphenylethene (TPE) unit (TPE-GDY in (figure 2.2(f)) [23], pyrediyne (PDY) [24], porphyrin-GDY (Por-GDY) [25] and allotrope-graphtetrayne (GTY) [26]. As is also the case for pristine GDY, these GDY derivatives are typically conjugated 2D large plane carbon structures composed of sp-hybridized butadiyne units. Due to the presence of a π orbit and electrons on the vertical orientation of the carbon large plane, the electrons tend to delocalize between electron-rich butadiyne and electronegative boron in BGDY (figure 2.2(g)) or benzene in GDY/GTY (figure 2.2(h)), leading to a narrow bandgap energy and a relatively high conductivity of 10^{-4}–10^{-2} S m^{-1} [16, 26].

2.2.2 Interfacial synthesis and the high-temperature metal-catalyst-free strategy

In addition to the aforementioned synthesis route, a couple of new strategies have been recently developed for the preparation of ultrathin GDY derivatives; for instance, interface reaction engineering and the high-temperature metal-catalyst-free strategy [27, 28]. The interfacial strategy was recently employed to prepare super-thin GDY sheets with the help of the restricted growth of materials on the interface between two phases. In a typical case, the reaction takes place on the liquid/liquid interface of a dichloromethane

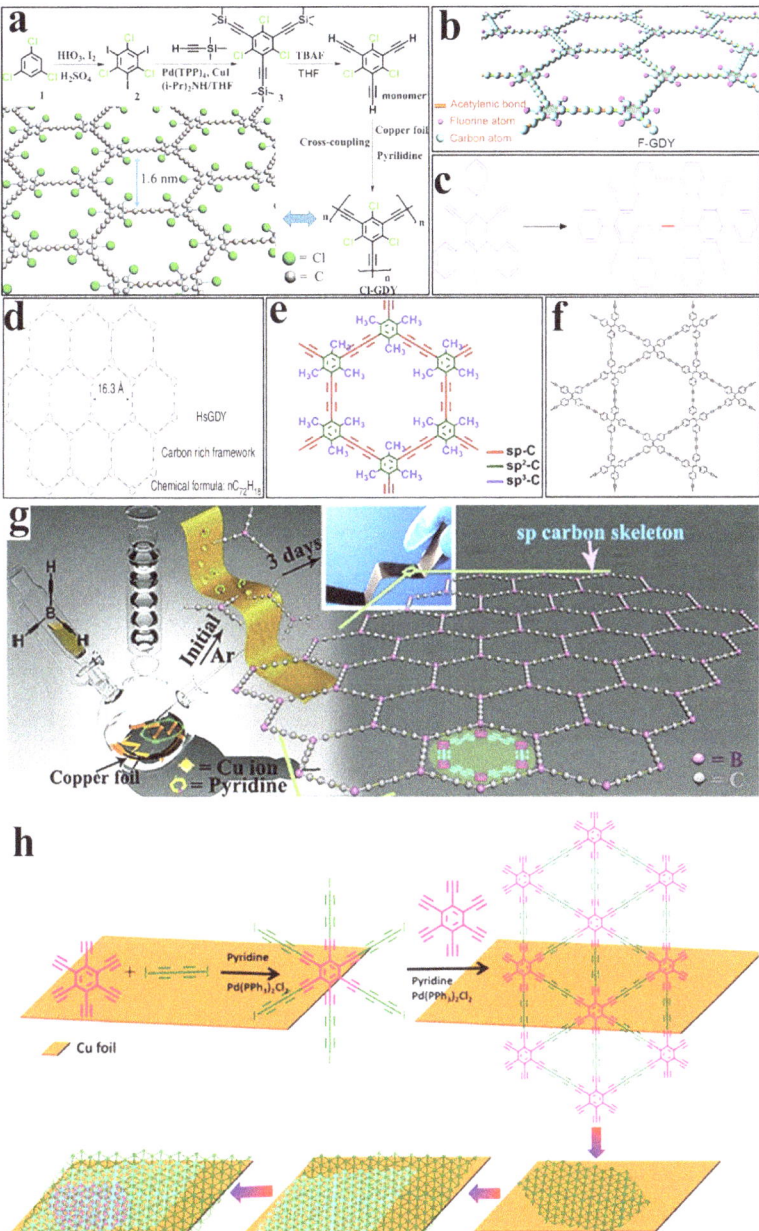

Figure 2.2. (a) Structural schematic of Cl-GDY [17], reproduced with permission © John Wiley and Sons. (b) Structural schematic of F-GDY [18], reproduced with permission © Royal Society of Chemistry. (c) Structural schematic of Ben-GDY [19], reproduced with permission © American Chemical Society. (d) Structural schematic of HsGDY [20], reproduced with permission © Springer Nature. (e) Structural schematic of Me-GDY [21], reproduced with permission © Elsevier. (f) Structural schematic of TPE-GDY [23], reproduced with permission © John Wiley and Sons. (g) Schematic of the experimental setup of BGDY [16], reproduced with permission © John Wiley and Sons. (h) Synthesis and structural schematic of GTY [26], reproduced with permission © Elsevier.

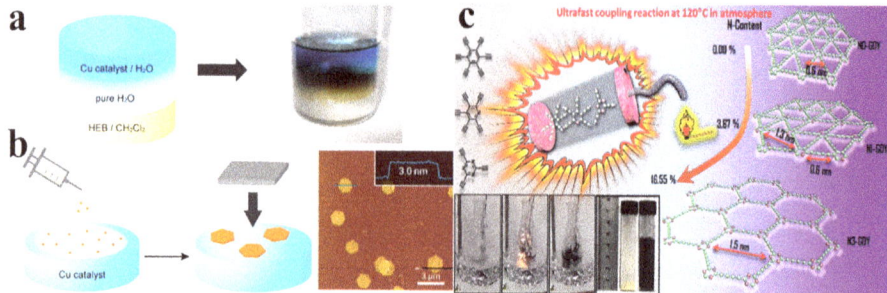

Figure 2.3. (a) Schematic diagram and a photograph of the liquid/liquid interfacial synthesis procedure; (b) schematic diagram of gas/liquid interface synthesis and topography of hexagonal GDY as determined by atomic force microscopy (AFM). Source for (a), (b): [28], reproduced with permission © American Chemical Society. (c) Schematic diagram of the preparation processes and structures of N-doped GDY [29], reproduced with permission © Elsevier.

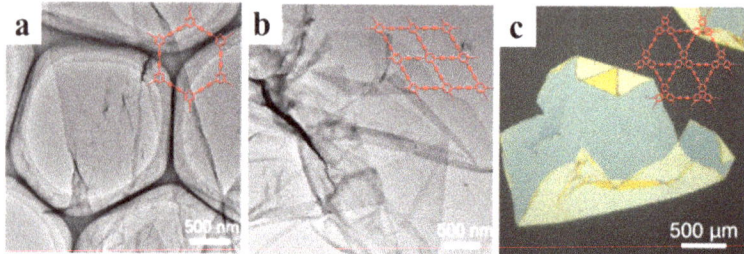

Figure 2.4. (a and b) N-doped GDY films [30], reproduced with permission © American Chemical Society. (c) Optical microscopic image of a TP-GDY film on an HMDS/Si (100) substrate [31], reproduced with permission © American Chemical Society.

phase with HEB and an aqueous phase containing a copper catalyst. An alkyne–alkyne coupling reaction in dichloromethane is initiated by the Cu on the interface, producing transparent ultrathin GDY (figure 2.3(a)). When further extended to a gas/liquid interface, 3 nm thick hexagonal GDY nanosheets were obtained by dropping an organic solution of HEB onto the surface of an aqueous solution of the catalyst (figure 2.3(b)) [28].

Taking another direction, an interesting solvent-free and catalyst-free coupling reaction was demonstrated for the high-yield preparation of N-doped GDY and GDY, which proceeds at a slightly increased temperature of 120 °C in air; oxygen was found to accelerate the coupling reaction (figure 2.3(c)) [29].

The abovementioned synthesis strategies were successfully used to prepare various GDY derivatives containing different aromatic cores connected by diacetylene linkers. These aromatic cores varied from the benzene of Ben-GDY to pyrazine, triazine, triphenylene, carbon ene–yne, etc. (figure 2.4) [31] leading to the formation of 2D GDY derivatives with different heteroatom contents, cavity sizes and distributions of conjugated sp–sp^2 bonds. Employing the liquid/liquid interface strategy, the thickness of these GDY derivative sheets can be controlled to an accuracy of less than 20 nm. Since there is no high-temperature treatment in the preparation process, the molecular structure and configuration, such as the shapes of cavities and the relative positions of heteroatoms and diacetylene units, can be exactly predicted and are not changed during synthesis, supporting the development of molecular designs for specific functions.

2.2.3 Template method

Based on the classical Cu-surface synthesis, Li and coworkers prepared GDY nanotube arrays using anodic aluminum oxide (AAO) as a template catalyzed by Cu foil (figure 2.5(a)) [32]. A Cu envelope catalysis strategy was then developed to synthesize structure-controlled GDY nanowalls on different substrates including 1D Si nanowires, 2D Au foils, 3D Ni foam, and metal oxides (MOs)0236+9 [33, 34]. Inspired by 3D graphene preparation that employs diatomite as a template [35], it is possible to prepare 3D GDY in this way. Cu nanoparticles are primarily absorbed on the surface and in the holes of the diatomite through a simple metallic replacement reaction that produces Cu nanoparticle and diatomite composites (Cu@diatomite). The absorbed Cu nanoparticles and diatomite play the roles of catalyst source and substrate, respectively. Finally, the diatomite was wrapped in GDY flakes (GDY@Cu@diatomite). After the residual Cu and diatomite were removed using etching reagents, porous 3D GDY with a freestanding structure was obtained [36]. Another approach developed the *in situ* growth of GDY and its

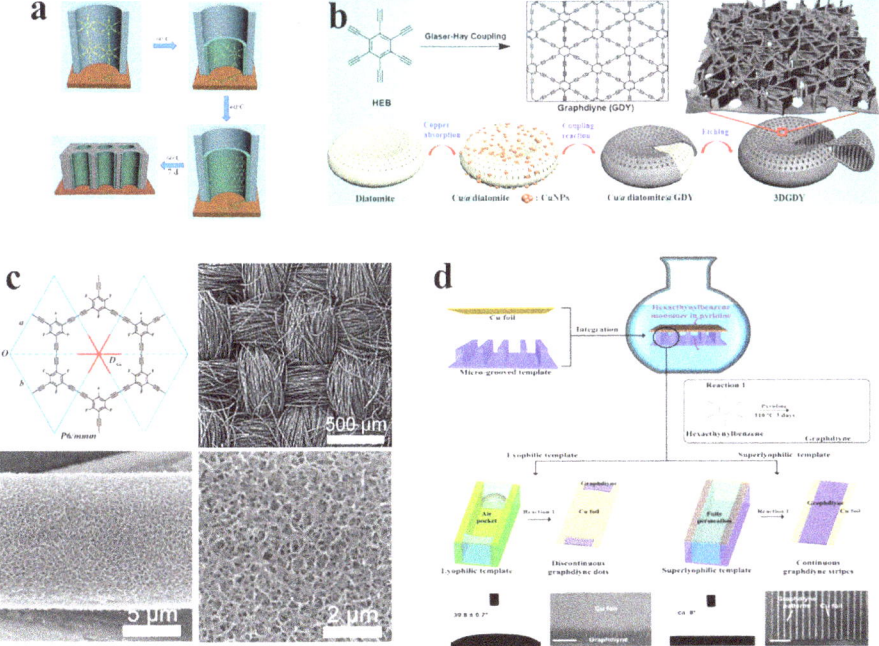

Figure 2.5. (a) Process used to fabricate GDY nanotube arrays [32], reproduced with permission © American Chemical Society. (b) 3D GDY synthesis using diatomite as a template [36], reproduced with permission © John Wiley and Sons. (c) F-GDY structure and morphologies of F-GDY/carbon cloth [40], reproduced with permission © Royal Society of Chemistry. (d) Schematic illustration of the *in situ* synthesis of GDY patterns [41], reproduced with permission © John Wiley and Sons.

analogs on a carbon cloth template [37–40, 48]. Precise patterning of GDY was also achieved on a predefined silicon template with different shapes at the microlevel [41].

2.2.4 Chemical vapor deposition method

Based on the experience of preparing graphene by the chemical vapor deposition (CVD) method, it was thought very significant to prepare high-quality GDY with fewer layers via this method. Zhang, Liu, and coworkers synthesized GDY on silver foil by atmospheric pressure chemical vapor deposition (APCVD) [42]. The silver-foil enclosure was put in the center of the heating zone and annealed at 890 °C under a specific atmosphere for a specific time in a quartz tube. After the furnace was cooled to the desired growth temperature (150 °C), 150 sccm Ar was introduced into the quartz tube. HEB vapor was then introduced into the furnace by bubbling 150 sccm Ar. After 2 h of growth, the silver enclosure was opened and the film grown on the inner surface was confirmed to be a GDY film with a thickness of 0.6 nm.

2.3 Applications of GDY in catalysis

2.3.1 Theoretical progress in GDY-based catalysts

Catalysts are essential in energy storage and conversion. The catalysts widely used today are mainly those based on precious metals. However, their high costs, scarcity, and instability in electrolytes severely restrict their commercial application. GDY is a novel 2D carbon material, which can grow *in situ* on any substrate surface under mild conditions. Moreover, GDY possess rich carbon chemical bonds including both sp- and sp^2-hybridization, a high electron mobility that can achieve to 10^5 cm^2 (V s)$^{-1}$ at 300 K [43], a moderate theoretical electronic bandgap in the range of 0.46–1.20 eV[44], excellent chemical stability, etc. These natural advantages and intrinsic characteristics endow GDY with great potential to be applied in the engineering of highly efficient catalysts. In recent years, numerous theoretical efforts have been made to study the catalytic properties and electrochemical applications of GDY-based catalysts.

2.3.1.1 GDY-based metal-atom catalysts
Atom catalysts (ACs) have been the leading edge of research and a hotspot in the field of catalysis [45–47]. In the case of the traditional single-atom catalysts (SACs), most of substrates destabilize isolated metal atoms and make them prone to aggregation during the fabricative and catalytic process due to the weak metal–support interaction. GDY is considered to be one of the most suitable substrates to be stably loaded with metal atoms; this is mainly attributed to its unique electronic properties, its porous structure with rich acetylenic bonds, and its 2D π-conjugated network composed of sp- and sp^2-hybridization. On one hand, the p$_x$–p$_y$ π/π* and p$_z$ π/π* states that coexist in GDY enable the π/π* orbitals to be rotated in any direction perpendicular to the –C≡C–C≡C– chain, inducing incomplete charge transfer between the metal atoms and the GDY. On the other hand, the widespread uniformly distributed triangular porous structure of GDY creates suitable anchoring environments for metal atoms. ACs

formed using GDY as a substrate can offer the benefits of infinitely high dispersion of active sites, clear electronic structures, and definite valence states—even zero valence.

2.3.1.1.1 Catalytic properties of ACs for the hydrogen evolution reaction

Hydrogen is an ideal green and renewable energy source with which to replace traditional fossil fuels due to its high energy density, renewability, and zero carbon emission. The hydrogen evolution reaction (HER, $2H^+ + 2e^- \rightarrow H_2$) driven by electricity is a promising approach for the production of hydrogen. However, the high cost and low abundance of today's most effective Pt-based catalysts severely limit their practical application. To overcome this bottleneck, it is necessary to develop stable, inexpensive, and highly efficient HER catalysts as alternatives.

GDY-supported zero-valence Ni/Fe ACs represent a great breakthrough in the field of catalysis; they were first synthesized by Li's group in 2018 [48] and exhibit superior HER activities (figure 2.6). Ni/Fe atoms can be stably adsorbed at the corner site of the truncated triangular pore structures of GDY. The active sites for HER are intuitively determined by the charge density distributions of the highest occupied molecular orbital (HOMO) and the lowest unoccupied molecular orbital (LUMO). Taking Ni/GDY as an example, the Ni-3d orbital exhibits a closed-shell

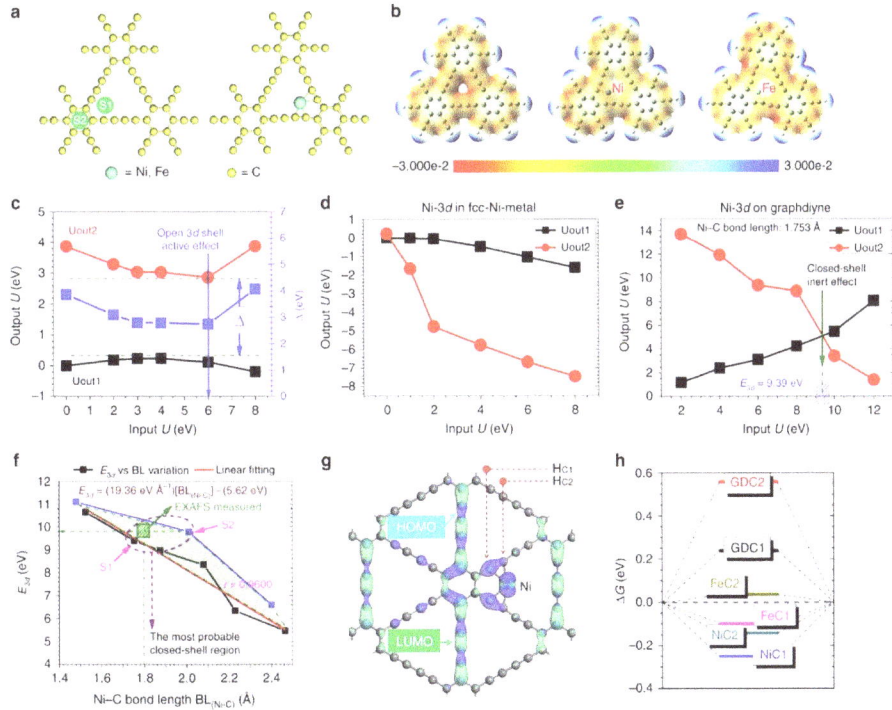

Figure 2.6. (a) Configuration of GDY-supported Ni/Fe atom catalysts. (b) Electrostatic potential diagrams for GDY, Ni/GDY, and Fe/GDY. Ni-3d orbital energies of (c) NiO, (d) Ni metal, and (e) Ni/GDY. (f) Calculated and measured Ni-3d orbital energies of Ni/GDY. (g) Charge distributions of the HOMO and the LUMO in Ni/GDY. (h) Free energies available for the HER (ΔG_H) at different carbon sites in GDY, Ni/GDY, and Fe/GDY. Source: [48], reproduced with permission © Springer Nature.

effect (crossing) which is completely different from the open-shell effect (non-crossing) in NiO. In addition, the corresponding Ni-3d orbital energy obtained by self-consistent calculation is much higher than that of fcc-Ni. These facts imply that there is a strong orbital overlap effect between the Ni atom and adjacent C atoms in Ni/GDY.

Li and coworkers successively synthesized a series of GDY-based ACs for the HER, such as Pd/GDY [49], Mo/GDY [50], Ru/GDY [51], etc. These ACs demonstrated superior long-term stabilities and electrocatalytic activities such as small Tafel slopes, small overpotentials, and large mass activities that were comparable to most state-of-the-art HER catalysts, which paved a new path for the development of HER catalysts. The outstanding HER activities are attributed to the strong chemisorption properties and charge transfer between the metal atoms and the GDY, which facilitate charge exchange and adsorption/desorption during the HER.

He *et al* [52] further examined the origins of AC activities that support the HER from the theoretical perspective (figure 2.7). Density functional theory (DFT) calculations revealed that the HER on ACs is site dependent and preferentially

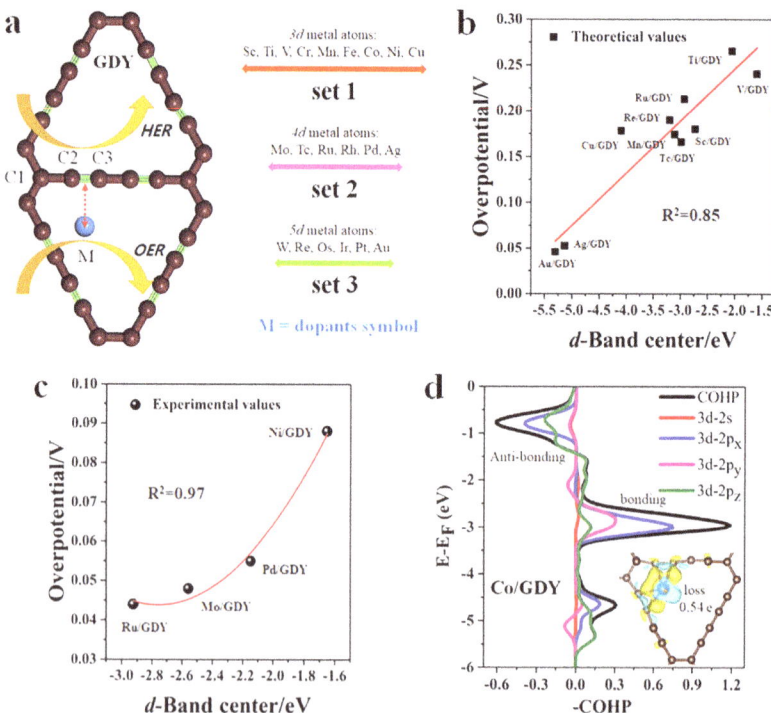

Figure 2.7. (a) Schematic diagram of GDY-supported metal atoms. Dependence of (b) theoretically calculated and (c) experimentally measured overpotentials (η) for the HER on the *d*-center of ACs. (d) Crystal orbital Hamilton population (COHP) and electron density differences in Co/GDY. The blue and yellow areas represent electron depletion and accumulation, respectively. Isosurface = 0.005 e/bohr^3. Source: [52], reproduced with permission © Springer Nature.

proceeds on the carbon site that binds with the metal atom. Moreover, lower *d*-centers of metal atoms correspond to smaller overpotentials, i.e. higher HER activities. It has been proposed that metal atoms with much deeper *d*-center positions induce much stronger d-p_x (p_y) orbital coupling in ACs according to crystal orbital Hamilton population (COHP) analysis; this would activate much obvious charge transfer between the metal site and the nearby bonded C site, thereby enhancing the HER activity of ACs. The results are of great significance in guiding the precise synthesis of highly efficient atom catalysts supported by GDY for the HER.

2.3.1.1.2 Catalytic properties of ACs for the oxygen evolution/reduction

The oxygen evolution reaction (OER) and the oxygen reduction reaction (ORR) are two important chemical reactions for sustainable energy supplies such as water splitting, fuel cells, and rechargeable metal–air batteries. There is an urgent need to search for high-efficiency and low-cost electrocatalysts to replace the widely used noble-metal-based catalysts for the OER (e.g. Ir or Ru oxides) and the ORR (e.g. Pt/C). Recently, it has been reported that GDY-based metal ACs have remarkable catalytic performance for the OER and the ORR [48, 51, 53, 54]. Du *et al* [55] systematically studied the OER/ORR activities of a series of GDY-based ACs using first-principles calculations (figure 2.8). The metal atoms can stably

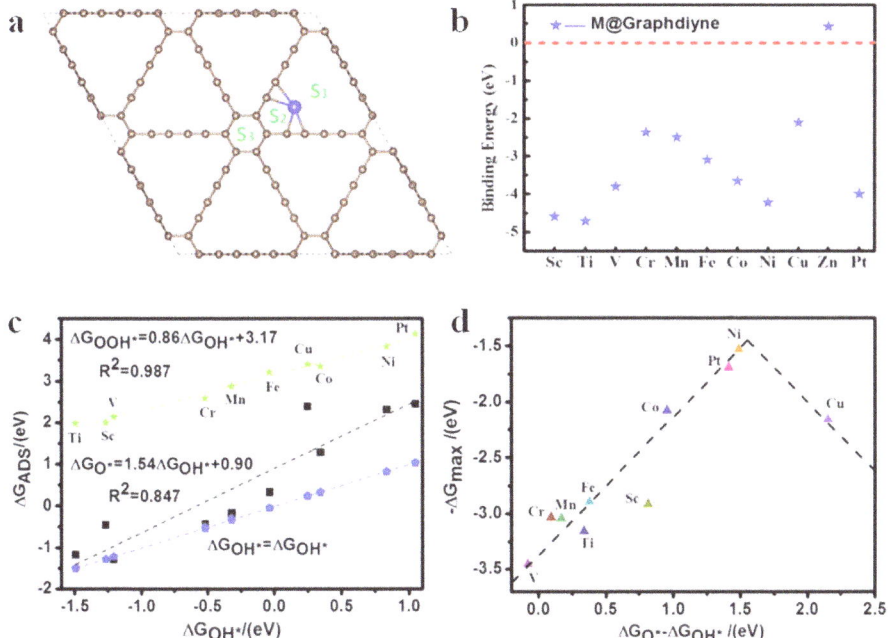

Figure 2.8. (a) Various possible anchoring sites for metal atoms on GDY. (b) Binding energies between metal atoms and GDY. (c) Scaling relation between the Gibbs free energies of oxygen-containing intermediates. (d) Trends of activity supporting the OER. Source: [55], reproduced with permission © John Wiley and Sons.

anchor to GDY without aggregating into metal clusters due to the high binding energies. The OER/ORR activities of ACs are evaluated by the overpotential η ($\Delta G_{max}/e - U_{equilibrium}$ or $U_{equilibrium} - \Delta G_{min}/e$), where ΔG_{max} is the Gibbs free energy of the potential-determining step during the OER/ORR. All the studied ACs exhibit excellent OER/ORR activities. Among them, the Ni/GDY catalyst exhibits the lowest overpotential of 0.29 V for the OER and a low overpotential of only 0.40 V for the ORR. Therefore, Ni/GDY is a promising highly efficient bifunctional electrocatalyst for both the OER and the ORR.

2.3.1.1.3 *Catalytic properties of ACs for carbon monoxide oxidation*

The oxidation of carbon monoxide (CO) is vital in order to enhance the durability of catalysts and solve the growing environmental problems caused by the emission of CO by automobile exhausts and inadequate industrial combustion. To efficiently activate the CO oxidation reaction, it is necessary to find appropriate catalysts. DFT calculations and molecular dynamics simulations suggested that GDY-based catalysts could be a good candidate for high efficient CO oxidation due to the low energy barrier in the rate-limiting step [56].

First-principles calculations show that GDY-supported Rh metal atoms (Rh/GDY) have excellent CO oxidation activity due to their strong adsorption ability for CO and O_2 molecules [57]. By considering all the possible pathways for CO oxidation, it can be found that the Langmuir–Hinshelwood (LH) and trimolecular Eley–Rideal (TER) mechanisms are preferable on Rh/GDY. The energy barriers in the rate-determining steps are only 0.54 eV and 0.49 eV via these two mechanisms, which are much lower than that of 2.32 eV via the Eley–Rideal (ER) mechanism. Zou *et al* [58] systematically examined the effect of N doping on a series of GDY-supported 3D metal atoms. The results revealed that N doping can effectively affect the adsorption behavior of O_2 and CO molecules on the metal sites. The adsorption energy of O_2 decreases with increasing N contents, while the opposite is true for the adsorption of CO, suggesting that the introduction of N dopants reduces O_2 adsorption but benefits CO adsorption. Among the combinations examined, the Fe@2NGDY catalyst exhibited the best catalytic activity for CO oxidation.

GDY-supported Ir atoms (Ir/GDY) are predicted to be a potential catalyst for efficient CO oxidation [59]. It has been found that CO oxidation on Ir/GDY prefers to occur in a new Eley–Rideal (NER) mechanism rather than the traditional ones; this discovery was made by comparing the reaction energy barriers (figure 2.9). In the traditional ER mechanism, the cleavage of the C–O bond to form the physisorbed CO_2 has to overcome a high energy barrier of 1.31 eV. In the traditional LH and TER mechanisms, the rate-limiting steps also need high barriers of 0.77 eV and 0.70 eV, respectively. In comparison, the adsorbed O_2 molecule can easily be activated by two physisorbed CO molecules in the NER mechanism, in which the electrons are transferred from the Ir/GDY substrate to the O_2 and the CO. The OOC–COO intermediate thus formed is then synchronously dissociated into two CO_2 molecules via C–C bond cleavage, accompanying by the return of the electrons to the Ir/GDY substrate. The energy barrier in the rate-limiting step for these two processes is only 0.37 eV.

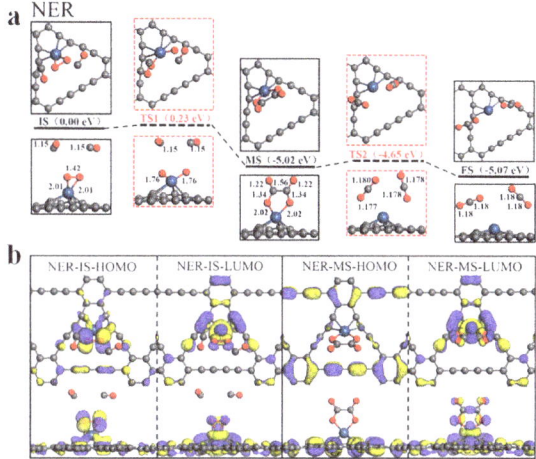

Figure 2.9. (a) CO oxidation catalyzed by Ir/GDY via the NER mechanism. (b) The HOMO and the LUMO of configurations in the initial state (IS) and the carbonate-like intermediate state during the NER pathway. The yellow and blue areas represent electron depletion and accumulation, respectively. Source: [59], reproduced with permission © American Chemical Society.

2.3.1.1.4 Catalytic properties of ACs for carbon dioxide reduction

The electrocatalytic carbon dioxide reduction reaction (CO_2RR) is one of the most attractive techniques for converting CO_2 into valuable chemical feedstocks and fuels. It is critical for the development of the CO_2RR to search for highly efficient catalysts. GDY can effectively promote the CO_2RR-supporting activities of catalysts. For example, Lu *et al* [60] reported a GDY-decorated bismuth subcarbonate catalyst (denoted by BOC@GDY), which showed a high level of activity in CO_2-to-formate conversion. Its outstanding performance was attributed to the electron-rich nature of the GDY used to decorate the BOC surface, which greatly elevated its CO_2 adsorption ability, reaction kinetics, and selectivity for formate. GDY can also be used as the substrate material to form dual-atom catalysts (DACs). In dual-atom GDY catalysts, the dual metal sites are adjacently embedded in the GDY support, inducing strong synergistic effects between two metal atoms, which play a critical role in promoting CO_2RR catalytic activity. It is a facile way to design highly efficient homo/heteronuclear dual-atom GDY catalysts for the CO_2RR using theoretical calculations. For example, theoretical studies predicted that Cu_2@GDY was a potential candidate to be a highly efficient CO_2RR electrocatalyst [61].

2.3.1.1.5 Catalytic properties of ACs for nitrogen reduction

Ammonia (NH_3) is a raw material for fertilizer production and a promising clean energy carrier. At present, the large scale of ammonia synthesis in industry is mainly due to the harsh Haber–Bosch process ($N_2+3H_2 \leftrightarrow 2NH_3$), which requires high temperature (400 °–600 °) and pressure (20–40 MPa). The electrocatalytic nitrogen reduction reaction (ECNRR) is an efficient and environmentally friendly ammonia synthesis technology that uses ambient conditions. However, most of the reported

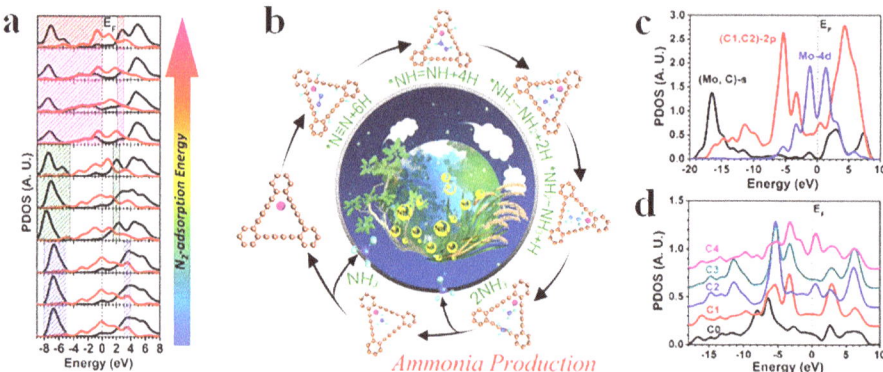

Figure 2.10. (a) Projected density of states (PDOS) evolution of the site-dependent N_2 adsorption energy on Mo/GDY. (b) Configuration evolution during the ammonia production process on Mo/GDY. (c) PDOS of Mo–(C1, C2) bonding motifs in Mo/GDY. (d) PDOS of C0–C4 sites in Mo/GDY. C0 is the C site in the benzene ring, and C1–C4 are the C sites that participate in sequential acetylenic chain bonding with Mo. Source: [50], reproduced with permission © American Chemical Society.

nitrogen reduction reaction (NRR) electrocatalysts have problems such as low catalytic activity and poor selectivity. Therefore, it is increasingly becoming urgent to develop highly efficient, stable, and cheap electrocatalysts for NRR.

Mo/GDY has been reported to be a promising atom catalyst for the NRR [50]. Its exact chemical/electronic structure, active sites, and catalytic mechanism have been studied (figure 2.10). The electron-enriched metal Mo atom is the active site, which can effectively modify the charge distribution of the surrounding C sites. There is an energy difference of nearly five times between the most energetically preferred nitrogen adsorption configuration of Mo–N≡N and the most energetically unfavorable adsorption configuration of Mo–N=N–C. Mo/GDY displays strong NRR catalytic activity that is preferentially conducted via the associative mechanism and has a low potential of $U = -0.71$ V. The high activity originating from the (C1, C2)-2p orbitals strongly preserves the valence electronic states of Mo-4d in various reaction steps during the NRR, and the moderate p-electron population of the C1 and C2 sites effectively promotes electron transfer between the Mo and the GDY.

The NRR activities of other metal-atom catalysts with GDY as the substrate were evaluated by calculating the overpotential (η) [62]. The results revealed that most GDY-based metal-atom catalysts (metal atom = Ti, V, Fe, Co, Zr, Rh, and Hf) exhibit much better NRR catalytic activities than the benchmark catalyst, which is a stepped Ru(0001) surface. Among the catalysts examined, the V/GDY catalyst possesses the best NRR catalytic activity with the lowest overpotential of 0.51 V. The potential-limiting step is $^*N_2 \rightarrow {}^*NNH$ for both alternating and distal mechanisms. In addition, there is an obvious linear correlation between the limiting potential and the N adsorption energy, which can be used as a simple descriptor for evaluating the NRR catalytic activities of GDY-based atom catalysts.

Multiple homonuclear metal atoms can be stabilized on a GDY substrate [63]. Ma *et al* [64] systematically studied the NRR catalytic activities of Mn, Fe, Co, and

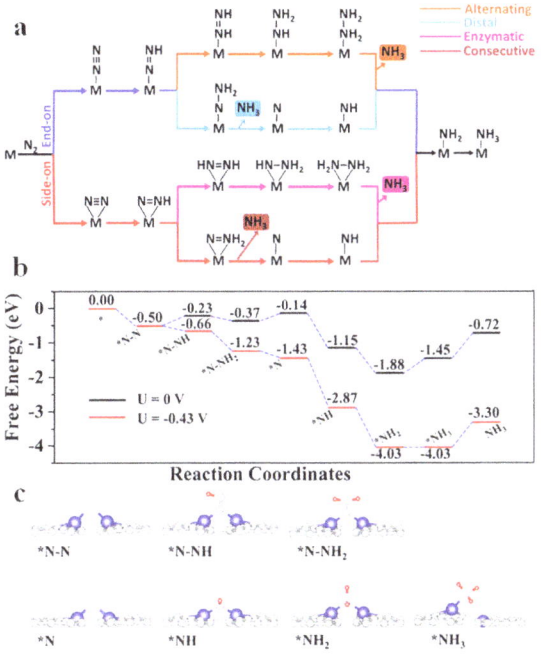

Figure 2.11. (a) Schematic depiction of different pathways for the NRR. (b) Free energy diagrams for the NRR on Co₂/GDY taking place via the distal mechanism at different applied potentials. (c) The corresponding structures of the reaction intermediates during the NRR. Source: [64], reproduced with permission © American Chemical Society.

Ni monomer- (SACs), dimer- (DACs), and trimer- (TACs) anchored GDY using first-principles calculations (figure 2.11). The results showed that most of the SACs and DACs have enhanced NRR catalytic activity compared with the Ru(0001) catalyst. In particular, Co_2/GDY exhibits the best NRR activity with a low $U_{limiting}$ of −0.43 V. The high NRR catalytic activity of Co_2/GDY is mainly attributed to the localized electronic states near the Fermi level induced by the Co dimer and the strong electron-donating ability of GDY. Furthermore, the predicted limiting potentials ($U_{limiting}$) have an approximately linear relation with the N adsorption energy, which may act as a simple descriptor for evaluating the NRR catalytic activity of such GDY-based multiple-metal-atom catalysts.

DFT calculations [65] have indicated that GDY-supported double heteronuclear metal atoms (FeM/GDY) are also potential NRR catalysts with high activity and selectivity (figure 2.12). The limiting potentials of FeM/GDY display a well-defined volcano-shaped relationship with the adsorption energy of the *NH₂ intermediate. The FeM/GDY catalysts (M = Ir, Hf, Rh, Ru, Mo, Nb, Cu, Ni, Co, Fe, and Cr) show improved catalytic activities compared to the Ru(0001) catalyst. Using the free energy differences between the *N₂ and *H intermediates (i.e., $\Delta G_{N2*} - \Delta G_{H*}$) as a descriptor, it has been demonstrated that the FeM/GDY catalysts (M = Ni, Mo, and Cr) exhibit a strong ability to suppress the competitive HER reaction. Therefore, GDY-based FeM atoms are promising NRR catalysts with both high activity and selectivity.

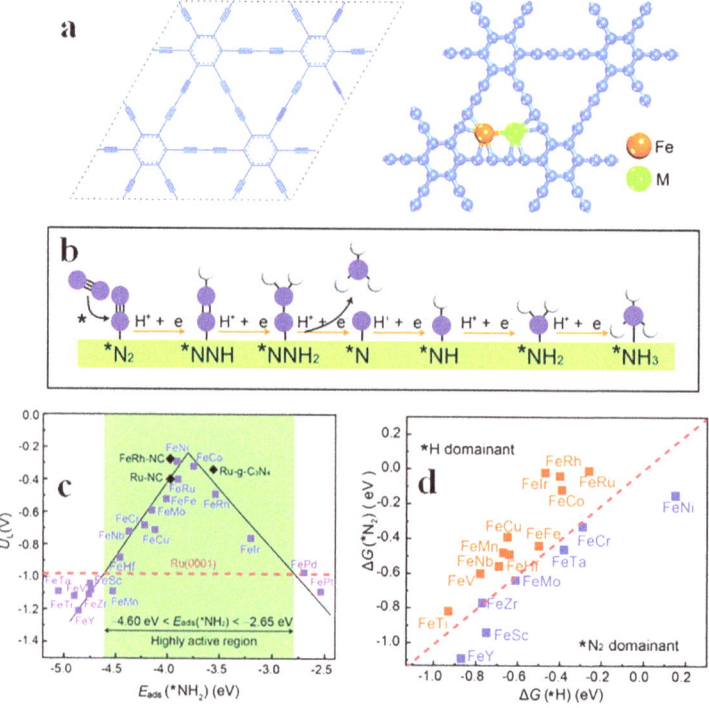

Figure 2.12. (a) Optimized structure of double heteronuclear metal atoms anchored by GDY. (b) Distal pathway for NRR. The purple and white spheres represent nitrogen and hydrogen atoms, respectively. (c) Volcano-shaped relationship between U_L and $E_{ads(*NH2)}$. (d) Comparison between $\Delta G(*N_2)$ ($N_2(g) \rightarrow *N_2$) and $\Delta G(*H)$ ($H^+ + e^- \rightarrow *H$) on FeM/GDY. The dashed line represents $\Delta G(*N_2) = \Delta G(*H)$. Source: [65], reproduced with permission © Royal Society of Chemistry.

2.3.1.2 GDY-based metal-free catalysts

The unique electron-rich sp-carbon chains endow GDY with inherent catalytic activity [66]. However, pristine GDY is not an excellent catalyst because the charge on sp-carbon is not large enough to facilitate the catalytic process. DFT calculations reveal that the poor catalytic activity of GDY is mainly limited by its weak ability to adsorb the intermediates [67]. An attractive strategy for improving the activity of pristine GDY is to dope it with nonmetal atoms to build GDY-based metal-free catalysts.

2.3.1.2.1 P/S-doped GDY catalysts for the hydrogen evolution reaction

DFT calculations have revealed that GDY-based metal-free electrocatalysts produced by P or S heteroatom doping can effectively activate the HER [68]. These P- or S-doped GDY catalysts have good conductivity and a small bandgap or even a zero bandgap; if the doping site is carefully regulated, they can exhibit superior HER catalytic activities with near-zero ΔG_{H*}. Such strong catalytic activities mainly originate from the charge transfer and the moderate p-band center that is modulated by the doped P or S atoms [69].

2.3.1.2.2 N/B-doped GDY catalysts for the carbon dioxide reduction reaction

Nitrogen-doped GDY (NGDY) is a promising catalyst for the CO_2RR [70]; it can improve the activity and selectivity of GDY to produce methanol (CH_3OH) and methane (CH_4). The lowest limiting potential for CH_3OH production on NGDY is only -0.46 V. This can be ascribed to its strong ability to adsorb the key intermediates, *COOH/*CHO. According to electronic structure analysis, its strong adsorption ability originated from the increased charge density of the active sites induced by N doping.

Zhao *et al* [71] systematically studied the electrocatalytic activity of N/B-doped GDY for the CO_2RR by calculating the various possible pathways that can generate C1 and C2 products (figure 2.13). The DFT calculation results indicate that these GDY-based metal-free catalysts have excellent catalytic activity and selectivity for converting CO_2 into CH_4 and C_2H_4, which strongly depend on the doping site. For both N-doped GDY and B-doped GDY, the potential-limiting step is the first step in the formation of the COOH* intermediate. It is noteworthy that all the CO_2RR intermediates are adsorbed at the B site of BGDY, but they are adsorbed at the sp-C site adjacent to the doped N of NGDY. This difference mainly originates from the different electronegativities of the heteroatoms. For example, the electronegativity of N (3.04) is larger than that of C (2.55), reflecting the positive charge density of adjacent C atoms. However, the electronegativity of B (2.04) is smaller than that of a C atom, resulting in the positive charge density of the B site.

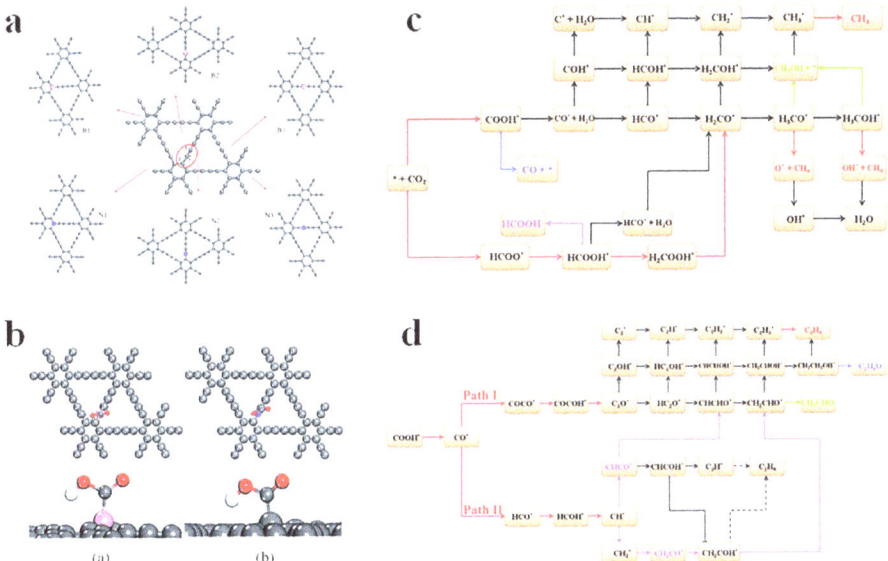

Figure 2.13. (a) Optimized structures of pristine and N/B-doped GDY at different sites. The gray, pink, and blue spheres represent C, B, and N atoms, respectively. (b) Optimized adsorption configurations of COOH* on N/B-doped GDY. The possible pathways to the (c) C1 products and (d) C2 products of the CO_2RR on N/B-doped GDY. Source: [71], reproduced with permission © Royal Society of Chemistry.

2.3.1.2.3 Nonmetal-doped GDY catalysts for the oxygen evolution reaction/oxygen reduction reaction

Numerous studies have reported that GDY-based metal-free catalysts, such as NGDY[72], B-doped GDY [73], B, N co-doped GDY [74], and N, F co-doped GDY [75] have high OER/ORR catalytic activities with low overpotential, small crossover effect, and long-term stability. However, the actual active site that determines the ORR/OER activity of heteroatom-doped GDY is unclear because nonmetal doping introduces two types of dopant, depending on which hybridized carbon is replaced. For example, replacing sp^2-C yields an sp^2-dopant, while replacing sp-C yields two different sp-dopants in the acetylenic chain. To facilitate the design of highly efficient GDY-based metal-free OER/ORR catalysts, it is necessary to deeply understand the influence of different dopants on the electronic structures and catalytic activities of GDY at the atomic level.

Chen *et al* [76] comprehensively studied the electrocatalytic activities of B, N, P, or S-doped GDY in the ORR/OER using DFT calculations (figure 2.14). Among the candidates, the graphitic-S1 dopant with a low overpotential (η) of 0.42 V endowed GDY with the best ORR catalytic activity, and the graphitic-P1 dopant with a low η of 0.35 V endowed GDY with the best OER catalytic activity. In addition, the sp-N3-doped GDY with a specific doping style (–C≡N–C≡C–) was the most promising bifunctional catalyst for both the ORR and the OER due to its overpotentials of only 0.49 V and 0.40 V, respectively. From the theoretical

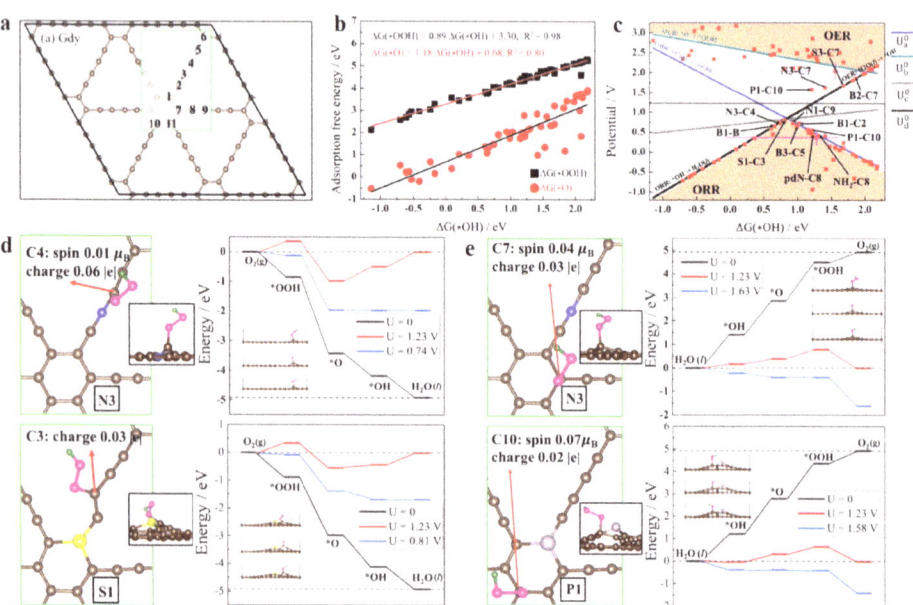

Figure 2.14. (a) Three possible doping sites (C1, C2, C3) on GDY. (b) Scaling relations between the adsorption free energies of oxygen-containing intermediates. (c) Volcano plot of the potentials of nonmetal-doped GDY in the ORR/OER. (d) Free energy diagrams for (d) the ORR and (e) the OER at various potentials. Source: [76], reproduced with permission © John Wiley and Sons.

perspective, the high ORR activities of N- or S-doped GDY arise from the active C sites with large positive charges (Q) induced by the nearby N and S dopant atoms, while the high OER activities of N- or P-doped GDY originate from the active C sites with high spin densities induced by the nearby N and P dopant atoms.

2.3.1.2.4 *Nonmetal-doped GDY catalysts for the nitrogen reduction reaction*

Nonmetal doping can lead to a charge redistribution in GDY, which induces much better electron transfer efficiency for the NRR. It has been reported that O-doped GDY (O-GDY) is a promising metal-free NRR electrocatalyst [77], in which the sp-C atom nearest to the O atom provides the catalyst with an enhanced ability to capture an N_2 molecule. In addition, the surface tensile strain (0% to +5%) can increase the NRR activity of O-GDY by decreasing the limiting potentials. The potential-limiting step for the NRR, namely $*NH_2 \rightarrow *NH_3$ via the preferentially distal mechanism, does not very with strain. The limiting potential via the distal mechanism is only 0.57 V when the surface strain of O-GDY has a value of up to +5%. It is noteworthy that overlarge surface strains (greater than +5%) are unfavorable for NRR because the carbon chains are disrupted when they adsorb the reaction intermediates. Overlarge compressive surface strain (less than −1%) also disfavors the capture of N_2 molecules because the O atom binds with four C atoms, forming a five-membered ring structure.

Double-boron-atom-doped GDY (2B-GDY) is an ideal metal-free NRR electrocatalyst [78]. In particular, 2B-GDY, in which two B atoms replace two equivalent sp-C atoms far from the benzene ring, exhibits excellent NRR catalytic activity, which is attributed to the acceptance–donation mechanism and the 'pull–pull' effect (figure 2.15).

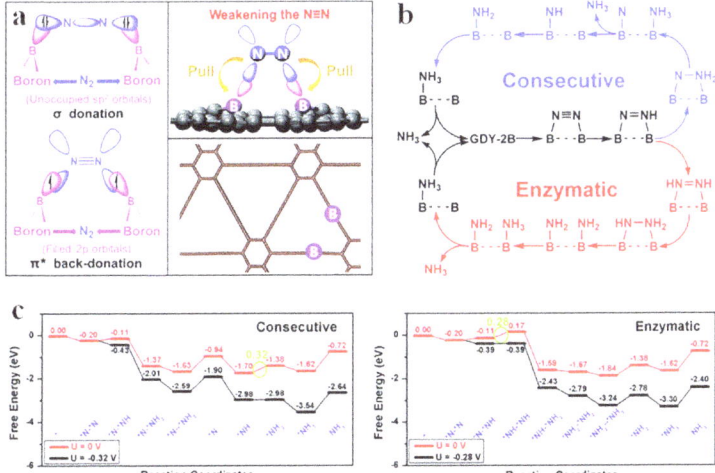

Figure 2.15. (a) Schematics of the acceptance–donation process and the 'pull–pull' effect in double-boron-atom-doped GDY (2B-GDY). (b) Schematic depiction of the consecutive pathway and the enzymatic pathway for NRR on 2B-GDY via the side-on N_2 adsorption configuration. (c) Free energy for NRR on 2B-GDY via the consecutive and enzymatic mechanisms at different applied potentials. Source: [78], reproduced with permission © Royal Society of Chemistry.

In the former, the orbitals of the B atom can be hybridized to generate sp^2 orbitals when is replaces the sp-carbon. The unoccupied sp^2 orbitals of B accept the lone-pair electrons of N_2, while the filled 2p orbitals of B donate electrons to the antibonding p-orbitals of N_2. The acceptance–donation process weakens the triple bond of the N_2 molecule. In the latter, two B atoms at a suitable distance can trap both lone pairs of electrons at the end of the N_2 molecule using a side-on adsorption configuration. The 'pull–pull' effect imposed by the Lewis pair of doped double B atoms effectively weakens the triple bond of N_2. The calculated overpotential (η) of 2B-GDY for NRR is as low as 0.12 V via the enzymatic mechanism. In addition, the 2B-GDY exhibits good NRR selectivity by greatly suppressing the competitive HER with a large ΔG_{H*} value of 1.25 eV.

B, N co-doped defective GDY (BN@GDY) exhibits excellent NRR catalytic activity (figure 2.16), which is superior to that of B-doped defective GDY (B@GDY) [79]. The overpotential of BN@GDY is only 0.32 V via the enzymatic mechanism, which is lower than that of B@GDY (0.61 V). The PDOS and Bader charge analysis show that the incorporation of N atoms into BN@GDY can convert the B hybridization from sp^2 to sp^3, which not only keeps the acceptance of lone-pair electrons from the adsorbed N_2 molecular to the empty orbital of the B atom, but also promotes the back-donation of electrons from the occupied B-p orbital to the unoccupied π^* orbital of N_2. The 'acceptance–donation' process plays a vital role in the activation of N_2 and the following hydrogenation steps. Additionally, BN@GDY exhibits high selectivity by substantially suppressing the competing HER during the NRR.

2.3.2 Experimental progress in GDY-based catalysts

Developing new catalysts with high activity, stability, and selectivity is of great importance for the efficient conversion and utilization of green energy under moderate conditions. The unique chemical and electronic structures of GDY endow it with infinite natural active sites, a large surface area, inhomogeneous surface charge, high conductivity and hole/electron transfer ability, and excellent intrinsic activity. In recent years, an abundance of original research achievements related to GDY-based catalysts has been reported, which evidenced their excellent performances in many fields. This section discusses the properties and experimental progress of GDY-based materials in the field of catalysis.

2.3.2.1 GDY-based atomic catalysts

As mentioned in section 2.3.1.1, ACs are ideal catalytic models because of their unique chemical and electronic structures, high atomic utilization efficiency, high reaction selectivity, high activity, and long-term stability. GDY-based ACs are stabilized by the incomplete charge transfer between the metal atoms and the GDY and exhibit many unique and well-defined chemical and electronic structures, which lead to high catalytic performance in many reactions. Remarkably, these novel ACs have effectively resolved the easy migration and aggregation of traditional SACs, and have been regarded as the next generation of catalysts.

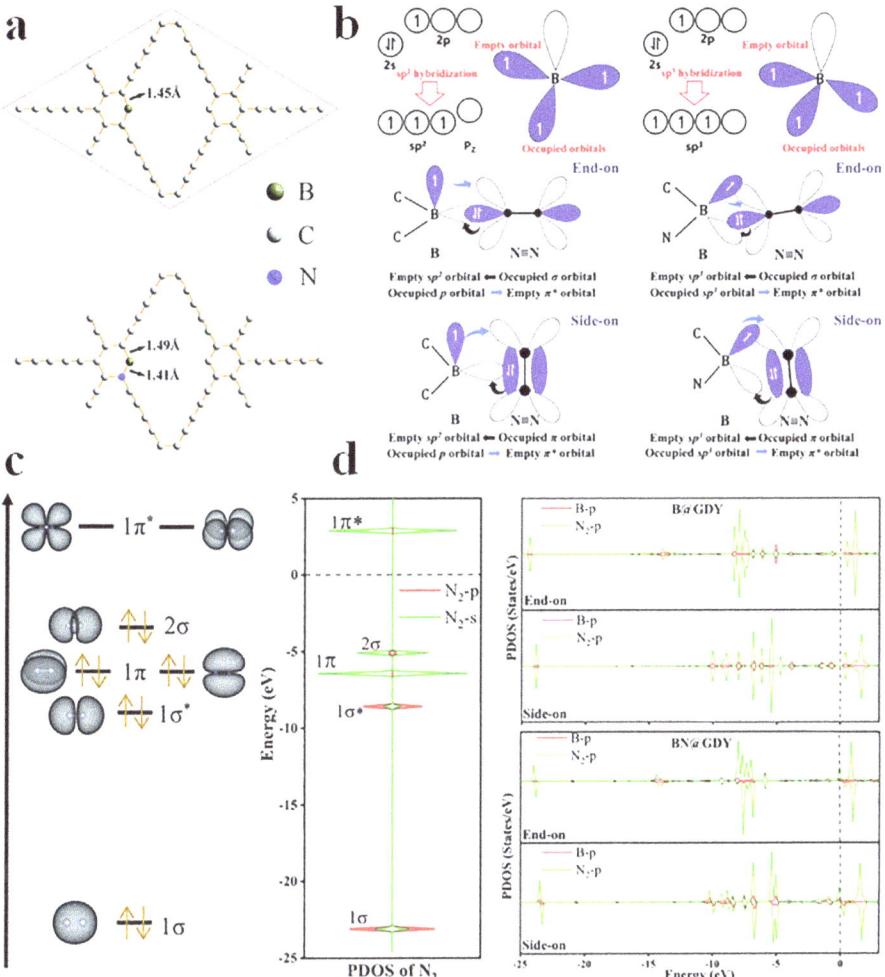

Figure 2.16. (a) Top view of B@GDY and BN@GDY structures. The celadon, white, and blue spheres represent B, C, and N atoms, respectively. (b) Simplified schematic diagram of N_2 adsorbed on a B site with sp^2 and sp^3 hybridization by means of the end-on and side-on modes. (c) Molecular orbitals of free N_2. (d) Spin-polarized PDOS of free N_2 and N_2 adsorbed on B@GDY and BN@GDY by means of the end-on and side-on modes. The dashed line represents the Fermi level. Source: [79], reproduced with permission © Royal Society of Chemistry.

2.3.2.1.1 Hydrogen evolution reaction

Electrochemical water splitting has been considered to be one of the most efficient and green approaches for the generation of hydrogen. However, the sluggish kinetics of water splitting largely reduce the reduces the overall efficiency and increases the cost of extra energy. In recent decades, high-performance hydrogen production has been reportedly been achieved. Unfortunately, only noble-metal-based catalysts (e.g. Pt-based ones for the HER and Ru-/Ir-based ones for the OER), which suffer from high cost and scarcity, are the most efficient electrocatalysts for water splitting at

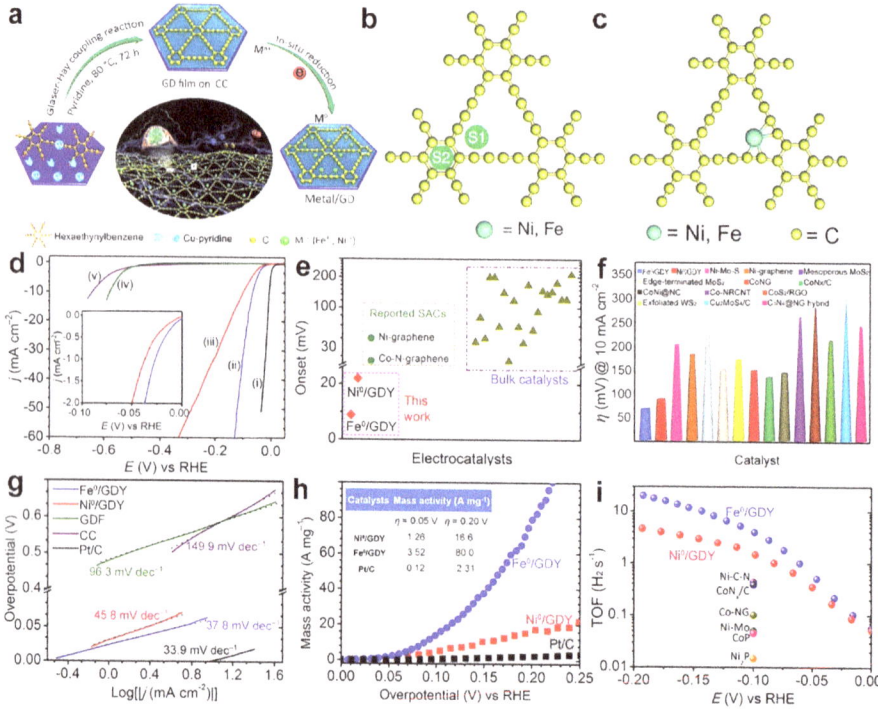

Figure 2.17. (a) Protocols for the synthesis of Ni/GDY and Fe/GDY. (b and c) Adsorption of single metal atoms on GDY (left: possible adsorption sites; right: optimized configuration). (d) Polarization curves of (i) Pt/C, (ii) Fe/GDY, (iii) Ni/GDY, (iv) graphdiyne foam (GDF), and (v) carbon cloth (CC) (inset: enlarged view of the linear sweep voltammetry (LSV) curves for Fe/GDY and Ni/GDY near the onset region). (e) Onset values of Ni/GDY and Fe/GDY (red squares) along with those of other nonprecious single-atom HER catalysts (green circles) and several bulk catalysts (olivine triangles); and (f) catalyst overpotentials at 10 mA cm⁻². (g) Tafel slopes of the catalysts. (h) Mass activities of catalysts at overpotentials of 0.05 and 0.20 V. (i) Turnover frequency (TOF) values of Fe/GDY and Ni/GDY together with those of several state-of-the-art HER electrocatalysts. Source: [48], reproduced with permission © Springer Nature.

present. The exploration of novel electrocatalysts with high reaction selectivity, activity, and stability is still critical in order to improve overall water-splitting performance.

In 2018, Professor Li [48] reported the first zerovalent metal ACs, produced by anchoring zerovalent transition-metal atoms to GDY (Ni^0/GDY and Fe^0/GY, figure 2.17). Detailed characterizations, such as high-angle annular dark-field (HAADF) imaging, high-resolution transmission electron microscopy (HR-TEM) imaging, extended x-ray absorption fine structure (EXAFS) imaging, and x-ray absorption near edge spectroscopy (XANES), unambiguously verified that the metal atoms were separately and highly dispersed on the surface of the GDY and were in their valent states. Theoretical calculations results showed that all the metal atoms were more energetically inclined to anchor at the corner of the acetylenic ring between two adjacent acetylenic bonds. The results also showed that the metal–C bonding in GDY-based ACs was due to orbital charge overlaps rather than the conventional covalent/ionic bond, reflecting the unique incomplete charge transfer between single metal atoms and GDY in GDY-based ACs. Thanks to the rich acetylenic bonds, the spatial confinement effects of the natural pores of GDY, and the incomplete charge transfer between metal atoms and GDY, GDY-based ACs possess excellent catalytic activity and stability, and eliminate the shortcomings of easy migration and aggregation of traditional SACs. Both experimental and theoretical calculations results demonstrated the excellent catalytic activity and long-term stability of Fe^0/GDY and Ni^0/GDY for the HER under acidic conditions. For example, they showed outstanding HER activity with lower over-potentials, smaller Tafel slopes, and higher stabilities than most state-of-the-art bulk catalysts [48]. Remarkably, the Ni^0/GD and Fe^0/GDY showed excellent long-term stability without any aggregation of Ni and Fe atoms on the surface of the GDY during practical operational processes. More recently, another zerovalent atom, Cu, was synthesized using GDY as the supporting material (Cu^0/GDY) [80]. The theoretical calculation results showed that the HOMO charge densities were concentrated within the Cu–C bonding area and that the strong p–d coupling connected localized charge along nearly all edges of the acetylenic rings. The special rapid charge-transfer behavior and the formation of zerovalent single Cu atoms by strong p–d coupling induced charge compensation. HAADF–scanning transmission electron microscope (STEM) images showed high dispersion of individual Cu atoms on GDY. Cu LMM Auger x-ray photoelectron spectroscopy confirmed the zerovalent states of the GDY-based Cu ACs. Experimental results further revealed the higher conductivity of Cu^0/GDY than GDY. These advantages of Cu^0/GDY endowed it with excellent HER performance, which was better than that of commercial 20 wt% Pt/C.

Noble-metal-based (Pt, Pd, Ru, etc) materials are considered to be the most efficient electrocatalysts for water splitting. However, the design and synthesis of zerovalent noble-metal ACs is still a great challenge. Using GDY, Li and coworkers [54] reported the first zerovalent palladium ACs (Pd^0/GDY), in which the single zerovalent Pd atoms were uniformly dispersed on the surface of GDY without any aggregation. When it was used as an electrocatalyst for hydrogen production, the Pd^0/GDY showed better HER activity, with a smaller overpotential (55 mV at 10 mA cm^{-2}), higher mass activity (61.5 A mg$_{metal}^{-1}$), and a higher turnover frequency (16.7 s^{-1}) than those of 20 wt% Pt/C and other reported electrocatalysts. In addition, the Pd^0/GDY also featured excellent long-term stability due to the mechanically and chemically inert properties of GDY and the intimate interactions between the anchored Pd atoms and the GDY.

2.3.2.1.2 *Nitrogen reduction reaction*

Artificial nitrogen fixation for NH_3 synthesis at ambient temperatures and pressures has been considered a promising approach for efficient NH_3 production. To this end, extensive efforts have been made to synthesize new catalysts with high selectivity and activity, such as noble metals [81, 82], metal nitrides [83], metal oxides [84–89] and metal-organic frameworks (MOFs) [90, 91]. Developing highly active and selective catalysts for efficient NH_3 production at ambient pressures and temperatures is still a giant challenge.

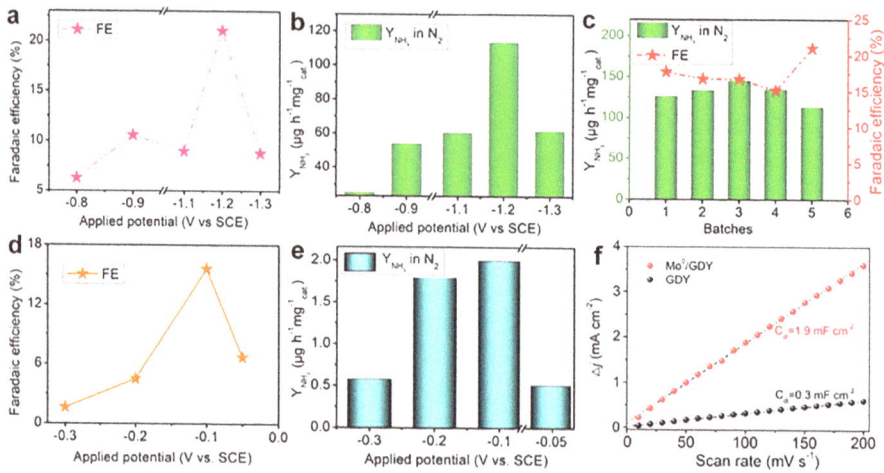

Figure 2.18. (a) Faradaic efficiencies (FEs) and (d) Y_{NH3} at different applied potentials in 0.1 M Na$_2$SO$_4$. (b) Y_{NH3} and FEs of NH$_3$ produced using different batches of Mo^0/GDY samples. (c) FEs and (d) Y_{NH3} at different applied potentials in 0.1 M HCl. (e) Amounts of NH$_3$ generated using pure GDY and a Mo^0/GDY electrode after 2 h of electrolysis at −0.1 V under ambient conditions. (f) Current density versus scan rates. Source: [39], reproduced with permission © American Chemical Society.

Molybdenum nitrogenase has been reported for most biological nitrogen fixation, and zerovalent molybdenum (Mo) atoms also play important roles in many reactions. However, zerovalent Mo atomic catalysts cannot be obtained through traditional methods. To this end, Professor Li [39] reported the first zerovalent Mo ACs by controlled anchoring single Mo atoms on GDY (Mo^0/GDY, figure 2.18). The Mo^0/GDY ACs thus obtained possess a high mass content of Mo atoms (up to 7.5 wt%). Experimental and DFT calculations revealed that the incomplete charge transfer between Mo atoms and GDY preserves the structural stability and the rapid Mo–C charge transfer, which ensures extremely strong activity in the synthesis of ammonia at room temperatures and ambient pressures. As expected, the Mo^0/GDY had very high selectivity and catalytic activities in both neutral (0.1 M Na$_2$SO$_4$) and acidic (0.1 M HCl) conditions. For example, a maximum ammonia yield and faradaic efficiency of Mo^0/GDY could reach up to 113.4–145.4 µg h^{-1} mg$_{cat}^{-1}$ and 15.2%–21.0% in 0.1 M N$_2$-saturated Na$_2$SO$_4$. Experiments using ^{15}N$_2$ isotope labeling confirmed the occurrence of ammonia electroproduction. The well-defined chemical/electronic structure and valence states show that atomically anchored Mo^0 on GDY can create numerous active sites, fast reversible Mo–C1 charge transfer, and an enlarged electrochemical active surface area, ensuring the high catalytic performance of Mo^0/GDY. More recently, a novel self-reduction strategy was proposed for the efficient anchoring of zerovalent Pd atoms on GDY (Pd-GDY, figure 2.19) [92]. The catalyst exhibited high ECNRR activity and selectivity with the greatest N$_2$ reduction activity under ambient conditions. The results of experimental and theoretical calculations demonstrated the unique distribution of separated zerovalent Pd atoms on GDY. The strong orbital interactions between Pd atoms and neighboring C sites led to a strong

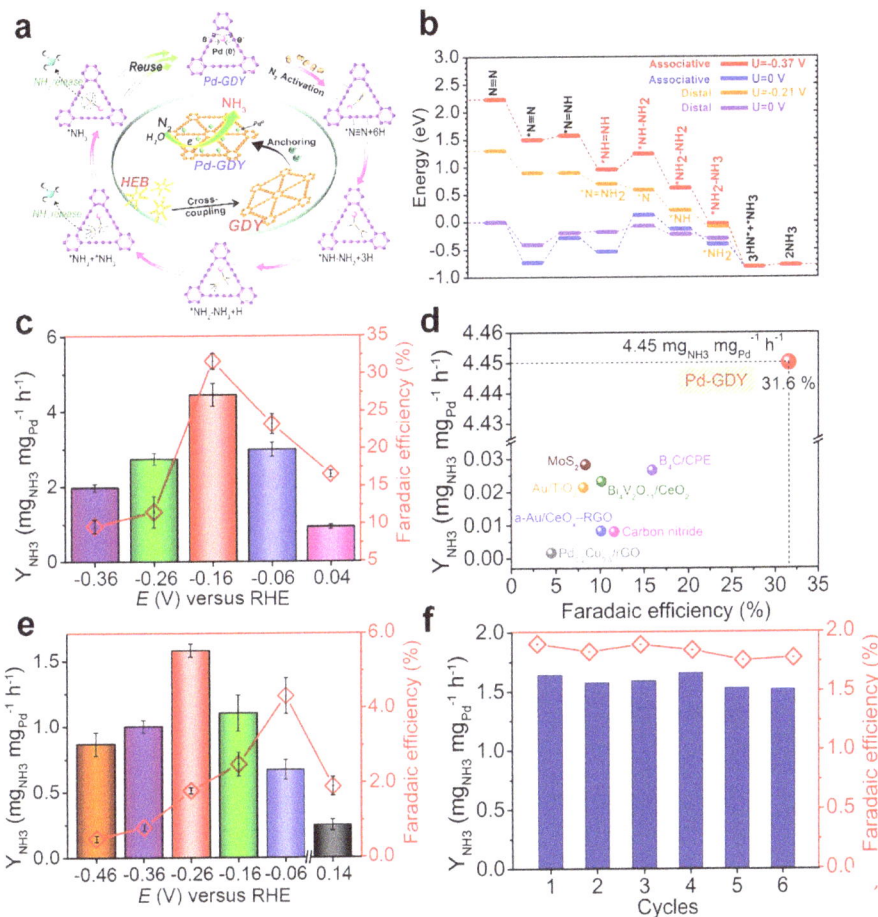

Figure 2.19. (a) Schematic of the synthesis (central green circle) and resuability of the Pd-GDY electrocatalyst for ammonia production. (b) ECNRR energetic pathway when the Pd-GDY catalyst is used. (c) Y_{NH3} and FEs at applied potentials in 0.1 M Na_2SO_4. Error bars represent standard deviation calculated using data from independent experiments (at least three values). (d) Comparison of the ECNRR performance of Pd-GDY and those of other catalysts. The error bars represent the standard deviation calculated using data from independent experiments (at least three values). (e) Y_{NH3} and FEs at applied potentials in 0.1 M HCl. (f) Stability test of Pd-GDY at −0.26 V versus a reversible hydrogen electrode (RHE) under ambient conditions. Source: [92], reproduced with permission © 2020, Oxford University Press.

electronegative reductive character for NH_3 production. Experimentally, the catalyst exhibited an ultrahigh NH_3 yield of 4.45 ± 0.30 mg_{NH3} mg_{Pd}^{-1} h^{-1} (much higher than the values reported for electrocatalysts under ambient conditions and even favorable compared to those used at higher temperatures/pressures), 100% reaction selectivity, and high faradaic efficiency in 0.1 M Na_2SO_4 aqueous solutions. Moreover, there was almost no decrease in the NH_3 yield rates and faradaic efficiencies of the catalyst during long-term stability tests. Duan *et al* [93] reported stereoconfinement-induced densely populated zerovalent metal single atoms (Rh, Ru, Co) on GDY for the

ECNRR, which were intended for NH_3 production at 55 atm. Their results showed that the HER of the prepared GDY-based ACs was greatly suppressed under the pressurized environment and the NRR activities were significantly enhanced. Among the prepared catalysts, the Rh^0/GDY exhibited the best performance and had an NH_3 yield rate of 74.15 µg h^{-1} cm^{-2}. The driving force for the formation of end-on N_2* on Rh^0/GDY dramatically increases under high pressure.

2.3.2.1.3 Oxygen reduction reaction

Using $NaBH_4$, Wu, Cai, et al [53] achieved the in situ reduction of Fe^{3+} on the surface of GDY to gain a GDY-based Fe atom catalyst (Fe-GDY). The experimental results showed that Fe-GDY could promote the 4e$^-$ pathway of ORR while limiting the 2e$^-$ pathway, which was very consistent with theoretical predictions. Under basic conditions (0.1 M KOH), the onset potential (U_{onset} = 0.21 V), half-wave potential ($U_{1/2}$ = 0.10 V), kinetic current density (i_k = 6.70 mA cm^{-2} under 0.1 V), and rate constant (k = 1.47 × 10^{-2} cm s^{-1}) of Fe-GDY were close to those of commercial Pt/C. Simultaneously, the half-wave potential and rate constant of Fe-GDY after a 5000-cycle accelerated stability test (accelerated degradation testing (ADT)) hardly exhibited any variation compared to the initial values. However, the same parameters showed obvious attenuation for commercial Pt/C, clearly demonstrating the excellent long-term stability of Fe-GDY. This study revealed the unique advantages of GDY-based metal-atom catalysts in the rational design and preparation of novel highly active ORR catalysts.

2.3.2.1.4 Organic reactions

Recently, Liu, Zhang, et al [94] reported the anchoring of zerovalent Pd atoms on a GDY/graphene (GDY/G) heterostructure through the van der Waals (vdW) epitaxy method in solution (figure 2.20). The mass loading of Pd was determined to be 0.855 wt%. Their results showed that the Pd_1/GDY/G catalyst thus obtained had high activity and selectivity for the conversion of 4-nitrophenol (4-NP) to 4-aminophenol (4-AP), with a rate constant of 0.953 min^{-1} and a turnover frequency (TOF) of 1762.17 min^{-1}. The catalyst also exhibited excellent long-time stability with 99% retention after ten repeated cycles. Lu and coworkers [95] reported the synthesis of Pd^s–GDY catalysts via a facile wet-chemistry route. Their experimental results showed that the Pd^s–GDY exhibited the strongest catalytic activity, with a TOF of up to 6290 h^{-1} at 100% conversion and 99.3% selectivity in the hydrogenation of phenylacetylene to styrene, compared to that of GDY-supported Pd nanoparticle (NP) catalysts, namely, Pd^{NP1}–GDY (with 2 nm Pd NPs) and Pd^{NP2}–GDY (with 12 nm Pd NPs). Their results revealed that Pd^s–GDY had the weakest adsorption of styrene, which was responsible for its high performance. These reports demonstrated the application potential of GDY-based metal atomic catalysts for organic reactions with desirable activity and selectivity.

2.3.2.2 GDY-based heterojunction catalysts

Heterostructured catalysts are another important type of catalyst due to their unique properties, such as easy fabrication, high stability, and the abundance of different

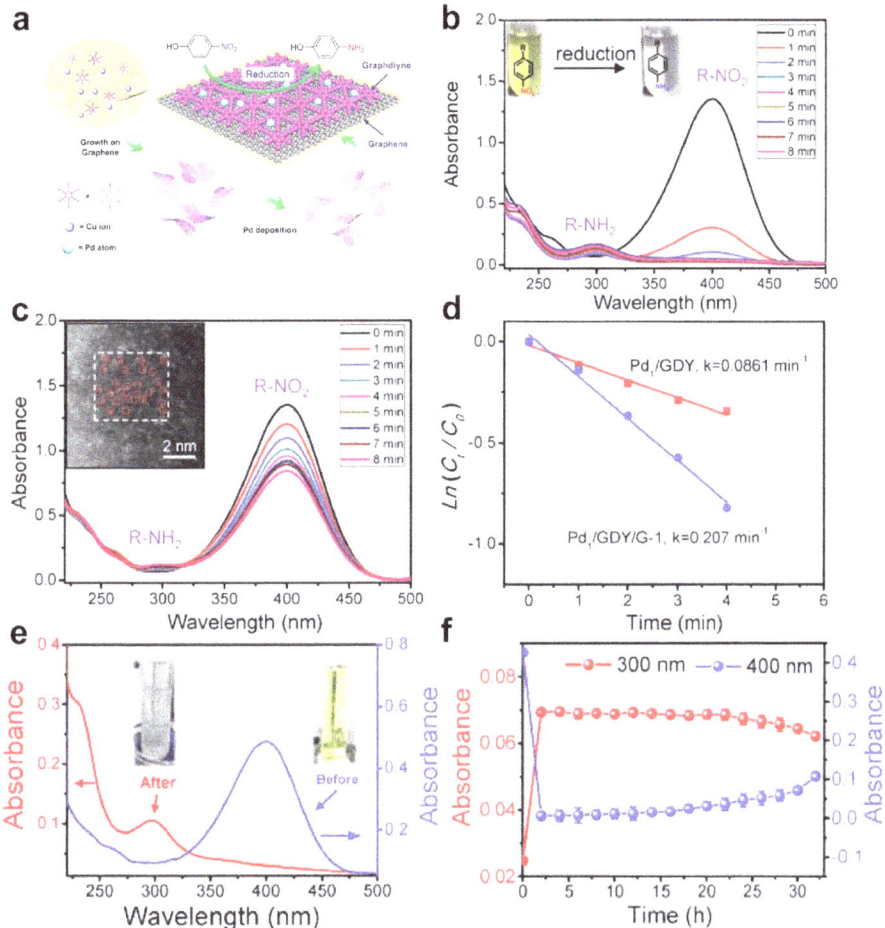

Figure 2.20. (a) Schematic illustration of the synthesis of a GDY/G heterostructure through solution-based van der Waals epitaxy, the preparation of Pd1/GDY/G, and the catalytic process of 4-nitrophenol (4-NP) reduction. (b) Time-dependent UV–Vis absorption spectra recorded during the 4-NP reduction catalyzed by (b) Pd1/GDY/G and (c) Pd1/GDY. The corresponding atomic-resolution HAADF-STEM image of Pd1/GDY is shown in the inset. (d) Plots of ln (C_t/C_0) as a function of the reaction time for the reduction of 4-NP catalyzed by Pd1/GDY/G-1 and Pd1/GDY. (e) UV–Vis absorption spectra recorded before (blue curve) and after (red curve) reduction in the continuous mode. (f) The absorbances of the lower solution at 400 nm and 300 nm after reduction as a function of the reaction time. Source: [94], reproduced with permission © John Wiley and Sons.

catalytically active sites. Benefitting from the superior characteristics of carbon materials, the incorporation of bulk catalysts into carbon materials has been widely used to fabricate high-performance catalysts. Among the reported carbon materials, GDY is the only sp-/sp^2-cohybridized carbon material that possesses excellent intrinsic activity and can be grown on various substrate surfaces, which suggests a new strategy for the preparation of novel catalysts with high catalytic activity and reaction selectivity, long-term stability, and high production yields [96, 97].

A synthesized GDY-based heterojunction exhibited more electrochemically active sites, more mass/ion transport and diffusion channels, and therefore more efficient electrocatalytic performance.

2.3.2.2.1 Electrocatalytic water splitting

Li and coworkers [96] reported a simple method for the synthesis of a 3D self-supported GDY/MoS$_2$ HER cathode by growing ultrathin MoS$_2$ nanosheets on the surface of GDY (eGDY/MDS). It was found that the combination of two semi-conductor materials, GDY and MoS$_2$, could produce metallic conductor properties. The synergistic interactions between eGDY and MDS endow the catalyst with facilitated electron transfer kinetics, higher conductivity, an increased number of active sites, and an optimal ΔG_H value that is close to thermoneutral. These advantages make it a highly active and stable electrocatalyst for efficient hydrogen production over a wide pH range (from acidic to alkaline conditions), with small overpotentials of 128 mV in 0.5 M H$_2$SO$_4$ and 99 mV in 1.0 M KOH, Tafel slopes of 46 mV dec^{-1} in 0.5 M H$_2$SO$_4$ and 89 mV dec^{-1} in 1.0 M KOH at 10 mA cm^{-2}, and long-term stability. Considering the fact that the best OER catalysts can only work well under alkaline or neutral conditions, GDY-nanosheet-supported cobalt nano-particles (CoNC/GDY) that function over a wide pH range were synthesized by a simple, scalable synthesis route involving the reduction of Co^{2+} and the simultaneous decomposition of dicyandiamide [98]. The synthesized CoNC/GDY exhibited unpre-cedented durability, which was far superior to that of commercial Pt/C (10 wt%). Yu et al [99] reported a 3D porous heterostructure fabricated through the in situ growth of 2D MoS$_2$ nanosheets on the surface of 2D N-doped GDY nanolayers (MoS$_2$/NGDY). The creative hybridization of MoS$_2$ and NGDY endowed this heterostruc-ture with structural and compositional advantages that boosted its catalytic activity (a low overpotential of 186 mV at 10 mA cm^{-2} and a Tafel slope of 63 mV dec^{-1}) and provided extraordinary stability ($\approx$15 200 cycles, which is higher than the values for all reported MoS$_2$-based materials and even better than those of commercial Pt and almost all benchmarked electrocatalysts). Wu et al [100] prepared a 3D Cu@GDY/Co electrode that exhibited high OER electrocatalytic activity with a small overpotential of nearly 0.3 V and a large unit mass activity of 413 A g^{-1} at 1.60 V vs RHE. In the course of 4 h of electrolysis, the electrode maintained a relatively constant current. Such performance is far from practical requirements. Moreover, to meet the require-ments of practical applications, there is an urgent demand for the development of an efficient HER/OER bifunctional electrocatalyst that can be operated under the same conditions to achieve highly efficient and stable overall water splitting (OWS). Li and coworkers [101] synthesized a 3D GDY network as a substrate for the controlled growth of NiCo$_2$S$_4$ nanowire arrays, and obtained the first GDY-based OER catalyst —NiCo$_2$S$_4$ NW/GDF. The NiCo$_2$S$_4$ NW/GDF electrode exhibited excellent catalytic activity and extraordinary long-term stability for both the HER and the OER, as well as OWS in 1.0 M KOH. When assembled into an alkaline water electrolyzer, it required only 1.53 V and 1.56 V to reach current densities of 10 and 20 mA cm^{-2}, respectively, and showed robust long-term stability over 140 h during continuous water splitting at 20 mA cm^{-2}. These observations show that the synergistic coupling interaction at the interfaces of $NiCo_2S_4$ and GDY can greatly enhance the performance of the catalyst. The results demonstrated that GDY can function as the key material for fabricating high-performance electrocatalysts for practical application in energy conversion.

Hui *et al* [103] employed a two-step strategy for preparing ultrathin GDY-wrapped iron carbonate hydroxide nanosheets on nickel foam (FeCH@GDY/NF) as efficient catalysts for electrical water splitting. The introduction of naturally porous GDY nanolayers on an FeCH surface endowed the pristine catalyst with structural advantages for enhanced catalytic activity and high long-term durability for the OER and the HER in 1.0 M KOH. Remarkably, this electrocatalyst can drive 10 mA cm^{-2} and 100 mA cm^{-2} at 1.49 V and 1.53 V, respectively. In addition, Liu *et al* [102] employed a simple method for the controllable synthesis of ultrathin charge-transfer complexes (CTs) of nickel with terephthalic acid nanosheets on a GDY surface (CTNS/GDY, figure 2.21). Benefiting from its superior morphological

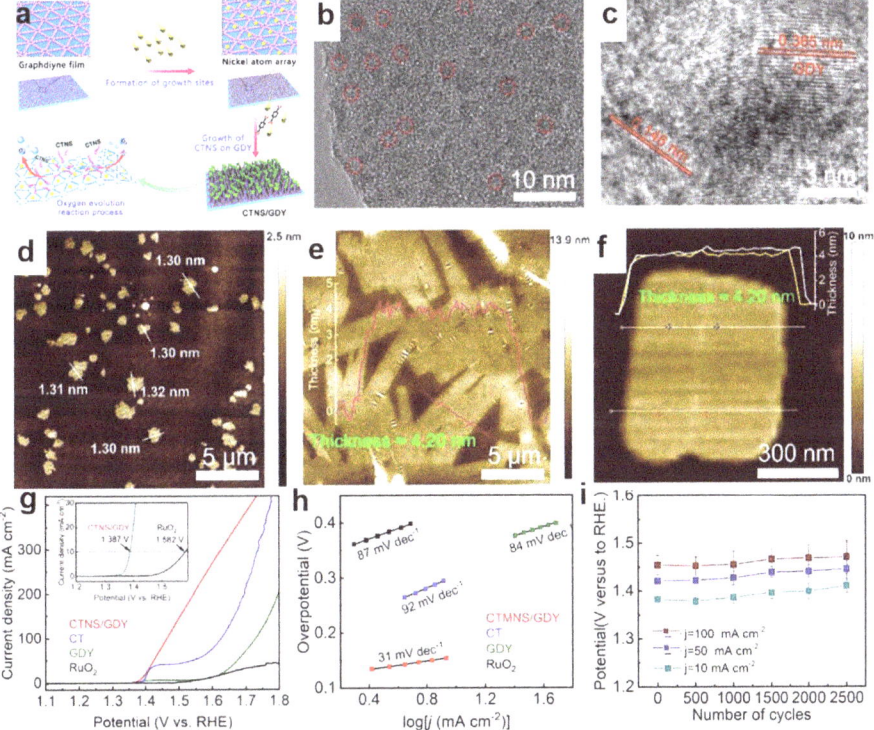

Figure 2.21. (a) Proposed mechanism for the growth of ultrathin CTNS/GDY nanosheets. (b and c) transmission electron microscopy (TEM) and HR-TEM images of CTNS/GDY. (d) AFM images of ultrathin GDY with a thickness of ≈1.30 nm. (e–f) AFM images of ultrathin CTNS/GDY nanosheets with a thickness of ≈4.20 nm. (g) Polarization curves of CTNS/GDY, CT, GDY, and RuO$_2$ in oxygen-saturated 1.0 M KOH (inset: an enlargement of the low overpotential area). (h) Tafel plots of the CTNS/GDY, CT, GDY, and RuO$_2$. (i) Stability measurements of CTNS/GDY, as determined using the potentials measured at 10, 50, and 100 mA cm^{-2} over 2500 cycles. The error bars show the standard deviation of the measurements. Source: [102], reproduced with permission © John Wiley and Sons.

and electronic properties, CTNS/GDY is a promising OER electrocatalyst, exhibiting an ultralow overpotential of 155 mV for the OER in alkaline electrolytes. DFT calculations revealed that this catalyst possesses a very low electron transfer barrier due to a strong p–d coupling effect. The enriched electronic distribution on the GDY surface not only promotes fast reversible redox switching but also enhances site selectivity due to its electron-rich character. These characteristics are key to minimizing the overall OER barrier. This work opens up new avenues for the design and optimization of future catalysts with applications in renewable energy technologies for sustainable energy conversion or storage.

Although the abovementioned GDY-supported HER electrocatalysts exhibit strong catalytic activity over a wide pH range from acidic to alkaline, such supported catalysts are gradually corroded during catalysis, leading to changes to their structure and morphology and a reduction of their service life. One of the unique properties of GDY is that it is the only carbon material that can be grown on any substrate surface at a low temperature and in a controlled manner. Therefore, it can be easily and effectively grown on the surface of a catalyst, protecting the catalyst from corrosion, and enhancing its long-term stability. Moreover, GDY has high intrinsic activity, an abundant number of active sites, and a large surface area, which are all beneficial to maximizing the catalytic activity of a catalyst.

Based on these considerations, Li and coworkers [104] were the first to report the *in situ* exfoliation and modification of bulk iron–cobalt layered double hydroxide (LDH) nanosheets through a GDY-induced intercalation/exfoliation/decoration strategy (e-ICLDH@GDY/NF, figure 2.22). Theoretical and experimental data revealed that the formation of a heterojunction structure between GDY and LDH can significantly reduce the resistance of the solution and the resistance to charge transfer, improve the charge-transfer ability, increase the number of catalytically active sites, and prevent corrosion, leading to high catalytic activity and stability for the OER and the HER in 1.0 M KOH. Furthermore, this structure only requires low cell voltages of 1.43, 1.46, and 1.49 V to provide current densities of 10, 100, and 1000 mA cm^{-2}, respectively, for OWS. These research results confirm that GDY can not only greatly enhance the catalytic activity of catalysts but also effectively improve the catalytic stability. This offers new insights for the design and synthesis of novel non-noble-metal-based electrocatalysts with strong catalytic activity and high stability.

During subsequent exploration and development, a series of GDY-coated electrocatalysts with superior catalytic properties were reported. For instance, GDY-encapsulated cobalt nitride (CoN$_x$@GDY NS, figure 2.23) nanosheets exhibit strong catalytic activity and stability in the HER, OER, and OWS processes[105]. In the HER and OER processes, the catalyst can reach 10 mA cm^{-2} at just 70 mV and 260 mV, respectively. When used both as the anode and the cathode for OWS, a very low cell voltage of 1.48 V is required to reach 10 mA cm^{-2}. Yu *et al* [106] reported a self-supported cubic heterostructure (NiO-GDY NC) with high catalytic activity and robust stability for OWS produced via surface engineering. As a result of its unique structure and synergistic interactions between the GDY and NiO species, the synthesized NiO-GDY NC exhibited greatly improved charge-transfer kinetics and

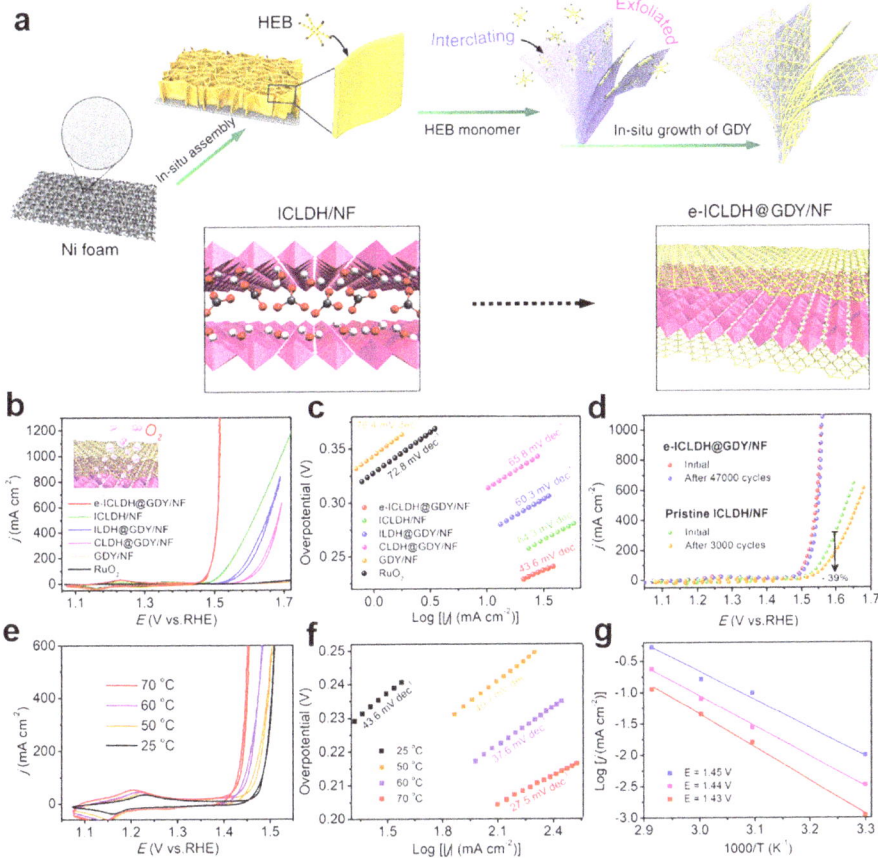

Figure 2.22. (a) Schematic representation of the synthesis strategy for the preparation of e-ICLDH@GDY/NF structures. (b) OER cyclic voltammetry (CV) curves and (c) corresponding Tafel plots of the synthesized samples. (d) Polarization curves of e-ICLDH@GDY/NF and pristine ICLDH/NF, recorded before and after 47 000 and 3000 cycles, respectively, of the OER. (e) OER CV curves and (f) corresponding Tafel plots for e-ICLDH@GDY/NF, recorded at various temperatures. (g) Arrhenius plots for the OER performed using e-ICLDH@GDY/NF at various potentials. Source: [104], reproduced with permission © Springer Nature.

hydrogen/oxygen-containing species adsorption abilities, which greatly enhanced its HER/OER catalytic performance.

2.3.2.2.2 Photoelectrocatalytic water splitting

In addition to electrocatalytic hydrogen and oxygen production, water splitting by sunlight to produce hydrogen and oxygen is another promising solution that addresses the problem of environmental issues and energy shortages. For the first time, GDY has been introduced into a photoelectrochemical (PEC) water-splitting cell as the hole transfer layer. Zhang *et al* [33] reported that the growth of GDY nanowalls on a $BiVO_4$ electrode could achieve excellent photoelectrochemical water-splitting performance, with a hole injection yield of up to 60% at 1.23 V versus RHE, which is higher than that of bare $BiVO_4$ (less than 30%). Wu *et al* [107] reported that

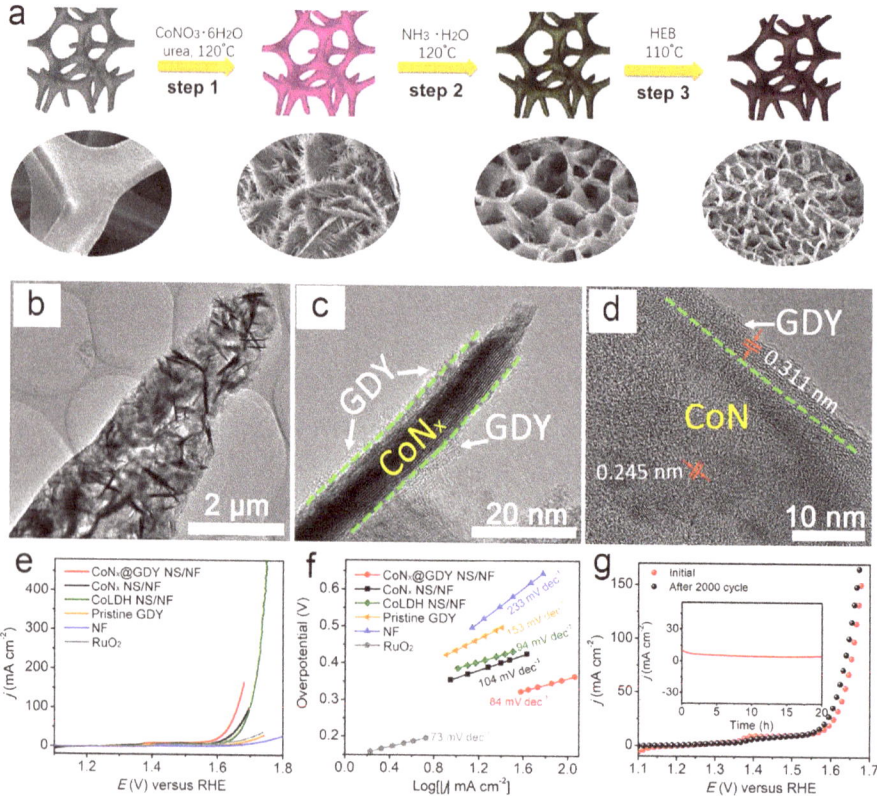

Figure 2.23. (a) Schematic of the synthesis of CoN$_x$@GDY NS/NF via an in-site growth strategy. (b) Low-resolution TEM image of CoN$_x$@GDY NS. (c-d) High-resolution TEM images of CoN$_x$@GDY NS. (e) Polarization curves and (f) Tafel plots of the prepared catalysts for the OER in 1.0 M KOH. (g) Polarization curves of CoN$_x$@GDY NS/NF before and after 2000 cycles (inset: time-dependent current-density curve of CoN$_x$@GDY NS/NF at an overpotential of 280 mV versus RHE for 20 h). Source: [105], reproduced with permission © Elsevier.

the GDY-based photoanode CoAl-LDH/GDY/BiVO$_4$ had a high level of photo-electrocatalytic activity in 0.1 M Na$_2$SO$_4$, producing a maximum photocurrent of ≈3.15 mA cm^{-2} at 1.23 V versus RHE. In addition, their results indicated that GDY has advantages that can overcome the sluggish kinetics and large overpotentials of the OER and thus significantly enhances catalytic activity. More recently, Li *et al* [108] constructed GDY-coated CuI composite materials (GDY-CuI) by growing GDY on the surface of CuI. The synthesized GDY-CuI displayed enhanced photocatalytic hydrogen production activity (465.95 mol/5 h), better than that of pure GDY (29.42 mol/5 h) and CuI (156.49 mol/5 h). Zhang and coworkers [109] constructed a SiHJ/GDY/NiO$_x$ heterojunction material by growing GDY on the surface of a Si substrate followed by controllably plating NiO$_x$ film onto GDY at a certain thickness. As a result of the superiorities of GDY and NiO$_x$, the optimized 10 nm SiHJ/GDY/NiO$_x$ exhibited a saturated photocurrent density that reached up to 39.1 mA cm^{-2}. An Ag$_3$PO$_4$/GDY-based emulsion that benefited from the fascinating properties of GDY and Ag$_3$PO$_4$ [110, 111], such as reduction ability, high stability, mediation of hole transfer (due to its low work function), highly conjugated structure, and high electron mobility was fabricated by Mao *et al* [112]. The catalyst was reported to have a high apparent rate constant (0.477 min^{-1}) for the degradation reaction of methylene blue, which was better than those of Ag$_3$PO$_4$/ CNT and Ag$_3$PO$_4$/graphene.

2.3.2.2.3 Nitrogen reduction reaction

The development of synthetic ammonia chemistry is a great opportunity to advance the progress of human society. Converting inert atmospheric N$_2$ molecules into NH$_3$ at room temperature and ambient pressure has been regarded as a promising alternative to the conventional Haber–Bosch processes. To this end, the rational design and synthesis of new catalysts with high activity, stability, and selectivity are in great demand.

Li and coworkers [113] reported a GDY-based heterostructure of GDY-cobalt nitride (GDY/Co$_2$N), which had a highly active and selective interface for the ECNRR (figure 2.24). Experimentally, at ambient pressures and temperatures, the

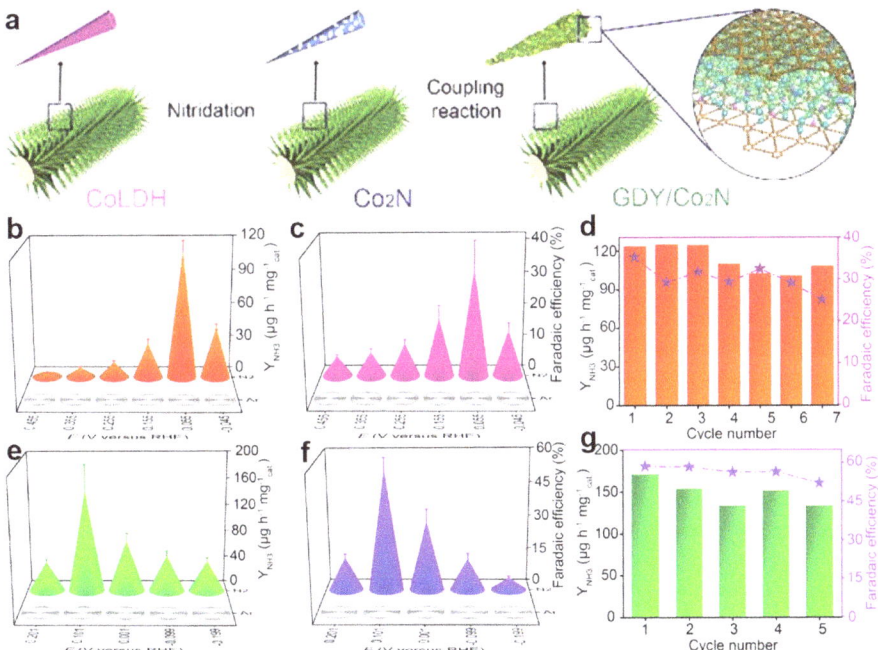

Figure 2.24. (a) The preparation of self-supporting GDY/Co$_2$N. (b) Y_{NH3} of GDY/Co$_2$N at different potentials in N$_2$- and Ar-saturated 0.1 M Na$_2$SO$_4$ (bars represent standard deviation). (c) FEs of GDY/ Co$_2$N at different potentials in N$_2$- and Ar-saturated 0.1 M Na$_2$SO$_4$ (bars represent standard deviation). (d) Cycling test of GDY/Co$_2$N at 0.055 V versus RHE in 0.1 M Na$_2$SO$_4$. (e) Y_{NH3} of GDY/Co$_2$N at different potentials in N$_2$- and Ar-saturated 0.1 M HCl (bars represent standard deviation). (f) FEs of GDY/Co$_2$N at different potentials in N$_2$- and Ar-saturated 0.1 M HCl (bars represent standard deviation). (g) Cycling test of GDY/Co$_2$N at 0.101 V versus RHE in 0.1 M HCl. Source: [113], reproduced with permission © John Wiley and Sons.

GDY/Co$_2$N reached a new record ammonia yield rate (Y_{NH3}) of 219.72 µg h^{-1} mg$_{cat.}$$^{-1}$ and an FE of 58.60% in acidic conditions, which are better than the values reported for ECNRR electrocatalysts. Moreover, their DFT calculations revealed that the interface-bonded GDY contributes a unique p-electronic character that optimally modifies the Co–N compound surface bonding, which generates the observed superior electronic activity for NRR catalysis at the interface region. In addition to the production of ammonia via the electrocatalytic nitrogen reduction, the conversion of atmospheric nitrogen molecules into ammonia by photocatalysts is another promising solution to the problems of environmental issues and energy shortages. Li and coworkers [114] reported a highly active GDY heterojunction (GDY@Fe-A and GDY@Fe-B, figure 2.25). The coordination environment and valence state of the Fe atoms in magnetite can be

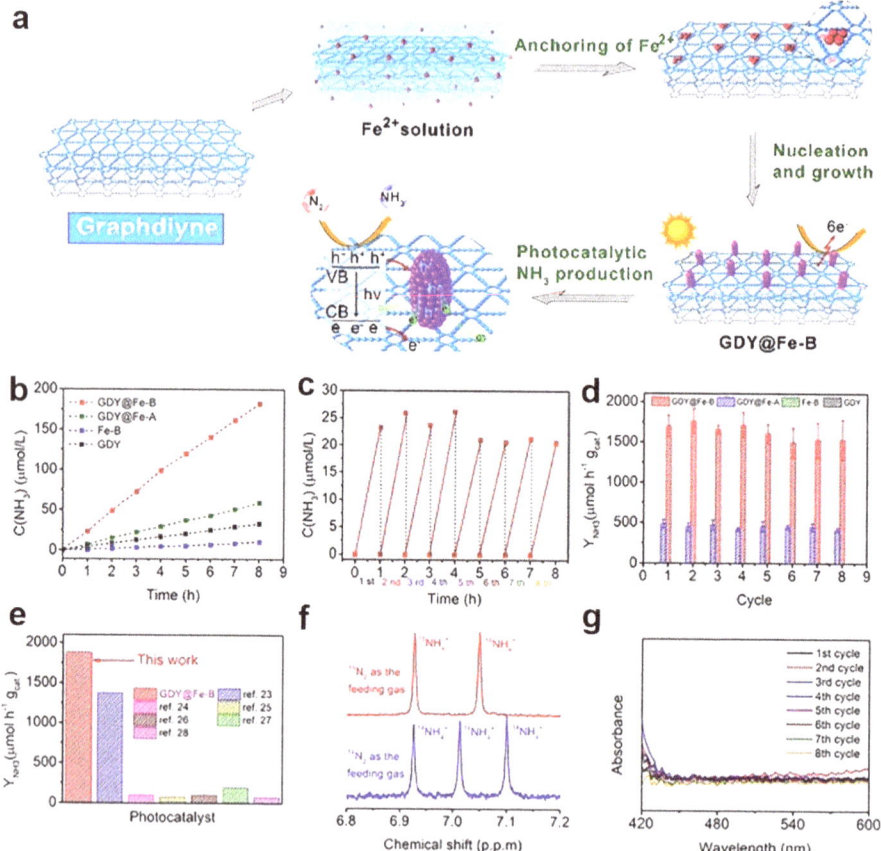

Figure 2.25. (a) Synthesis of GDY@Fe-B and its reaction route. (b) Ammonia concentrations as a function of time for catalysts under irradiation. (c) Cycling test for PCNRR with GDY@Fe-B. (d) Y_{NH3} obtained after the stability test. (e) Comparison between the Y_{NH3} of GDY@Fe-B and those of reported PCNRR catalysts. (f) Isotope labeling experiment. ^{1}H NMR spectra of ammonia obtained via the PCNRR using ^{15}N$_2$ (red line) and ^{14}N$_2$ (blue line) as the feeding gases, respectively. (g) Detection of N$_2$H$_4$ after the cycling test for PCNRR in N$_2$-saturated 0.1 M Na$_2$SO$_4$. Source: [114], reproduced with permission © John Wiley and Sons.

effectively changed by incorporating them into GDY, thereby enabling the magnetite/GDY heterojunctions to participate in highly photocatalytic nitrogen reduction reactions (PCNRRs). When it is used for the photoelectrodes, the ammonia yield (Y_{NH3}) due to the transformative photocatalytic activity of GDY@Fe-B can reach an unprecedented level of 1762.35 ± 153.71 µmol h^{-1} g$_{cat.}^{-1}$ (the maximum Y_{NH3} reached up to 1916.06 µmol h^{-1} g$_{cat.}^{-1}$), which is much higher than those of GDY@Fe-A (486.13 ± 48.96 µmol h^{-1} g$_{cat.}^{-1}$), Fe-B (4.75 ± 0.69 µmol h^{-1} g$_{cat.}^{-1}$), GDY (9.76 ± 0.69 µmol h^{-1} g$_{cat.}^{-1}$), and recently reported PCNRR catalysts. These excellent properties are achieved because the incorporation of Fe site-specific magnetite in GDY results in a valence state transition within the catalyst. These results demonstrate the advantages of GDY in effectively regulating magnetite activity and coordination environments and also indicate that magnetite can selectively form two different valences, specifically, those of tetrahedral-coordination Fe and octahedral-coordination Fe.

2.3.2.2.4 Other catalytic reactions

Unique carbon hybridization (sp^2 and sp) endows GDY with unevenly distributed surface charges as compared to carbon nanotubes and graphene. This has promoted the advancement of GDY as a highly active electrode material for various applications including (photo)electrochemical water splitting, the NRR, the ORR, batteries, etc. Recently, the application of GDY/GDY-based materials for other catalytic reactions has been explored. For example, Mao et al [115] showed that the oxygen-containing groups on the graphdiyne oxides (GDYO) surface play an important role in controlling the formation of Pd clusters by increasing the anchoring ability of Pd nuclei and avoiding Ostwald ripening in nuclei. The same groups exhibited strong reaction activity for 4-NP reduction, resulting in rate constants about 40, 11, and 5 times higher than those of Pd/MWNT, Pd/GO, and commercial Pd/C, respectively. Lu and coworkers [116] reported a catalyst comprising Pd sub-nanometric catalysts (SNCs, with an average particle size of 0.83 nm) stabilized on pyrenyl GDY (Pyr-GDY) ultrafine nanofibers (3–10 nm) for the reduction of nitro-arenes to arylamines and Suzuki coupling reactions. Its catalytic activities for the reduction of 4-NP were 300 and 25 times higher than those of commercial Pd/C and Pd/GO, respectively. Tan et al [117] reported a 2D heterogeneous hybrid nanomaterial (P5A-Au-GD) based on GDY and pillar[5]arene (P5A)-reduced Au nanoparticles (P5AAu) for the efficient reduction of 4-NP and methylene blue, which had a higher catalytic performance than that of a commercial Pd/C catalyst. Li and Shi et al [118] utilized the strong interactions between GDY and Pt nanoparticles that can prevent the thermal migration of Pt nanoparticles on the GDY surface to fabricate an ultrastable Pt-GDY catalyst. Their results showed that Pt NPs with sizes of 2–3 nm exhibited high performance in the hydrogenation of aldehydes and ketones to the corresponding alcohols compared with commercial Pt–C.

GDY has also been applied for photocatalyzed organic reactions. Wang et al [12] successfully chemically incorporated titania nanoparticles (P25) into GDY nanosheets through a facile hydrothermal treatment, forming a novel P25-GDY nanocomposite which was directly used for the photocatalytic degradation of methyl blue. P25-GDY

(0.6 wt% GDY) exhibited the best catalytic activity compared to those of P25-carbon nanotubes and P25-graphene. Their later results showed that the combination of GDY and the TiO_2 (001) crystal plane could contribute to an efficient carrier separation process and a longer photogenerated carrier lifetime; it also facilitated photocatalytic oxidation [119]. Ramakrishnan et al [120] further showed that the strong interaction between GDY/GDYO and TiO_2 could effectively optimize the hole transport process and increase the photocurrent density of the catalyst, resulting in an enhancement of the catalytic performance. Recently, Lu and coworkers [121] further combined GDY and CdS nanoparticles to obtain a CdS/GDY heterojunction for photocatalytic hydrogen production. Their results showed that GDY can not only stabilize CdS, but can also effectively transfer photogenerated holes and effectively prevent photogenerated electron–hole recombination. When the GDY content was 2.5 wt%, the catalytic performance of a CdS/GDY heterojunction was 2.6 times that of pristine CdS; it also exhibited high recyclability.

2.3.2.3 GDY-based metal-free catalysts

Although metal-based materials have been widely used for catalysis, energy conversion and storage, and many other important industrial processes, their high cost and poor stability have seriously limited their practical uses [122]. Carbon-material-based metal-free catalysts have been shown to have many unique properties, including rich and adjustable chemical or electronic structures and excellent resistance to acidic/alkaline conditions, which are beneficial for efficient catalysis. Among all the reported carbon materials, GDY possesses many unique properties and has many promising advantages for metal-free catalysis.

2.3.2.3.1 Water splitting

Li and coworkers [123] reported the utilization of self-active metal-free GDY as a model carbon-based metal-free electrocatalyst for exploring the origin of electrocatalytic performance at the atomic level. Their results showed that the unusual electrocatalytic properties of GDY originate from its unique nanostructure, as a benefit of which, GDY acts as a metal-free efficient HER electrocatalyst with Pt-like HER activity, but with long-term durability superior to that of Pt/C over a wide pH range (from acidic to basic). Its HER performances were much better than those of other reported metal-free electrocatalysts and even Pt-based ones.

Heteroatom-doped GDY exhibits excellent properties in the field of water electrolysis; however, molecularly designed and carefully doped GDY is also of great significance in promoting the kinetics of the ORR. Li and coworkers [124] controlled the growth of a film of fluorine-substituted GDY (F-GDY) on carbon cloth (p-FGDY/CC, figure 2.26). As expected, the obtained p-FGDY/CC electrode gained high HER, OER, and OWS performances and stability in both acidic and alkaline environments. The results revealed that the strong C-F bond could lead to the redistribution of p-electron orbitals, improve the electron-rich characteristics of the C2 site, and increase the electron transfer ability, which finally endowed the electrode with excellent OWS performances over a wide pH range. Zhao et al [125] reported the N- and S-co-doped GDY exhibited better OER activity than that of only N- or

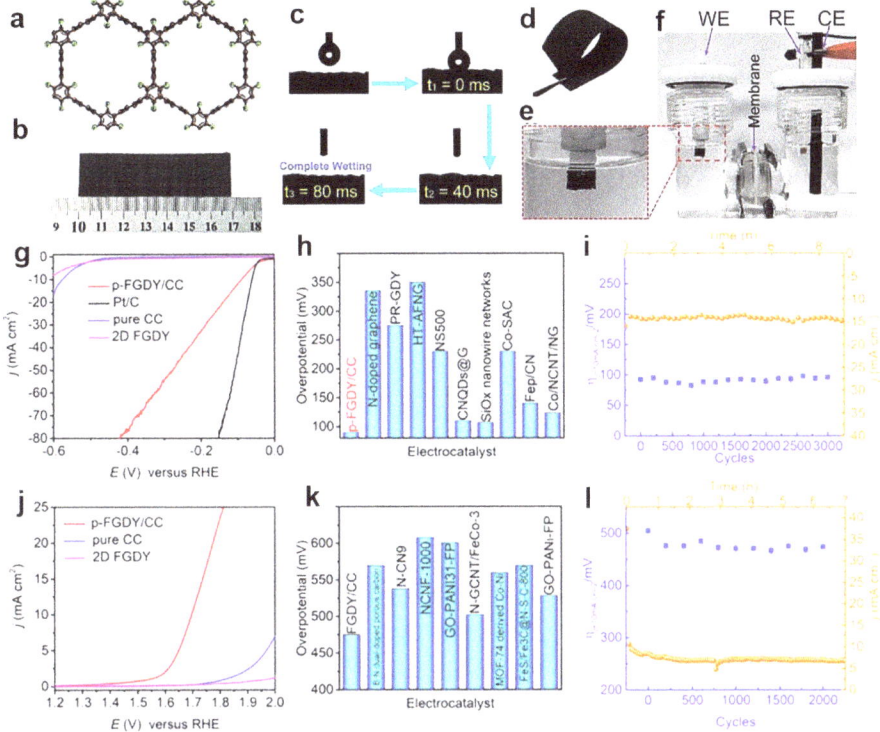

Figure 2.26. (a) Schematic of the structure of F-GDY. (b) Photograph of the p-FGDY/CC material. (c) Contact angle measurements of p-FGDY/CC. (d) Photograph of a bent p-FGDY/CC electrode, demonstrating its flexibility. (e) Photograph of the three-electrode system (WE: working electrode; RE: reference electrode; CE: counter electrode) and (f) an enlarged image of the WE in (f). (g) Polarization curves of the catalysts for HER in 0.5 M H_2SO_4. (h) Comparison of the HER performances of p-FGDY/CC and reported catalysts in 0.5 M H_2SO_4. (i) Long-term stability tests of p-FGDY/CC in 0.5 M H_2SO_4. (j) Polarization curves of catalysts for the OER in 1.0 M KOH. (k) Comparison of the OER performances of the p-FGDY/CC and the reported catalysts. (l) Long-term stability tests of p-FGDY/CC in 1.0 M KOH during the OER process. Source [124], reproduced with permission © John Wiley and Sons.

S-doped GDY. This study also revealed that the synergistic effect of double doping and 3D positioning could effectively enhance the overall catalytic performance.

2.3.2.3.2 Oxygen reduction reactions

Recently, Huang *et al* [126] synthesized a GDY-like carbon material, in which one carbon atom in every benzene ring of GDY was replaced by pyridinc N (PyN-GDY). The synthesized PyN-GDY with its 'defined' molecular structure is an ideal model for addressing the intrinsic activity of active sites at the molecular level. It exhibits excellent performance in both alkaline and acidic media as electrochemical catalyst for the ORR. DFT calculations have been used to analyze and determine which sites of PyN-GDY may be active in the ORR. The precise construction of specific nitrogen-doped carbon materials is an effective method of producing efficient catalysts with improved electrocatalytic performance for the

ORR. Li *et al* [75] reported that an N, F co-doped GDY catalyst exhibited catalytic activity equivalent to that of a Pt/C electrode and also had excellent stability in an alkaline environment (1.0 M KOH). Zhao *et al* [127] recently reported that sp-N doping promoted the adsorption and electron transfer of O_2 on a catalyst's surface, so that the material had great electrocatalytic ORR performance. In an alkaline environment, sp-N-doped GDY had a smaller Tafel slope (60 mV dec^{-1}) than Pt/C (76 mV dec^{-1}). In addition, Zhang *et al* [72] prepared N-doped GDY. An electro-catalyst test showed that the limiting current density of N 550-GD at 0.05 V vs. RHE in 0.1M KOH reached about 4.5 mA cm^{-2}, which was comparable to that of 20% Pt/C, and N 550-GD exhibited excellent stability.

2.3.2.3.3 Photocatalysis

As a direct-gap natural semiconductor, GDY has great electron and hole transport capability and has been used as an excellent metal-free material for photocatalytic processes. A 2D/2D heterojunction made from g-C3N4/GDY was applied for photocatalytic H_2 production by Si *et al* [128]. They found that only 1% GDY could realize a high H_2 evolution rate of (454.28 μmol h^{-1}). Recently, Xu *et al* [129] combined graphitized carbon nitride and graphyne to form a heterostructure which improved the separation efficiency of charge carriers and prolonged the life of charge carriers. The electron flow in the photocatalyst was accelerated and reduced. The overpotential of the HER was analyzed. Li and coworkers [130] recently reported their work on the synthesis and self-assembly of a crystalline dehydrobenzoannu-lene-based 3D GDY (PDBA), and its application in photocatalytic hydrogen production. DFT calculations revealed that the C5 site (0.10 eV, carbon atom on the twisted butadiyne of the DBA unit) and the C2 site (0.16 eV, carbon atom on the connecting butadiyne) possessed higher activities and were favorable for H_2 evolution, confirming the role of alkynes in photocatalysis. The energy level for the reduction of H_2O to H_2 was below the E_{cb} of PDBA and the energy level for the oxidation of H_2O to O_2 was above the E_{vb} of PDBA. In a three-electrode setup, the PDBA photocathode gave a prompt and reproducible photoresponse under on/off cycles of visible light excitation and exhibited a strong ability to produce H_2 at a rate of 340 mmol h^{-1} g^{-1} and an apparent quantum efficiency of 4.68% at 420 nm.

2.3.2.4 GDY-based quantum dot catalysts

Quantum dots (QDs), which have the unique advantages of excellent light-harvest-ing ability, high surface-to-volume ratios, and a large number of exposed active sites have been expected to find a place in the field of electrocatalysis. Compared with traditional carbon materials, GDY-based QD catalysts exhibit excellent catalytic performance due to their larger specific surface areas, increased numbers of active sites, improved corrosion resistance, etc.

2.3.2.4.1 Photocatalytic water splitting

GDY, for the first time, has been introduced into a PEC water-splitting cell as the hole transfer layer. Zhang, Wu, *et al* [131] prepared a film of CdSe QDs/GDY nanosheets through π–π stacking between mercaptopyridine and GDY. The GDY acted as a hole transfer layer for hydrogen production in neutral water. Their experimental results showed that the strong π–π interactions between the GDY and the mercaptopyridine surface-functionalized CdSe QDs facilitated hole transportation and photocurrent enhancement, which promoted photogenerated electron-charge separation efficiency and improved the photocatalytic hydrogen evolution performance. As a result, when exposed to illumination produced by a Xe lamp, the CdSe QDs/GDY photocathodes exhibited enhanced photoactivity and nearly 90% ± 5% FE for hydrogen production during a 12 h test. These results were attributed to the higher hole mobility and stability of GDY.

Recently, Du *et al* [132] synthesized OsO_x QDs on the surface of GDY (OsO_x QDs/GDY) through a facile *in situ* growth approach (figure 2.27). The catalyst exhibited greatly improved catalytic activity and stability for the HER under light induction. The calculated and experimental results indicated that GDY was not only used as the hole transfer layer to prevent hole–electron recombination but also induced charge transfer for more high-coordination osmium (Os^{4+}). For example, under alkaline conditions, OsO_x QDs/GDY exposed to light has the lowest over-potential and attains a current density of 100 mA cm^{-2} at 42.5 mV, which is much lower than the value achieved by OsO_x QDs/GDY without light exposure (210 mV, at 100 mA cm^{-2}). The Tafel slope of OsO_x QDs/GDY is 39.3 mV dec^{-1} when illuminated, which is also significantly lower than those of other reference samples. In addition, GDY can promote good distribution and stability of the metal-oxide QDs. The fundamental advantages of the OsO_x QDs/GDY catalyst are that it promotes hole transport, generates a large number of active sites, and has superior photocurrent performance and photo/electrocatalytic activity.

2.3.2.4.2 *Electrocatalytic water splitting*

In Wang's work [133], a controllable GDY-induced growth strategy was established; a highly uniform size distribution of oxidized iridium quantum dots was prepared on the surface of GDY (IrO_xQD/GDY). GDY has unique porous and acetylene-rich structural units; this suitable space and incomplete charge transfer between the metal atoms and the GDY can precisely anchor the metal atoms, improving the aggregation and stability of the QDs and producing a surface structure that has multiple active sites. The catalyst exhibits a strong acidic OER with a small overpotential of 236 mV versus the RHE at a current density of 10 mA cm^{-2} and a Tafel slope of 70 mV dec^{-1}, which are better than those of other reported acidic OER electrocatalysts. Furthermore, the IrO_x QD/GDY catalyst was found to be much more stable than IrO_x QDs self-supported on CC. When it was used as both cathode and anode in an acidic electrolyzer, the catalyst exhibited a much lower cell voltage of 1.49 V (vs RHE) than a Pt/C||RuO$_2$ combination and other reported catalysts. The results show the superior advantages of GDY in effectively increasing the numbers of catalytically active sites in order to improve the charge-transfer behavior and protect the metal catalysts from corrosion.

Recently, Li *et al* [134] reported a series of porous 2D GDY-loaded bimetallic vanadium-ruthenium oxides (VRuO$_x$/GDY) produced through the in situ growth of VRuO$_x$ on a GDY surface. GDY can guide the formation of an optimum interface

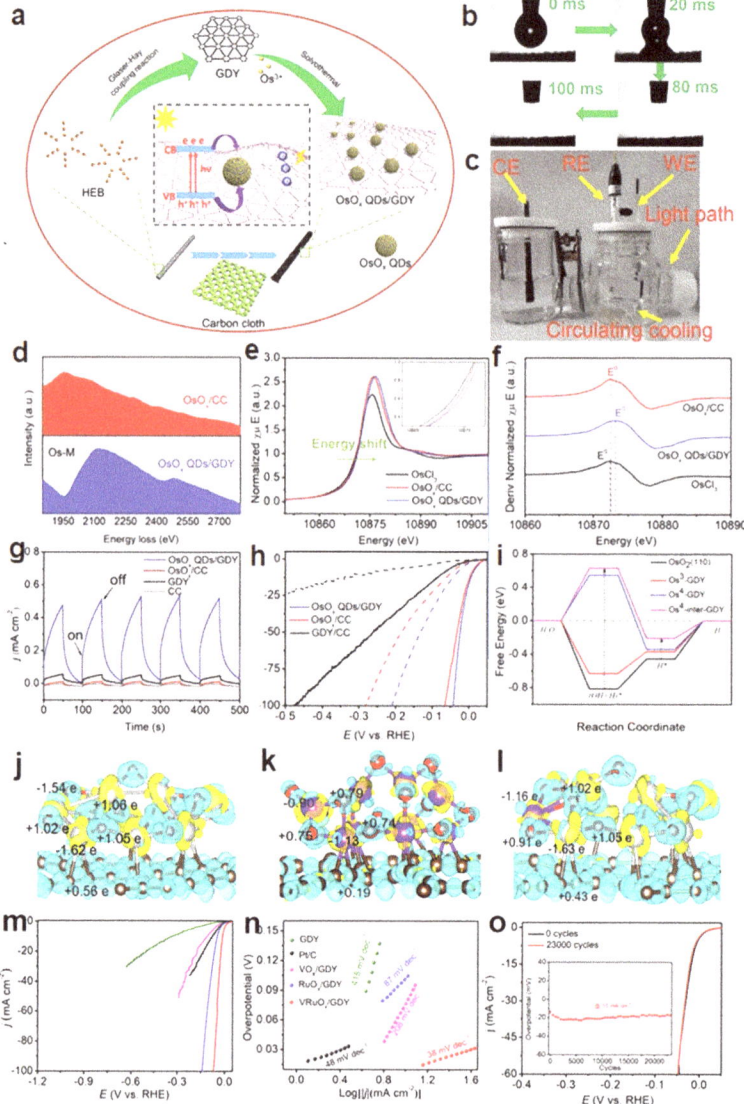

Figure 2.27. (a) Schematic representation of the strategy for the synthesis of OsO_x QDs/GDY. (b) Contact angle measurement on OsOx QDs/GDY. (c) Photograph of the three-electrode system equipped with a light path and a circulating cooling water system (CE: the counter electrode, RE: the reference electrode, WE: the working electrode). (d) Electron energy loss spectroscopy (EELS) results of OsO_x QDs/GDY and OsO_x/CC. (e) Normalized Os L-edge XANES spectra and (f) the first derivative of XANES of samples. (g) Transient photocurrent responses of the samples. (h) LSV curves of catalysts recorded before (solid lines) and after (dash–dotted lines) illumination in 1.0 M KOH. (i) Free energy diagrams of OsO_x QDs/GDY and OsO_2 (110) for the alkaline HER under equilibrium potential. Charge distribution in (j) VO_x/GDY, (k) RuO_x/GDY, and (l) $VRu_{0.027}O_x$/GDY (brown, silver, purple, and red spheres represent C, V, Ru, and O atoms, respectively). (m) Polarization curves and (n) corresponding Tafel slopes of the samples for the HER in 1.0 M KOH. (o) Polarization curves of $VRu_{0.027}O_x$/GDY before and after 23 000 CV cycling tests in 1.0 M KOH (inset: CV measurements of $VRu_{0.027}O_x$/GDY in 1.0 M KOH). Source: [132], reproduced with permission © John Wiley and Sons.

structure with highly catalytic activity and durability for the HER under both alkaline and neutral conditions, and the introduction of bimetal species could effectively tune the catalyst composition, electronic structure, and the number of the active sites, finally improving the intrinsic electrocatalytic activity. When the Ru/V ratio was 0.027, the catalyst ($VRu_{0.027}O_x$/GDY) possessed the largest electrocatalytic activity with minimum overpotentials of 13 mV and 12 mV at 10 mA cm^{-2} under alkaline and neutral conditions, respectively. These values are also much better than those of all prepared electrocatalysts and almost all reported HER electrocatalysts under their respective conditions. The superior HER activity of $VRu_{0.027}O_x$/GDY could be attributed to the optimized electronic structure and enriched active sites. In addition, the stability of $VRu_{0.027}O_x$/GDY for HER was also tested and showed negligible activity loss after long-term stability tests. The results reveal that the synergism of different components can efficiently improve the electron/mass transport properties, reduce the energy barrier, and increase the active site number for high catalytic performance (figure 2.28).

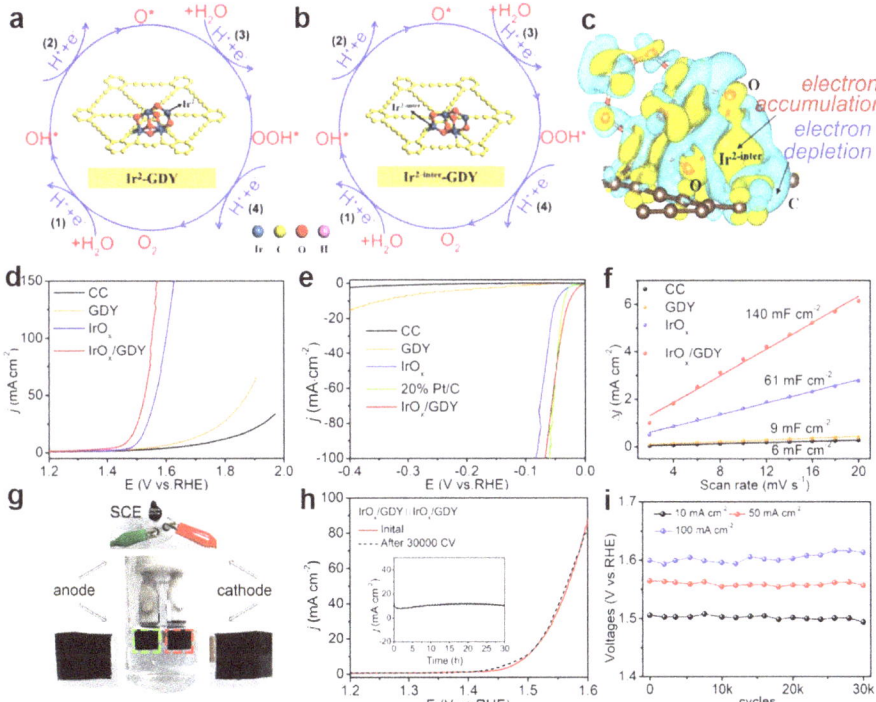

Figure 2.28. Four consecutive elementary electron steps on (a) non-heterogeneous and (b) heterogeneous interfacial iridium sites. (c) Charge density distribution of heterogeneous interfacial iridium atoms. (d) Polarization curves of catalysts for the OER in 0.5 M H_2SO_4 at a scan rate of 2 mV s^{-1}. (e) Polarization curves of catalysts for the HER in 0.5 M H_2SO_4 at a scan rate of 2 mV s^{-1}. (f) Cyclic voltammograms of an IrO_x QD/GDY electrode in 0.5 M H_2SO_4 at different scan rates from 2 to 20 mV s^{-1}. (g) Photograph of acidic electrolyzer for OWS in 0.5 M H_2SO_4. (h) Long-term durability measurements of IrO_xQD/GDY‖IrO_xQD/GDY. (i) Cell voltages at 10 (black dots), 50 (red dots), and 100 mA cm^{-2} (blue dots) recorded during continuous cycling tests. Source: [133], reproduced with permission © John Wiley and Sons.

2.4 Application to energy storage

Currently, graphite is used as the dominant anode material for lithium-ion batteries; it has a capacity of 372 mAh g^{-1} in the form of LiC$_6$. It not only exhibits high coulombic efficiency, but also a flat lithiation–delithiation voltage plateau. Different kinds of carbon material have been investigated as possible lithium storage materials, such as hard carbon, carbon nanotubes, and 2D graphene. As in the case of the widely studied graphene, GDY, as an emerging 2D-conjugated planar structure, can also be used as an anode for storing Li$^+$, and Li$^+$ can reversibly intercalate into and de-intercalate out of the interlayer space of GDY, due to the electron delocalization from electron-rich di-yne and benzene to Li$^+$. Theoretical investigations have revealed that the Li$^+$ can be stored on top of the benzene rings and in the triangular cavities, thus both Li$^+$ storage models lead to the formation of LiC$_3$ together with a high theoretical Li$^+$ storage capacity of 744 mAh g^{-1} [135]. To date, the reported GDY materials have had stable long-term Li$^+$ capacities from 450 to 550 mAh g^{-1}, based on different morphologies and stacking modes. As a result of the rich in-plane triangular cavities of GDY, which are large enough for the free transport of Li$^+$ across the cavities (figure 2.29), Li$^+$ can migrate fast both parallel and perpendicular to the 2D GDY sheet with a low energy barrier, which contributes to high C-rate performance.

Since the cavities in a GDY sheet not only enrich the positions for Li$^+$ storage, but also facilitate Li$^+$ transport, a further expansion in the cavity size should theoretically improve the electrochemical performance. In this respect, H-substituted GDY (HsGDY) film has been fabricated from a triethynylbenzene precursor

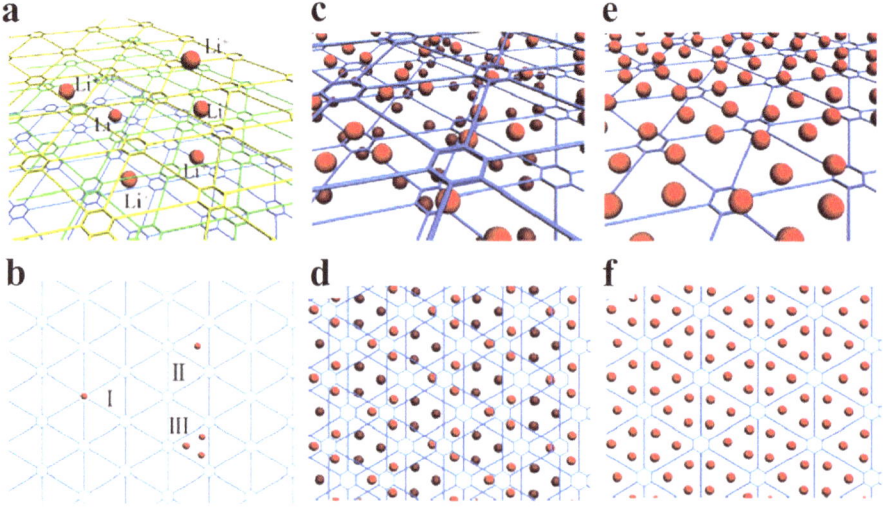

Figure 2.29. GDY for Li storage in a Li-ion cell. (a) Li-intercalated GDY. (b) Three different sites that can be occupied by Li atoms in GDY. Absorption of Li atoms on (c, d) both sides and (e, f) one side of a GDY plane; and (c, e) angled and (d, f) top views of Li absorption geometries. Source: [136], reproduced with permission © Elsevier.

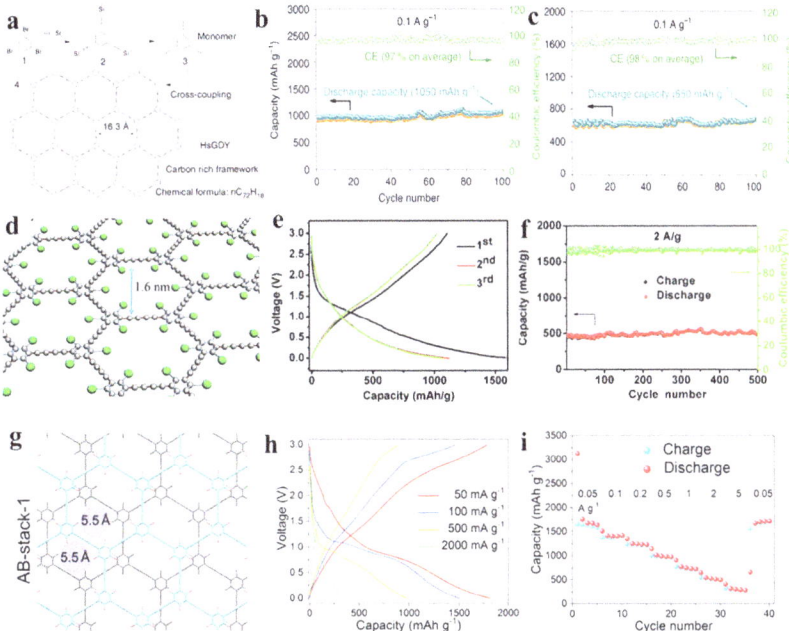

Figure 2.30. (a) Structural schematic of HsGDY. The cycling performance of HsGDY for (b) LIBs and (c) SIBs. Source for (a), (b), and (c): [20], reproduced with permission © Springer Nature. (d) The structure of Cl-GDY, (e) charge–discharge curves of a Cl-GDY electrode at a current density of 50 mA g^{-1}, (f) cycle performance of a Cl-GDY electrode. Source for (d), (e), and (f) [137], reproduced with permission © John Wiley and Sons. (g) Stacking configuration of F-GDY calculated by the DFT method, (h) charge–discharge curves and (i) the rate performance of F-GDY in a Li metal half-cell. Source for (g), (h), and (i): [18], reproduced with permission © Royal Society of Chemistry.

(figure 2.30(a)), and large capacities of 1050 mAh g^{-1} for lithium-ion batteries (LIBs) and 650 mAh g^{-1} for sodium-ion batteries (SIBs) have been achieved (figures 2.30(b) and (c)) [20]. Additionally, heteroatom-doped GDYs produced using trichlorinebenzene (Cl-GDY) and trifluorobenzene precursors (figures 2.30(d)–(f)) (F-GDY) (figures 2.30(g)–(i)) also exhibited a high Li$^+$ storage capability of over 1000 mAh g^{-1}, about twice that of pristine GDY. This increased capacity can be attributed to the enlarged cavity size (1.6 nm in Cl-GDY) and the heteroatom doping strategy, which further enrich the positions available for anchoring the Li$^+$ [18, 137].

GDY has a wider electrochemical window than those of current electrode materials; it can be used as an artificial protection layer to modify the surface of the cathode or the anode to inhibit the continuous reaction between a liquid electrolyte and the electrode. Using the *in situ* method of growing GDY on a Cu nanowire, it has been found that Cu@GDY can deliver high performance in terms of capacity, rate performance, and stability. The obtained Cu@GDY composite had an excellent rate performance of 1000 mAh g^{-1} at 5 A g^{-1} [27]. Its 3D continuous network greatly improved the reaction kinetics of the electrodes. Inspired by this method, metal-oxide anode and silicon anode materials have been modified by the seamless growth of an all-carbon GDY layer to improve stability [34, 138, 139]. High-energy-density silicon anodes are pulverized by the ultra-large volume expansion (300% for silicon) that occurs during the lithiation process. This process severely disintegrates the conductive network and solid-state electrolyte interfaces (SEIs), thus reducing the capacity retention. When an *in situ* coating of GDY is applied, the voids in the flexible GDY nanosheets become conducive to accommodating large volume expansions. The seamless GDY coating layer not only maintains the electronic conductivity with the current collector, but also holds the pulverized particles together and maintains the accessibilities of both electrons and ions to the Si particles. Under the protection of GDY, a silicon electrode showed significant enhancements in its capacity (2300 mAh g^{-1}) and stability [139]. As in the case of Si anodes, the promising metal-oxide (MO) anodes are also facing intense interfacial polarization and volumetric variation due to the conversion process involved in Li$^+$ storage. A seamless GDY interfacial layer can efficiently protect the secondary architecture of MOs, and the electrodes thus obtained can deliver high performances in terms of both stability and power performance (figures 2.31(e) and (f)).

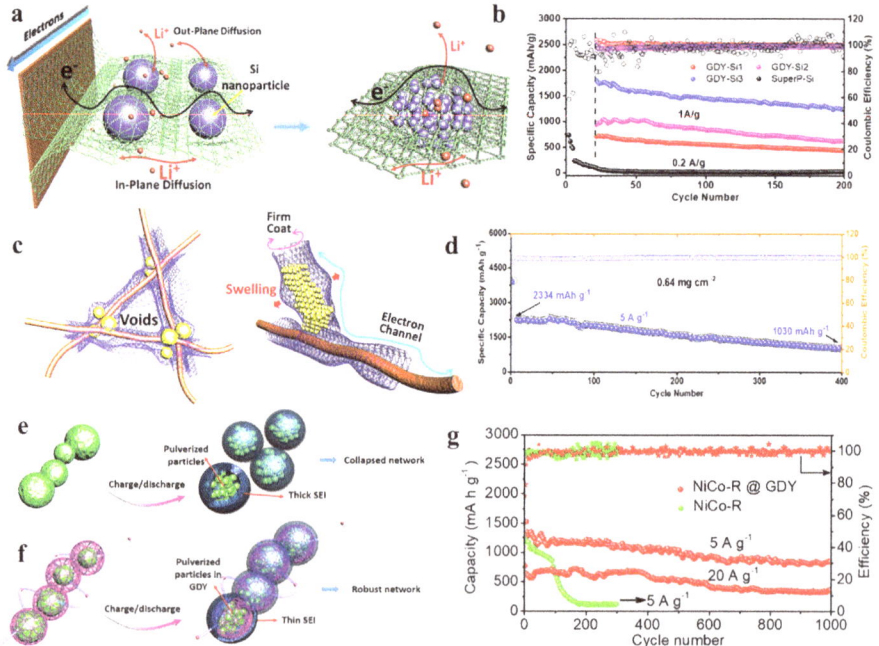

Figure 2.31. (a) Protective mechanism of GDY on a silicon anode. (b) Cycling performance of super P Si and GDY-Si anodes. Source for (a), (b): [138], reproduced with permission © Elsevier. (c) The mechanism of the GDY-wrapped Si particles on Cu wires with a high performance, (d) cycling performance. Source for (c), (d): [139], reproduced with permission © John Wiley and Sons. Schematic diagram for showing (e) the pulverization of pristine NiCo$_2$O$_4$ electrode with thick SEI and (f) the NiCo$_2$O$_4$@GDY electrode with thin SEI. (g) Cycling performance of pristine NiCo$_2$O$_4$ electrode and NiCo$_2$O$_4$@GDY electrode. Source for (e–g): [34], reproduced with permission © John Wiley and Sons.

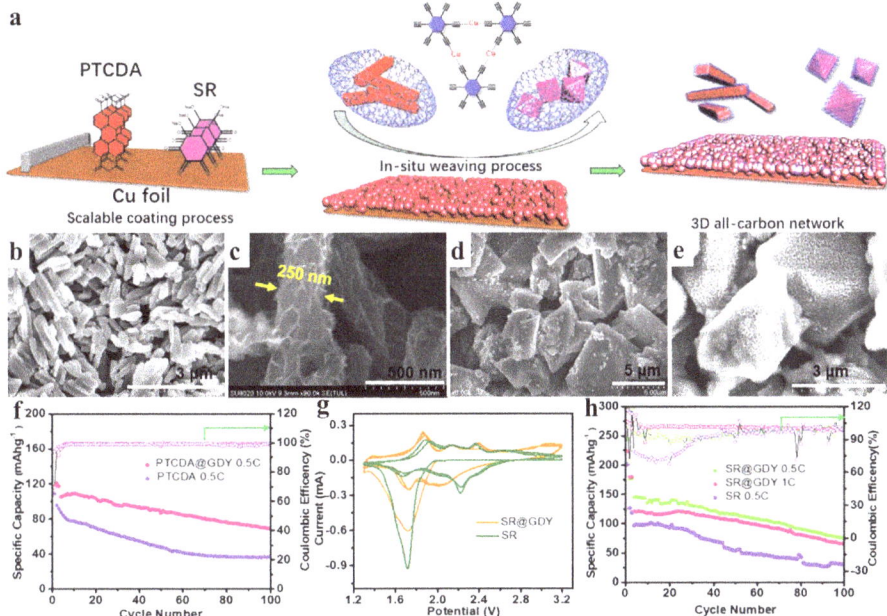

Figure 2.32. (a) Schematic illustration of GDY coating approaches used to improve the electrochemical performance of PTCDA. Super-resolution scanning electron microscopy (SEM) images of (b) PTCDA, (c) PTCDA@GDY, (d) SR, and (e) SR@GDY. (f) Cycling performances of PTCDA and PTCDA@GDY in SIBs. (g) CV curves of SR and SR@GDY in SIBs. (h) Cycling performances of SR and SR@GDY in SIBs. Source: [140], reproduced with permission © John Wiley and Sons.

Without the GDY protection, pristine $NiCo_2O_4$ has a sharp capacity decay during long-term cycling. With the addition of GDY protection, a $NiCo_2O_4$@GDY electrode showed a robust capacity of 837.3 mAh g^{-1} after 1000 cycles (figure 2.31(g)) [34].

A GDY coating also can be applied to organic cathodes to solve their issues of low conductivity and molecular dissolution. The GDY forms an all-carbon nano-coat *in situ* on small-molecule organic cathodes (figure 2.32(a)). This is the first time that an all-carbon GDY coating layer has been constructed on small-molecule organic cathodes. As a result of this modification, the presence of GDY on the organic cathode improves both the electron and ion conductivities at the interface. In long-term cycling tests, pristine 3,4,9,10-perylenetetracarboxylic dianhydride (PTCDA) could only supply 29.5 mAh g^{-1} after 300 cycles at 0.5 C, but the PTCDA@GDY retained high capacities of 84.9 mAh g^{-1} at 5 C and 73.3 mAh g^{-1} at 10 C in LIBs, respectively. After GDY coating, the effective mass loading of the organic cathodes was significantly increased to 93%, which was comparable to that of current LIBs and much better than previously reported values. This is beneficial for achieving a high-energy-density organic battery. Moreover, GDY-modified PTCDA and sodium rhodizonate (SR) dibasic organic cathodes were also tested in SIBs. SR@GDY delivered large capacities of 74.6 mAh g^{-1} and 64.6 mAh g^{-1} at rates of 0.5 C and 1 C after 100 cycles, while pristine SR only provided 30.1 mAh g^{-1}

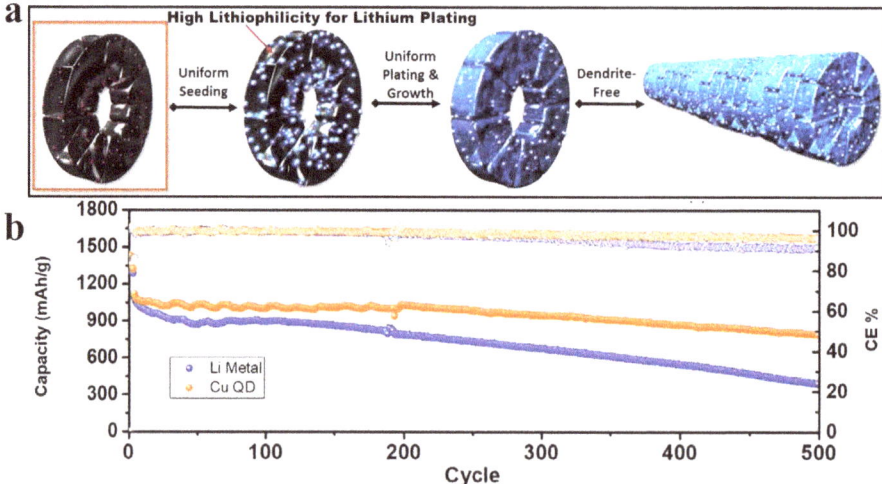

Figure 2.33. (a) Schematic illustration of the Li plating process and the suppression of Li dendrites on the CuQDs@GDY. (b) Cycling performance of the Li–S battery with and without the CuQDs@GDY at 1 C. Source: [141], reproduced with permission © John Wiley and Sons.

at 0.5 C (figure 2.32(h)) [140]. This report confirmed that the main issues of molecular dissolution were completely solved using the GDY coating strategy.

The Cu substrate plays a key role in catalyzing the growth of GDY. It is very interesting to find that, during the growth of GDY, polycrystalline Cu can be split into copper quantum dots (CuQDs). The CuQDs thus formed are well dispersed on the GDY nanosheets (CuQDs@GDY). According to theoretical calculations, the CuQDs further increase the lithiophilicity of the GDY. Thus, the uniformly dispersed CuQDs on GDY offer many active spots, which is very beneficial for improving the plating/ stripping processes in lithium metal batteries. According to tests, the presence of CuQDs@GDY caused the lithium metal to be evenly plated and effectively inhibited lithium dendrites (figure 2.33(a)). Moreover, CuQDs@GDY were tested in Li–S full cells. After 500 cycles at 1 C, a full cell fabricated using CuQDs@GDY retained 73% of its capacity. In contrast, the Li metal only retained 38% (figure 2.33(b)) [141]. This performance improvement was ascribed to the uniform Li plating and stripping that resulted from the use of the CuQDs@GDY.

As a conductive and mechanical framework, GDY can be used as the host to store sulfur and improve the conductivity of elemental S. Importantly, the polyanion (Nafion) can be seamlessly embedded in the GDY shell due to the mild growth conditions compared with those of traditional carbon materials. In the heterostructure of S@Nafion@GDY, the function of the GDY is to increase the electron and ion conductivities at the interface, and its atomic-level selectivity can be advantageous for suppressing the shuttle effect of the polysulfide. The inner Nafion can enhance mass transfer behavior in the primary nanostructure, and is thus beneficial for improving the reaction kinetics inside the heterostructure. In long-term tests of

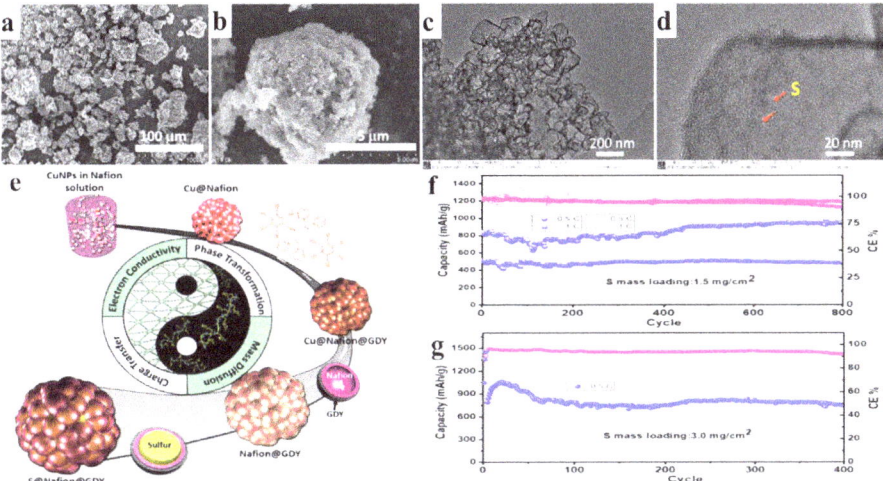

Figure 2.34. SEM images of (a, b) S@Nafion@GDY. TEM images of (c, d) S@Nafion@GDY. (e) Schematic illustration of the preparation of the Nafion@GDY core–shell nanostructure. (f) Cycling performance of S@Nafion@GDY. (g) Cycling performance of S@Nafion@GDY that has a high mass loading of 3 mg cm^{-2}. Source: [142], reproduced with permission © Elsevier.

Li–S batteries, the capacity of S@Nafion@GDY exhibited no obvious reduction after 800 cycles at 0.5 and 1 C (figure 2.34(f)) [142].

As in the case of Li$^+$ storage in GDY, the Na$^+$ could also be inserted into the interlayer of GDY to form NaC$_{5.14}$, which has a theoretical capacity of 316 mAh g^{-1}. This value is lower than that of Li storage, because the Na$^+$ ion is larger than Li$^+$. Until now, the experimental capacity of GDY powder for Na storage has been about 200 mAh g^{-1} [143, 144]. Better electrochemical performance could be obtained through the expansion of the in-plane cavities and the introduction of heteroatoms. As examples, boron-doped GDY (BGDY) and HsGDY both showed a capacity of over 600 mAh g^{-1} for Na$^+$ (figure 2.35) [16, 20]. In the case of BGDY, the improved electrochemical performance was attributed to the introduction of negatively charged boron and an appropriate cavity size. According to a theoretical simulation, the Na$^+$ was shown to be located in the angle of the butadiyne linkage and boron in the molecular plane (sites A1, A2, and A3 in figure 2.35(e)), and also in the middle of the butadiyne linkage (sites A4 and A5 in figure 2.35(e)).

The extraordinary performance of supercapacitors is also achieved by taking advantage of a GDY property. In a study, GDY powder provided a specific capacitance of 71.4 F g^{-1} at a current density of 3.5 A g^{-1}, and the specific capacitance of GDY nanowalls was 189 F g^{-1} [145]. Remarkably, improved energy density (8.66 Wh kg^{-1}) and specific capacitance (250 F g^{-1}) were obtained by an adjustable N-doping configuration [29]. When GDY was used as an ultrafine nanochain coating with a high specific surface area and 3D connectivity on the substrate, the synergistic effect of the GDY coating and the substrate itself created a promising area capacitance of 53.66 mF cm^{-2} with robust long-term retention (99% after 1300 cycles) [146].

The appearance of 2D GDY greatly enriches the members of the carbon family, and opens the door to the artificial synthesis of new carbon allotropes in solution.

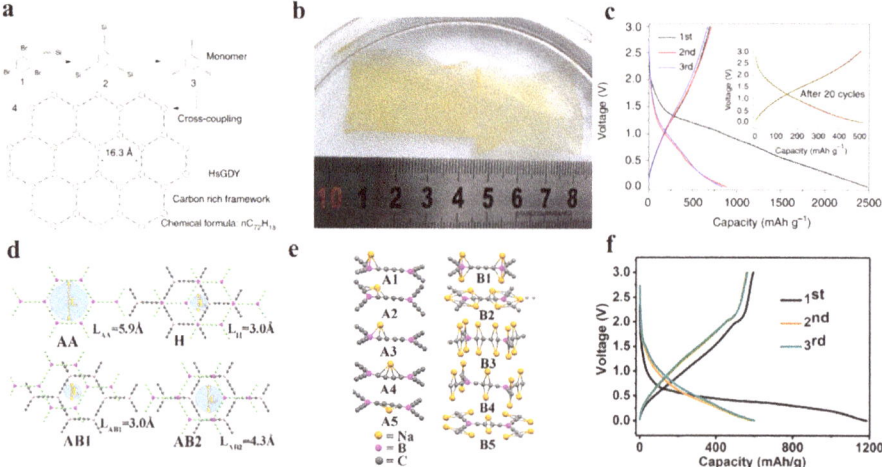

Figure 2.35. (a) Structural schematic of HsGDY. (b) Photograph of freestanding HsGDY films. (c) Charge–discharge curves of HsGDY in SIBs at a current density of 0.1 A g^{-1}. Source for (a), (b), and (c): [20], reproduced with permission © Springer Nature. (d) The calculated stacking configuration of bilayer BGDY, L = simulated pore size, shown as blue disks. (e) The geometries of fully occupied mode nNa$^+$+C$_{16}$B$_2$ (B1–B5) and optimized singly occupied mode Na$^+$+C$_{16}$B$_2$ (A1–A5). (f) Charge–discharge curves of BGDY at a current density of 50 mA g^{-1}. Source for (d), (e), and (f): [16], reproduced with permission © John Wiley and Sons.

The carbon structure, namely a 2D π-conjugated carbon network with sp and sp^2 hybridization, offers many novel and attractive properties to scientists working in various areas, and has inspired innovation in the fields of catalysis, energy conversion, and intelligent devices. In the field of energy conversion, its exceptional applications have revealed many original understandings of the scientific problems confronted in this system. The unique properties of GDY, such as its highest-areal-density in-plane atomic-level cavities, tunable molecular structure, and ultra-mild preparation, represent unparalleled advantages compared to traditional carbon materials, and provide new insights that will help us to systematically solve the bottleneck problems of electrodes at both the microscale and the macroscale. Although the studies of GDY are at an early stage, and more effort is required in order to realize these promising advantages, more and more recent extensive progress has demonstrated that the GDY has great potential to become one of the materials required to construct the next-generation high-efficiency energy conversion devices. In future studies, the GDY mechanisms that can be used in these next-generation energy conversions should be systematically investigated; the realization of low-defect or defectless GDY for energy conversion devices remains a major challenge.

2.5 Conclusions

As the first chemically synthesized all-carbon material, GDY has played a significant role in the development of carbon materials, particularly artificial carbon materials. Therefore, there is a vital need to review GDY in the light of its concept, properties, and applications. In this chapter, we comprehensively discussed recent progress in the GDY field in terms of its synthesis methods and applications, in conjunction with its structure and properties. First, the approaches used to synthesise GDY and its derivatives were summarized, in which both classical Cu-surface synthesis and new methods were critically discussed. Subsequently, the applications of GDY in catalysis were scrutinized from the perspective of theoretical and experimental progress. The designed strategies and behaviors of GDY-based metal-atom, metal-free, and heterojunction catalysts in various catalytic fields (e.g. the HER, OER, ORR, and CO_2RR) were analyzed. GDY-based catalysts with well-designed structures were demonstrated to have superior catalytic performance. Finally, the applications of GDY in energy storage were presented. After structural tuning, GDY exhibits an excellent ability to reversibly store metal ions (e.g. Li^+, Na^+) at high capacities. Additionally, GDY can also be used to modify and protect electrodes.

In summary, GDY possesses great potential in various applications, which is attributed to its 2D-conjugated structure with sp and sp^2 hybridization, and subsequent special features (e.g. in-plane atomic-level cavities, tunable molecular structure). Although further work is still needed in the areas of the synthesis and the structural design of GDY as well as the characterization and simulation of GDY-related chemical/electrochemical behavior, GDY is believed to remain a promising area of exploration for next-generation catalysts and energy storage devices.

References

[1] Diederich F and Thilgen C 1996 Covalent fullerene chemistry *Science* **271** 317–23

[2] Franklin A D 2013 Electronics: the road to carbon nanotube transistors *Nature* **498** 443–4

[3] Geim A K and Novoselov K S 2007 The rise of graphene *Nat. Mater.* **6** 183–91

[4] Inagaki M and Kang F 2014 Graphene derivatives: graphane, fluorographene, graphene oxide, graphyne and graphdiyne *J. Mater. Chem.* A **2** 13193–206

[5] Li G X, Li Y L, Liu H B, Guo Y B, Li Y J and Zhu D B 2010 Architecture of graphdiyne nanoscale films *Chem. Commun.* **46** 3256–8

[6] Zuo Z, Wang D, Zhang J, Lu F and Li Y 2019 Synthesis and applications of graphdiyne-based metal-free catalysts *Adv. Mater.* **31** 1803762

[7] Li Y, Xu L, Liu H and Li Y 2014 Graphdiyne and graphyne: from theoretical predictions to practical construction *Chem. Soc. Rev.* **43** 2572–86

[8] Jia Z Y, Li Y J, Zuo Z C, Liu H B, Huang C S and Li Y L 2017 Synthesis and properties of 2D carbon-graphdiyne *Acc. Chem. Res.* **50** 2470–8

[9] Huang C, Li Y, Wang N, Xue Y, Zuo Z, Liu H and Li Y 2018 Progress in research into 2D graphdiyne-based materials *Chem. Rev.* **118** 7744–803

[10] Gao X, Liu H, Wang D and Zhang J 2019 Graphdiyne: synthesis, properties, and applications *Chem. Soc. Rev.* **48** 908–36

[11] Xue Y, Huang B, Yi Y, Guo Y, Zuo Z, Li Y, Jia Z, Liu H and Li Y 2018 Anchoring zero valence single atoms of nickel and iron on graphdiyne for hydrogen evolution *Nat. Commun.* **9** 1460

[12] Wang S, Yi L, Halpert J E, Lai X, Liu Y, Cao H, Yu R, Wang D and Li Y 2012 A novel and highly efficient photocatalyst based on P25-graphdiyne nanocomposite *Small* **8** 265–71

[13] Xiao J Y, Shi J J, Liu H B, Xu Y Z, Lv S T, Luo Y H, Li D M, Meng Q B and Li Y L 2015 Efficient $CH_3NH_3PbI_3$ perovskite solar cells based on graphdiyne (GD)-modified P3HT hole-transporting material *Advanced Energy Materials* **5** 1401943

[14] Jia Z, Zuo Z, Yi Y, Liu H, Li D, Li Y and Li Y 2017 Low temperature, atmospheric pressure for synthesis of a new carbon ene–yne and application in Li storage *Nano Energy* **33** 343–9

[15] Li G, Li Y, Liu H, Guo Y, Li Y and Zhu D 2010 Architecture of graphdiyne nanoscale films *Chem. Commun.* **46** 3256–8

[16] Wang N, Li X, Tu Z, Zhao F, He J, Guan Z, Huang C, Yi Y and Li Y 2018 Synthesis and electronic structure of boron-graphdiyne with an sp-hybridized carbon skeleton and its application in sodium storage *Angew. Chem. Int. Ed. Engl.* **57** 3968–73

[17] Wang N *et al* 2017 Synthesis of chlorine-substituted graphdiyne and applications for lithium-ion storage *Angew. Chem. Int. Ed. Engl.* **56** 10740–5

[18] He J, Wang N, Yang Z, Shen X, Wang K, Huang C, Yi Y, Tu Z and Li Y 2018 Fluoride graphdiyne as a free-standing electrode displaying ultra-stable and extraordinary high Li storage performance *Energy Environ. Sci.* **11** 2893–903

[19] Zhou W *et al* 2018 Direct synthesis of crystalline graphdiyne analogue based on supra-molecular interactions *J. Am. Chem. Soc.* **141** 48–52

[20] He J *et al* 2017 Hydrogen substituted graphdiyne as carbon-rich flexible electrode for lithium and sodium ion batteries *Nat. Commun.* **8** 1172

[21] Li X, Wang N, He J, Yang Z, Tu Z, Zhao F, Wang K, Yi Y and Huang C 2020 Designing the efficient lithium diffusion and storage channels based on graphdiyne *Carbon* **162** 579–85

[22] Zhang Z, Wu C, Pan Q, Shao F, Sun Q, Chen S, Li Z and Zhao Y 2020 Interfacial synthesis of crystalline two-dimensional cyano-graphdiyne *Chem Commun (Camb)* **56** 3210–3

[23] Liu H, Zhang Z, Wu C, Pan Q, Zhao Y and Li Z 2019 Interfacial synthesis of conjugated crystalline 2D fluorescent polymer film containing aggregation-induced emission unit *Small* **15** e1804519

[24] Prabakaran P, Satapathy S, Prasad E and Sankararaman S 2018 Architecting pyrediyne nanowalls with improved inter-molecular interactions, electronic features and transport characteristics *J. Mater. Chem.* C **6** 380–7

[25] Pan Q Y, Chen X S, Liu H, Gan W J, Ding N X and Zhao Y J 2021 Crystalline porphyrin-based graphdiyne for electrochemical hydrogen and oxygen evolution reactions *Mater. Chem. Front.* **5** 4596–603

[26] Gao J, Li J, Chen Y, Zuo Z, Li Y, Liu H and Li Y 2018 Architecture and properties of a novel two-dimensional carbon material-graphtetrayne *Nano Energy* **43** 192–9

[27] Shang H, Zuo Z, Li L, Wang F, Liu H, Li Y and Li Y 2018 Ultrathin graphdiyne nanosheets grown *in situ* on copper nanowires and their performance as lithium-ion battery anodes. *Angew. Chem. Int. Ed. Engl.* **57** 774

[28] Matsuoka R, Sakamoto R, Hoshiko K, Sasaki S, Masunaga H, Nagashio K and Nishihara H 2017 Crystalline graphdiyne nanosheets produced at a gas/liquid or liquid/liquid interface *J. Am. Chem. Soc.* **139** 3145

[29] Shang H, Zuo Z, Zheng H, Li K, Tu Z, Yi Y, Liu H, Li Y and Li Y 2018 N-doped graphdiyne for high-performance electrochemical electrodes *Nano Energy* **44** 144

[30] Kan X, Ban Y, Wu C, Pan Q, Liu H, Song J, Zuo Z, Li Z and Zhao Y 2018 Interfacial synthesis of conjugated two-dimensional N-graphdiyne *ACS Appl. Mater. Interfaces* **10** 53

[31] Matsuoka R *et al* 2019 Expansion of the graphdiyne family: a triphenylene-cored analogue *ACS Appl. Mater. Interfaces* **11** 2730

[32] Li G, Li Y, Qian X, Liu H, Lin H, Chen N and Li Y 2011 Construction of tubular molecule aggregations of graphdiyne for highly efficient field emission *J. Phys. Chem.* C **115** 2611–5

[33] Gao X *et al* 2017 Direct synthesis of graphdiyne nanowalls on arbitrary substrates and its application for photoelectrochemical water splitting cell *Adv. Mater.* **29** 1605308

[34] Wang F, Zuo Z, Li L, He F, Lu F and Li Y 2019 A universal strategy for constructing seamless graphdiyne on metal oxides to stabilize the electrochemical structure and interface. *Adv. Mater.* **31** e1806272

[35] Chen K *et al* 2016 Growing three-dimensional biomorphic graphene powders using naturally abundant diatomite templates towards high solution processability *Nat. Commun.* **7** 13440

[36] Li J, Xu J, Xie Z, Gao X, Zhou J, Xiong Y, Chen C, Zhang J and Liu Z 2018 Diatomite-templated synthesis of freestanding 3D graphdiyne for energy storage and catalysis application *Adv. Mater.* **30** e1800548

[37] Yu H, Xue Y, Hui L, Zhang C, Li Y, Zuo Z, Zhao Y, Li Z and Li Y 2018 Efficient hydrogen production on a 3D flexible heterojunction material *Adv. Mater.* **30** e1707082

[38] Xing C, Xue Y, Huang B, Yu H, Hui L, Fang Y, Liu Y, Zhao Y, Li Z and Li Y 2019 Fluorographdiyne: a metal-free catalyst for applications in water reduction and oxidation. *Angew. Chem. Int. Ed. Engl.* **58** 13897–903

[39] Hui L, Xue Y, Yu H, Liu Y, Fang Y, Xing C, Huang B and Li Y 2019 Highly efficient and selective generation of ammonia and hydrogen on a graphdiyne-based catalyst *J. Am. Chem. Soc.* **141** 10677–83

[40] Xing C *et al* 2020 A highly selective and active metal-free catalyst for ammonia production *Nanoscale Horiz.* **5** 1274–8

[41] Wang S S, Liu H B, Kan X N, Wang L, Chen Y H, Su B, Li Y L and Jiang L 2017 Superlyophilicity-facilitated synthesis reaction at the microscale: ordered graphdiyne stripe arrays *Small* **13** 1602265

[42] Liu R, Gao X, Zhou J, Xu H, Li Z, Zhang S, Xie Z, Zhang J and Liu Z 2017 Chemical vapor deposition growth of linked carbon monolayers with acetylenic scaffoldings on silver foil *Adv. Mater.* **29** 1604665

[43] Long M, Tang L, Wang D, Li Y and Shuai Z 2011 Electronic structure and carrier mobility in graphdiyne sheet and nanoribbons: theoretical predictions *ACS Nano* **5** 2593–600

[44] Luo G *et al* 2011 Quasiparticle energies and excitonic effects of the two-dimensional carbon allotrope graphdiyne: theory and experiment *Phys. Rev.* B **84** 075439

[45] Lei Y *et al* 2010 Increased silver activity for direct propylene epoxidation via subnanometer size effects *Science* **328** 224–8

[46] Thomas J M 2015 Catalysis tens of thousands of atoms replaced by one *Nature* **525** 325–6

[47] Kyriakou G, Boucher M B, Jewell A D, Lewis E A, Lawton T J, Baber A E, Tierney H L, Flytzani-Stephanopoulos M and Sykes E C H 2012 Isolated metal atom geometries as a strategy for selective heterogeneous hydrogenations *Science* **335** 1209–12

[48] Xue Y R, Huang B L, Yi Y P, Guo Y, Zuo Z C, Li Y J, Jia Z Y, Liu H B and Li Y L 2018 Anchoring zero valence single atoms of nickel and iron on graphdiyne for hydrogen evolution *Nat. Commun.* **9** 1460

[49] Yu H *et al* 2019 Ultrathin nanosheet of graphdiyne-supported palladium atom catalyst for efficient hydrogen production *iScience* **11** 31

[50] Hui L, Xue Y, Yu H, Liu Y, Fang Y, Xing C, Huang B and Li Y 2019 Highly efficient and selective generation of ammonia and hydrogen on a graphdiyne-based catalyst *J. Am. Chem. Soc.* **141** 10677–83

[51] Yu H *et al* 2020 2D graphdiyne loading ruthenium atoms for high efficiency water splitting *Nano Energy* **72** 104667

[52] Wu H and He F 2021 Activity origins of graphdiyne based bifunctional atom catalysts for hydrogen evolution and water oxidation *Chem. Res. Chin. Univ* **37** 1334–40

[53] Gao Y, Cai Z, Wu X, Lv Z, Wu P and Cai C 2018 Graphdiyne-supported single-atom-sized fe catalysts for the oxygen reduction reaction: DFT predictions and experimental validations *ACS Catal.* **8** 10364–74

[54] Yu H D *et al* 2019 Ultrathin nanosheet of graphdiyne-supported palladium atom catalyst for efficient hydrogen production *iScience* **11** 31

[55] He T W, Matta S K, Will G and Du A J 2019 Transition-metal single atoms anchored on graphdiyne as high-efficiency electrocatalysts for water splitting and oxygen reduction *Small Methods* **3** 1800419

[56] Chen Z W, Wen Z and Jiang Q 2017 Rational design of Ag-38 cluster supported by graphdiyne for catalytic CO oxidation *J. Phys. Chem.* C **121** 3463–8

[57] Xu G L, Liu F X, Lu Z S, Talib S H, Ma D W and Yang Z X 2021 Design of promising single Rh atom catalyst for CO oxidation based on graphdiyne sheets *Physica E-Low-Dimensional Systems Nanostructures* **130** 114676

[58] Zou L, Zhu Y, Cen W, Jiang X and Chu W 2021 N-doping in graphdiyne on embedding of metals and its effect in catalysis *Appl. Surf. Sci.* **557** 149815

[59] Xu G, Wang R, Ding Y, Lu Z, Ma D and Yang Z 2018 First-principles study on the single Ir atom embedded graphdiyne: an efficient catalyst for CO oxidation *J. Phys. Chem.* C **122** 23481–92

[60] Tang S F, Lu X L, Zhang C, Wei Z W, Si R and Lu T B 2021 Decorating graphdiyne on ultrathin bismuth subcarbonate nanosheets to promote CO_2 electroreduction to formate *Sci. Bull.* **66** 1533–41

[61] Feng Z, Tang Y, Ma Y, Li Y, Dai Y, Chen W, Su G, Song Z and Dai X 2021 Theoretical computation of the electrocatalytic performance of CO_2 reduction and hydrogen evolution reactions on graphdiyne monolayer supported precise number of copper atoms *Int. J. Hydrogen Energy* **46** 5378

[62] Feng Z, Tang Y N, Chen W G, Li Y, Li R Y, Ma Y Q and Dai X Q 2020 Graphdiyne coordinated transition metals as single-atom catalysts for nitrogen fixation *Phys. Chem. Chem. Phys.* **22** 9216–24

[63] Lin Z-Z 2015 Graphdiyne as a promising substrate for stabilizing Pt nanoparticle catalyst *Carbon* **86** 301–9

[64] Ma D, Zeng Z, Liu L, Huang X and Jia Y 2019 Computational evaluation of electrocatalytic nitrogen reduction on TM single-, double-, and triple-atom catalysts (TM = Mn, Fe, Co, Ni) based on graphdiyne monolayers *J. Phys. Chem.* C **123** 19066–76

[65] Xu Y K, Cai Z W, Du P, Zhou J X, Pan Y H, Wu P and Cai C X 2021 Taming the challenges of activity and selectivity in the electrochemical nitrogen reduction reaction using graphdiyne-supported double-atom catalysts *J. Mater. Chem.* A **9** 8489–500

[66] Diederich F 1994 Carbon scaffolding – building acetylenic all-carbon and carbon-rich compounds *Nature* **369** 199–207

[67] Kang B, Wu S, Ma J, Ai H and Lee J Y 2019 Synergy of sp-N and sp(2)-N codoping endows graphdiyne with comparable oxygen reduction reaction performance to Pt *Nanoscale* **11** 16599–605

[68] Wang S Y, Jiao D X, Liu J W, Shang Y C and Zhao J X 2021 P- or S-Doped graphdiyne as a superior metal-free electrocatalyst for the hydrogen evolution reaction: a computational study *New J. Chem.* **45** 8101–8

[69] Ku R Q, Yu G T, Gao J, Huang X R and Chen W 2020 Embedding tetrahedral 3D transition metal TM4 clusters into the cavity of two-dimensional graphdiyne to construct highly efficient and nonprecious electrocatalysts for hydrogen evolution reaction *Phys. Chem. Chem. Phys.* **22** 3254–63

[70] Liu T F, Wang Q, Wang G X and Bao X H 2021 Electrochemical CO_2 reduction on graphdiyne: a DFT study *Green Chem.* **23** 1212–9

[71] Zhao J, Chen Z and Zhao J 2019 Metal-free graphdiyne doped with sp-hybridized boron and nitrogen atoms at acetylenic sites for high-efficiency electroreduction of CO_2 to CH_4 and C_2H_4 *J. Mater. Chem.* A **7** 4026–35

[72] Liu R, Liu H, Li Y, Yi Y, Shang X, Zhang S, Yu X, Zhang S, Cao H and Zhang G 2014 Nitrogen-doped graphdiyne as a metal-free catalyst for high-performance oxygen reduction reactions *Nanoscale* **6** 11336–43

[73] Das B K, Sen D and Chattopadhyay K K 2016 Implications of boron doping on electrocatalytic activities of graphyne and graphdiyne families: a first principles study *Phys. Chem. Chem. Phys.* **18** 2949–58

[74] Das B K, Sen D and Chattopadhyay K K 2016 Nitrogen doping in acetylene bonded two dimensional carbon crystals: ab-initio forecast of electrocatalytic activities vis-a-vis boron doping *Carbon* **105** 330–9

[75] Zhang S *et al* 2016 Heteroatom doped graphdiyne as efficient metal-free electrocatalyst for oxygen reduction reaction in alkaline medium *J. Mater. Chem.* A **4** 4738–44

[76] Gu J X, Magagula S, Zhao J X and Chen Z F 2019 Boosting ORR/OER activity of graphdiyne by simple heteroatom doping *Small Methods* **3** 1800550

[77] Feng Z, Tang Y A, Chen W G, Wei D, Ma Y Q and Dai X Q 2020 O-doped graphdiyne as metal-free catalysts for nitrogen reduction reaction *Molecular Catalysis* **483** 110705

[78] Fu C, Li Y F and Wei H Y 2021 Double boron atom-doped graphdiynes as efficient metal-free electrocatalysts for nitrogen reduction into ammonia: a first-principles study *Phys. Chem. Chem. Phys.* **23** 17683–92

[79] Cao J G, Li N and Zeng X 2021 Exploring the synergistic effect of B-N doped defective graphdiyne for N−2 fixation dagger *New J. Chem.* **45** 6327–35

[80] Hui L, Xue Y, Yu H, Zhang C, Huang B and Li Y 2020 Loading copper atoms on graphdiyne for highly efficient hydrogen production *ChemPhysChem* **21** 2145–9

[81] Deng G, Wang T, Alshehri A A, Alzahrani K A, Wang Y, Ye H, Luo Y and Sun X 2019 Improving the electrocatalytic N_2 reduction activity of Pd nanoparticles through surface modification *J. Mater. Chem.* A **7** 21674–7

[82] Lv J, Wu S, Tian Z, Ye Y, Liu J and Liang C 2019 Construction of PdO–Pd interfaces assisted by laser irradiation for enhanced electrocatalytic N_2 reduction reaction *J. Mater. Chem.* A **7** 12627–34

[83] Yang X, Nash J, Anibal J, Dunwell M, Kattel S, Stavitski E, Attenkofer K, Chen J G, Yan Y and Xu B 2018 Mechanistic insights into electrochemical nitrogen reduction reaction on vanadium nitride nanoparticles *J. Am. Chem. Soc.* **140** 13387–91

[84] Wu T *et al* 2019 Greatly improving electrochemical N_2 reduction over TiO_2 nanoparticles by iron doping *Angew. Chem. Int. Ed.* **58** 18449–53

[85] Chu K, Liu Y-p, Wang J and Zhang H 2019 NiO nanodots on graphene for efficient electrochemical N_2 reduction to NH_3 *ACS Appl. Energy Mater.* **2** 2288–95

[86] Liu Y-p, Li Y-b, Huang D-j, Zhang H and Chu K 2019 ZnO quantum dots coupled with graphene toward electrocatalytic N_2 reduction: experimental and DFT investigations *Chemistry* **25** 11933–9

[87] Zhang S, Zhao C, Liu Y, Li W, Wang J, Wang G, Zhang Y, Zhang H and Zhao H 2019 Cu doping in CeO_2 to form multiple oxygen vacancies for dramatically enhanced ambient N_2 reduction performance *Chem. Commun.* **55** 2952–5

[88] Li Y, Chen X, Zhang M, Zhu Y, Ren W, Mei Z, Gu M and Pan F 2019 Oxygen vacancy-rich MoO_{3-x} nanobelts for photocatalytic N_2 reduction to NH_3 in pure water *Catalysis Sci. Technol.* **9** 803–10

[89] Xing Z, Kong W, Wu T, Xie H, Wang T, Luo Y, Shi X, Asiri A M, Zhang Y and Sun X 2019 Hollow Bi_2MoO_6 sphere effectively catalyzes the ambient electroreduction of N_2 to NH_3 *ACS Sustainable Chem. Eng.* **7** 12692–6

[90] Gao Y, Han Z, Hong S, Wu T, Li X, Qiu J and Sun Z 2019 ZIF-67-derived cobalt/nitrogen-doped carbon composites for efficient electrocatalytic N_2 reduction *ACS Appl. Energy Mater.* **2** 6071–7

[91] Cui Q, Qin G, Wang W, Geethalakshmi K R, Du A and Sun Q 2019 Mo-based 2D MOF as a highly efficient electrocatalyst for reduction of N_2 to NH_3: a density functional theory study *J. Mater. Chem.* A **7** 14510–8

[92] Yu H X Y, Hui L, Zhang C, Fang Y, Liu Y, Chen X, Zhang D, Huang B and Li Y 2021 Graphdiyne-based metal atomic catalysts for synthesizing ammonia *Nat. Sci. Rev.* **8** nwaa213

[93] Zou H, Rong W, Wei S, Ji Y and Duan L 2020 Regulating kinetics and thermodynamics of electrochemical nitrogen reduction with metal single-atom catalysts in a pressurized electrolyser *Proc. Natl Acad. Sci.* **117** 29462–68

[94] Li J *et al* 2019 Atomic Pd on graphdiyne/graphene heterostructure as efficient catalyst for aromatic nitroreduction *Adv. Funct. Mater.* **29** 1905423

[95] Yin X-P, Tang S-F, Zhang C, Wang H-J, Si R, Lu X-L and Lu T-B 2020 Graphdiyne-based Pd single-atom catalyst for semihydrogenation of alkynes to alkenes with high selectivity and conversion under mild conditions *J. Mater. Chem.* A **8** 20925–30

[96] Yu H, Xue Y, Hui L, Zhang C, Li Y, Zuo Z, Zhao Y, Li Z and Li Y 2018 Efficient hydrogen production on a 3D flexible heterojunction material *Adv. Mater.* **30** 1707082

[97] Xue Y, Guo Y, Yi Y, Li Y, Liu H, Li D, Yang W and Li Y 2016 Self-catalyzed growth of Cu@graphdiyne core–shell nanowires array for high efficient hydrogen evolution cathode *Nano Energy* **30** 858–66

[98] Xue Y, Li J, Xue Z, Li Y, Liu H, Li D, Yang W and Li Y 2016 Extraordinarily durable graphdiyne-supported electrocatalyst with high activity for hydrogen production at all values of pH *ACS Appl. Mater. Interfaces* **8** 31083–91

[99] Yu H, Xue Y, Hui L, Zhang C, Zhao Y, Li Z and Li Y 2018 Controlled growth of MoS2 nanosheets on 2D N-doped graphdiyne nanolayers for highly associated effects on water reduction *Adv. Funct. Mater.* **28** 1707564

[100] Li J, Gao X, Jiang X, Li X-B, Liu Z, Zhang J, Tung C-H and Wu L-Z 2017 Graphdiyne: a promising catalyst–support to stabilize cobalt nanoparticles for oxygen evolution *ACS Catal.* **7** 5209–13

[101] Xue Y, Zuo Z, Li Y, Liu H and Li Y 2017 Graphdiyne-supported $NiCo_2S_4$ nanowires: a highly active and stable 3D bifunctional electrode material *Small* **13** 1700936

[102] Liu Y, Xue Y, Yu H, Hui L, Huang B and Li Y 2021 Graphdiyne ultrathin nanosheets for efficient water splitting *Adv. Funct. Mater.* **31** 2010112

[103] Hui L, Jia D, Yu H, Xue Y and Li Y 2019 Ultrathin graphdiyne-wrapped iron carbonate hydroxide nanosheets toward efficient water splitting *ACS Appl. Mater. Interfaces* **11** 2618–25

[104] Hui L *et al* 2018 Overall water splitting by graphdiyne-exfoliated and -sandwiched layered double-hydroxide nanosheet arrays *Nat. Commun.* **9** 5309

[105] Fang Y, Xue Y, Hui L, Yu H, Liu Y, Xing C, Lu F, He F, Liu H and Li Y 2019 *In situ* growth of graphdiyne based heterostructure: toward efficient overall water splitting *Nano Energy* **59** 591–7

[106] Yu H *et al* 2019 Graphdiyne-engineered heterostructures for efficient overall water-splitting *Nano Energy* **64** 103928

[107] Li J *et al* 2019 Superhydrophilic graphdiyne accelerates interfacial mass/electron transportation to boost electrocatalytic and photoelectrocatalytic water oxidation activity *Adv. Funct. Mater.* **29** 1808079

[108] Li Y, Yang H, Wang G, Ma B and Jin Z 2020 Distinctiveimproved synthesis and application extensions graphdiyne for efficient photocatalytic hydrogen evolution *ChemCatChem* **12** 1985–95

[109] Zhang S, Yin C, Kang Z, Wu P, Wu J, Zhang Z, Liao Q, Zhang J and Zhang Y 2019 Graphdiyne nanowall for enhanced photoelectrochemical performance of si heterojunction photoanode *ACS Appl. Mat. Interfaces* **11** 2745–9

[110] Yi Z *et al* 2010 An orthophosphate semiconductor with photooxidation properties under visible-light irradiation *Nat. Mater.* **9** 559–64

[111] Si H-Y *et al* 2018 Z-scheme Ag_3PO_4/graphdiyne/g-C_3N_4 composites: enhanced photocatalytic O_2 generation benefiting from dual roles of graphdiyne *Carbon* **132** 598–605

[112] Guo S, Jiang Y, Wu F, Yu P, Liu H, Li Y and Mao L 2019 Graphdiyne-promoted highly efficient photocatalytic activity of graphdiyne/silver phosphate pickering emulsion under visible-light irradiation *ACS Appl. Mater. Interfaces* **11** 2684–91

[113] Fang Y *et al* 2020 Graphdiyne interface engineering: highly active and selective ammonia synthesis *Angew. Chem. Int. Ed.* **59** 13021–7

[114] Fang Y, Xue Y, Hui L, Yu H and Li Y 2021 Graphdiyne@Janus magnetite for photocatalytic nitrogen fixation *Angew. Chem. Int. Ed.* **60** 3170–4

[115] Qi H, Yu P, Wang Y, Han G, Liu H, Yi Y, Li Y and Mao L 2015 Graphdiyne oxides as excellent substrate for electroless deposition of Pd clusters with high catalytic activity *J. Am. Chem. Soc.* **137** 5260–3

[116] Yang L-L, Wang H-J, Wang J, Li Y, Zhang W and Lu T-B 2019 A graphdiyne-based carbon material for electroless deposition and stabilization of sub-nanometric Pd catalysts with extremely high catalytic activity *J. Mater. Chem.* A **7** 13142–8

[117] Tan X, Xu J, Huang T, Wang S, Yuan M and Zhao G 2019 Graphdiyne bearing pillar[5]arene-reduced Au nanoparticles for enhanced catalytic performance towards the reduction of 4-nitrophenol and methylene blue *RSC Adv.* **9** 38372–80

[118] Shen H, Li Y and Shi Z 2019 A novel graphdiyne-based catalyst for effective hydrogenation reaction *ACS Appl. Mater. Interfaces* **11** 2563–70

[119] Yang N, Liu Y, Wen H, Tang Z, Zhao H, Li Y and Wang D 2013 Photocatalytic properties of graphdiyne and graphene modified TiO_2: from theory to experiment *ACS Nano* **7** 1504–12

[120] Ramakrishnan V, Kim H and Yang B 2019 Improving the photo-cathodic properties of TiO_2 nano-structures with graphdiynes *New J. Chem.* **43** 12896–9

[121] Lv J-X, Zhang Z-M, Wang J, Lu X-L, Zhang W and Lu T-B 2019 *In situ* synthesis of CdS/graphdiyne heterojunction for enhanced photocatalytic activity of hydrogen production *ACS Appl. Mat. Interfaces* **11** 2655–61

[122] Liu X and Dai L 2016 Carbon-based metal-free catalysts *Nat. Rev. Mater* **1** 16064

[123] Hui L, Xue Y, Liu Y and Li Y 2021 Efficient hydrogen evolution on nanoscale graphdiyne *Small* **17** 2006136

[124] Xing C, Xue Y, Huang B, Yu H, Hui L, Fang Y, Liu Y, Zhao Y, Li Z and Li Y 2019 Fluorographdiyne: a metal-free catalyst for applications in water reduction and oxidation *Angew. Chem. Int. Ed.* **58** 13897–903

[125] Zhao Y, Yang N, Yao H, Liu D, Song L, Zhu J, Li S, Gu L, Lin K and Wang D 2019 Stereodefined codoping of sp-N and S atoms in few-layer graphdiyne for oxygen evolution reaction *J. Am. Chem. Soc.* **141** 7240–4

[126] Lv Q, Wang N, Si W, Hou Z, Li X, Wang X, Zhao F, Yang Z, Zhang Y and Huang C 2020 Pyridinic nitrogen exclusively doped carbon materials as efficient oxygen reduction electro-catalysts for Zn-air batteries *Appl. Catalysis* B **261** 118234

[127] Zhao Y *et al* 2018 Few-layer graphdiyne doped with sp-hybridized nitrogen atoms at acetylenic sites for oxygen reduction electrocatalysis *Nat. Chem.* **10** 924–31

[128] Si H, Deng Q, Yin C, Zhou J, Zhang S, Zhang Y, Liu Z, Zhang J, Zhang J and Kong J 2020 Gas exfoliation of graphitic carbon nitride to improve the photocatalytic hydrogen evolution of metal-free 2D/2D g-C3N4/graphdiyne heterojunction *J. Alloys Compd.* **833** 155054

[129] Xu Q, Zhu B, Cheng B, Yu J, Zhou M and Ho W 2019 Photocatalytic H_2 evolution on graphdiyne/g-C_3N_4 hybrid nanocomposites *Appl. Catalysis* B **255** 117770

[130] Shen H, Zhou W, He F, Gu Y, Li Y and Li Y 2020 A dehydrobenzoannulene-based three dimensional graphdiyne for photocatalytic hydrogen generation using Pt nanoparticles as a co-catalyst and triethanolamine as a sacrificial electron donor *J. Mater. Chem.* A **8** 4850–5

[131] Li J, Gao X, Liu B, Feng Q, Li X-B, Huang M-Y, Liu Z, Zhang J, Tung C-H and Wu L-Z 2016 Graphdiyne: a metal-free material as hole transfer layer to fabricate quantum dot-sensitized photocathodes for hydrogen production *J. Am. Chem. Soc.* **138** 3954–7

[132] Du Y, Xue Y, Zuo Z, Li Y, Liu H and Li Y 2021 Photoinduced electrocatalysis on 3D flexible OsO_x quantum dots *Adv. Energy Mater.* **11** 2100234

[133] Wang Z, Zheng Z, Xue Y, He F and Li Y 2021 Acidic water oxidation on quantum dots of IrOx/graphdiyne *Adv. Energy Mater.* **11** 2101138

[134] Gao Y, Xue Y, Liu T, Liu Y, Zhang C, Xing C, He F and Li Y 2021 Bimetallic mixed clusters highly loaded on porous 2D graphdiyne for hydrogen energy conversion *Adv. Sci. (Weinh)* **8** e2102777

[135] Sun C and Searles D J 2012 Lithium storage on graphdiyne predicted by DFT calculations *J. Phys. Chem.* C **116** 26222

[136] Huang C, Zhang S, Liu H, Li Y, Cui G and Li Y 2015 Graphdiyne for high capacity and long-life lithium storage *Nano Energy* **11** 481–9

[137] Wang N *et al* 2017 Synthesis of chlorine-substituted graphdiyne and applications for lithium-ion storage *Angew. Chem. Int. Ed. Engl.* **56** 10740

[138] Li L, Zuo Z, Shang H, Wang F and Li Y 2018 *In-situ* constructing 3D graphdiyne as all-carbon binder for high-performance silicon anode *Nano Energy* **53** 135–43

[139] Shang H, Zuo Z, Yu L, Wang F, He F and Li Y 2018 Low-temperature growth of all-carbon graphdiyne on a silicon anode for high-performance lithium-ion batteries *Adv. Mater.* **30** e1801459

[140] Li L, Zuo Z, Wang F, Gao J, Cao A, He F and Li Y 2020 *In situ* coating graphdiyne for high-energy-density and stable organic cathodes *Adv. Mater.* **32** e2000140

[141] Zuo Z, He F, Wang F, Li L and Li Y 2020 Spontaneously splitting copper nanowires into quantum dots on graphdiyne for suppressing lithium dendrites *Adv. Mater.* **32** e2004379

[142] Wang F, Zuo Z, Li L, He F and Li Y 2020 Graphdiyne nanostructure for high-performance lithium-sulfur batteries *Nano Energy* **68** 104307

[143] Farokh Niaei A H, Hussain T, Hankel M and Searles D J 2017 Sodium-intercalated bulk graphdiyne as an anode material for rechargeable batteries *J. Power Sources* **343** 354

[144] Zhang S, He J, Zheng J, Huang C, Lv Q, Wang K, Wang N and Lan Z 2017 Porous graphdiyne applied for sodium ion storage *J. Mater. Chem.* A **5** 2045

[145] Wang K, Wang N, He J, Yang Z, Shen X and Huang C 2017 Graphdiyne nanowalls as anode for lithium—ion batteries and capacitors exhibit superior cyclic stability *Electrochim. Acta* **253** 506

[146] Wang F, Zuo Z, Shang H, Zhao Y and Li Y 2019 Ultrafastly interweaving graphdiyne nanochain on arbitrary substrates and its performance as a supercapacitor electrode *ACS Appl. Mater. Interfaces* **11** 2599–607

Chapter 3

Carbon fibers and nanofibers

Daxiong Wu, Shichun Yang, Xinhua Liu and Jianmin Ma

Carbon-based fibers have received a great deal of attention in various fields because of their light weight, high conductivity, high mechanical strength, good flexibility, excellent creep, and chemical resistance. This chapter summarizes the properties and fabrication methods of carbon-based fibers, especially polymer/pitch-based carbon fibers (CFs), graphene-based fibers and lignin-based CFs, as well as various strategies for improving their mechanical, electrical, and electrochemical properties. It focuses on the advanced design of these carbon-based fibers and their applications in supercapacitors, alkali-metal ion batteries (AMBs), catalysts, field-effect transistor (FET) sensors, biomedicine, airline industries, sporting industries, automotive industries, etc. Finally, the challenges and future opportunities of carbon-based fibers are discussed. We expect the reader to gain knowledge and information from this chapter.

3.1 History

Fibers occur universally in nature as continuous filaments or elongated objects, such as spiderwebs, natural silk, etc. which provided an important source of inspiration for the development of synthetic fibers [1]. Around 1300, people obtained fibers from wool and cotton for use in fabrics and clothing and fabricated the spindle. By the 1880s, this practice was slowly evolving into the textile industry. With the development of chemistry and polymer science, fiber materials such as CFs, aramid fibers, polyketone fibers, etc. are applied in industrial production in various fields and have grown greatly [2, 3].

Among these, CF is defined as having at least 92 to 100 weight% (wt%) of carbon content with a length of '1 mm and is prepared from a polymeric precursor or produced from carbon allotrope building blocks [4]. CFs have excellent mechanical strength with very good creep resistance, high flexibility, low density ($\rho = 1.75$–2.00 g cm^{-3}), outstanding electrical conductivity, chemical stability, high temperature resistance, and small coefficient of thermal expansion. Benefitting from these good properties, CFs have attracted the attention of many researchers. Therefore, the CF industry has been developing continuously in various research fields including flexible and wearable electronics [5, 6], aerospace and aviation [4], automobiles, military hardware, civil engineering, and medical and sporting goods. CFs have been developed for over 150 years (figure 3.1). The first CF was produced by Swan, who carbonized paper filaments for incandescent light bulbs in 1860. In the late 19th century, Edison also prepared CF filaments for incandescent lamps using carbonized cotton threads and bamboo slivers as raw materials [7]. In the early 1960s, CF was one of the attractive industrial materials for developing modern science and technology, which was extracted from different carbon precursors. For example, Ford and Mitchell realized commercial CFs by heat treating rayon strands at up to 3000 °C [8]. Shindo from Japan was the first to produce commercial PAN-based CFs using polyacrylonitrile (PAN) fibers as precursors [9]. Otani was considered to be the first person to propose the construction method of isotropic pitch-based CFs (IPCFs) extracted from polyvinyl chloride (PVC) [10]. In the decades that followed, the CFs has widely applicated in various fields owing to the low cost of the product [11]. According to the latest report published by Carbon Composites e.V., the global CF demand will increase from 58 000 tons in 2015 to 120 000 tons by 2022, indicating a bright future for the development of the CF market (figure 3.2) [12].

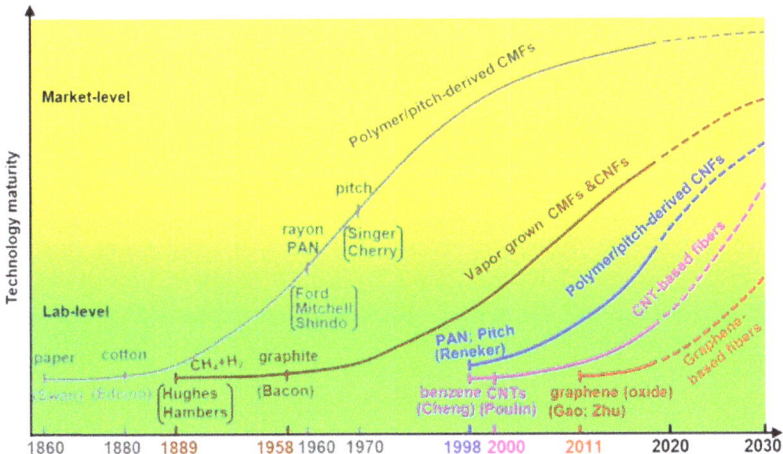

Figure 3.1. The past and future development trend of CFs [11], reproduced with permission © American Chemical Society.

Figure 3.2. Global demand for CFs from 2010 to 2022 [12], reproduced with permission © Royal Society of Chemistry.

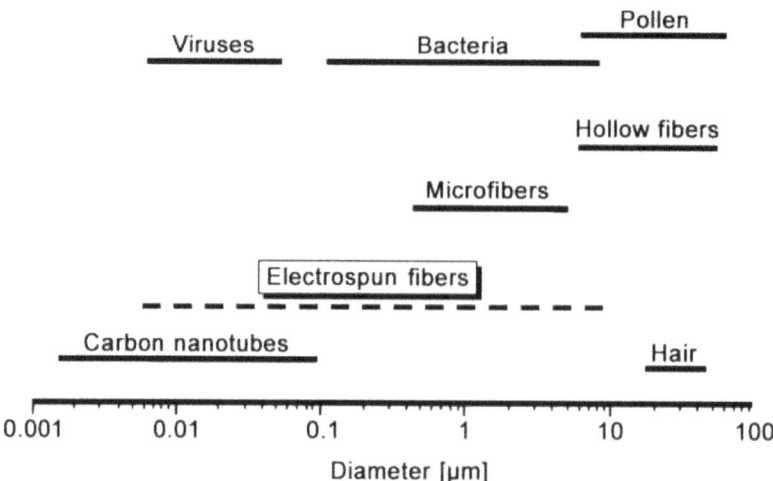

Figure 3.3. Comparison of the diameters of different CFs [13], reproduced with permission © John Wiley and Sons.

3.2 Classification of CFs

CFs can be divided into many different types. Depending on the diametric sizes of CFs, they can be divided into carbon nanofibers (CNFs, diameter <1 μm) and carbon microfibers (CMFs, diameter >1 μm). The reduction of the dimensions of a CF can result in new properties. A comparison of the diameters of different CFs is shown in figure 3.3. As the diameter decrease down to the nanoscale, some new properties can be observed due to the change of Fermi level and the reduction potential [13]. According to their microscopic structures, CFs can also classified as compact, porous, and even hollow-structure CFs. These special structures play an important role in different fields. In addition, depending on the precursor used, CFs mainly consist of polymer/pitch-based CFs, graphene-based fibers, and lignin-based CFs.

3.2.1 Polymer/pitch-based CFs

These CFs are produced using a polymer as the precursor, but is very limited due to a lack of suitable polymers. To date, commonly used polymers include polyacrylonitrile (PAN) [14–16], poly(vinyl alcohol) (PVA) [17, 18], poly(vinyliden fluoride) (PVDF) [19, 20], polyvinyl pyrrolidone (PVP) [21, 22], and pitch [7, 23, 24]. Among them, PAN has commonly been used as a raw material to fabricate commercial CFs with various diameters due to its high carbon output, low cost, and the good performance of the product. The polymerization of the polymer is crucial, especially in the production of CFs, where the properties of the CFs that result from spinning, stabilization, and carbonization of the primary filament are largely dependent on the characteristics of the polymer precursor used. Polymer/pitch-based CFs have played an important role in different fields including catalysis [25, 26], energy [27, 28], environmental protection [29, 30], and so on.

3.2.2 Graphene-based fibers

Graphene-based fibers are typical one-dimensional (1D) linear structural compo-nents, and they also constitute a new type of CF material. Graphene-based fibers successfully inherit both the excellent physical properties of graphene and the flexibility and wearability of the fiber structure, which is suitable for the production of multifunctional electronic textiles, sensing/monitoring, artificial skin and muscles, and wearable electronic products [31, 32]. However, graphene-based fibers are extracted from graphene oxide (GO), which inevitably introduces some of the disadvantages of abundant oxygen-containing functional groups (e.g., –COOH, –OH) and large numbers of defects [33]. So far, considerable efforts have been dedicated to improving the mechanical, physical, and electrical properties of graphene-based fibers, which have also achieved significant progress.

3.2.3 Lignin-based CFs

Lignin is the second most abundant aromatic monomer in nature; it is widely found in cork, hardwood, grass, and other plants, and is the only renewable raw material [34]. In addition, the carbon yield of lignin is more than 60%, and it can be used as a good precursor for the preparation of CFs [4]. Therefore, lignin-based CFs have attracted widespread attention because of their low production cost and high yield. Recently, a lot of research has shown that CFs prepared directly from lignin or precursors containing lignin are successful and they have been widely applied to produce electrical devices (such as supercapacitors and batteries) [35–37].

3.3 Fabrication of CFs

The fabrication of CFs has two major methods, the carbonization of precursor fibers (method 1) and the spinning of carbon nanomaterials (method 2). In method 1, the preparation of CFs generally requires the basic fabrication steps of the precursor fiber, spinning, stabilization, and high-temperature carbonization of the precursor fiber (temperature >500 °C). In method 2, nanocarbons (such as carbon nanotubes and graphene) are spun into fibers from their aerosol, wet dispersion, or solid film [11].

Generally, CF properties mainly depend on the characteristics of the precursor fibers [38]. Therefore, the precursor fibers play a crucial role in obtaining strong CFs with high performance. The preparation of the precursor fibers needs to stretch the colloidal solution or melt to further form the fiber. There are two main spinning methods, namely, solution spinning and melt spinning. The melt spinning method can be used with precursors that have good thermal stability. In addition, the spinning technologies also include stretch spinning, blow spinning, electrostatic spinning, and centrifugal spinning, which vary according to the different forces exerted on the colloid, which rely on mechanical stretching, an air jet, an electrostatic field force, and a centrifugal force to obtain the fibers, respectively. The next section introduces the universal solution spinning, melt spinning, and electrospinning techniques in detail.

3.3.1 Solution spinning

Solution spinning is a method in which a concentrated polymer solution is quantitatively extruded from a spinneret hole and the solution flow is solidified into fibers by a coagulation bath, hot air, or a hot inert gas. Solution spinning can be divided into wet spinning and dry spinning [11]. Wet spinning involves the following processes: (1) the spinning stock is prepared; (2) the raw colloid is pressed out from the spinneret hole to form a trickle; (3) the raw colloid trickle solidifies into primary fibers; (4) the raw fibers are rolled or directly post processed. Wet spinning is an important process that affects the physical, chemical, and mechanical properties of the fibers. Common CFs, such as polymer-based CFs and lignin-based CFs, are usually produced by wet spinning technology. GO fibers can be prepared by wet spinning technology; a schematic diagram is shown in figure 3.4(a) [39]. First, GO sheets are uniformly dispersed in a water solution, then injected into a coagulation bath to produce GO fibers. Importantly, rotating the bath or drawing the fiber with a collection unit should maintain a certain movement speed, thereby obtaining uniform and continuous GO gel-state fibers [39]. In addition, graphene-based fibers are also produced by dry spinning technology. Dry spinning is different from wet spinning; in dry spinning from a spinneret capillary hole, the pressure of the spinning liquid does not carry the trickle into a coagulation bath, but into a spinning tunnel. Under the influence of a hot air flow in the passage, the solvent in the thin stream of the original liquid evaporates rapidly, and the solvent vapor thus volatilized is taken away by the hot air flow. The original solution solidifies while the solvent is gradually removed; it elongates and becomes thin under the action of the winding tension to form the primary fiber. Gao and his colleagues first proposed a homemade dry-spinning apparatus for the preparation of graphene-based fibers; a schematic illustration of the dry spinning process is displayed in figure 3.4(b) [40]. A nozzle was used to extrude the GO coating. With the help of infrared lamps, the solvent in the fiber was evaporated, and the fibers were dried for further collection.

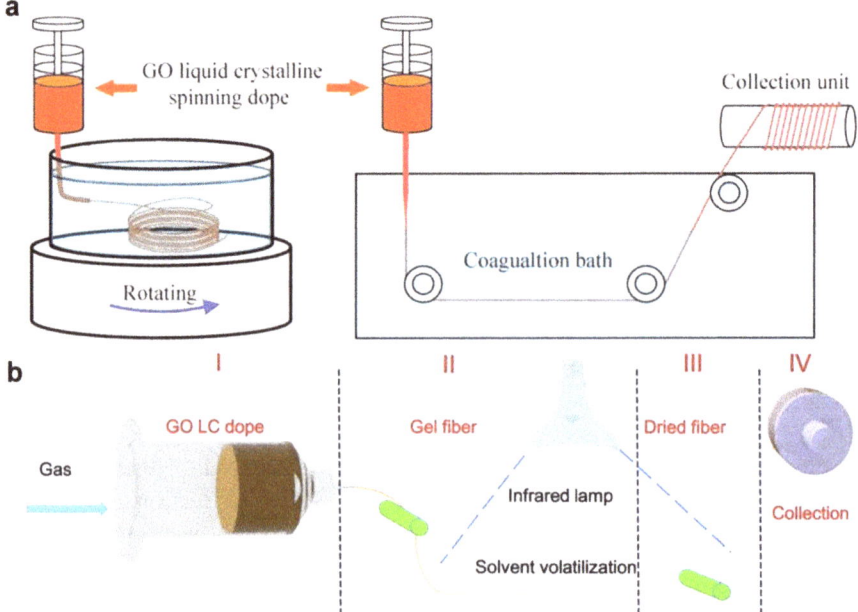

Figure 3.4. (a) Schematic diagram of the preparation of graphene oxide fiber via a wet spinning method with a rotating coagulation bath and a collection unit [39], reproduced with permission © John Wiley and Sons. (b) Schematic illustration of the dry spinning process for the preparation of GO fibers [40], reproduced with permission © Royal Society of Chemistry.

3.3.2 Melt spinning

Melt spinning, is a kind of molding method which takes a polymer melt as the raw material (similar to extrusion after plastic melting) and uses a melt spinning machine. Any polymer that can be melted or transformed into a viscous fluid state without significant degradation when heated, i.e. materials with good thermal stability, can be spun by melt spinning. During melt spinning, the bulk polymer is melted in a screw extruder and then fed into the spinning site. It is quantitatively fed into the spinning assembly by the spinning pump. After filtration, it is extruded from the pores of the spinneret plate. The liquid filament is gradually solidified as it passes through the cooling medium and is then drawn at high speed by the winding device below. The filament is the primary fiber, and the primary fiber is processed into the finished fiber by further processing. The typical preparation method of pitch-based fibers is by melt spinning owing to the good thermal stability of pitch; schematic diagrams of the melt spinning of pitch-based fibers are shown in figure 3.5(a) [41]. Wang and his colleagues utilized a miniature twin-screw extruder to produce polysulfone-block-poly(ethylene glycol) (PSF-b-PEG) hollow fibers; a schematic is shown in figure 3.5(b) [42]. In addition, Pia Willberg-Keyriläinen *et al* used the melt spinning method to prepare cellulose octanoate fibers (figure 3.5(c)) [43]. Melt spinning is generally more efficient than solution spinning because it does not require the use of solvents or coagulation baths during the preparation process [11].

In addition, the melt spinning technique can be used to produce unidirectional, patterned, continuous CFs with controlled surface geometries and properties via applying a bicomponent fiber in combination with elimination of the fugitive component, adequate thermo-chemical stabilization, and carbonization of the desired component. As show, in figure 3.6, Amit K Naskar *et al* proposed a flexible

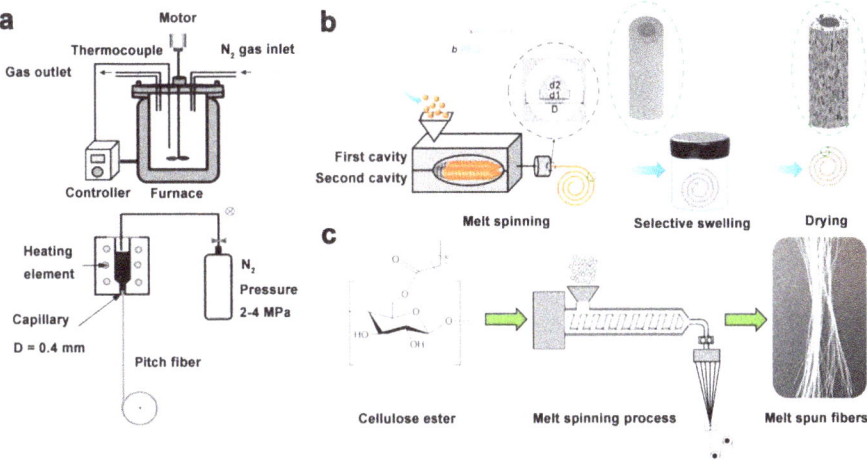

Figure 3.5. (a) Schematic diagrams of the melt spinning of pitch-based fibers [41], reproduced with permission © Elsevier. (b) Schematic of the preparation of hollow-fiber membranes of block copolymers [42], reproduced with permission © Elsevier. (c) Schematic of the preparation of cellulose octanoate fibers [43], reproduced with permission © John Wiley and Sons.

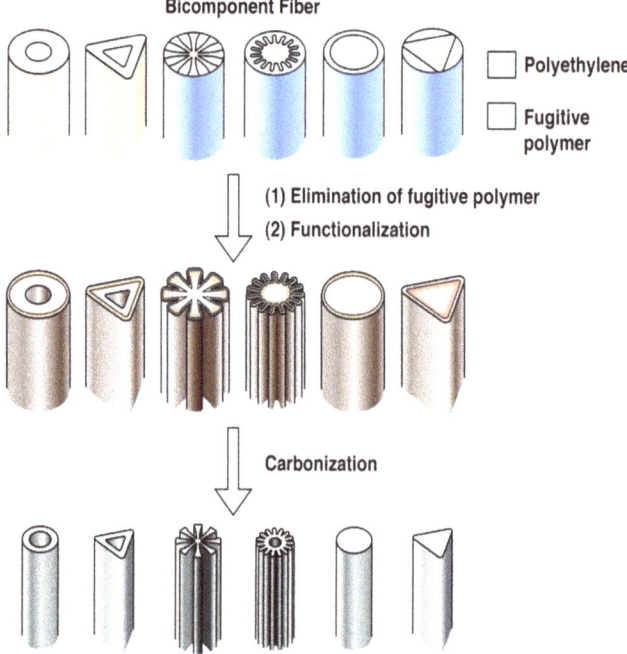

Figure 3.6. The principal processing path for creating novel CF structures from polylactic acid (PLA)/ polyethylene (PE) precursors by sulfonation and carbonization. Reproduced with permission from [108] © John Wiley and Sons.

technique for producing large-volume, technologically innovative fibers in myriad configurations such as fiber bundles and nonwoven mat assemblies. This technology makes it easy to design and manufacture fibers with customized surface contours, and at the same time, it can control the fiber diameter to the submicron level, and the porosity can be adjusted through the diffusion control functionalization of the precursor.

3.3.3 Electrospinning

Electrospinning is a special fiber manufacturing process in which polymer solution or melt is sprayed while spinning under the action of strong electric field forces. A schematic illustration of a basic electrospinning setup is shown in figure 3.7, which mainly consists of three components: a high-voltage power supply (1–100 kV), a spinneret (a needle nozzle with inner diameter of 0.1–1.0 mm), and a collector [44].

The electrospinning concept was proposed by William Gilbert in an early study in 1600 [45]. It was not until 1934 that Anton Forhals invented an experimental device for preparing polymer fibers by electrostatic force and applied for a patent. His patent described how the polymer solution formed a jet between the electrodes, and was the first patent to describe in detail a device for preparing fibers using high-voltage electrostatic electricity [46]. This is recognized as the beginning of fiber preparation by the electrospinning technique. In addition, electrospinning theory also experienced a comparatively slow development period. In 1964, Taylor proposed the concept of the Taylor cone, which was not tested until 1969. Over the next few decades, the

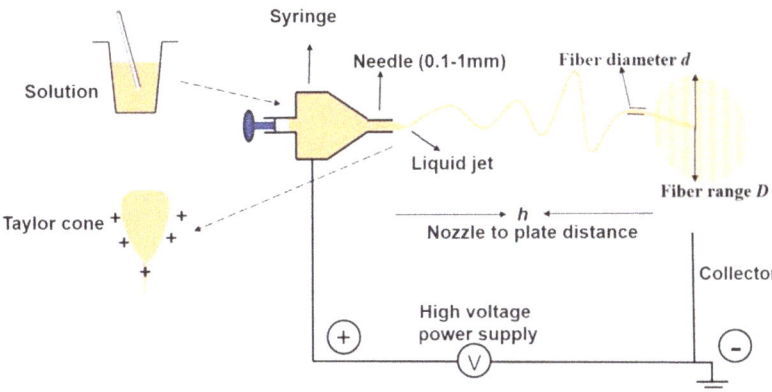

Figure 3.7. Schematic illustration of a basic electrospinning setup [44], reproduced with permission © John Wiley and Sons.

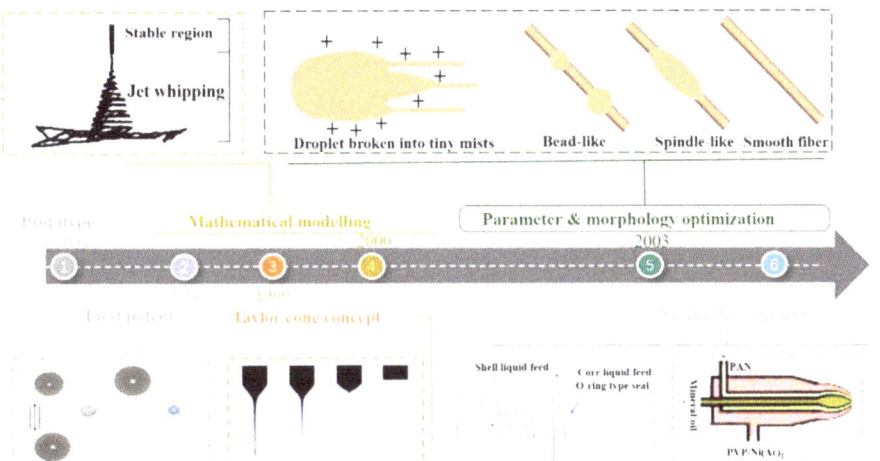

Figure 3.8. Schematic of the development history of the electrospinning technique [44], reproduced with permission © John Wiley and Sons.

electrospinning technique developed very rapidly. Reneker, Yarin *et al* established the experimental phenomenology and theory of the mechanical electrospinning process by analyzing the bending instability (figure 3.8). The electrospinning technique has played a very important role in the field of constructing 1D nanostructured materials with different structures including solid and hollow structures, a core–shell microfiber structure for applications in catalysis, conversion, storage, photoelectricity, food engineering, cosmetics, and so on. By designing different collection devices, single fibers, bundles, highly oriented fibers or randomly oriented fiber membranes can be obtained. However, the electrostatic spinning technique also faces some challenges related to control of the fiber structure: first, to realize the industrial application of electrostatic fiber spinning, it is necessary to obtain a staple or continuous nanofiber bundle. The preparation of the fiber orientation is an effective way to solve this problem, but there are many gap distance goals. In the future, the objective should be to make the fibers as straight and oriented as possible by improving the nozzle and receiving device and adding an auxiliary electrode, so as to obtain an oriented fiber array with excellent overall performance. Second, as a new research field of electro-spinning nanofibers, research into nano-spiderwebs is still in its early stages, and the theoretical analysis and modeling of the formation process of the nano-spiderweb still need to be further studied. In addition, in order to improve the performance of electrospun fiber membranes in the field of ultrafine filtration, it is necessary to reduce the diameter of the fiber. Ways of reducing the average diameter of the fiber to less than 20 nm are a challenge for the electrospinning technique.

The preparation of nanofiber materials by the electrospinning technique has been one of the most important academic and technical activities in the field of materials science and technology in the world in recent years. Electrospinning has become one of the main ways to effectively prepare nanofiber materials owing to its advantages such as simple manufacturing equipment, low spinning cost, wide variety of spinnable materials, and controllable process. Electrospinning has produced a wide variety of nanofibers, including organic fibers, organic/inorganic composites, and inorganic nanofibers. However, there are still some problems to be solved in the preparation of nanofibers by electrospinning. Firstly, in the preparation of organic nanofibers, the natural polymer varieties used in electrospinning are very limited, while the structures and properties of the resulting products are not perfect enough, and the final products are mostly only in the experimental stage, especially the industrial production of these products. Secondly, the properties of electrospun organic/inorganic composite nanofibers are not only related to the structure of the nanoparticles, but also related to the aggregation and synergistic properties of the nanoparticles, the structural properties of the polymer matrix, the structural properties of the interface between the particles and the matrix, and the processing and composite technology. Methods for the preparation of suitable, high-perform-ance, multifunctional composite nanofibers form a key area of research.

3.4 Applications of CFs

CFs, including polymer/pitch-based CFs, graphene-based fibers, and lignin-based CFs, have mainly been applied in supercapacitors, rechargeable alkali-metal batteries (such as Li, Na, and K ion batteries), and catalysis; these applications benefit from their high mechanical strength (high tensile strength and high modulus), low density, excellent creep, chemical resistance, and good electrical conductivity. Some potential applica-tions have been developed in the areas of FET sensors, biomedicine, airline industries, sporting industries, automotive industries, etc. have been developed.

3.4.1 Supercapacitors

Supercapacitors, which have high power densities, fast charge/discharge rates, and long cycle lives have been widely used in electronics and electric vehicles [47–49]. Supercapacitors are composed of a negative electrode, a separator, an electrolyte and a positive electrode [50]. Both electrodes consist of capacitive materials, and the electrode materials mainly determine the electrochemical performance of super-capacitors [51]. Typically, carbon materials are mostly used as electrode materials, and commercially available capacitors utilize activated carbons as electrodes. As society's needs have continued to develop, supercapacitors have suffered from issues caused by their insufficient energy density, which severely hamper their application scope, in particular, their high-energy output. CFs with highly porous or hollow nanostructures have a high specific surface area and good conductivity are better electrodes for supercapacitors than activated carbons. Prospective capacitive mate-rials made from CFs can provide an extremely large surface area accessible to the electrolyte, fast electron and ion transport, and good structural stability that mitigates volume changes [52–54]. These properties significantly help the develop-ment of high-performance supercapacitors.

To date, many strategies have been used to improve the performance of super-capacitors. Typically, porous carbon nanofibers are designed to be used as high-capacitance materials, and higher capacitance can be easily realized by changing the pore size distribution of the carbon nanofibers to increase the ion conductivity and specific surface area, which further promotes an increased interface between the electrolyte and the electrode [55–57]. It is worth noting that the connecting pore structure within a single nanofiber can introduce new electroactive sites and diffusion pathways to accelerate the dynamics of ion transport. In addition, heteroatom dopants (N, P, B, S, F, etc.) can help to improve the electrochemical properties of CF by modifying the bandgap and/or changing the surface characteristics [58, 59]. Poly-heteroatom doping (co-doping) will probably inherit the merits from the overall synergistic properties of both carbon and the dopants [60, 61]. Furthermore, heter-oatom doping can not only change the electron donor/acceptor properties of carbon materials, thus generating additional pseudocapacitance, but also maintain excellent rate performance and cycle stability [62]. Among these heteroatoms, N-atom doping seems to be the most effective way to increase capacity in addition to improving conductivity and surface wettability [63, 64]. Therefore, many researchers have devoted great efforts to synthesizing N-doped CFs for high-performance supercapacitors.

Lou and his colleagues prepared hollow particle-based nitrogen-doped carbon nanofibers (HPCNFs-N) with a hierarchical porous structure by embedding ultra-fine zeolitic imidazolate framework (ZIF-8) nanoparticles into electrospun PAN [58]. As shown in figures 3.9(a)–(c), the resultant HPCNFs-N sample has very long and uniform carbon nanofibers with a rough surface and numerous hollow nano-particles interconnected with each other and exhibits good flexibility. Furthermore, the C and N elements are uniformly distributed on the carbon nanofibers over the entire porous structure (figure 3.9(d)). Benefiting from its hierarchical porous structural feature and desirable chemical composition, an assembled supercapacitor device with electrodes made from this material exhibited a remarkable capacitive performance of 307.2 F g^{-1} and 193.4 F g^{-1} at 1.0 and 50.0 A g^{-1}, respectively, while offering high energy/power density and long cycling stability with only a 1.8% capacitance loss over 10 000 cycles (figures 3.9(e) and (f)).

As shown in figure 3.10, Shi *et al* designed a boron and nitrogen co-doped carbon (BCN), which was manufactured by immersing active carbon fiber cloth (ACC) in a

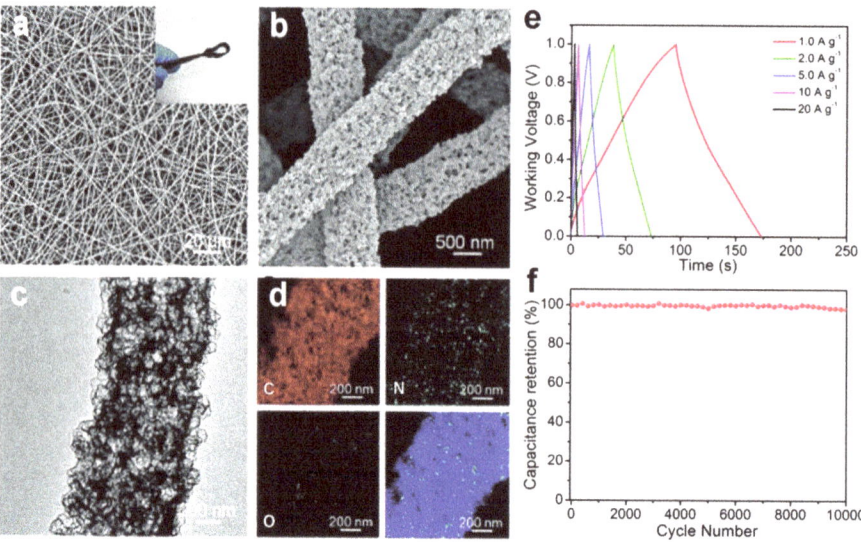

Figure 3.9. Morphological characterization (a–c) and elemental mapping images (d) of the HPCNFs-N sample. Charge–discharge curves at different current densities (e) and cycling performance of the HPCNFs-N device at 5.0 A g^{-1} (f) [58], reproduced with permission © Royal Society of Chemistry.

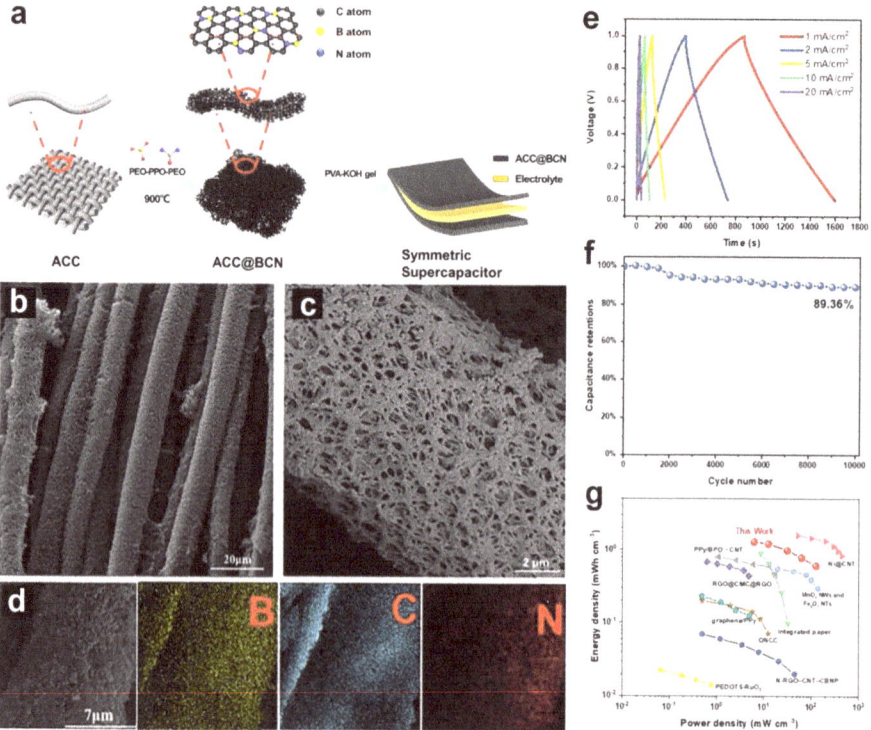

Figure 3.10. (a) Schematic diagram of the preparation of the ACC@BCN electrode and the structure of the symmetric supercapacitor. SEM (b and c) and elemental mapping images (d) of the ACC@BCN. The electrochemical performances of ACC@BCN FSSCs (e)–(g) [65], reproduced with permission © Elsevier.

mixed solution of urea, boric acid, and polyethylene oxide–propylene oxide which was then calcined [65]. The morphology of the prepared ACC@BCN surface become rough and porous compared to the smooth surface of the ACC, and the surface was uniformly covered by the porous BCN which formed a complex 3D cross-linked network; this increased the hydrophilicity of the electrode. In addition, the distribution of elements in ACC@BCN was examined, and the C, B, and N elements were uniformly distributed on the ACC@BCN. The prepared BCN with its porous 3D network structure provides a large specific surface area, good hydrophilicity, and abundant active sites; when it was used in the electrodes of flexible all-solid-state supercapacitors, they exhibited an excellent volume specific capacitance of 9.212 F cm^{-3} at a current density of 1 mA cm^{-2} and an ultrahigh capacitance retention of 89.5% even after 10 000 cycles. This research provides a simple and effective strategy for the large-scale low-cost preparation of high-performance flexible capacitor electrode materials for high-performance energy storage devices.

3.4.2 Rechargeable alkali-metal batteries

With the rapid development of portable consumer electronics, power grids, and electric vehicles, the societal requirements for advanced energy storage technologies are also increasing [66–68]. AMBs based on Li, Na, and K are potential candidates for the next-generation energy technologies and have practically or potentially been used as power sources in a multitude of application fields due to their high energy density, long lifespan, portability, energy conservation, and environmental friendliness [69–71]. At present, their low specific capacity, structural instability, slow redox kinetics, and poor electronic/ionic conductivity are the main bottlenecks of energy storage systems, which can result in poor electrochemical performance and severely limit their practical application [44]. The properties of the electrode materials play a crucial role in AIB performance. Therefore, a new type of AIB electrode material is needed to enhance the specific capacity, structural stability, redox kinetics, and electronic/ionic conductivity, thus improving the overall performance of these batteries. However, most electrode materials suffer from huge volume changes during the charge–discharge process, leading to structural destruction and electronic or ionic transport degradation, further causing rapid capacity decay and short service life. The preparation of carbon-based composites with active high-capacity materials is be considered to be one of the most effective strategies for addessing these problems. Using this approach, it is easy to control the composition, structure, and topology of electrodes, while improving their specific capacity and conductivity.

According to research, CFs have high mechanical strength, flexibility, good electrical conductivity, and good electrochemical properties, and they have long been used as reinforcements in metal matrices [11, 27]. They have advantages that no other material can match in AIB systems [27]. However, there are major technical problems that need to be solved in the prior art. The problems of dispersion technology and reinforcement technology of CF need to further solved, so that the base material has the characteristics of uniformity and excellent strength. In addition, in order to ensure that the substrate meets the required tightness and has good surface flatness, the hot-pressing process should be studied to solve the problems of resin migration, uniform thickness, and surface flatness. In addition, a suitable high-temperature carbonization process should be explored, which would improve the strength of the substrate and solve the problem of easy fracture.

In recent years, different types of CFs-based material have been investigated for use in various electrochemical energy storage systems [72–77]. For example, facile synthesis of carbonaceous fibers using *Tyromyces fissilis* wild fungus was proposed via a facile carbonization step, as displayed in figure 3.11(a) [78]. These CFs were used as the anode and the conventional $LiCoO_2$ was used as the cathode with 1 M $LiPF_6$ in a 1:1:1 volume ratio of ethylene carbonate (EC): dimethyl carbonate (DMC): diethyl carbonate (DEC) with 1 % vinylene carbonate (VC) additive as the electrolyte of rechargeable lithium-ion batteries, which powered 2 V and 3 V yellow and green LEDs, respectively. In addition, the electrochemical performances of lithium half-cell configurations were also measured (figures 3.11(b)–(d)). The anode delivered a reversible capacity of 340 mAh g^{-1} at a rate of C/10 rate, while achieving a high capacity of 300 mAh g^{-1} at a rate of C/5 for 150 cycles without significant attenuation of capacity.

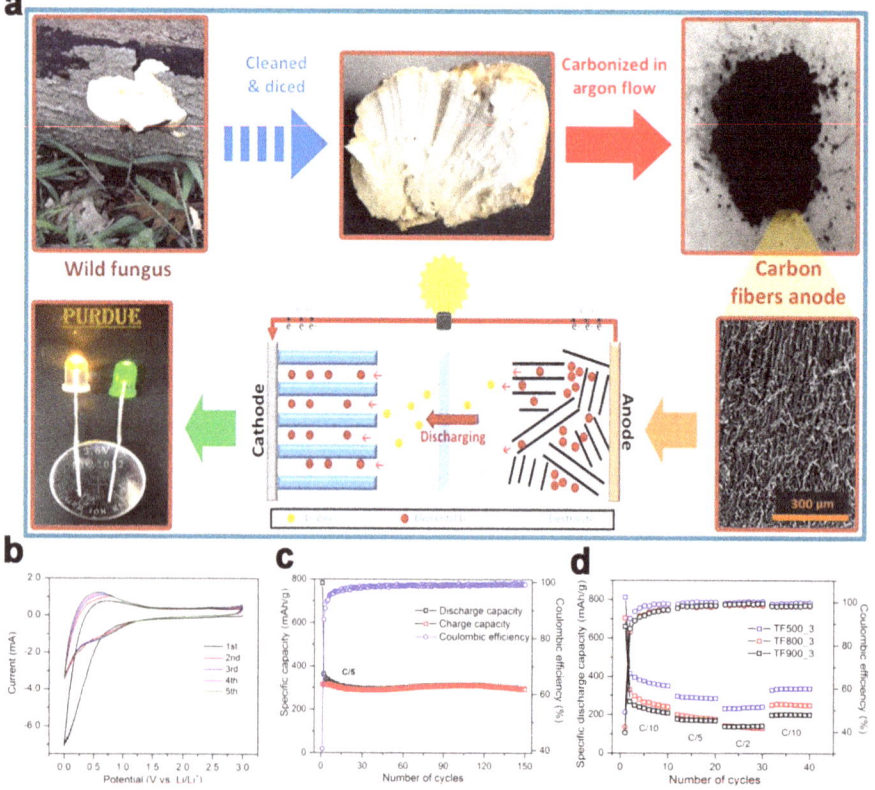

Figure 3.11. (a) Preparation of a CF anode and the construction of a full cell to power light-emitting diodes. The CV curve, (b) cycling, (c) and (d) rate performance of a half-cell [78], reproduced with permission © American Chemical Society.

In addition, micro/mesopore multihole CFs have been regarded as good sulfur hosts that can improve the electrochemical performance of lithium–sulfur batteries owing to their high electrical conductivity; they provide an effective conductive network on the cathode, a porous structure with a high sulfur loading, and good mechanical properties to buffer the volume change during the charge–discharge process [79–81]. For example, sulfur-embedded activated multichannel carbon nanofibers were proposed, which consist of parallel mesoporous channels connected by micropores for sulfur containment, as shown in figure 3.12(a) [82]. First, multicore polymer nanofibers (MPNFs) were prepared by the co-electrospinning technology with both PAN and poly(methyl methacrylate) (PMMA) polymer solutions. The MPNFs were then carbonized and further activated by a KOH activation process to obtain activated multichannel carbon nanofibers (a-MCNF). A uniform carbon fibrous morphology with a rough surface and 3D networked small

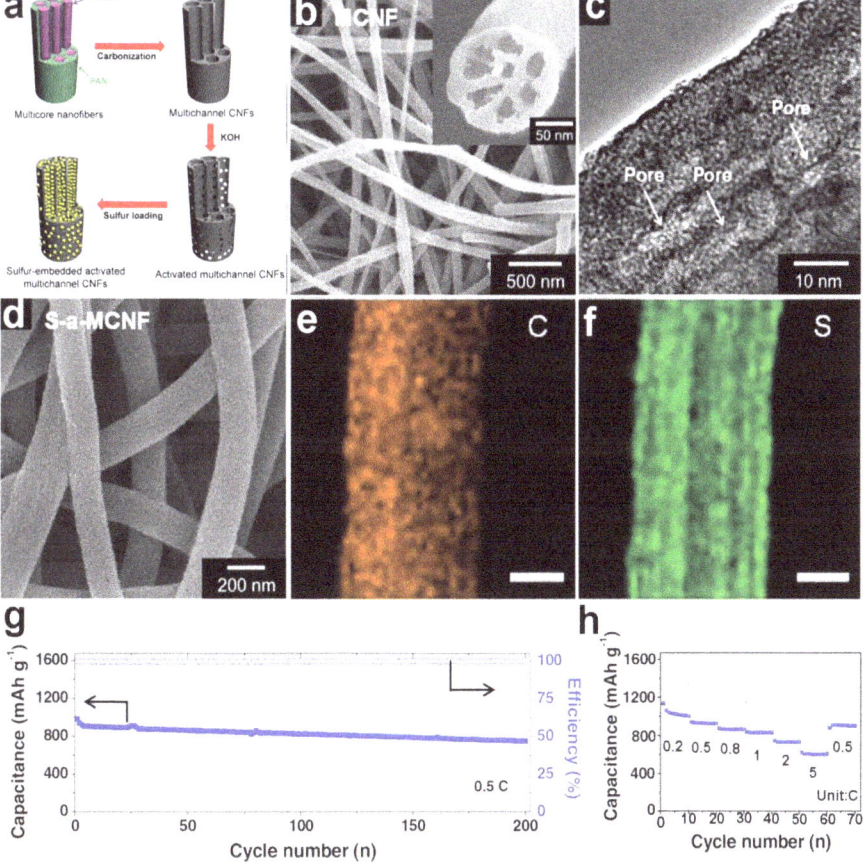

Figure 3.12. (a) Illustration of the sequential fabrication steps used for the sulfur-embedded activated multichannel carbon nanofibers. SEM (b) and TEM (c) images of a-MCNF. SEM and element mapping analysis images of S-a-MCNF (d-f). The cycling (g) and rate (h) performances of a S-a-MCNF-based electrode [82], reproduced with permission © John Wiley and Sons.

pores can be observed in figures 3.12(b)–(d). In addition, C and S element mapping analysis showed that the S component filled the pores instead of attaching to the outer carbon surface and was uniformly distributed on the CFs (figures 3.12(e) and (f)). Therefore, the KOH activation process acts not only to etch CFs but also to generate a 3D nanoscale porous network, which effectively increases the loading amount of sulfur (80 wt% in the composite) and forms a nanoscale sulfur decoration. The S-a-MCNF-based electrode achieved a high capacity of 1138 mAh g^{-1} at 0.2 C, a good rate performance (623 mAh g^{-1} at 5 C), and a good cycling stability with 76% capacity retention (753 mAh g^{-1}) after 200 cycles at 0.5 C at a higher sulfur loading of 4.6 mg cm^{-2}. This study provides an effective way to deal with the issues of the rapid capacity fade and low sulfur utilization in Li–S batteries.

Generally, CFs are recognized as one of the most promising materials and have been widely applied to high-performance SIBs due to their unique structure, low potential, low cost, eco friendliness, and chemical and cycling stability [83]. As in the case of LIBs, heteroatom doping can further increase the energy density and overall capacity of the Na ion battery [84, 85]. N atoms, which are used for doping, have a similar covalent radius to those of carbon materials; such doping can produce more defects and improve electronic conductivity [86]. S atoms have a large covalent radius, and the introduction of S atoms facilitates an expansion of the distance between the carbon layers so that they accommodate more Na ions; it also facilitates the Na insertion/extraction process [87]. Therefore, it can be expected that flexible independent electrodes will be prepared based on N/S co-doped CFs to synergistically promote the Na ion transport and insertion processes for high sodium storage performance. Yu *et al* fabricated a sulfur-enriched N-doped multichannel hollow carbon nanofiber (S-NCNF) film via an electrospinning and heat treatment technique with sublimed S as the flexible anode for SIBs, as displayed in figure 3.13 [88]. The SEM and TEM images show a multichannel hollow 3D structure, and the C, N, and S elements are uniformly distributed throughout the whole CF. S-NCNF, which has abundant defects and a large interlayer spacing, can promote Na ion diffusion and electron transportation, and the S-NCNF electrode achieved superior rate performance (132 mAh g^{-1} at 10 A g^{-1}) and outstanding cycling stability (a reversible specific capacity of 187 mA h g^{-1} at 2 A g^{-1} over 2000 cycles).

CFs also are ideal candidates for use as anode materials for potassium-ion batteries (PIBs) because of their unique physicochemical characteristics [89–91]. Similarly, heteroatom doping can also improve the energy density and specific capacity of PIBs [92, 93]. As shown in figure 3.14, the red P nanoparticles are uniformly distributed within electrospun N-doped porous hollow carbon nanofibers (labeled red P@N-PHCNFs) by an encapsulating technology [94]. The unique structural properties effectively alleviate the huge volume change of red P without destroying the solid-state electrolyte interfaces (SEI) formed by the carbon matrix and the external surface, while greatly enhancing the K$^+$ diffusion kinetics and rapid charge transfer. Benefiting from these advantages, the red P@N-PHCNF anode realized an excellent cycle life with a high charge capacity of 636 mAh g^{-1} at the second cycle, while maintaining a high capacity retention of 84% after 200 cycles at a higher current density of 1 A g^{-1}, and an outstanding rate capability of 342 mAh g^{-1} at 5 A g^{-1} for PIBs.

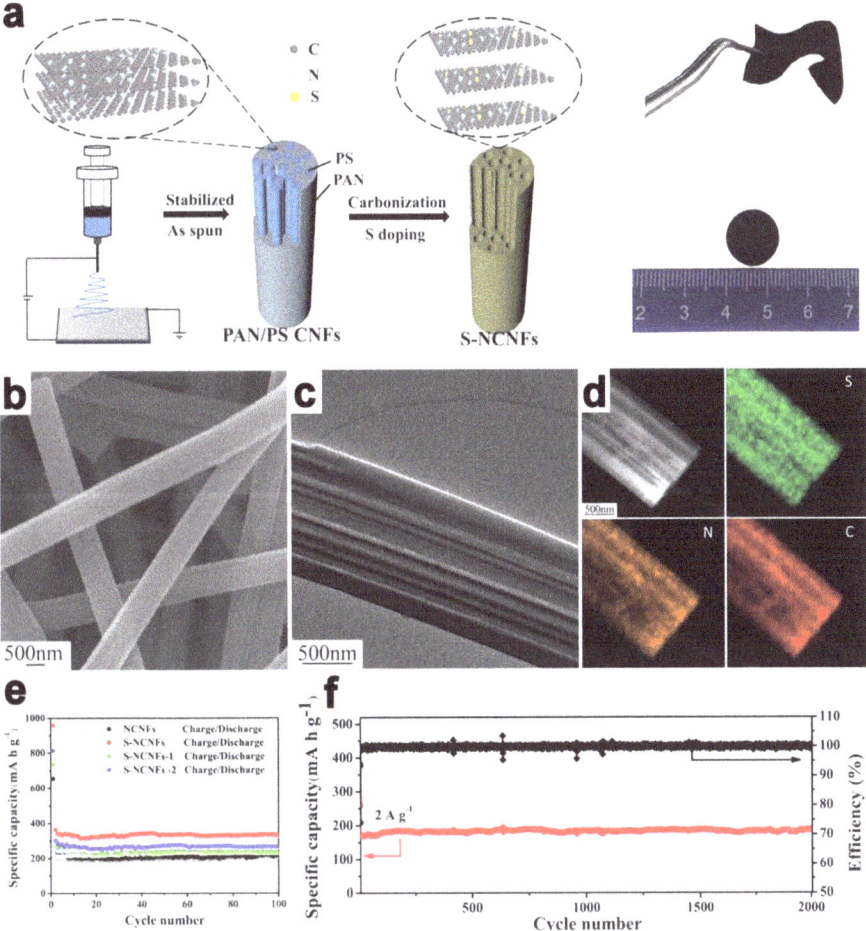

Figure 3.13. (a) The fabrication process and digital images of S-NCNFs. (b and c) SEM and (d) element mapping analysis images of S-NCNF. (e and f) The cycling performance of the S-NCNF-based electrode [88], reproduced with permission © John Wiley and Sons.

In addition, pure CFs are used as anodes of AMBs, yielding a low specific capacity. To increase the capacity density, CFs composited with alkali-metal anode materials with high theoretical capacities, such as CFs/silicon, CFs/metallic oxide, CFs/metallic sulfide, and CFs/metallic selenide etc. have been investigated; many great successes have been achieved, which demonstrates that these strategies are highly desirable in order to improve AMB performance.

3.4.3 Catalysis

The development of clean energy is considered to be one of the most promising strategies for the alleviation of environmental pollution and energy crises. Regardless of whether we consider a high-energy-density and zero-emission hydrogen energy source, or a fuel cell with high energy density, environmental friendliness, and high

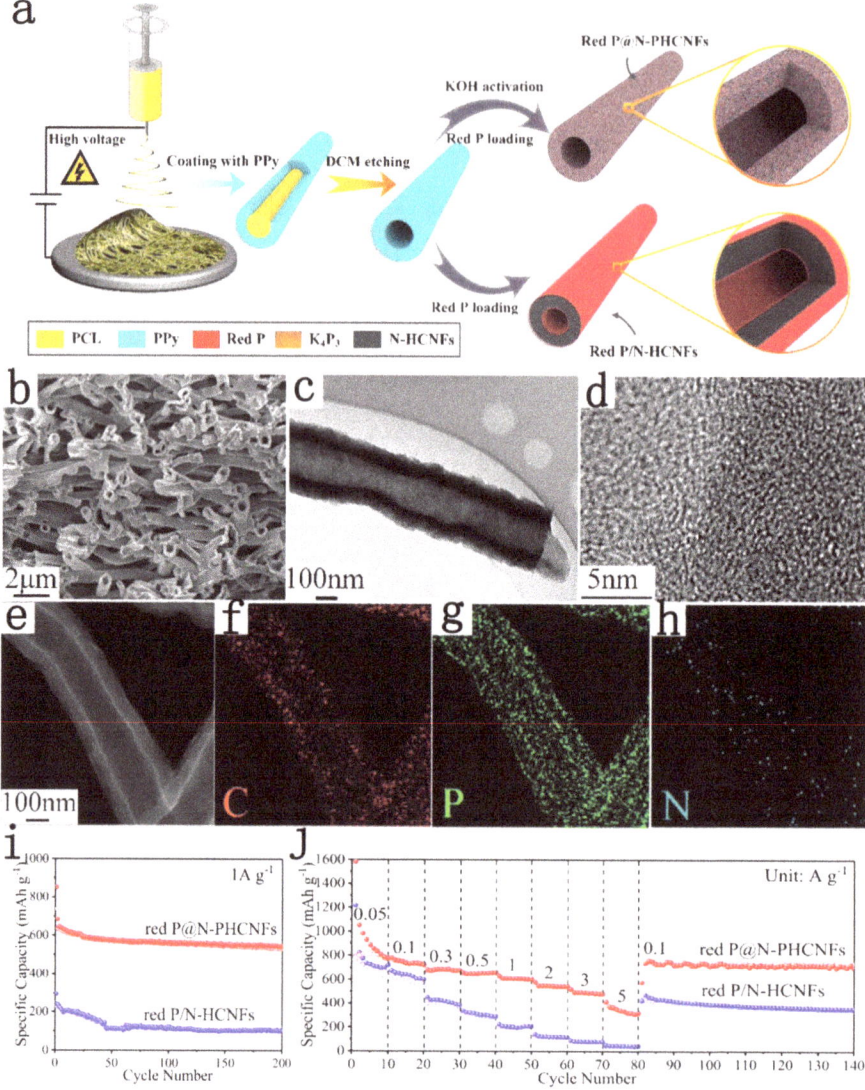

Figure 3.14. (a) Images of the synthesis process of red P@N-PHCNFs and the red P/N-HCNF sample. SEM (b), TEM (c and d), and element mapping analysis (e–h) images of the red P@N-PHCNF. (e and f) The cycling performance of these electrodes [94], reproduced with permission © American Chemical Society.

efficiency, an efficient catalyst is required [95]. Therefore, catalysts are vital to the development of clean energy. To date, platinum (Pt)-group metals are still the most efficient catalysts for oxygen reduction reactions (ORR) and the oxygen evolution reaction (OER) [96], while they are the most advanced low-overpotential hydrogen evolution reaction (HER) electrocatalysts [97]. However, the high cost, limited storage, and poor durability of Pt have severely impeded the practical application of fuel cells and water splitting technology. In this regard, researchers have devoted themselves to exploring non-precious-metal and metal-free materials to achieve high-efficiency ORR catalytic performance or HER catalytic performance. Among these materials, porous CFs have low preparation cost, good catalytic activity, and long cycle stability, and are considered to be a new generation of catalysts [98]. Furthermore, heteroatom-doped porous CF materials with large specific surface areas, high porosity, and fully accessible pores on/in individual nanofibers that effectively improve the transport kinetics can provide many defects and abundant active sites to increase the essential activity of CFs [99, 100]. In addition, the 3D network skeleton of CFs can endow immobilized catalysts with increased activity, selectivity, durability, and thermal stability [1].

N-doped CFs derived from the carbonization of an electron-spun PAN membrane were proposed as a high-efficiency and stable metal-free electrocatalyst for HER catalysis in both acidic and alkaline media [101]. The electron-spun PAN membrane was carbonized at 700 °C, 800 °C, and 900 °C (denoted as NCFs-700, NCFs-800, and NCFs-900, respectively), corresponding to the SEM images in figures 3.15(a)–(c). The energy dispersive spectrometer (EDS) mappings of the NCFs-800 electrocatalyst maintained a completely fibrous structure and a uniform distribution of nitrogen atoms before and after a durability test (figures 3.15(d) and (e)). Therefore, NCFs-800 requires only low overpotentials of 114.3 mV and 198.6 mV vs. RHE to realize a cathodic current density of 10 mA cm^{-2} in 0.5 M H$_2$SO$_4$ and 1 M KOH electrolytes, respectively. Furthermore, the NCFs-800 electrocatalyst had 95.2 mV dec^{-1} and 131.3 mV dec^{-1} Tafel slopes, and exhibited insignificant degradation in HER activity after 2000 potential cycles, while the HER activity of commercial Pt/C was seriously deteriorated. This work indicates that the preparation of N-doped CF electrocatalysts can probably replace Pt/C in acidic and alkaline water splitting.

As displayed in figure 3.16, metal-free N, P co-doped porous CFs (NPCFs) were synthesized by the simple pyrolysis of inexpensive polypyrrole as the nitrogen source and phosphoric acid as the phosphorus source [102]. First, a nitrogen source polypyrrole precursor was synthesized in large quantities through an improved oxidative template assembly method. The polypyrrole carbon nanofibers were then immersed in phosphoric acid (H$_3$PO$_4$: 0.5 M, 1.0 M, 1.5 M) aqueous solutions at different molar concentrations and finally thermally cracked, which successfully synthesized N and P atom co-doped porous CFs (denoted by NPCF-1, NPCF-2, and NPCF-3, respectively). The porous structure of the NPCF-2 catalyst, which had a large surface area, effectively increased the diffusion of oxygen molecules to the active sites and the exposure of active sites, while the N and P atoms regulated the charge distribution, further making the N, P, and adjacent C atoms effective ORR active sites. Thus, the NPCF-2 catalyst achieved excellent alkaline ORR catalytic activity, ultralong durability, and high methanol tolerance.

3.4.4 Field-effect transistor sensors

The rapid identification of low-concentration chemicals and biomolecules is essential for chemical analysis, healthcare, environmental monitoring, and disease

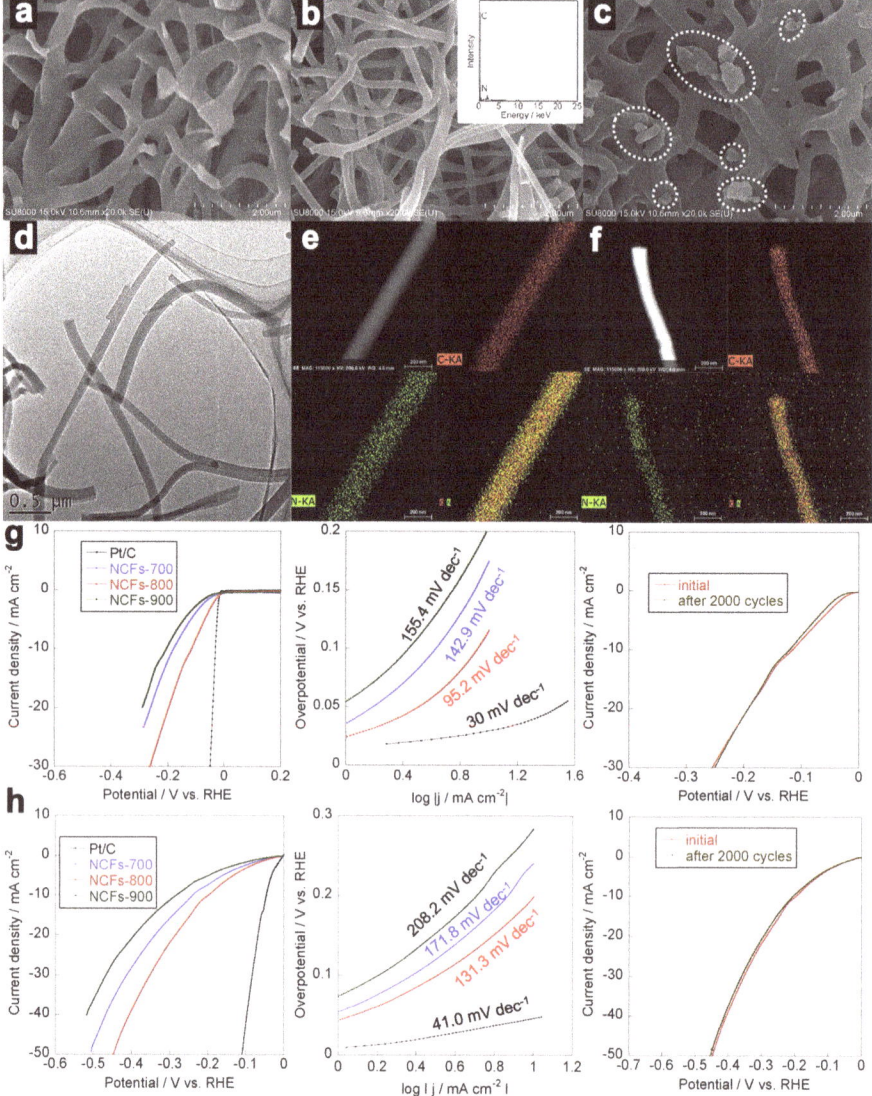

Figure 3.15. SEM images of (a) NCFs-700, (b) NCFs-800, and (c) NCFs-900. (d) TEM and the EDS mappings before (e) and after (f) a durability test of NCFs-800. (g) HER performance in 0.5 M H_2SO_4 and (h) 1 M KOH electrolyte [101], reproduced with permission © Elsevier.

diagnosis. FET sensors have attracted extensive attention from researchers due to their high sensitivity, fast detection speed, and simple test procedures. The working principle of the FET sensor is based on a change in the conductance of the FET channel that occurs when the target molecule is adsorbed [103]. The FET sensor is a resistive sensor with a gate voltage that is used to adjust the current in the FET channel. The concept of chemical sensing using FET devices has been demonstrated through the use of bulk semiconductor materials as channels, such as gas-sensitive

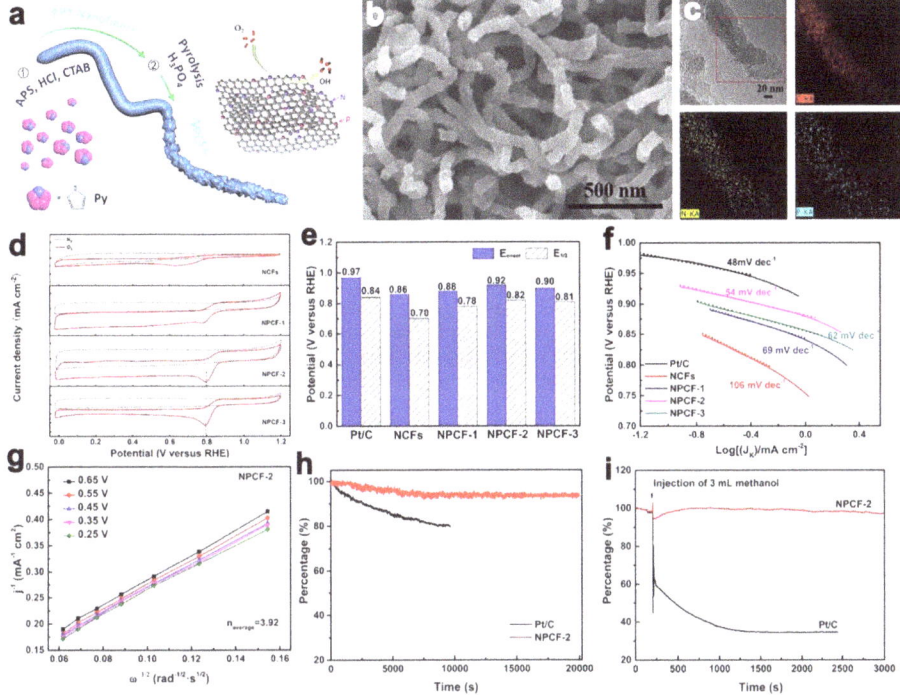

Figure 3.16. (a) The fabrication process used for NPCFs. (b) SEM and (c) element mapping images of the NPCF-2 sample. (d) CV curves and (e) bar plots of E_{onset} and $E_{1/2}$. (f) The Tafel slopes of these samples and (g) the Koutecky–Levich plots of NPCF-2. (h) The long-term stabilities and (i) methanol crossover effects of NPCF-2 and commercial Pt/C electrodes [102], reproduced with permission © Institute of Physics.

metal oxides and ion-sensitive polymer films. However, due to the inherent electronic properties of bulk materials and the limited interaction between the target molecule and the channel surface, FET sensors based on bulk materials generally have lower sensitivity or require specific operating conditions. In contrast, FET sensors that use semiconductor nanomaterials as the sensing channel show better performance. One-dimensional nanostructures, such as CNTs and CFs, have long been studied as channel materials in FET sensors [103]. For example, multiscale porous CFs were used as a field-effect transistor (FET) biosensor to detect Nesfatin-1; the biosensor exhibited high sensitivity for levels as low as 0.1 fM of Nesfatin-1, even in the presence of other interfering biomolecules (figure 3.17(a)) [104]. Kim *et al* proposed an FET sensor using aptamer-modified multichannel carbon nanofibers (MCNFs) to detect bisphenol A. The FET sensor exhibited a high sensitivity and specificity for bisphenol A at an extremely low concentration of 1 fM (figure 3.17(b)) [105].

3.4.5 Biomedicine

In the last two decades, carbon nanofibers have been widely used in various biomedical applications. By regulating their structure and properties, including

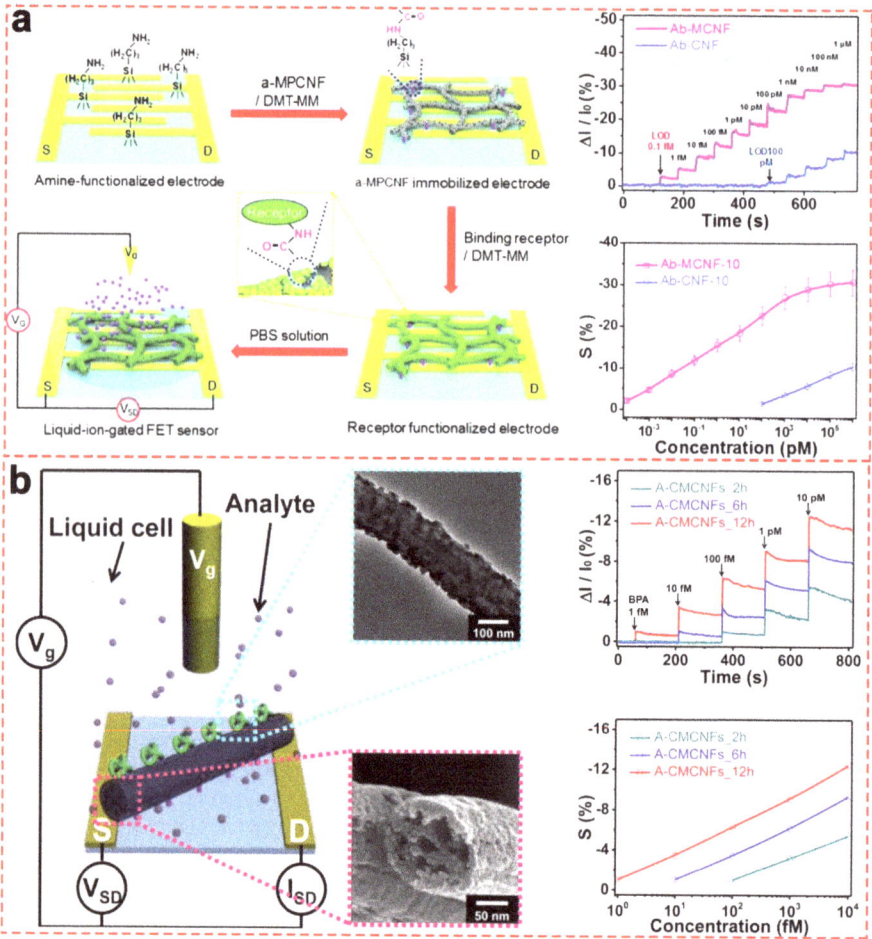

Figure 3.17. (a) The fabrication steps of the multiscale porous CF-based liquid-ion gated FET-type biosensor and real-time responses of the normalized current and the calibration curves [104], reproduced with permission © Royal Society of Chemistry. (b) Schematic diagram of the FET sensor constructed using MCNFs and the real-time responses of the normalized current and the calibration curves [105], reproduced with permission © American Chemical Society.

their diameter, porosity, arrangement, stacking, patterning, surface functional groups, mechanical properties, and biodegradability, both 2D and 3D scaffolds have been designed and manufactured to control cell migration or stem cell differentiation in order to enhance the repair or regeneration of nerves, skin, heart, blood vessels, musculoskeletal systems, and tissue interfaces. In addition, electrospun carbon nanofibers have been actively explored for cancer diagnosis and the construction of *in vitro* 3D tumor models for cancer research. They have also been reported to treat cancer by manipulating the migration of cancer cells using uniaxially arranged nanofibers and/or by incorporating drugs into nanofibers for controlled release. In addition, electrospun carbon nanofibers have been used as implant coatings, barrier membranes, filter membranes, etc. to improve or develop biomedical equipment [1]. Sun and his colleagues used a simple and feasible strategy to convert commercial fluorinated carbon (FC) fiber into nanoscale FC with good solubility in water and culture media, good biocompatibility, a high drug loading capacity, and enhanced thermal performance (figure 3.18) [106]. Moreover, cell experiments showed that through a suitable combination of chemotherapy and photothermal therapy, the constructed carbon nanofibers could easily be transferred into cells through endocytosis, exhibiting low toxicity and excellent cancer treatment effects.

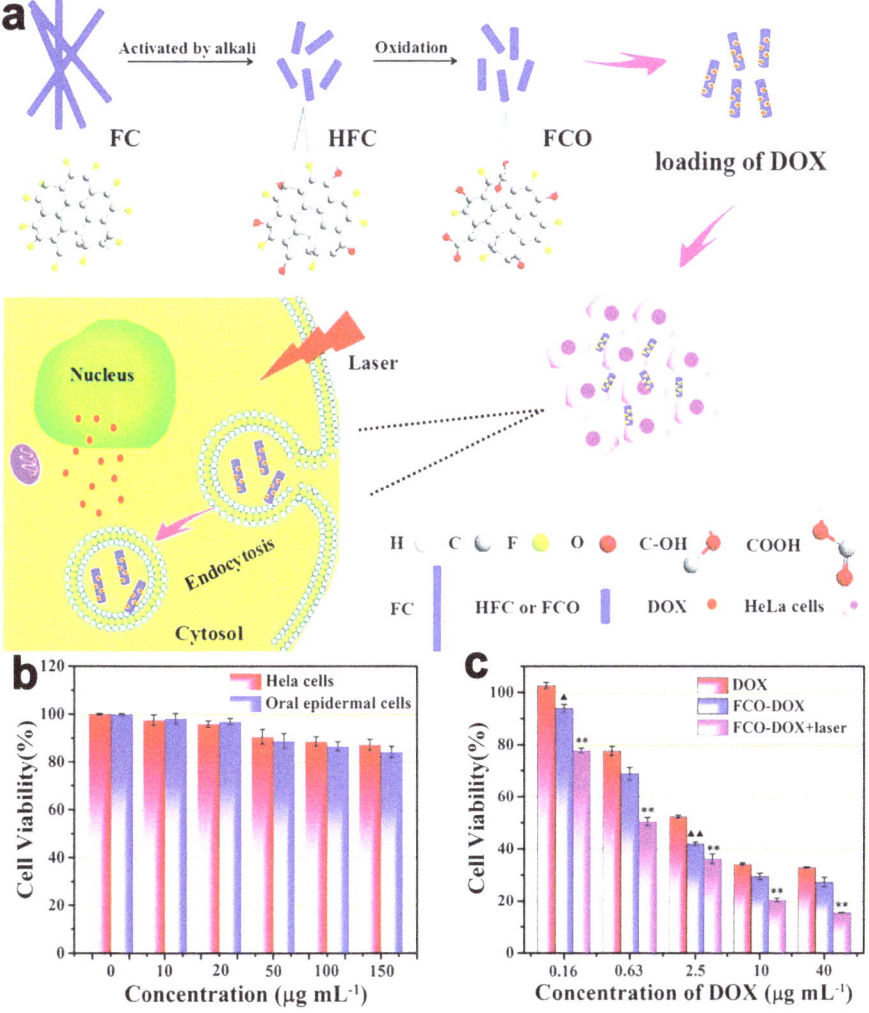

Figure 3.18. (a) Illustration of the synthesis of a sample and intracellular release combined with photothermal therapy of cancer cells. (b and c) Cell viabilities of HeLa cells and oral epidermal cells at different concentrations [106], reproduced with permission © Royal Society of Chemistry.

3.4.6 Airlines, sporting goods, and the automotive industry

CF-based composites have been applied in the fairings, flight control surfaces, landing gear doors, floor beams, floor boards, primary wing and fuselage structures of next-generation aircraft, etc. and account for about 50% or more of the materials derived from CF-based composites (figure 3.19(a)). For example, the Boeing 787 Dreamliner and Airbus A350 XWB contain nearly 50% and 53% of CF-based composite materials, respectively, while increasing the usage of CF-based composites significantly improves the fuel efficiency—by 25% according to Airbus [107].

In addition, 18%–20% of the CF market is accounted for by the sporting goods industry, as shown in figures 3.19(b) and (c). In terms of weight, about 46 000 tons of CFs are used in various sports equipment around the world; they are widely applied in golf club shafts, hockey sticks, rackets, fishing rods, ski poles, snowboards, sailboard masts, marine hulls, backpack frames, tent poles, bicycle frames, etc. The CFs are surface reacted with sodium hypochlorite and nitric acid, which is done to protect the fiber. The fibers are further impregnated with resin or epoxy resin, partially cured at 121 °C–176 °C to form the desired shape, and fully cured at 482 °C. The CFs are used in various sports equipment, which not only offers light weight but also improves sports performance [107].

CFs are also commonly used in the automotive industry. PAN-based CFs can be made in molds of different shapes and installed in cars. For example, lightweight CFs are used in the BMW-i3 electric vehicle. Compared with the conventional aluminum structure, the weight is reduced by nearly 550 pounds, which greatly improves the vehicle's mileage. In addition, the production time of cars can also be shortened due to the use of CF materials, which saves costs and lowers prices for merchants. Other commercially available CF-based marques include Aston Martin, Ferrari, Lamborghini, etc [107].

In addition to the abovementioned applications, there are some other interesting applications of CFs, such as micro/flexible/wearable/structured power supplies, electric actuators, electrochromic devices, capacitive desalination/deionization, memory storage, zinc/aluminum ion batteries, zinc–air batteries, solar cells, fuel cells, transistors, light-emitting devices, tissue repair or regeneration, cancer research, etc. which are also widely used [1, 11, 107].

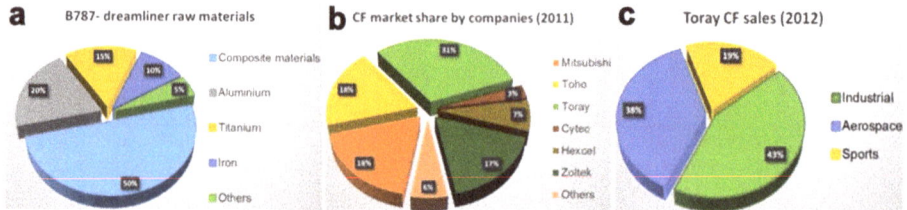

Figure 3.19. (a) Statistical data for the composition of the Boeing 787 Dreamliner. (b) CF market share by company and (c) market shares of Toray in various industries [107].

3.5 Perspectives

CF will be the main component of high-strength and high-performance materials in the future due to its unique physical and chemical properties, which can improve energy efficiency and payload size/range in different application fields. However, because the use of these CFs and CF composite materials is still in its early stages, they are still expensive. Therefore, the price of CFs and their composite materials is the focus of attention of researchers, companies, and end users in various fields.

Current research is focused on the development of new precursor materials, new processing technologies, and the efficient use of recycling to produce high-quality, low-cost CFs in large quantities. At present, the preparation of the most commonly used PAN-based CF precursor material accounts for almost 50% of the cost, and its price is quite high. Therefore, it is very important in the CF industry to develop new precursor materials and reduce the manufacturing cost of the precursors. At present, many research teams are studying new precursors of CFs, such as lignin, cellulose, rayon, and other polymers with high carbon contents. In addition, it is not only necessary to reduce the cost of fiber processing, but also to optimize fiber structures to support high-end applications. For the preparation of the precursor fibers and the carbonization and graphitization processes, it is necessary to develop convenient methods that offer easier control of the CF microstructures and mechanical and electrochemical properties while more effectively saving time and energy, reducing costs, and even allowing large-scale production. This also requires a more systematic and in-depth study of the effects of temperature and atmosphere on the evolution of the CF microstructure and its various mechanical properties. In addition, for demanding applications, stronger and harder CFs are needed, and it is hoped that through innovative technologies, the strengths and moduli of the fibers can be made much higher than those of current commercial fibers.

CFs have achieved great success in various fields because of their unique physical and chemical properties, such as energy storage, photonics, electronics, biomedicine, etc. In the future, with the rapid development of society, the requirements for CF use in different fields will also increase, so more efforts are needed to update devices and meet social needs, while controlling device sizes, weights, functions, and performances.

References

[1] Xue J, Wu T, Dai Y and Xia Y 2019 Electrospinning and electrospun nanofibers: methods, materials, and applications *Chem. Rev.* **119** 5298–415

[2] Rajak D K, Pagar D D, Kumar R and Pruncu C I 2019 Technology, recent progress of reinforcement materials: a comprehensive overview of composite materials *J. Mater. Res. Technol.* **8** 6354–74

[3] Zhang R, Gao B, Ma Q, Zhang J, Cui H and Liu L 2016 Directly grafting graphene oxide onto carbon fiber and the effect on the mechanical properties of carbon fiber composites *Mater. Design* **93** 364–9

[4] Frank E, Steudle L M, Ingildeev D, Spörl J M and Buchmeiser M R 2014 Carbon fibers: precursor systems, processing, structure, and properties *Angew. Chem. Int. Ed.* **53** 5262–98

[5] Yu D, Goh K, Wang H, Wei L, Jiang W, Zhang Q, Dai L and Chen Y J 2014 Scalable synthesis of hierarchically structured carbon nanotube–graphene fibres for capacitive energy storage *Nat. Nanotechnol.* **9** 555–62

[6] Liu L, Yu Y, Yan C, Li K and Zheng Z J 2015 Wearable energy-dense and power-dense supercapacitor yarns enabled by scalable graphene–metallic textile composite electrodes *Nat. Commun.* **6** 7260

[7] Liu J, Chen X, Liang D and Xie Q J 2020 Recovery, utilization; effects, E., Development of pitch-based carbon fibers: a review *Energy Source* A **2020** 1–21

[8] Ford C E and Mitchell C V 1963 *Fibrous graphite* US3107152A

[9] Nakamura O, Ohana T, Tazawa M, Yokota S, Shinoda W and Itoh J 2009 Study on the PAN carbon-fiber-innovation for modeling a successful R&D management–An excited-oscillation management model *Synthesiology Eng. Ed.* **2** 154–64

[10] Ōtani S J 1965 On the carbon fiber from the molten pyrolysis products *Carbon* **3** 31–8

[11] Chen S, Qiu L and Cheng H-M 2020 Carbon-based fibers for advanced electrochemical energy storage devices *Chem. Rev.* **120** 2811–78

[12] Fang W, Yang S, Wang X-L, Yuan T-Q and Sun R-C 2017 Manufacture and application of lignin-based carbon fibers (LCFs) and lignin-based carbon nanofibers (LCNFs) *Green Chem.* **19** 1794–827

[13] Greiner A and Wendorff J H 2007 Electrospinning: a fascinating method for the preparation of ultrathin fibers *Angew. Chem.* **46** 5670–703

[14] Wu D, Zhang W, Feng Y and Ma J 2020 Necklace-like carbon nanofibers encapsulating V_3S_4 microspheres for ultrafast and stable potassium-ion storage *J. Mater. Chem.* A **8** 2618–26

[15] Ye Q, Wu Y, Qi Y, Shi L, Huang S, Zhang L, Li M, Li W, Zeng X and Wo H 2019 Effects of liquid metal particles on performance of triboelectric nanogenerator with electrospun polyacrylonitrile fiber films *Nano Energy* **61** 381–8

[16] Bao Y, Tay Y S, Lim T-T, Wang R, Webster R D and Hu X 2019 Polyacrylonitrile (PAN)-induced carbon membrane with *in situ* encapsulated cobalt crystal for hybrid peroxymo-nosulfate oxidation-filtration process: preparation, characterization and performance evaluation *Chem. Eng. J.* **373** 425–36

[17] He J, He Y, Chen Y, Zhang X, Hu C, Zhuang J, Lei B and Liu Y 2018 Construction and multifunctional applications of carbon dots/PVA nanofibers with phosphorescence and thermally activated delayed fluorescence *Chem. Eng. J.* **347** 505–13

[18] Li Y, Li S and Sun J 2021 Degradable poly (vinyl alcohol)-based supramolecular plastics with high mechanical strength in a watery environment *Adv. Mater.* **33** 2007371

[19] Sheikh F A, Zargar M A, Tamboli A H and Kim H 2016 A super hydrophilic modification of poly(vinylidene fluoride) (PVDF) nanofibers: by *in situ* hydrothermal approach *Appl. Surf. Sci.* **385** 417–25

[20] Yang Z, Jia Y, Niu Y, Yong Z and Li Q 2020 Wet-spun PVDF nanofiber separator for direct fabrication of coaxial fiber-shaped supercapacitors *Chem. Eng. J.* **400** 125835

[21] Luo M, Ming Y, Wang L, Li Y, Li B, Chen J and Shi S 2018 Local delivery of deep marine fungus-derived equisetin from polyvinylpyrrolidone (PVP) nanofibers for anti-MRSA activity *Chem. Eng. J.* **350** 157–63

[22] Han M, Ge Y, Liu J, Cao Z, Li M, Duan X and Hu J 2020 Mixed polyvinyl pyrrolidone hydrogel-mediated synthesis of high-quality Ag nanowires for high-performance transparent conductors *J. Mater. Chem.* A **8** 21062–9

[23] Lee H-M, Kwac L-K, An K-H, Park S-J and Kim B-J 2016 Management, electrochemical behavior of pitch-based activated carbon fibers for electrochemical capacitors *Energy Convers. Manage.* **125** 347–52

[24] Naito K, Tanaka Y and Yang J-M 2017 Transverse compressive properties of polyacrylonitrile (PAN)-based and pitch-based single carbon fibers *Carbon* **118** 168–83

[25] Lee K H, Lee S H and Ruoff R S 2020 Synthesis of diamond-like carbon nanofiber films *ACS Nano* **14** 13663–72

[26] Zhang W, Cai G, Wu R, He Z, Yao H B, Jiang H L and Yu S H 2021 Templating synthesis of metal–organic framework nanofiber aerogels and their derived hollow porous carbon nanofibers for energy storage and conversion *Small* **17** 2004140

[27] Yang S, Cheng Y and Xiao X *et al* 2020 Development and application of carbon fiber in batteries *Chem. Eng. J.* **384** 123294

[28] Wang L, Zhang G, Zhang X, Shi H, Zeng W, Zhang H, Liu Q, Li C, Liu Q and Duan H 2017 Porous ultrathin carbon nanobubbles formed carbon nanofiber webs for high-performance flexible supercapacitors *J. Mater. Chem.* A **5** 14801–10

[29] Yan J, Dong K, Zhang Y, Wang X, Aboalhassan A A, Yu J and Ding B 2019 Multifunctional flexible membranes from sponge-like porous carbon nanofibers with high conductivity *Nat. Commun.* **10** 5584

[30] Qin W, Zhang B, Xu R and Wen X 2020 Process safety and environmental protection: an optimized solid phase denitrification filter by using activated carbon fibers for secondary effluent treatment *Process Saf. Environ.* **142** 99–108

[31] Zeng W, Shu L, Li Q, Chen S, Wang F and Tao X M 2014 Fiber-based wearable electronics: a review of materials, fabrication, devices, and applications *Adv. Mater.* **26** 5310–36

[32] Qin Y, Wang X and Wang Z L 2008 Microfibre–nanowire hybrid structure for energy scavenging *Nat.* **451** 809–13

[33] Noh S H, Eom W, Lee W J, Park H, Ambade S B, Kim S O and Han T H 2019 Joule heating-induced sp2-restoration in graphene fibers *Carbon* **142** 230–7

[34] Kouris P D, Huang X, Boot M D and Hensen E J 2018 Scaling-up catalytic depolymerisation of lignin: performance criteria for industrial operation *Top. Catal.* **61** 1901–11

[35] Zhu M, Liu H, Cao Q, Zheng H, Xu D, Guo H, Wang S, Li Y and Zhou J 2020 Engineering, electrospun lignin-based carbon nanofibers as supercapacitor electrodes *ACS Sustain. Chem. Eng.* **8** 12831–41

[36] García-Mateos F J, Ruiz-Rosas R, Rosas J M, Morallon E, Cazorla-Amorós D, Rodríguez-Mirasol J and Cordero T 2020 Activation of electrospun lignin-based carbon fibers and their performance as self-standing supercapacitor electrodes *Sep. Purif. Technol.* **241** 116724

[37] Liu H, Xu T, Liu K, Zhang M, Liu W, Li H, Du H and Si C 2021 Products, lignin-based electrodes for energy storage application *Ind. Crop. Prod.* **165** 113425

[38] Morgan P 2005 *Carbon Fibers and Their Composites* (Boca Raton, FL: CRC Press)

[39] Meng F, Lu W, Li Q, Byun J H, Oh Y and Chou T W 2015 Graphene-based fibers: a review *Adv. Mater.* **27** 5113–31

[40] Tian Q, Xu Z, Liu Y, Fang B, Peng L, Xi J, Li Z and Gao C 2017 Dry spinning approach to continuous graphene fibers with high toughness *Nanoscale* **9** 12335–42

[41] Wazir A H and Kakakhel L 2009 Preparation and characterization of pitch-based carbon fibers *New Carbon Mater.* **24** 83–8

[42] Zhong D, Zhou J and Wang Y 2021 Hollow-fiber membranes of block copolymers by melt spinning and selective swelling *J. Membrane Sci.* **632** 119374

[43] Willberg-Keyriläinen P, Rokkonen T, Malm T, Harlin A and Ropponen J 2020 Melt spinnability of long chain cellulose esters *J. Appl. Polym. Sci.* **137** 49588

[44] Li X, Chen W, Qian Q, Huang H, Chen Y, Wang Z, Chen Q, Yang J, Li J and Mai Y W 2021 Electrospinning-based strategies for battery materials *Adv. Energy Mater.* **11** 2000845

[45] Gilbert W 1958 *De Magnete* (North Chelmsford, MA: Courier Corporation)

[46] Formhals A 1934 Process and apparatus for preparing artificial threads *US Patent Specification* 1975504

[47] Zhang Y-Z, Wang Y, Cheng T, Yao L-Q, Li X, Lai W-Y and Huang W 2019 Printed supercapacitors: materials, printing and applications *Chem. Soc. Rev.* **48** 3229–64

[48] Choudhary N, Li C, Moore J, Nagaiah N, Zhai L, Jung Y and Thomas J 2017 Asymmetric supercapacitor electrodes and devices *Adv. Mater.* **29** 1605336

[49] Zhou Y *et al* 2021 Two-birds-one-stone: multifunctional supercapacitors beyond traditional energy storage *Energy Environ. Sci.* **14** 1854–96

[50] Wang F, Wu X, Yuan X, Liu Z, Zhang Y, Fu L, Zhu Y, Zhou Q, Wu Y and Huang W 2017 Latest advances in supercapacitors: from new electrode materials to novel device designs *Chem. Soc. Rev.* **46** 6816–54

[51] Simon P and Gogotsi Y 2010 Materials for electrochemical capacitors *Nat. Mater.* **7** 320–9

[52] He S, Hu Y, Wan J, Gao Q, Wang Y, Xie S, Qiu L, Wang C, Zheng G and Wang B 2017 Biocompatible carbon nanotube fibers for implantable supercapacitors *Carbon* **122** 162–7

[53] Zhang K, Zhang L L, Zhao X S and Wu J 2010 Graphene/polyaniline nanofiber composites as supercapacitor electrodes *Chem. Mater.* **22** 1392–401

[54] Ma W, Li W and Li M *et al* 2021 Unzipped carbon nanotube/graphene hybrid fiber with less 'dead volume' for ultrahigh volumetric energy density supercapacitors *Adv. Funct. Mater.* **31** 2100195

[55] Ma C, Wu L, Dirican M, Cheng H and Zhang X 2021 Carbon black-based porous sub-micron carbon fibers for flexible supercapacitors *Appl. Surf. Sci.* **537** 147914

[56] Zhang X, Li H and Qin B *et al* 2019 Direct synthesis of porous graphitic carbon sheets grafted on carbon fibers for high-performance supercapacitors *J. Mater. Chem.* A **7** 3298–306

[57] Jia D, Yu X and Tan H *et al* 2017 Hierarchical porous carbon with ordered straight micro-channels templated by continuous filament glass fiber arrays for high performance super-capacitors *J. Mater. Chem.* A **5** 1516–25

[58] Chen L-F, Lu Y, Yu L and Lou X W D 2017 Designed formation of hollow particle-based nitrogen-doped carbon nanofibers for high-performance supercapacitors *Energy Environ. Sci.* **10** 1777–83

[59] Gp A, Feng C A, Yz A and Xx B 2020 N-doped carbon nanofibers arrays as advanced electrodes for supercapacitors *J. Mater. Sci. Technol.* **55** 144–51

[60] Li Y, Wang G, Wei T, Fan Z and Yan P 2016 Nitrogen and sulfur co-doped porous carbon nanosheets derived from willow catkin for supercapacitors *Nano Energy* **19** 165–75

[61] Fu M, Lv R, Lei Y and Terrones M 2021 Ultralight flexible electrodes of nitrogen-doped carbon macrotube sponges for high-performance supercapacitors *Small* **17** 2004827

[62] Miao F, Shao C, Li X, Wang K and Liu Y 2016 Flexible solid-state supercapacitors based on freestanding nitrogen-doped porous carbon nanofibers derived from electrospun poly-acrylonitrile@ polyaniline nanofibers *J. Mater. Chem.* A **4** 4180–7

[63] Chen L-F, Zhang X-D, Liang H-W, Kong M, Guan Q-F, Chen P, Wu Z-Y and Yu S-H 2012 Synthesis of nitrogen-doped porous carbon nanofibers as an efficient electrode material for supercapacitors *ACS Nano* **6** 7092–102

[64] Liu C, Liu J, Wang J, Li J, Luo R, Shen J, Sun X, Han W and Wang L 2018 Electrospun mulberry-like hierarchical carbon fiber web for high-performance supercapacitors *J. Colloid Interf. Sci.* **512** 713

[65] Shi L, Ye J, Lu H, Wang G and Ning G 2021 Flexible all-solid-state supercapacitors based on boron and nitrogen-doped carbon network anchored on carbon fiber cloth *Chem. Eng. J.* **410** 128365

[66] Bai Y, Muralidharan N and Sun Y K *et al* 2020 Energy and environmental aspects in recycling lithium-ion batteries: concept of battery identity global passport *Mater. Today* **41** 304–15

[67] Vaalma C, Buchholz D, Weil M and Passerini S 2018 A cost and resource analysis of sodium-ion batteries *Nat. Rev. Mater.* **3** 18013

[68] Hosaka T, Kubota K and Hameed A S *et al* 2020 Research development on K-ion batteries *Chem. Rev.* **120** 6358–466

[69] Xie X, Mao M and Qi S *et al* 2019 ReS$_2$-based electrode materials for alkali-metal ion batteries *CrystEngComm* **21** 3755–69

[70] Wang J, Feng L, Qu Y, Yang L, Yang Y, Li W and Zhao M 2018 PNTCDA: a promising versatile organic electrode material for alkali-metal ion batteries *J. Mater. Chem.* A **6** 24869–76

[71] Li F, Qu Y and Zhao M 2016 Germanium sulfide nanosheet: a universal anode material for alkali metal ion batteries *J. Mater. Chem.* A **4** 8905–12

[72] Ghosh S, Bhattacharjee U and Patchaiyappan S *et al* 2021 Multifunctional utilization of pitch-coated carbon fibers in lithium-based rechargeable batteries *Adv. Energy Mater.* **11** 2100135

[73] Savignac L, Danis A S and Charbonneau M *et al* 2021 Valorization of carbon fiber waste from the aeronautics sector: an application in Li-ion batteries *Green Chem.* **23** 2464–70

[74] Ma J L, Meng F L, Xu D and Zhang X B 2016 Co-Embedded N-Doped carbon fibers as highly efficient and binder-free cathode for Na-O$_2$ batteries *Energy Stor. Mater.* **6** 1–8

[75] Wang H F, Tang C, Wang B, Li B Q, Cui X and Zhang Q 2018 Defect-rich carbon fiber electrocatalysts with porous graphene skin for flexible solid-state zinc–air batteries *Energy Stor. Mater.* **15** 124-30

[76] Kong Y, Nanjundan A K and Liu Y *et al* 2019 Modulating ion diffusivity and electrode conductivity of carbon nanotube@ mesoporous carbon fibers for high performance aluminum–selenium batteries *Small* **15** 1904310

[77] Liu D, Yang L, Chen Z, Zou G, Hou H and Ji X 2020 Ultra-stable Sb confined into N-doped carbon fibers anodes for high-performance potassium-ion batteries *Sci. Bulletin* **65** 1003–12

[78] Tang J, Etacheri V and Pol V G 2016 Wild fungus derived carbon fibers and hybrids as anodes for lithium-ion batteries *ACS Sustain. Chem. Eng.* **4** 2624–31

[79] Yang Y, Sun W, Zhang J, Yue X, Wang Z and Sun K 2016 High rate and stable cycling of lithium–sulfur batteries with carbon fiber cloth interlayer *Electrochim. Acta* **209** 691–9

[80] Lin L, Pei F, Peng J, Fu A, Cui J, Fang X and Zheng N 2018 Fiber network composed of interconnected yolk-shell carbon nanospheres for high-performance lithium–sulfur batteries *Nano Energy* **54** 50–8

[81] Zhang X, Zhong Y, Xia X, Xia Y, Wang D, Zhou C, Tang W, Wang X, Wu J B and Tu J 2018 Metal-embedded porous graphitic carbon fibers fabricated from bamboo sticks as a novel cathode for lithium–sulfur batteries *ACS Appl. Mater. Interf.* **10** 13598

[82] Lee J S, Kim W and Jang J *et al* 2017 Sulfur-embedded activated multichannel carbon nanofiber composites for long-life, high-rate lithium–sulfur batteries *Adv. Energy Mater.* **7** 1601943

[83] Han H, Chen X and Qian J *et al* 2019 Hollow carbon nanofibers as high-performance anode materials for sodium-ion batteries *Nanoscale* **11** 21999–2005

[84] Liu Y, Liu Q and Jian C *et al* 2020 Red-phosphorus-impregnated carbon nanofibers for sodium-ion batteries and liquefaction of red phosphorus *Nat. Commun.* **11** 2520

[85] Ma L, Cao M and song Zhao C *et al* 2021 The novel N-rich hard carbon nanofiber as high-performance electrode materials for sodium-ion batteries *Ceram. Int.* **47** 9118–24

[86] Wang Z, Qie L, Yuan L, Zhang W, Hu X and Huang Y 2013 Functionalized N-doped interconnected carbon nanofibers as an anode material for sodium-ion storage with excellent performance *Carbon* **55** 328–34

[87] Wang *et al* 2015 A high performance sulfur-doped disordered carbon anode for sodium ion batteries *Energy Environm. Sci.* **8** 2916–21

[88] Sun X, Wang C and Gong Y *et al* 2018 A flexible sulfur-enriched nitrogen doped multichannel hollow carbon nanofibers film for high performance sodium storage *Small* **14** 1802218

[89] Hu X, Zhong G and Li J *et al* 2020 Hierarchical porous carbon nanofibers for compatible anode and cathode of potassium-ion hybrid capacitor *Energy Environm. Sci.* **13** 2431–40

[90] Zhao X, Xiong P and Meng J *et al* 2017 High rate and long cycle life porous carbon nanofiber paper anodes for potassium-ion batteries *J. Mater. Chem.* A **5** 19237–44

[91] Wu D, Zhang W, Feng Y and Ma J 2020 Necklace-like carbon nanofibers encapsulating V$_3$S$_4$ microspheres for ultrafast and stable potassium-ion storage *J. Mater. Chem.* A **8** 2618–26

[92] Zhu Y, Wang M, Zhang Y, Wang R and Wang C 2021 Nitrogen/oxygen dual-doped hierarchically porous carbon/graphene composite as high-performance anode for potassium storage *Electrochim. Acta* **377** 138093

[93] Zheng F, Niu P, Xu Y, Li Z, Wei L, Yao G and Wang J 2021 Tuning the electronic conductivity of porous nitrogen-doped carbon nanofibers with graphene for high-performance potassium-ion storage *Inorg. Chem. Front.* **8** 3926–33

[94] Wu Y, Hu S and Xu R *et al* 2019 Boosting potassium-ion battery performance by encapsulating red phosphorus in free-standing nitrogen-doped porous hollow carbon nanofibers *Nano Lett.* **19** 1351–58

[95] Zhao G, Rui K, Dou S X and Sun W 2018 Heterostructures for electrochemical hydrogen evolution reaction: a review *Adv. Funct. Mater.* **28** 1803291

[96] Sato Y, Kowalski D and Aoki Y *et al* 2020 Long-term durability of platelet-type carbon nanofibers for OER and ORR in highly alkaline media *Appl. Catal. Gen.* **597** 117555

[97] Strmcnik D, Lopes P P, Genorio B, Stamenkovic V R and Markovic N M 2016 Design principles for hydrogen evolution reaction catalyst materials *Nano Energy* **29** 29–36

[98] Zhang L, Xiao J, Wang H and Shao M 2017 Carbon-based electrocatalysts for hydrogen and oxygen evolution reactions *ACS Catal.* **7** 7855–65

[99] Zhang J, Qu L, Shi G, Liu J, Chen J and Dai L 2016 N, P-codoped carbon networks as efficient metal-free bifunctional catalysts for oxygen reduction and hydrogen evolution reactions *Angew. Chem. Int. Edit.* **128** 2270–4

[100] Wang M 2017 Plasma-induced nanoporous metal oxides with nitrogen doping for high-performance electrocatalysis *Nanotechnology* **28** 242501

[101] Sun J, Ge Q, Guo L and Yang Z 2020 Nitrogen doped carbon fibers derived from carbonization of electrospun polyacrylonitrile as efficient metal-free HER electrocatalyst *Int. J. Hydrogen Energy* **45** 4035–42

[102] Xie Y, Yang Y, Zhou Q, Wang L, Yao Y, Zhu X, Fang L, Zhao H and Zhang J 2019 Metal-free N, P-codoped porous carbon fibers for oxygen reduction reactions *J. Electrochem. Soc.* **166** H549–55

[103] Mao S, Chang J, Pu H, Lu G, He Q, Zhang H and Chen J 2017 Two-dimensional nanomaterial-based field-effect transistors for chemical and biological sensing *Chem. Soc. Rev.* **46** 6872–904

[104] Kim S G and Lee J S 2021 Multiscale pore contained carbon nanofiber-based field-effect transistor biosensors for nesfatin-1 detection *J. Mater. Chem.* B **9** 6076–83

[105] Kim S G, Lee J S, Jun J, Shin D H and Jang J 2016 interfaces, ultrasensitive bisphenol a field-effect transistor sensor using an aptamer-modified multichannel carbon nanofiber transducer *ACS Appl. Mater. Inter.* **8** 6602–10

[106] Sun L, Gong P, Liu X, Pang M, Tian M, Chen J, Du J and Liu Z 2017 Fluorinated carbon fiber as a novel nanocarrier for cancer chemo-photothermal therapy *J. Mater. Chem.* B **5** 6128–37

[107] Hiremath N, Mays J and Bhat G 2017 Recent developments in carbon fibers and carbon nanotube-based fibers: a review *Polym. Rev.* **57** 339–68

[108] Hunt M A, Saito T, Brown R H, Kumbhar A S and Naskar A K 2012 Patterned functional carbon fibers from polyethylene *Adv. Mater.* **24** 2386–9

Chapter 4

Carbon nanotubes and the related applications

Yongde Long[†], Fenghui Ye[†] and Chuangang Hu[*]

Carbon nanotubes (CNTs), consisted of carbon hexagons arranged in a concentric manner, usually own a diameter ranging from a few angstroms to tens of nanometers and a length of up to centimeters. The hollow geometry and the conjugated all-carbon structure allow CNTs to exhibit various new intriguing electrical, mechanical, and thermal properties, which make the CNTs are promising for many applications, ranging from composite materials to electronic and biomedical devices. Due to the high specific surface area, mesoporous structure, and good electrical property, CNTs can also be used as high-performance catalysts for the applications in energy-related areas, such as supercapacitors, fuel cells, batteries, and solar cells will be discussed in succeeding sections as appropriate.

4.1 Introduction

CNTs are tubes with a nanometer-scale outer diameter made of carbon atoms. In 1991, Iijima proposed the concept of 'helical microtubules of graphitic carbon' (figure 4.1(a)) [1] which aroused great interest in CNTs and opened a new era in materials science and technology. A little later, Iijima [2] and Bethune [3] described single-walled CNTs (SWCNTs, figure 4.1(b)) that were obtained using Fe or Co as a catalyst in an arc-discharge apparatus. As a result of their peculiar hollow geometry, coupled with a conjugated all-carbon structure, CNTs are endowed with unique and excellent electrical, mechanical, thermal, and chemical properties. Therefore, CNTs are widely used in fields ranging from condensed-matter physics to chemistry, and in both academia and industry, for example in nanoelectronic materials and devices, biology, chemical engineering, catalysts, energy storage and conversion, and catalysts. The structure, production, physical properties, functionalization, and applications of CNTs are discussed in this chapter.

[†] These authors contributed equally to this work.
[*] Corresponding author. Email: chuangang.hu@mail.buct.edu.cn

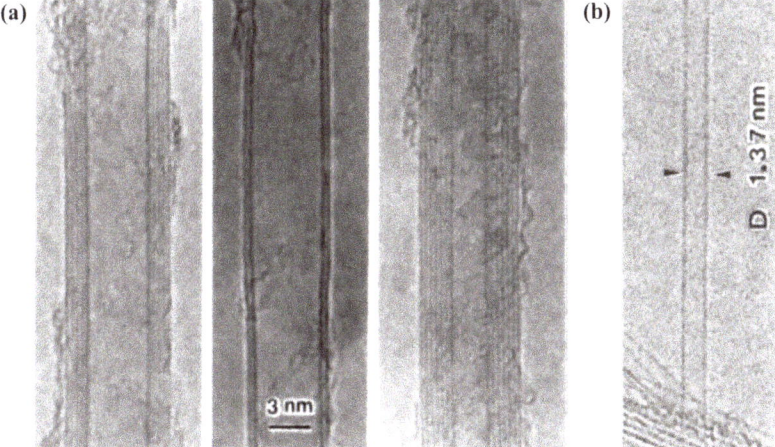

Figure 4.1. (a) High-resolution transmission electron microscope (HR-TEM) images of multiwalled CNTs (diameter 6.7, 5.5 and 6.5 nm, respectively) [1], reproduced with permission © Springer Nature. (b) HR-TEM images of single-walled CNTs [2], reproduced with permission © Springer Nature.

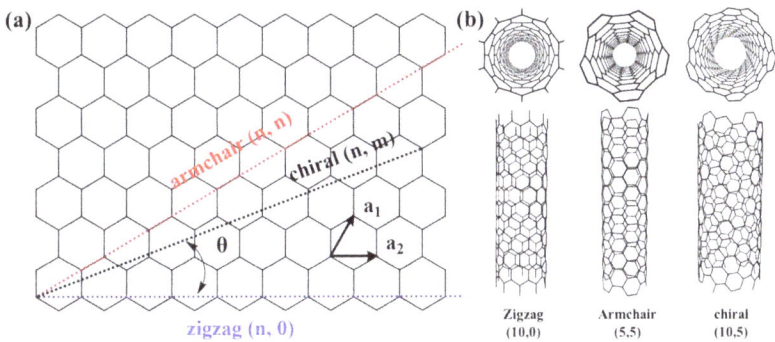

Figure 4.2. (a) A 2D graphene sheet is shown along with the unit vectors $\vec{a}_1$ and $\vec{a}_2$, the chiral vector $\vec{C}_h$, the chiral angle θ, and the descriptors n and m. (b) A schematic representation of CNTs: (10,0) zigzag nanotube; (5,5) armchair nanotube; (10,5) chiral nanotube.

4.2 Structure

CNTs can be viewed as rolled up graphene sheets with a honeycomb lattice of sp^2 hybridized carbon atoms. SWCNTs are rolled from a single layer of graphene, while MWCNTs consist of multiple rolled layers of graphene. SWCNTs can be completely described through the so-called chiral vector $\vec{C}_h$ (figure 4.2(a)), which is given by the equation $\vec{C}_h = n\vec{a}_1 + m\vec{a}_2$ ($n \geqslant m$), where $\vec{a}_1$ and $\vec{a}_2$ are unit vectors in the two-dimensional (2D) hexagonal lattice [4] and the integers (n,m) are the numbers of steps along the zigzag carbon bonds [5]. The angle between $\vec{C}_h$ and $\vec{a}_1$, θ is known as the chiral angle. There are three types of CNTs, including zigzag, armchair, and chiral CNTs, depending on the chiral vector $\vec{C}_h$ (figure 4.2(b)) [6]. When $n = m$ or $\theta = 30°$, the CNTs are known as armchair nanotubes, whereas zigzag CNTs correspond to either $m = 0$ or $\theta = 0°$. In addition, a large number of CNTs formed with $0 < \theta < 30°$ or $n > m > 0$ are known as chiral nanotubes. The chirality and diameter in turn depend on the integers (n,m), which have the greatest impact on the electronic band structure [4, 5]. Theoretical calculations have indicated that (n,m) CNTs are metallic if $2n + m = 3q$ where q is an integer, while the other CNTs are either small or moderate bandgap semiconductors [7].

The structures of MWCNTs are very complicated and cannot be completely described using just the chiral vector used for SWCNTs. Two models can be used to illustrate the structure of MWCNTs, i.e. the Russian doll model and the parchment model. In the Russian doll model [8], sheets of graphene are arranged in perfectly concentric cylinders, whereas in the parchment model, a single graphene sheet is rolled up on itself, like a scroll of parchment or a rolled newspaper [9]. Normally, MWCNTs are considered to be composed of concentric SWCNTs, because electron micrographs show that the space between individual tubes is usually 0.34 nm, which is approximately the same as in disordered graphite, and the reflexes of electron diffraction in the TEM also indicate that the composition of MWCNT is concentric and consists of independent tubes [10].

4.3 Preparation

In recent decades, a multitude of techniques have been developed to produce CNTs, including arc discharge, laser ablation, and chemical vapor deposition (CVD). Enormous efforts have been made to grow CNTs with the desired structure through high-purity, high-quality, large-scale production.

4.3.1 Arc discharge

The arc-discharge method developed by Kratschmer and his colleagues for the preparation of macroscopic quantities of fullerenes is suitable for the production of nanotubes as well [10, 11]. A typical arc-discharge apparatus [12, 13] is shown in figure 4.3(a). In 1991, during the process for preparing C_{60} by arc discharge, Iijima observed MWCNTs in the cathode product using HR-TEM [1]. In 1993, they

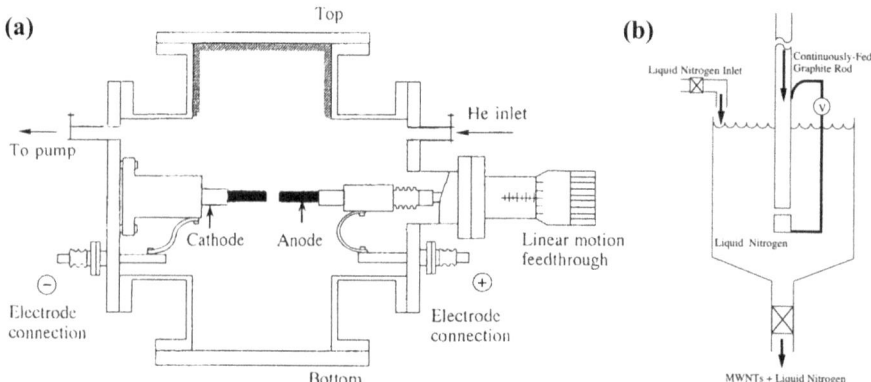

Figure 4.3. (a) Schematic representation of a typical arc-discharge apparatus [12], reproduced with permission © AIP Publishing. (b) Schematic of an arc-discharge generator with a liquid-nitrogen reaction chamber for the continuous synthesis of CNTs [14], reproduced with permission © Elsevier.

prepared SWCNTs by a similar method but with iron (Fe) as a catalyst [2]. In the same year of 1993, Bethune *et al* [3] obtained a small amount of SWCNTs by arc discharge using cobalt (Co) as the catalyst. Since then, the arc-discharge method has been extensively investigated for the preparation of CNTs.

In 1992, Ebbesen *et al* increased the yield of CNTs to ca. 75% by adjusting the type of inert gas, the direct or alternating current, the applied voltage, and the relative size of the electrodes; their report [15] was the first on the large-scale production of MWCNTs. To improve the quality of CNTs and reduce the sintering of CNTs on the cathode, Colbert *et al* [16] connected the graphite cathode to a water-cooled copper cathode base, which reduced the defects of CNTs greatly, permitting the production of CNTs longer than 40 micrometers. In addition to optimizing the process of the arc-discharge method, for example, by changing the type of inert gas (such as He, H_2, Ar, N_2), the current, the applied voltage, or the relative size and angle of the electrodes, researchers also simplified the device by changing the arc-discharge medium to be a liquid solution (e.g. N_2 or water). Ishigami *et al* [14] then reported the simplified arc-discharge method, which allows for the continuous synthesis of MWCNTs as shown in figure 4.3(b). This method only requires a DC power supply, a graphite electrode, and a container of liquid N_2; it does not require the pumps, seals, water-cooled vacuum chambers, or purge-gas handling systems which are necessary for CNT production by the conventional arc-discharge method. Subsequently, Zhu's group [17] and Amaratunga *et al* [18] have shown that CNTs can also be made by arc discharge under water. They found that the quality of CNTs produced under water was significantly improved compared to those produced under liquid N_2.

MWCNTs are commonly produced via the arc-discharge method without the use of catalysts, which are necessary for the growth of SWCNTs. The catalysts used include transition metals such as Fe, Co, Ni, rare-earth metals such as Y and Gd, and noble metals such as Pt, Pd, or metal mixtures (e.g., Co/Pt, Ni/Y). The products prepared by the arc method contained a large amount of impurities such as amorphous carbon, metal-based catalyst particles, and fullerenes, while the yield of CNTs was meager. In 1997, Journet *et al* [19] improved the traditional arc-discharge apparatus, using Ni/Y as a catalyst to prepare SWCNTs with a purity of 70% to 90%. This method is currently used worldwide for the high-yield production of SWCNTs. In order to further improve the yield and quality of SWCNTs, Liu and coworkers [20] developed a hydrogen arc-discharge method for the semicontinuous synthesis of SWCNTs. Using H_2 in the arc-discharge apparatus, the impurities of amorphous carbon can be selectively etched to improve the quality of the SWCNTs. Generally, due to the higher temperature in the arc zone, the CNTs prepared by the arc-discharge method have a high degree of crystallization. However, the electric arc-discharge method has poor control of the microstructure (e.g. diameter and chirality) of the CNTs produced, and the high-purity graphite electrodes, metal powders, and ultrahigh-purity inert gases involved in the process induce high costs.

4.3.2 Laser ablation

In the laser ablation method, a graphite target is irradiated by a focused laser beam to evaporate carbon atoms at high temperatures under an inert atmosphere to generate

CNTs. In 1985, Kroto and his colleagues [21] produced the first molecule of C_{60} composed of 60 carbon atoms in an experiment that evaporated graphite by laser ablation in a helium stream. SWCNTs were obtained by adding a certain amount of catalyst to the graphite, and MWCNTs could be formed when a pure graphite target was used [22, 23]. Meanwhile, Guo *et al* [22] reported that the yield of SWCNTs largely depended on the type of metal catalyst used and the temperature during the laser ablation processes. At 1200 °C, for the single metal catalysts, the highest yield of CNT was obtained by using Ni as the catalyst without forming MWCNTs, followed by Co as the catalyst. It was also found that Pt produced few CNTs, while no CNTs were observed when Cu or Nb were used separately as catalysts [22]. Moreover, for bimetallic catalysts, Co/Ni and Co/Pt mixtures produced SWCNTs at the same abundance, and an SWCNT yield 10–100 times as much as that obtained using the single metal counterparts. For example, in 1996, a double laser pulse was used to prepare SWCNTs with yields of more than 70% using a Co/Ni mixture as the catalyst at 1200 °C [24]. The resulting SWCNTs appeared to be continuous and free of defects without any associated metal particles. For the first time, a relatively large number of SWCNTs (up to 1 g per day [13]) were obtained via this method, which provided a material basis for studying the physical chemistry properties of SWCNTs. These SWCNTs were metallic, and (10,10) may have been the dominant component. In another independent experiment, Dai's group also reported that the laser ablation method preferentially grows metallic single-walled nanotubes (SWNTs) [25]. Unfortunately, the laser ablation method is not suitable for the synthesis of CNTs on a large scale as the equipment is complicated and expensive, has high energy consumption (powerful laser), and still produces low yields of CNTs.

4.3.3 Chemical vapor deposition

The chemical vapor deposition (CVD) technology involves the thermal decomposition of a carbon source (such as CH_4, C_2H_2, C_2H_4, C_6H_6, CO, CH_3OH, etc.) vapor in the presence of a catalyst (Co, Fe, Ni, etc), which then achieves the nucleation and growth of CNTs. Compared to the arc discharge and laser ablation methods, this method has the advantages of low cost, high yield and purity, and easy control of the experimental conditions; it can also define the structure of the CNTs produced.

The CVD method can produce both SWCNTs and MWCNTs whose structure can be precisely defined, including the diameter, length, number of walls, electrical behavior (metallic or semiconductor), chirality, and macrostructure (array, film, or fiber), by adjusting the preparation parameters such as the catalysts and their supports, carbon sources [26], gas flow rate, deposition temperature, growth time, etc. Among these parameters, the design of the catalyst plays an important role in the control of the CNT structure (figure 4.4). According to the experimental and theoretical results, Sinnott *et al* [27] suggested that the growth of SWCNTs or MWCNTs depended on the diameter of the catalytic particles. Liu's group found that there was an upper limit for the diameter of the catalytic particles (between 4 and 8 nm) that could be used to nucleate SWCNTs [28]. This phenomenon was also observed by Dai's group: small catalyst nanoparticles (<~1.8 nm) tended to be active in producing SWCNTs, while no

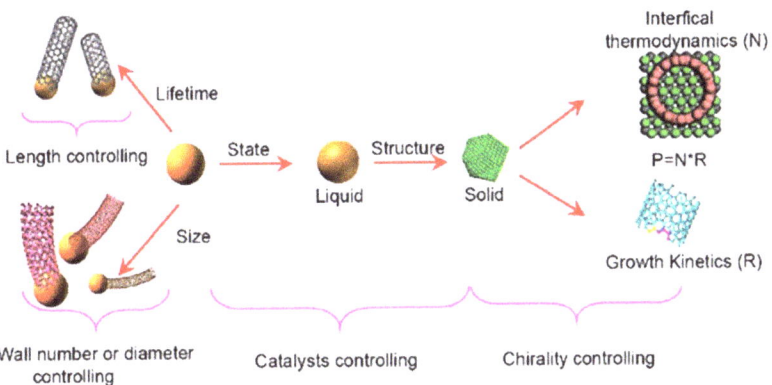

Figure 4.4. Relationship between catalyst design and CNT structure [33], reproduced with permission © Acta Phys. -Chim. Sin.

SWCNTs were formed by large catalyst particles with diameters centered around ~7 nm [29]. Cheung *et al* [30] prepared nearly monodisperse iron nanoclusters with average diameters of 3, 9, and 13 nm used for the growth of CNTs, the average diameters of which were 3, 7, and 12 nm, respectively. They also observed that both SWCNTs and double-walled carbon nanotubes (DWCNTs) were nucleated from small iron nanoclusters (3 nm), SWCNTs and thin-walled MWCNTs (two to four layers) were formed from nanoclusters with a diameter of 9 nm, whereas only MWNTs were observed to be formed from the 13 nm nanoclusters of metal catalysts. It was proposed that the partial pressure of the carbon source played an important role in controlling the diameter of CNT growth [30]. In addition, the catalytic activity that promoted CNT growth (e.g. the growth of ultra-long CNTs) could be tuned by varying the growth parameters, namely, the catalyst's composition and size, the growth temperature, the carbon feedstock, and even the substrate [31]. Under optimized conditions, 550 mm long CNTs with perfect structures were successfully synthesized by Wei and his colleagues [31]. The substrates of the catalysts mainly affect the growth mode, including the tip and/or base growth mechanism. If there is a strong interaction between the substrate and the catalyst, the particle remains anchored to the surface, resulting in the base growth of CNTs. If the interaction is between the substrate and the catalyst, tip growth occurs [13, 27, 32].

In addition to the diameter, length, and number of walls, the electrical behavior of CNTs can be controlled during their CVD growth. In 2009, by introducing a UV beam into a CVD system to destroy metallic SWCNTs (m-SWCNTs), Zhang *et al* [34] obtained a well-aligned semiconducting SWCNT (s-SWCNT) array. Later, this group also presented a simple way of preparing s-SWCNTs using water vapor as a weak oxidant to etch the m-SWCNTs during or after CNT growth [35]. In 2011, Cheng's group synthesized s-SWCNTs with an average diameter of 1.6 nm by *in situ* etching m-SWCNTs with oxygen [36]. However, O_2 can not only remove m-SWCNTs but also etch s-SWCNTs, causing defects in s-SWCNTs and reducing product quality. Therefore, they used H_2 to selectively etch m-SWCNTs *in situ* to

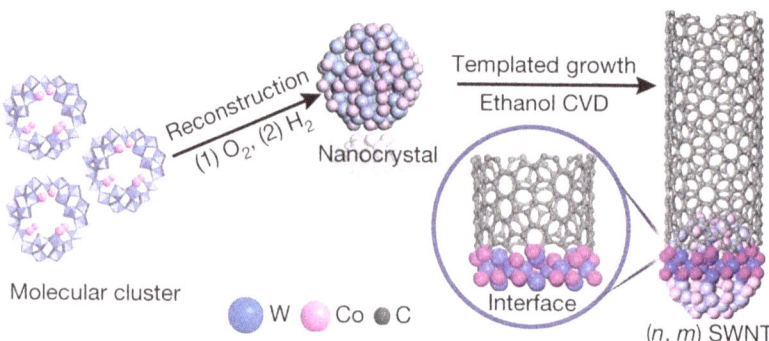

Figure 4.5. Preparation of the nanocrystal catalyst W_6Co_7 and the templated growth of an SWNT with specified (n,m) [42], reproduced with permission © Springer Nature.

prepare s-SWCNTs with high quality and high purity (~93.0%) [37]. Zhang *et al* [38] investigated the selectivity of different etchants, namely, O_2, H_2O, and CO_2, for SWCNTs through *in situ* polarized optical microscopy and found that the etching selectivity of m-SWCNTs followed the order: $H_2O > CO_2 > O_2$.

Furthermore, the CVD method can also control the chirality of CNTs. Traditional metal catalysts (Fe/Ru [39], Fe/Co [40], Co/Mo [41]) are called 'liquid catalysts', as their low melting points are caused by their small particle sizes (figure 4.4). Due to the structural instability of these traditional metal catalysts, there are many possibilities for the growth of SWCNTs during CVD deposition, with the result that the chiral selectivity of the prepared CNTs is unsatisfactory. Therefore, it is necessary to select metal or metal compounds with high melting points and good stability, such as W_6Co_7, WC, W/Co, and Mo_2C, to improve the selectivity of chiral CNTs. In 2014, Li *et al* [42] used alloy nanocrystals of tungsten and cobalt alloy nanoparticles (W_6Co_7) prepared from molecular clusters with specific structural characteristics and high structural stability as a catalyst and ethanol as a carbon source; they obtained (12,6) SWCNTs with high selectivity of more than 92% (figure 4.5), although the density was relatively low. Later, Zhang's group designed Mo_2C and WC solid catalysts with uniform sizes and melting points, using which, (8,4) and (12,6) SWNT arrays were grown with high density and selectivity [43].

Despite the many advantages of the CVD method mentioned above, there are still many growth-related challenges. The quality of CVD-produced CNTs, especially MWCNTs, is still inferior to those made by arc discharge, and there are many defects and impurities in the prepared CNTs, which hinder their further application.

4.3.4 Other techniques

In addition to the abovementioned methods for the production of CNTs, researchers have also developed many other preparation methods, such as high-pressure CO disproportionation (HiPco) [44], plasma-enhanced CVD (PECVD), the electrolytic method [45–47], ball milling [48], and diffusion flame synthesis [49–51], in which HiPco and PECVD are well studied and widespread.

In 1999, Smalley's group reported the HiPco method for the first time, which is an important technique for the catalytic synthesis of SWCNTs on a large scale [44]. The organometallic compound ($Fe(CO)_5$) is delivered by a high-pressure cold CO flow and meets hot CO in the reaction zone; it then decomposes to form small iron clusters that enable the catalytic growth of SWNTs from CO. The direct injection of the catalyst precursor into the reactant feedstock significantly promotes the continuous production of CNTs, which is one of the major advantages of the HiPco process. PECVD has been widely investigated for the synthesis of CNTs due to its advantages of lower growth temperatures with respect to regular CVD and the ability to control the growth direction of CNTs (vertical alignment and/or patterning [52, 53]). A typical PECVD apparatus consists of two parallel plate electrodes to which a high-frequency voltage is applied, and the substrate is placed on the grounded electrode for the synthesis of CNTs. MWCNTs were grown on nickel substrates by PECVD at temperatures below 666 °C, as reported by Ren *et al* in 1998. Then, in 2004, Dai's group described the production of high-quality s-SWCNTs at a good yield (nearly 90%) by the PECVD method at 600 °C [25]. In a later work, Qu *et al* [54] increased the yield of s-SWCNTs to 96%, using a thin film of Fe (0.5 nm) sputter-coated onto a 10 nm-thick Al layer precoated on the SiO_2 surface of a SiO_2/Si wafer as the catalyst.

4.4 Physical properties

4.4.1 Electrical properties

Studies [54–56] have shown that the transportation of electrons in CNTs takes place via ballistic transport, which means there is no energy loss, while joule heat is generated when electrons pass along a regular conductor (such as Cu). In other words, CNTs can conduct a large current without getting hot, which is ideal for building nanoscale circuits. SWCNTs can be either metallic or semiconducting in their electrical behavior, depending on their chirality and diameter. More attractively, the electrical properties of CNTs can be altered by introducing dopants, such as alkali metals, Si [57], B [58], C_{60} [59], and $FeCl_3$ [60], leading to p-type or n-type behavior.

4.4.2 Mechanical properties

CNTs, which have an inherent tensile strength of more than 100 GPa and a Young's modulus of more than 1 TPa, are considered to be one of the strongest and stiffest materials ever discovered [61, 62]. This results from the covalent sp^2 bonds formed between the individual carbon atoms. Individual CNTs also exhibit excellent fatigue resistance, which was measured for the first time by Wei's group using a noncontact acoustic resonance test system in 2020 [63]. Although the tensile strength of an individual CNT is extremely high (over 100 GPa), it is challenging to prepare CNTs in long lengths. For this reason, CNTs are often used in fibers or composite materials, which always leads to a significant reduction in effective strength due to their defects, impurities, random orientations, and discontinuous lengths. In 2017, Suhr *et al* [64] were the first to measure the tensile properties of macroscopically long and continuous CNTs; the results indicated that the tensile strength and Young's modulus are related to the lengths, diameters, and volumes of the CNTs used.

In 2018, Bai and coworkers reported an ultra-long defect-free CNT fiber with a tensile strength of 80 GPa, which is far higher than that of any other strong fiber [61]. In addition, because of the hollow structure and high aspect ratio of CNTs, Jensen *et al* [65] found that an individual MWCNT tends to undergo buckling when subjected to compressive or bending stress.

4.4.3 Thermal properties

CNTs are expected to have good thermal conductivity in the length direction, exhibiting a property known as 'ballistic conduction' over a certain range [66]. The thermal conductivity of an individual MWCNT produced by arc evaporation has been reported to be higher than 3000 W m^{-1} K^{-1} at room temperature [67], while the value measured for an individual SWCNT was nearly 3500 W m^{-1} K^{-1} [68]. However, ropes or films of SWCNTs exhibit relatively low thermal conductivity [69]. Recently, Hamasaki [70] reported an *in situ* nanoscale observation of the anisotropic thermal transport of a single bundle of SWCNTs without spacing between the tubes (figure 4.6). Their experimental results indicated that thermal transport along the bundle axis is significantly higher than that in the perpendicular direction. CNTs also show extreme thermal stability, i.e. atomic-scale stability up to 3200 K (MWCNTs), which implies that CNTs may be more robust than either graphite or diamond [71]. In addition, SWCNTs are stable at over 3000 K in vacuum and 1000 K in air [72].

4.4.4 Optical properties

The optical properties of CNTs strongly depend on their unique electronic structure, which also gives them peculiar Raman spectra, optical absorption, and photo-luminescence. The optical absorption of CNTs is a convenient way to investigate their quality (figure 4.7(a)) [73]. Based on their outstanding visible light absorption, CNTs (e.g. vertically aligned nanotube arrays) can act as black-body materials [74–77]. In addition, the type (s-SWCNTs and m-SWCNTs) and chirality ((*n,m*) index) of CNTs can be determined via photoluminescence (figure 4.7(b), (c)) [78–80]. Furthermore, resonance Raman scattering spectroscopy is a fast and non-destructive technique for characterizing and studying CNTs (figure 4.7(d)) [81, 82].

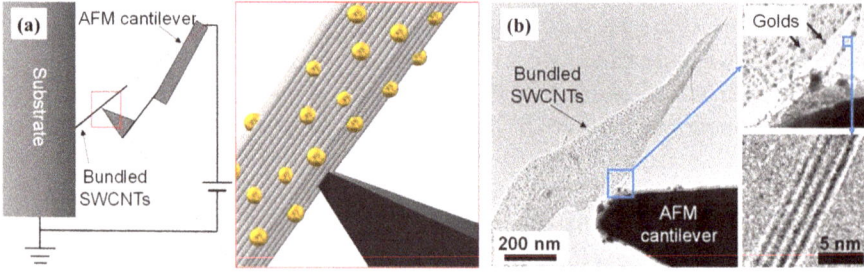

Figure 4.6. Experimental setup used for *in situ* observations of thermal transport in a bundle of SWCNTs. (a) Schematic of the experimental setup and (b) TEM images of a typical bundle in contact with the electrode probe tip [70], reproduced with permission © American Chemical Society.

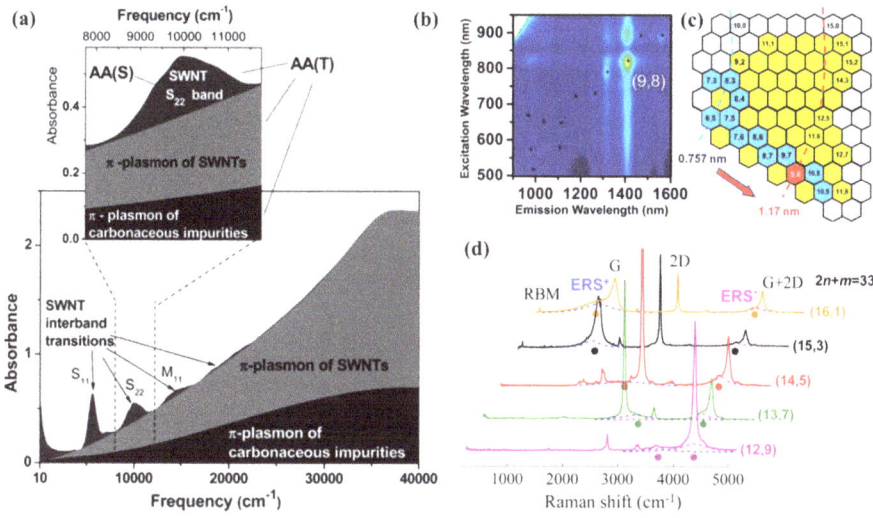

Figure 4.7. (a) Schematic illustration of the electronic spectrum of a typical SWCNT sample produced by the electric arc method [73], reproduced with permission © American Chemical Society. (b) Room-temperature Raman spectra of purified SWCNTs excited at five different laser frequencies [81].

4.5 Functionalization and chemical modification

The one-dimensional morphology of CNTs, which have a large aspect ratio and frequent bending caused by Stone–Wales defects, together with the strong van der Waals force in between CNTs, makes them tend to form bundles and even cross-linked networks. This complicated texture limits the good dispersion of CNTs in most solvents and their flexibility in further applications; the chemical functionalization of CNTs has been used to solve this problem.

In terms of chemical reactivity, CNTs can be divided into different parts: the caps shaped like half-fullerenes at the ends of CNTs, the sidewall with a hex atomic ring network, and inevitable defects formed in the synthesis and purification processes. If the carbon atom network stays intact, the planar graphitic sheets tend to exhibit inert chemical reactivity, while the fullerene possesses a much greater derivation tendency in chemical environments. The carbon skeleton of the spherical surface of fullerene presents a high degree of pyramidalization, and the conjugated coplanar π-orbitals are damaged, which endows the fullerene with high chemical reactivity. The caps of CNTs share comparable properties with fullerene, being the most reactive sites in CNTs [83]. When attacked by oxidizing reactants, the caps are easily broken or even completely removed, resulting in open ends and an exposed inner cavity. With regard to the tubular outer wall, the reactivity of the CNTs depends on the tube diameter. A smaller tube diameter means a bigger curvature and a higher degree of pyramidalization of the CNT carbon skeleton. Therefore, CNTs with smaller tube diameters are more vulnerable to oxidative reagents [84]. In addition, if CNTs are exposed to strongly oxidizing environments for a long enough time, the outer sidewall may also be damaged, and some holes or fractures may be generated. The rims of the holes and fractures are surrounded by defects with dangling bonds and/or heterocyclic carbon rings, which are likely to take part in further chemical reactions.

Many approaches have been developed to chemically modify the properties of CNTs, including covalent functionalization, non-covalent functionalization, endohedral filling, and heteroatomic doping functionalization [85–87]. Through chemical reactions between reactants and CNTs, covalent bonds, including –COOH, –OH, –F, –SO$_3$H, and –NH$_2$, etc. can be introduced into the carbon network. The formed covalent bonds not only change the surface chemical properties of CNTs, but also affect the conjugated π-orbitals through transforming the carbon hybridization type from sp^2 to sp^3, which may influence the electronic conductivity and mechanical integrity of CNTs [88, 89]. Non-covalent functionalization is another way to achieve surface property modification while simultaneously keeping the integrity of CNTs at the very highest level [90]. Various organic or inorganic molecules can be attached to the outer wall of CNTs by many kinds of non-covalent interaction such as π–π conjugation, the van der Waals force, electrostatic binding, or hydrophobic effects. In particular cases, some molecules may be introduced into the tube cavity and achieve the endohedral filling functionalization of CNTs. In addition, some carbon atoms in the skeletons of CNTs can be replaced by other atoms, i.e. by chemical heteroatomic doping, which can induce electron redistribution to take place on the surfaces of CNTs and thus endow them with special catalytic properties.

4.5.1 Chemistry on the open ends and defects of CNT

Currently, the CNTs in commercial use are mainly prepared by the CVD method, which requires the use of a large number of transition metal (Fe, Co, Ni, etc.) catalysts [91–93]. In subsequent applications, it is often necessary to undertake a purification process involving strong oxidative acid treatment to remove residual metallic particulate impurities and amorphous carbon fragments. Oxidative acids such as HNO$_3$ and/or H$_2$SO$_4$ have been used to purify CNTs and chemically modify them during the process [94, 95]. The acid treatment cuts the highly reactive caps and may also damage the sidewalls of CNTs, resulting in open ends and defective holes (figures 4.8(a) and (b)). These two sites both represent the terminals of the conjugated

Figure 4.8. (a) Typical defects in a CNT [103], reproduced with permission © John Wiley & Sons. (b) Reactivity of the open ends and defects of CNT during the purification process. (c) Typical Stone–Wales (or 7-5-5-7) defect on the sidewall of a nanotube [100], reproduced with permission © John Wiley & Sons.

graphite grid system, in which carbon atoms at the edges are rearranged and thus have special chemical reactivity [96, 97]. On the other hand, another type of defect called the Stone–Wales defect, which is not located at the terminal sites of CNTs, should be also noted. It is a carbon configuration that consists of a five-membered ring and an adjacent seven-membered ring (figure 4.8(c)), at which point CNTs tend to appear bent, and the outer wall presents an increased curvature [98–100]. Additionally, chemical functionalization is also favored at Stone–Wales defect sites [101, 102].

4.5.2 Sidewall functionalization

4.5.2.1 Covalent functionalization

Chemical functionalization of the ends and defects is not enough to ensure that CNTs have good dispersion, and it is thus necessary to functionalize their walls. To achieve this goal, reagents need to attack the π conjugation delocalization system all over the wall, which needs a special reaction environment. In recent years, a number of covalent functionalization methods have been developed for CNT walls, including but not limited to fluorination, sulfonation, surface addition, and polymer grafting [104, 105]. The use of these methods can induce the overall chemical functionalization of CNTs, and make them suitable for many specific applications [85, 106].

Fluorination and nucleophilic substitution: fluorine has the highest electronegativity and strong oxidation activity, so it can react with most elements to form bonds; it thus has the ability to modify the surface properties of CNTs. Mickelson and coworkers reported a CNT fluorination reaction in 1998 [107]. They found that the fluorine atoms could attack the sidewalls of CNTs directly and that the carbon atoms adjacent to fluorine would undergo a hybridization transition from sp^2 to sp^3. Further research revealed that CNTs could be fluorinated in the range from room temperature to 600 °C [107–109]. The fluorination of CNTs at room temperature by F_2 led to a black product with an F/C ratio of less than 0.4, while a temperature of 500 °C resulted in completely fluorinated white compounds with an F/C ratio of 1.0 [109]. On the other hand, in addition to F_2, other fluorine-containing monomers (e.g., CF_4, XeF_2 and HF) can also function as the fluorinating agent [109–111].

The electronic properties of fluorinated CNTs are very different from those of the original CNTs because of the destruction of the sidewall conjugated π-orbital. The conductivity of fluorinated CNT decreases, while the Fermi energy level and conduction band are lower. Therefore, fluorinated CNTs had better electron acceptance and were easier to replace with strong nucleophilic reagents than the original CNTs [112]. Grignard reagents, alkyllithium, alkoxides, amines, and peroxide are all available for sidewall alkylation of the fluorinated CNTs, thus providing additional route options for the dissolution and application of CNTs in various solvents (figure 4.9).

Sulfonation: many studies have been conducted on CNTs functionalized with sulfonic acid groups. The most commonly used reagent is sulfuric acid. Through a microwave-assisted treatment with a mixture of concentrated nitric and sulfuric acid (ratio of 1:1), Mitra and coworkers found coexistent carboxyl (–COOH) and

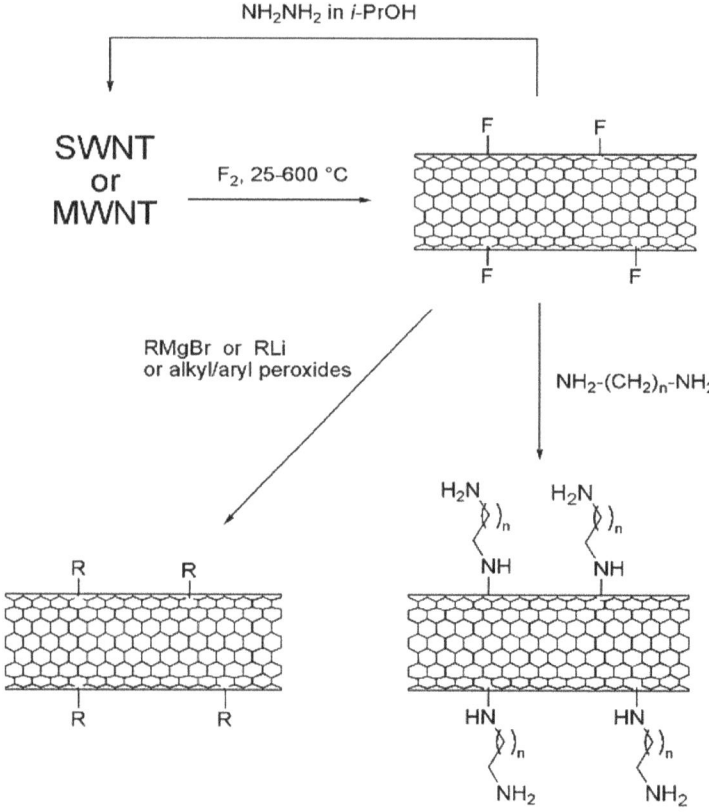

Figure 4.9. Reaction scheme for the fluorination of CNTs and further derivatization [105], reproduced with permission © American Chemical Society.

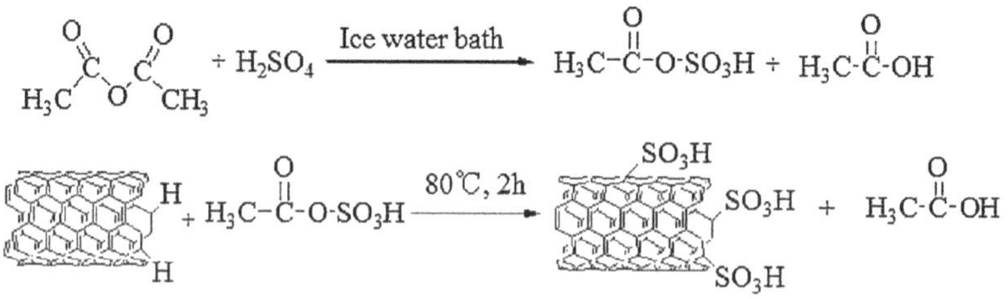

Figure 4.10. Procedure used to synthesize sulfonated CNTs using acetyl sulfate [115], reproduced with permission © Elsevier.

sulfonate ($-SO_3H$) groups on the surfaces of CNTs [113]. Liang and coworkers [114] reported a sulfonation process of CNTs in oleum (20% free SO_3), based on a preliminary step in the formation of phenylated CNTs. In addition, Wei and coworkers [115] revealed that acetyl sulfate, the reaction product of acetic anhydride and concentrated sulfuric acid, could also reserve as an efficient sulfonation reagent at a mild treatment temperature of 80 °C (figure 4.10).

Cycloaddition: Haddon and coworkers were the first to report [2+1] cycloaddition to CNTs, utilizing dichlorocarbene to attack C=C bonds connecting two adjacent six-membered carbon sings and thus produce 1,1-dichlorocyclopropane [116]. Another kind of cycloaddition reaction between CNTs and nitrene was conducted by Hirsh and coworkers. The alkoxycarbonylnitrene formed via nitrogen elimination of an organic azide, then performed [2+1] cycloaddition to the sidewalls of CNTs, producing alkoxycarbonylaziridino-CNTs (figure 4.11) [117]. The [2+1] cycloaddition reaction promotes the derivatization of CNTs and their solubility in dimethyl sulfoxide or 1,2-dichlorobenzene.

Polymer grafting: this is a significant route for enhancing the solubility of CNTs by covalent reactions between CNTs and polymers, as the long polymer chains contribute to the dissolution of the CNTs in a wide range of solvents, even at a low degree of functionalization [108]. The process of establishing a covalent linkage between polymers chain and CNTs is referred to as 'grafting' and involves two methods, i.e. 'grafting to' and 'grafting from' (figure 4.12). The 'grafting to' method involves a pre-synthesized polymer which is attached to the CNT surface via a certain interaction of its end groups, while the 'grafting from' method refers to covalent attachment of the polymer precursors to the CNT surface and subsequent *in situ* polymerization [118].

4.5.2.2 Non-covalent functionalization

As a promising alternative to covalent approaches that largely maintains the chemical structure and physical properties of CNTs, the non-covalent functionalization method has also been demonstrated to enhance the solubility and expand the applications of CNTs in the photovoltaic, electronic, and biomedical fields and beyond. In recent years, great progress has been made in the non-covalent functionalization of CNTs, using surfactants, small aromatic molecules, macrocyclic conjugated systems, coating polymers, or polypeptide molecules.

Dai and coworkers reported a general and attractive approach to the non-covalent functionalization of CNT sidewalls and the subsequent immobilization of biological molecules onto SWNTs with a high degree of control and specificity [119]. They found that pyrene-based molecules could be irreversibly adsorbed (figure 4.13(a)) onto the hydrophobic CNT sidewall via π–π stacking, leading to a highly stable self-assembled structure, which dispersed well in DMF, MeOH, and aqueous solutions due to the addition of hydrophilic functional groups. Richard and coworkers [120] solubilized

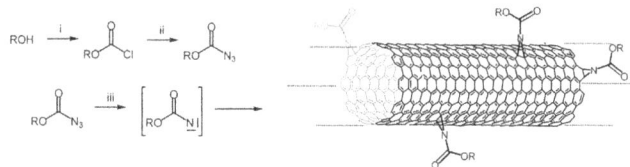

Figure 4.11. Sidewall functionalization of SWCNTs with azidocarbonates [117].

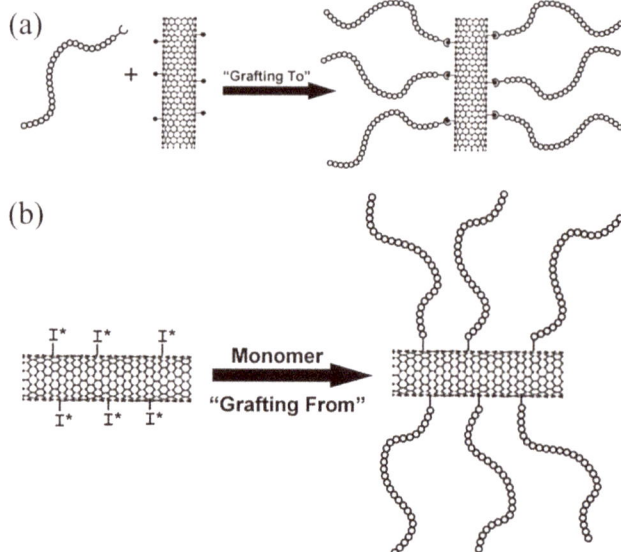

Figure 4.12. (a) 'Grafting to' and (b) 'grafting from' approaches for producing CNT–polymer composites [118], reproduced with permission © Elsevier.

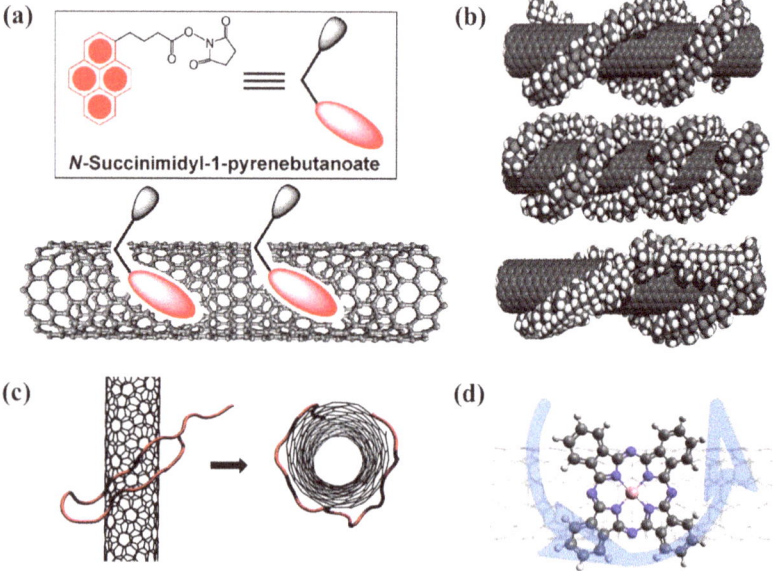

Figure 4.13. Non-covalent functionalization of a CNT sidewall by: (a) pyrene-based molecules [119], reproduced with permission © American Chemical Society; (b) PVP linear polymers [120], reproduced with permission © Elsevier; (c) a peptide chain [121], reproduced with permission © American Chemical Society; and (d) a heterocyclic aromatic molecule [122].

CNTs in an aqueous solution by non-covalently associating them with linear polymers (e.g. polyvinyl pyrrolidone (PVP)) with the aid of sodium dodecylsulfate (SDS, figure 4.13(b)). The polymer was able to wrap around nanotubes to form a stable self-assembled CNT/PVP complex, providing a route to the precise and reliable manipulation of CNTs by solution-phase techniques. In another independent experiment, Rajesh and coworkers [121] then described a rational design for a peptide recognition element (PRE) that was capable of non-covalently attaching to CNTs as well as binding to trinitrotoluene (TNT). The PRE contained two binding domains, one for TNT and another for CNTs; it was able to spontaneously decorate CNTs and provide target selectivity in a chemical sensor (figure 4.13(c)).

Heterocyclic aromatic molecules have been found to easily be immobilized on the sidewalls of CNTs through π–π stacking and intermolecular van der Waals forces, generating great potential for enhancing the CNTs' catalytic properties [122]. For example, Wang and coworkers showed that when it is immobilized on CNTs, cobalt phthalocyanine catalyzes the six-electron reduction of CO_2 to methanol with appreciable activity and selectivity (figure 4.13(d)) [122], reproduced with permission © Springer Nature.

4.5.3 Endohedral filling

Uncapped CNTs with open ends and an inner empty cavity have been demonstrated to be endohedrally filled by various materials, ranging from fullerenes and metal-lofullerenes to different metals and metal salts as well as biomolecules. The materials that fill CNTs are confined to the nanometer scale, and their structure and dynamic properties are radically changed. At the same time, the electrical, magnetic, and mechanical properties of CNTs with endohedral fillings also change.

In 1998, Smith and coworkers [123] accidentally discovered that C_{60} could fill the interior of a SWCNT while observing a CNT sample using HR-TEM. The C_{60} can fit well inside CNTs, aligning in a straight line. Ever since then, researchers have been dedicated to developing the fabrication of fullerene-filled CNTs and have succeeded in encapsulating fullerenes into CNTs by plasma irradiation, gas phase, and liquid-phase techniques (figure 4.14) [124]. In addition to C_{60}, some higher-order carbon spheres such as C_{70} and C_{80}, various different metallofullerenes including $Gd–C_{82}$, $Ti_2–C_{80}$, and $Sm–C_{82}$, and even some metals and inorganic salts have also been inserted into the cavity of nanotubes [105, 125–127].

4.5.4 Heteroatomic doping functionalization

Doping CNTs with other chemical elements could be an especially interesting way of tuning their physicochemical properties. The chemical doping of CNTs involves the replacement of some carbon atoms within the honeycomb graphitic structure by other heteroatoms, including N, B, S, F, P, and so on (figure 4.15(a)) [87]. These heteroatoms are different in size and/or electronegativity from carbon atoms, and the doping can cause charge transfer and redistribution to take place on the surface of CNTs, resulting in changes in their electronic structure and physical/chemical properties. In general, there are two pathways that can be used for the heteroatom

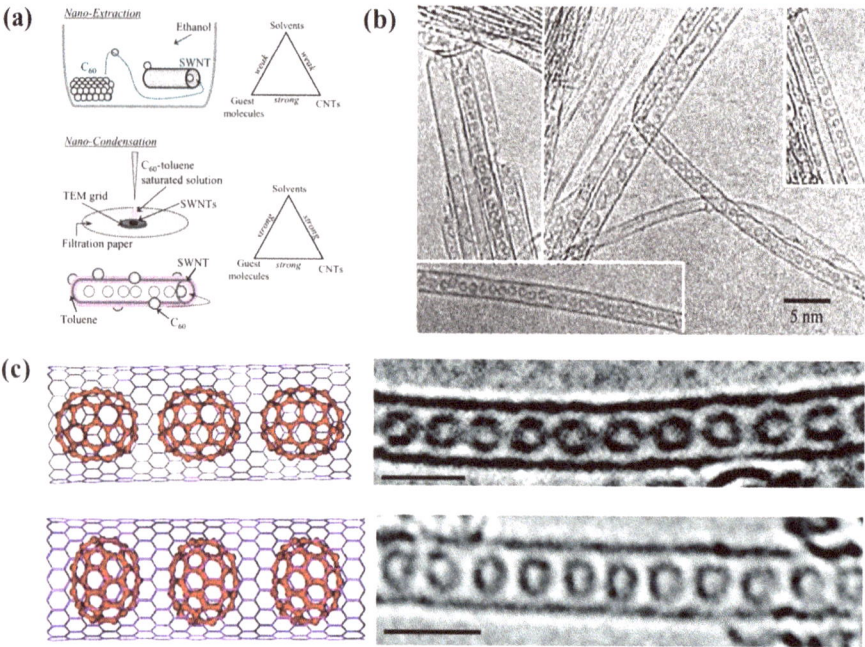

Figure 4.14. (a) Schematic illustration of the liquid-phase technique used for the endohedral filling CNTs with C_{60} and (b) the corresponding TEM images [124], reproduced with permission © Elsevier, (c) C_{70} nanopeapods with lying and standing alignments (scale bar: 2 nm) [127], reproduced with permission © American Chemical Society.

doping of functionalized CNTs: *in situ* doping during synthesis and post-doping of preformed CNTs with heteroatom-containing molecules.

Stephan and coworkers initiated synthesis that included nitrogen and boron co-doping in 1994 using the arc-discharge method [128]. Since then, much work has been done on the direct synthesis of nitrogen- and/or boron-doped CNTs (figure 4.15(b)) [129]. The simple but efficient approach to *in situ* N doping is the pyrolysis of a metal phthalocyanine to produce vertically aligned CNTs, since the metal phthalocyanine molecules contain all the key ingredients for doped CNT growth: the catalyst, the carbon source, and nitrogen heteroatoms [130]. Moreover, the CVD method was used by Yang *et al* to synthesize boron-doped CNTs with benzene, triphenylborane, and ferrocene as precursors. In 2010, Dai and coworkers developed a metal-free plasma-etching technology for the efficient growth of undoped and/or nitrogen-doped SWCNTs; they generated SiO_2 nanoparticles to function as catalysts (figure 4.15(c)) [131]. Post-doped CNTs have also been developed in the presence of heteroatom-containing precursors as dopants. The thermal treatment of prepared CNTs in an NH_3 atmosphere has become a widely accepted method for nitrogen doping, leading to a better dispersion of doped CNTs in an aqueous medium [132]. Likewise, a high-temperature post-treatment of CNTs wrapped with precursors containing N and S obtained N, S co-doped CNTs (figure 4.15(d)) [133].

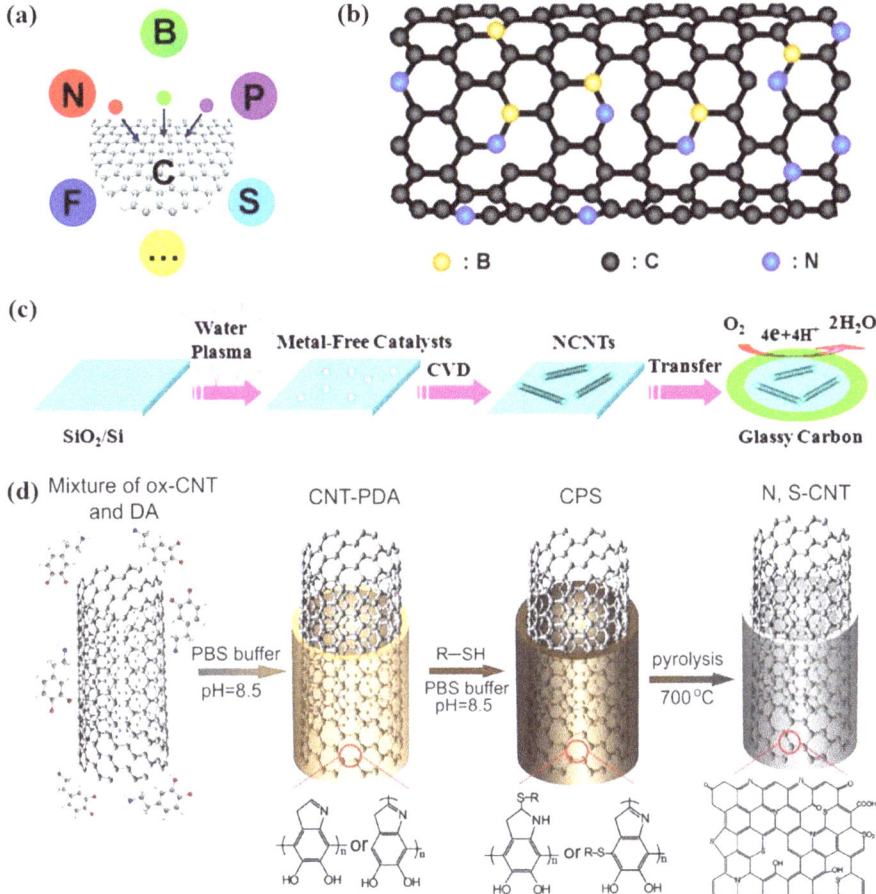

Figure 4.15. (a) The doping of a graphitic carbon structure with heteroatoms [87], reproduced with permission © American Chemical Society. (b) Illustration of B, N co-doped CNT [129], reproduced with permission © John Wiley & Sons. (c) Metal-free growth of N-doped CNT [131], reproduced with permission © American Chemical Society. (d) Fabrication of N,S-CNT following a two-step 'graft-and-pyrolyze' route [133], reproduced with permission © John Wiley & Sons.

4.6 Applications

As mentioned above, CNTs possess a unique set of electrical, mechanical, thermal, and chemical properties, which have been stimulating increasing interest in their applications. The applications of functionalized CNTs are more extensive and can be found in catalytic, energy-related, biological, and environmental fields.

4.6.1 Catalytic and energy-related applications

As a result of their high electronic conductivity, chemical and thermal stability, and capability for modification by functional groups, CNTs have been regarded as promising materials for catalytic applications. A large number of reactions have been catalyzed by CNTs or their derived materials. Thus, CNTs exhibit great advantages when they are used as electrodes or additive materials in energy-related applications, including batteries, solar cells, fuel cells, and supercapacitors.

4.6.1.1 Electrochemical catalysis

It has been demonstrated that CNTs with designable functionalization can catalyze various electrochemical reactions, such as the oxygen reduction reaction (ORR), water splitting, and the CO_2 reduction reaction (CO_2RR), which are traditionally catalyzed by metal-based catalysts, demonstrating the great promise of highly efficient metal-free catalysis. Since pristine CNTs have poor activities for electrochemical catalysis, two approaches are usually developed to endow CNTs with high catalytic activity, i.e. heteroatomic doping and surface modification.

Oxygen reduction reaction (ORR): the ORR is an important reaction in fuel cells and metal–air batteries. Dai and coworkers reported the first metal-free carbon-based catalyst made of nitrogen-doped vertically aligned CNTs (VA-NCNTs) for the electrochemical ORR in 2009 [134]. They found that the VA-NCNTs exhibited better electrocatalytic activity, long-term operation stability, and tolerance to the crossover effect than Pt/C in an alkaline medium. It was demonstrated that doping-induced charge transfer and redistribution facilitated the chemisorption mode of O_2 and further weakened the O-O bond, resulting in excellent ORR performance (figure 4.16(a)) [134, 135]. Since then, much research has been devoted to searching for high-efficiency, durable, and reliable metal-free ORR catalysts by doping various heteroatom(s) (such as N [135], P [136], B [137], S [138], etc.) into MWCNTs or SWCNTs [131]. For instance, Dai's group prepared a nitrogen-doped SWCNT from metal-free nanoparticles using the CVD method [131]. The metal-free nitrogen-doped SWCNTs, when used for the ORR, exhibited enhanced electrocatalytic activity and long-term stability compared with undoped SWCNTs in an acidic medium. Surface modification also contributes to the ORR, and various functionalized CNTs with different functional groups, including O-containing functional groups (e.g., –OH, –COOH, C=O) [139], S-containing functional groups [140, 141] (e.g. sulfonated CNTs), and polyelectrolytes [142], have also been developed that had significantly improved electrocatalytic activity in the ORR. Furthermore, poly(diallyldimethylammonium chloride) (PDDA) functionalized CNTs (PDDA-CNTs) have been shown to possess remarkable electrocatalytic activity for the ORR, which suggests that PDDA has a strong electron-accepting ability that withdraws electrons from the carbon atoms of CNTs to induce a net positive charge, facilitating ORR catalytic activity (figure 4.16(b)) [142]. As mentioned above, CNTs can be endohedral, i.e. filled with various materials. Carini *et al* [143] reported a new method for producing DWCNTs (N-SWNT@SWNT) as ORR catalysts with nitrogen doped on the inner wall. First, they filled the internal cavity of SWCNTs with dicyanopyrazophenanthroline, and then converted nitrogen-rich molecules into nitrogen-doped CNTs by exposing them to electron beams or annealing them at temperatures greater than 1300 °C (figure 4.16(c)). N-SWNT@SWNT had a better

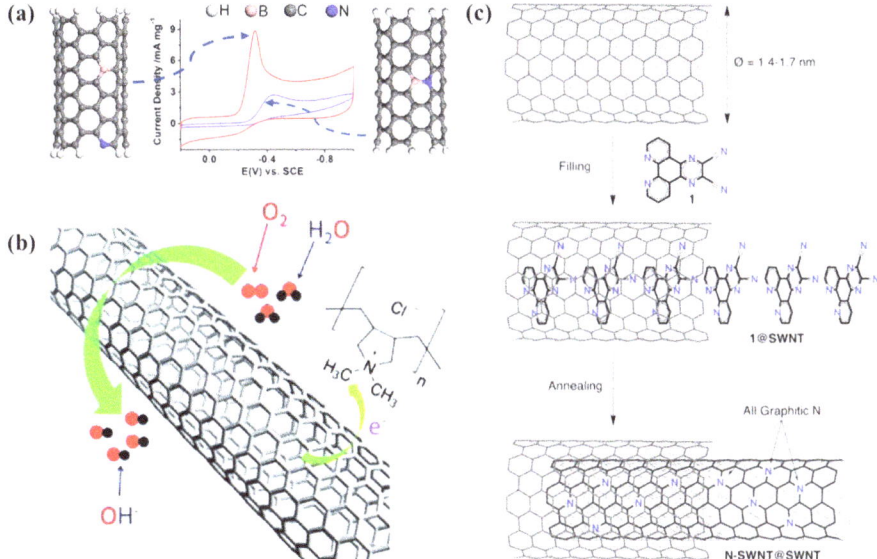

Figure 4.16. (a) Bonded and separated B and N co-doped CNT, and its ORR performance [135]. (b) Diagram of the charge-transfer process and the ORR on PDDA-CNT [142], reproduced with permission © American Chemical Society. (c) Synthetic approach for the preparation of coaxial N-SWNT@SWNT [143], reproduced with permission © John Wiley & Sons.

ORR performance than that of SWCNTs, confirming that the internal nitrogen-doped inner wall was able to transduce its properties across the external wall [143].

Water splitting: Water splitting involves two important electrochemical reactions, namely, the cathodic hydrogen evolution reaction (HER) and the anodic oxygen evolution reaction (OER). In the same way as for the ORR, heteroatom(s) doping is an efficient way to produce metal-free catalysts that improve OER and/or HER activity. Zhao *et al* [141, 144] demonstrated that oxygen-containing functional groups, such as hydroxyl (–OH), epoxy (–C–O–C–), carbonyl (–C=O), and carboxyl (–COOH), may be responsible for enhancing OER performance. Those groups can alter the electronic structures of the adjacent carbon atoms and facilitate the adsorption of OER intermediates [144]. Subsequently, Suib's group developed a sequential two-step strategy to dope sulfur into CNTs, which boosted the activity of OER in an alkaline medium [145]. To achieve a bifunctional catalyst for water splitting coupled with the OER and the HER, a co/multi-doping strategy with two or more heteroatoms has been conducted. Specifically, nitrogen and sulfur co-doped CNTs display superb bifunctional catalytic activities for both the HER and the OER in alkaline solutions (figures 4.17(a) and (b)) [133]. Experimental characterizations confirmed that the excellent performance could be attributed to the N and S dopants, while theoretical computations revealed the favorable effect of the secondary sulfur dopant in enhancing the spin density of dual-doped samples and consequently in forming highly electroactive sites for both the HER and the OER [133]. Rather than

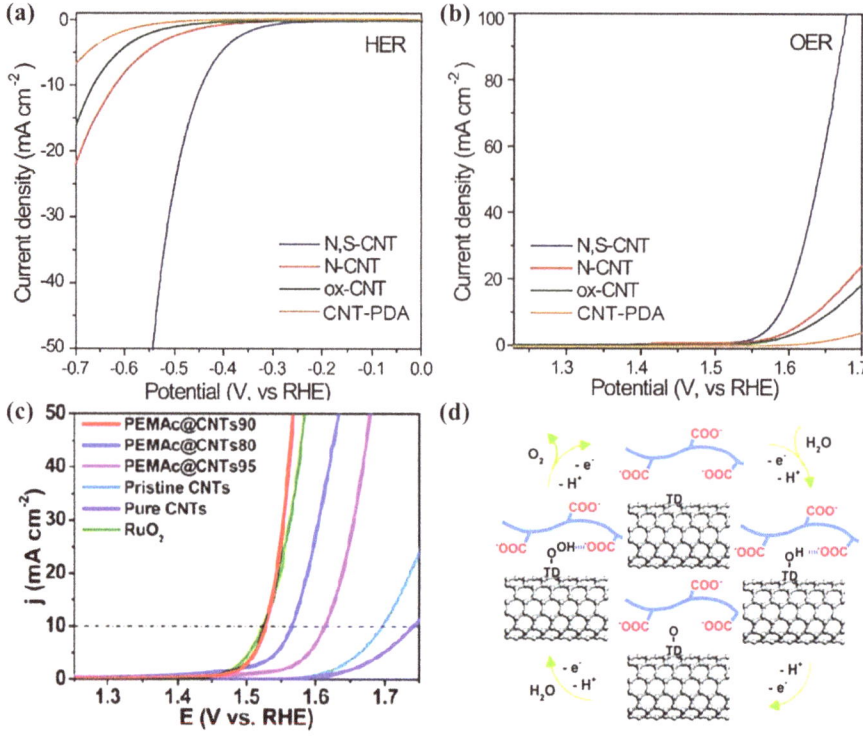

Figure 4.17. (a) HER and (b) OER polarization curves [133], reproduced with permission © John Wiley & Sons. (c) LSV curves of the OER and (d) proposed reaction mechanism in alkaline media [146], reproduced with permission © The Royal Society of Chemistry.

heteroatom(s) doping, Dai *et al* [146, 147] also developed another strategy that involved simply wrapping CNTs with polymer to improve the OER performance. This method can preserve the integrity of CNTs without damaging the lattices, facilitating electron transport during the OER process. A class of electrochemically inert polymers, such as poly(ethylene-alt-maleic acid) (PEMAc), poly(acrylic acid) (PAA), poly(vinyl alcohol) (PVA), poly(vinyl acetate) (PVAc), poly(ethylene glycol) (PEG), were then employed to functionalize the CNTs [146], which exhibited dramatically improved OER activities in alkaline media, even comparable to that of RuO_2 (figure 4.17(c)). The wrapped polymer was introduced as a 'kinetics accelerator' which could promote the absorption of intermediates (figure 4.17(d)) [146]. Furthermore, the electron-withdrawing polyelectrolyte PDDA-modified CNTs could induce interfacial intermolecular electron transfer in the positively charged PDDA on CNTs and thus capture OH^- via electrostatic interactions, which was beneficial to the reaction kinetics of the OER [147].

CO$_2$RR: As described above, nitrogen-doped CNTs (NCNTs) are effective catalysts for the electrocatalytic ORR. NCNTs also have been demonstrated to be promising electrocatalysts for the CO$_2$RR. In 2014, Meyer's group reported NCNTs functionalized by polyethylenimine (PEI) for the electrocatalytic reduction of CO_2 to formate [148]. Due to the important role of PEI in stabilizing the intermediate and concentrating CO_2, these combination catalysts were able to significantly reduce overpotential and increase the current density and efficiency [148]. Soon after, Ajayan *et al* reported that nitrogen-doped CNT arrays exhibited a low overpotential for the CO_2RR to CO with high selectivity (~80%) and long-term stability (over 10 h) [149]. It was also found that the most selective site for CO production was induced by pyridinic N [149]. On the contrary, Zheng and coworkers concluded that pyrrolic N-enriched NCNTs enhanced the selectivity of the CO_2RR to CO. Hot water steam was employed to etch pyridinic and graphitic nitrogen atoms in CNTs, resulting in an increased pyrrolic N level, which subsequently demonstrated high selectivity for the CO_2RR and HER suppression.

4.6.1.2 Energy storage and conversion applications

Lithium-ion batteries (LIBs): as a result of their unique mechanical and electrical properties and high aspect ratios, CNTs can be used as conductive additives or active electrode materials in LIBs, which are the dominant energy storage systems for electric vehicles (EVs) and portable energy devices in modern society [150–152]. These electrodes with CNTs not only exhibited high conductivity for rapid electronic transport, but also provided high capacity to boost the rate performance of high-energy-density batteries [153]. Ning *et al* [154] fabricated a conductive network using VA-CNTs in $LiFePO_4$ cathodes, which significantly promoted the specific capacities and low-temperature performance of LIBs. Wang and coworkers reported highly stable, crystalline 2D polyarylimide functionalized CNTs (2D-PAI@CNTs) which possessed abundant π-conjugated redox-active naphthalene diimide units, a high specific surface area, and a porous structure [153]. 2D-PAI@CNT was further used as a cathode material in LIBs, which exhibited ultrastability (100% capacity retention after 8000 cycles) and a high rate capability (figure 4.18). In addition, anodes made from a segregated network composite of CNTs with lithium storage materials (e.g. silicon, graphite, and metal oxide particles) were used in LIBs, which exhibited high conductivities and low charge-transfer resistances [155].

Supercapacitors: Since they benefit from high conductivity and structural stability, CNTs are also regarded as ideal electrode materials for supercapacitors. Dai and coworkers [156] fabricated large-area multicomponent hybrid films by the sequential self-assembly of PEI-modified graphene sheets and acid-oxidized CNTs via electrostatic interactions onto various substrates. The resultant hybrid films possessed an interconnected network and well-defined nanopores, leading to a nearly rectangular cyclic voltammogram even at a high scan rate of 1 V s^{-1} with an average specific capacitance of 120 F g^{-1} (figure 4.19(a)). Cao *et al* [157] developed a highly stretchable and reliable supercapacitor based on crumpled vertically aligned CNT forests on an elastomer substrate (figures 4.19(b)–(d)). The resulting supercapacitor was able to withstand high uniaxial (300%) and biaxial strains (300%×300%). After thousands of stretching–relaxation cycles, the electrode performance remained unchanged.

Solar cells: the conversion of solar energy into electricity is one of the most promising routes to meeting the increasing energy demands of future generations

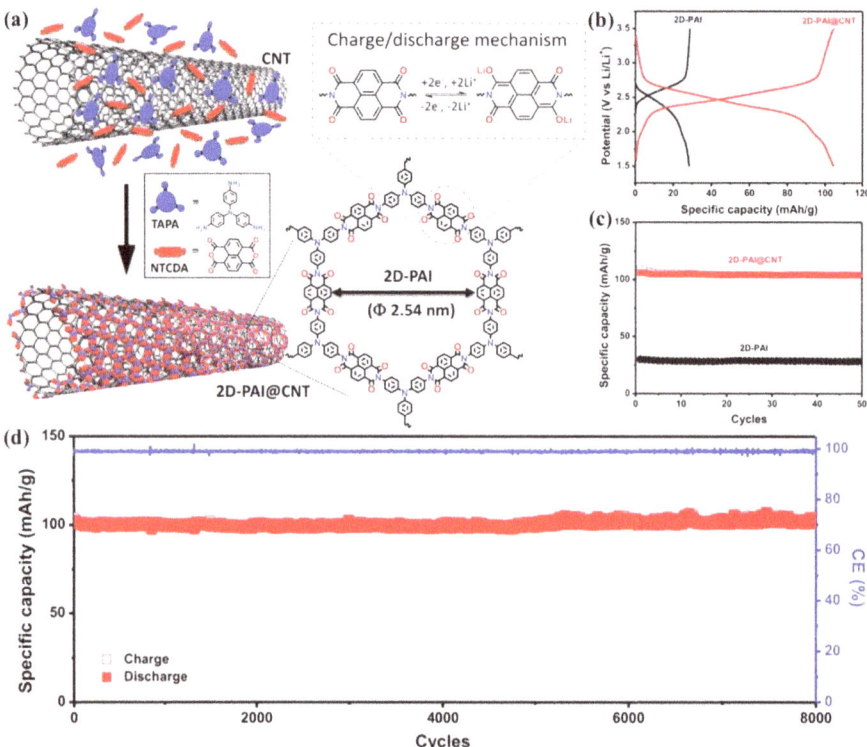

Figure 4.18. (a) Schematic illustration of the synthesis of crystalline 2D-PAI@CNT and its energy storage mechanism; (b) galvanostatic charge–discharge curves and (c) cycling performance of 2D-PAI- and 2D-PAI@CNT-based batteries at 0.1 A g^{-1}, and (d) long-term cycling stability of 2D-PAI@CNT at 0.5 A g^{-1} [153], reproduced with permission © John Wiley & Sons.

[158, 159]. Photovoltaics, more generally known as solar cells, have received much attention in recent years due to their promise as clean and efficient light-harvesting devices. When applied in different types of solar cell, such as organic, silicon, dye-sensitized, and perovskite cells, CNTs have been demonstrated to efficiently play the roles of not only electron acceptors and charge transporters, but also of light absorbers and electron donors, thereby improving both exciton generation and the transport of photoexcited carriers. Habisreutinger and coworkers [160] replaced an organic hole transport material with polymer-functionalized SWCNTs embedded in an insulating polymer matrix to mitigate the thermal degradation issue in perovskite solar cells (figures 4.20(a) and (b)). Power conversion efficiencies of up to 15.3% were observed, and the resilience of the cells against thermal stress and moisture ingress was remarkably enhanced. This CNT–polymer composite contributes to a drive toward achieving a 25-year operational lifetime for perovskite solar cells [160]. SWCNTs have also attracted significant attention for use in silicon heterojunction solar cells due to their high carrier mobility, high light transmittance, and low resistance. Cheng's group synthesized a SWCNT film using the floating catalyst CVD method which had a low sheet resistance of 180 Ω sq^{-1} and a high

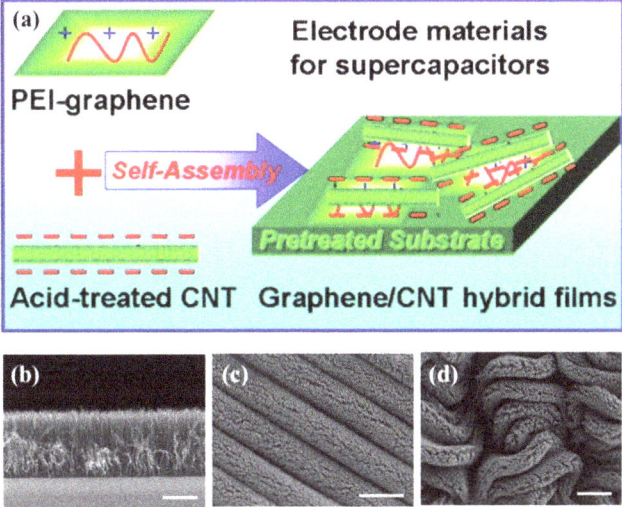

Figure 4.19. (a) Illustration of the processes of positively charged PEI-GN and negatively charged CNT film deposition on an appropriate substrate [156], reproduced with permission © John Wiley & Sons. (b) SEM image of a CNT forest (scale bar: 10 µm), (c) SEM image of the parallel ridge pattern formed by a CNT forest on an elastomer substrate after relaxation in one direction. (d) SEM image of the crumpled pattern formed by a CNT forest on a fully relaxed elastomer substrate (300% × 300%) (scale bar: 100 µm) [157], reproduced with permission © John Wiley & Sons.

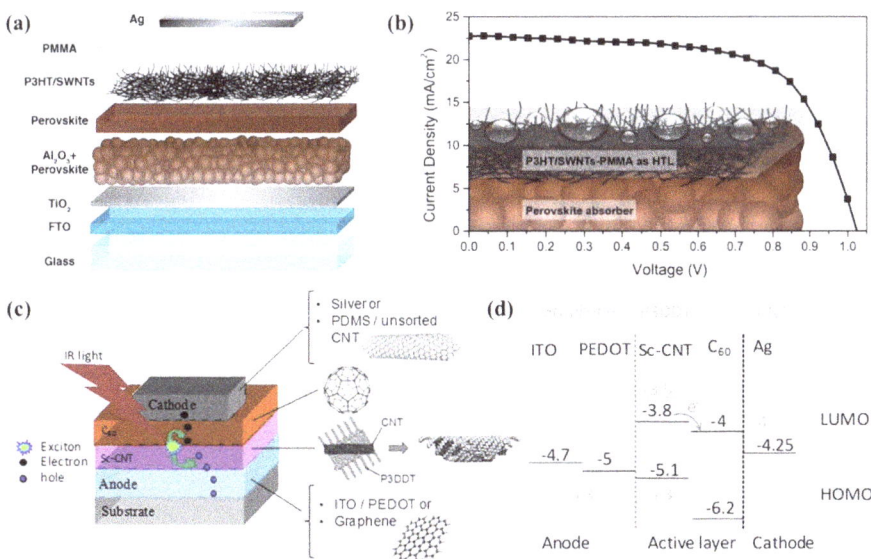

Figure 4.20. (a) Schematic illustration of a solar cell with a CNT/polymer composite as the hole-transporting structure. (b) Plots of the current–voltage characteristics [160], reproduced with permission © American Chemical Society. (c) Structure of all-carbon solar cells. (d) Band diagram of the photovoltaic device shown in (c) [163], reproduced with permission © American Chemical Society.

transmittance of 90% [161]. They then fabricated SWCNT/Si solar cells with a high photovoltaic performance that was attributed to the good optical and electrical properties of the SWCNT films. As a result of their bandgap, s-SWCNTs can act as active layers [162, 163]. In 2012, Bao's group reported the first all-carbon solar cell, in which s-SWCNTs were used as the light absorber and donor, and the maximum power conversion efficiency (PCE) was 0.46% under an illumination of AM1.5 sun (figures 4.20(c) and (d)) [163]. A PCE of 3.1% was described a short time later by Hersam, who used polychiral s-SWCNTs and PC71BM fullerene as active layers [162].

Field effect transistors (FET): CNTs have been shown to be promising semi-conducting materials for FETs due to their low electron scattering, compatibility with high-k dielectric materials, and mechanical flexibility [164]. CNT FETs (CNTFETs) are capable of operating at room temperature using single-electron digital switches. For instance, Borghetti and coworkers [165] demonstrated that polymer-functionalized CNTFETs, as very sensitive charge sensors, proved to be a good tools with which to study the charge distribution and dynamics of polymer thin-film transistors. Depending on the applied gate bias, the device could be optimized to function as a memory element or as an optical switch (figures 4.21(a)

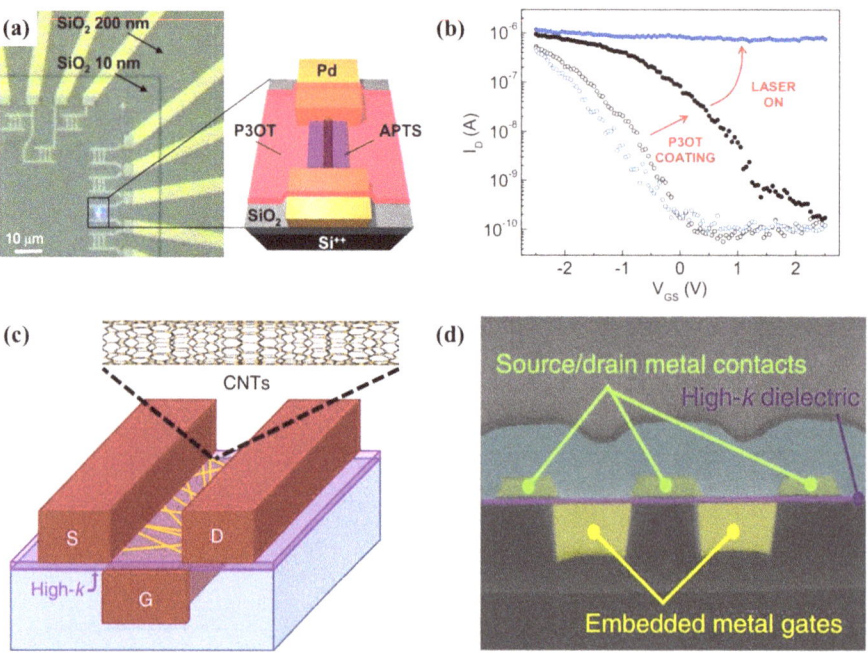

Figure 4.21. (a and b) Self-assembled CNTFETs and their response to polymer photoexcitation [165], reproduced with permission © John Wiley & Sons. (c and d) Schematic of a CNTFET with a bottom gate geometry and a cross-sectional SEM image of two series CNTFETs [166], reproduced with permission © Springer Nature.

and (b)). Moreover, Bishop and coworkers reported a method of CNT deposition through incubation that could fabricate CNTFETs within commercial silicon manufacturing facilities, demonstrating wafer-scale uniformity and reproducibility across multiple 200 mm wafers, which have displayed the potential to meet the requirements for a manufacturable, high-performance future CNTFET technology (figures 4.21(c) and (d)) [166].

4.6.2 Biological applications

In recent years, integrating and analyzing the signals of biological components has become a prerequisite for the early diagnosis of many diseases and attracted a lot of biomedical researchers. As a typical inorganic nanomaterial, CNTs have multifarious outstanding properties such as superior electrochemical properties, a high specific surface area, thermal stability, and in particular, exposure sensitivity and low toxicity to biomolecules [167, 168]. Moreover, the surface of CNTs can be schematically functionalized by the addition of various active substances, which benefit accurate targeting and the combination of functionalized CNTs and receptors [169]. These features have led to an increase in the investigations of CNTs for use in biological applications.

Biosensors: Biosensors that can accept and subsequently transform biological signals into measurable chemical or physical information are principally used to realize the specific detection of biomolecules. An ever-growing number of reports have recently emerged, showing that functionalized CNTs have the ability to address several concerns related to biocompatibility and toxicology and can serve as ideal materials for highly sensitive nanoscale biosensor devices [168, 170, 171]. CNT-based biosensors could work well in the detection of DNA biomarkers, cell-surface sugars, protein receptors, neurotransmitters, and enzymes [168]. For instance, Dai and coworkers reported single-strand DNA chains chemically grafted to aligned CNT electrodes as DNA sensors, which exhibited high sensitivity and selectivity when used to probe complementary DNA and target DNA chains with specific sequences [172]. In addition, an electrochemical, label-free method was developed by Zhu *et al* to detect folate-receptor-positive tumor cells via specific recognition by a polydopamine-coated CNT–folate nanoprobe of cell-surface folate receptors [173]. Furthermore, Peng *et al* [174, 175] presented flexible and miniaturized implantable fiber biosensors for stable *in vivo* interfaces to chemically monitor the signals deep in biotissues. The sensor was fabricated by twisting CNTs into hierarchical and helical bundles of fibers resembling muscle filaments. They matched the bending stiffnesses of tissues and cells and therefore exhibited good biocompatibility (figure 4.22).

Bioimaging: Bioimaging, an advanced and widely applied imaging approach in biomedical technology, can reflect the behaviors of cells, tissues, organs, or even human bodies. CNTs have been investigated in bioimaging because of their unique physical properties. Specifically, 1D semiconducting CNTs exhibit a narrow bandgap of about 1 eV, which allows fluorescent emission in the near-infrared (NIR) regions (700–900 nm and 1100–1400 nm). As one of the darkest materials,

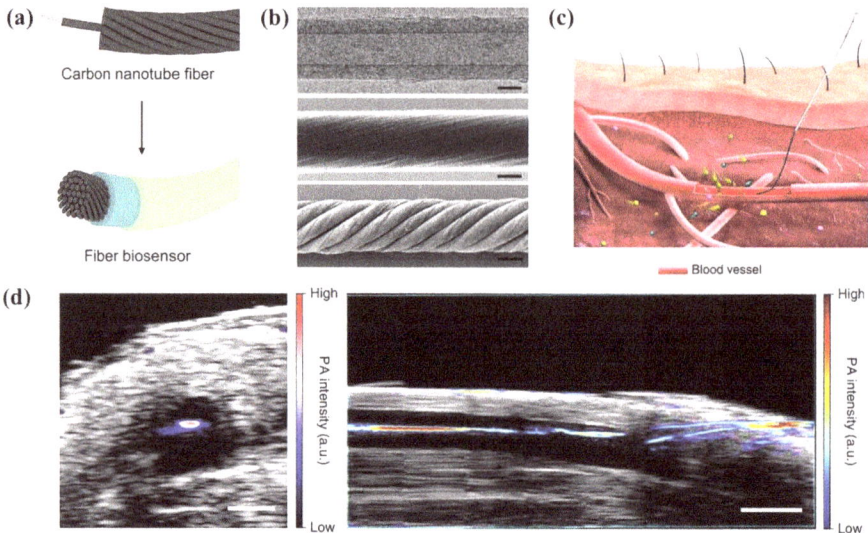

Figure 4.22. (a) Schematics of implantable fiber biosensors based on CNTs [174], reproduced with permission © American Chemical Society. (b) Representative TEM image of a multiwalled CNT (scale bar: 3 nm), an SEM image of a primary CNT fiber (scale bars, 6 µm), and a hierarchically helical CNT fiber assembled from primary CNT fibers (scale bar: 20 µm). (c) Schematic showing the injection of the fiber into a blood vessel. (d) Photoacoustic images of a feline vein after multiple sensing fibers were injected—top and side views (scale bars: 700 µm and 2 mm) [175], reproduced with permission © Springer Nature.

CNTs display strong absorption from the ultraviolet (UV) to the NIR and can thus be utilized as photoacoustic imaging contrast agents. One example is that CNTs coated with gold nanoparticles were capable of being used in photoacoustic imaging. They were used as new NIR contrast agents for non-invasive targeted *in vivo* mapping of the lymphatic system in live mice, and viability tests showed that gold-coated CNTs have minimal toxicity (figure 4.23) [176]. In addition, CNTs also exhibit strong resonance-enhanced Raman scattering. When used as Raman probes, CNTs are greatly advantageous in *in vitro* and *in vivo* bioimaging applications [177, 178].

4.6.3 Environmental applications

Today, environmental pollution has become a significant issue and causes great concern. Carbon-based materials are increasingly used to remediate the environment by monitoring or treating polluted air, water, and soil [179–181]. Their numerous advantages, such as large surface area, good thermal stability, wide pore size, different functional groups, and low mass transfer resistance, make CNTs competent in environmental applications.

CNT-based membranes have emerged as a promising water purification technology. Their hollow tubes can efficiently transport water molecules, benefitting the development of high-flux separation techniques. Appropriate pore diameters can

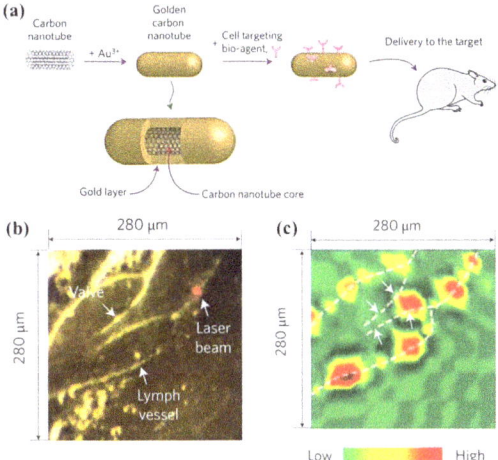

Figure 4.23. (a) Schematics of the synthesis of gold-coated CNTs and their delivery to the target [176], reproduced with permission © Springer Nature. (b) Fragment of mouse mesentery with mapping area. (c) Two-dimensional lymphatic mapping in selected mesenteric area after administration of gold-coated CNTs [176], reproduced with permission © Springer Nature.

constitute energy barriers at the channel entries, rejecting salt ions and permitting water through the nanotube. Flow rates can be further increased by fabricating aligned CNT membranes (figure 4.24(a)) [182]. When functionalized with positive, negative, or hydrophobic functional groups, CNT-based membranes can adjust the water influx and selectivity for the retention of particular pollutants (figure 4.24(b)) [182, 183].

In addition, adsorption has been demonstrated to be an economical and eco-friendly technique that is suitable for the efficient removal of emerging contaminants including dyes and heavy metals from aqueous systems [181] and undesired gas components from flue gas or by-product vapors [184, 185]. Functionalized CNTs are regarded as excellent sorbents due to their large surface areas and wide availability of surface adsorption sites. Magnetic MWCNTs have been studied for the removal of methylene blue dye from wastewater and achieved a removal efficiency of 178.57 mg g^{-1} [186]. In another report, CNTs modified with functional groups such as –COOH, –NH$_2$, and –SH were used for Hg(II) uptake [187]. The thiol functional CNTs exhibited the highest Hg(II) sorption performance, which was approximately three times that of pristine CNTs.

Acknowledgments

The authors are grateful to all the authors whose papers are cited here for their research work, and are sorry to the authors whose works are not cited here due to limited space. The authors acknowledge the financial support of the Fundamental Research Funds for the Central Universities (buctrc202118) and the National Natural Science Foundation of China (52172179).

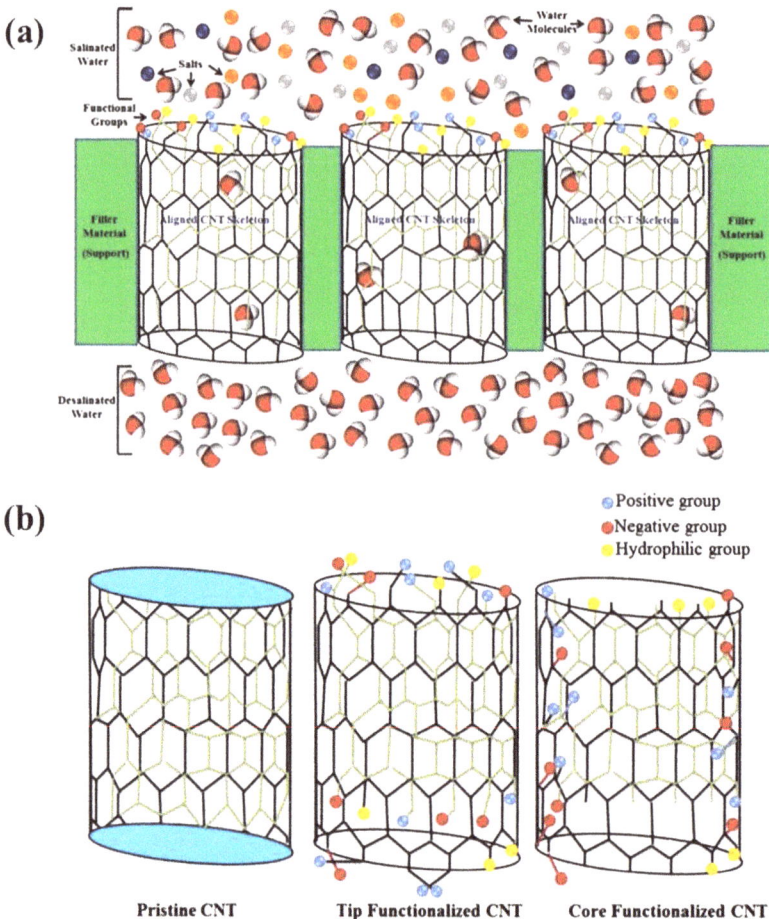

Figure 4.24. (a) A prototype of a CNT membrane, the trapping of salts, and the movement of water molecules. (b) Functionalization of CNT membranes [182], reproduced with permission © Elsevier.

References

[1] Iijima S 1991 Helical microtubules of graphitic carbon *Nature* **354** 56–8

[2] Iijima S and Ichihashi T 1993 Single-shell carbon nanotubes of 1-nm diameter *Nature* **363** 603–5

[3] Bethune D S, Kiang C H and De Vries M S *et al* 1993 Cobalt-catalysed growth of carbon nanotubes with single-atomic-layer walls *Nature* **363** 605–7

[4] Dai L 2004 *Intelligent Macromolecules for Smart Devices: From Materials Synthesis to Device Applications* (Berlin: Springer)

[5] Zhang F, Hou P-X and Liu C *et al* 2016 Epitaxial growth of single-wall carbon nanotubes *Carbon* **102** 181–97

[6] Dresselhaus M S, Dresselhaus G and Saito R 1995 Physics of carbon nanotubes *Carbon* **33** 883–91

[7] Saito R, Fujita M and Dresselhaus G *et al* 1992 Electronic structure of chiral graphene tubules *Appl. Phys. Lett.* **60** 2204–6

[8] Reznik D, Olk C H and Neumann D A *et al* 1995 X-ray powder diffraction from carbon nanotubes and nanoparticles *Phys. Rev.* B **52** 116–24

[9] Duclaux L 2002 Review of the doping of carbon nanotubes (multiwalled and single-walled) *Carbon* **40** 1751–64

[10] Krueger A 2010 *Carbon Materials and Nanotechnology.* (New York: Wiley)

[11] Krätschmer W, Lamb L D and Fostiropoulos K *et al* 1990 Solid C60: a new form of carbon *Nature* **347** 354–8

[12] Saito Y, Nishikubo K and Kawabata K *et al* 1996 Carbon nanocapsules and single-layered nanotubes produced with platinum-group metals (Ru, Rh, Pd, Os, Ir, Pt) by arc discharge *J. Appl. Phys.* **80** 3062–7

[13] Harris P J and Harris P J F 2009 *Carbon Nanotube Science: Synthesis, Properties and Applications* (Cambridge: Cambridge University Press)

[14] Ishigami M, Cumings J and Zettl A *et al* 2000 A simple method for the continuous production of carbon nanotubes *Chem. Phys. Lett.* **319** 457–9

[15] Ebbesen T W and Ajayan P M 1992 Large-scale synthesis of carbon nanotubes *Nature* **358** 220–2

[16] Colbert D T, Zhang J and Mcclure S M *et al* 1994 Growth and sintering of fullerene nanotubes *Science* **266** 1218

[17] Zhu H W, Li X S and Jiang B *et al* 2002 Formation of carbon nanotubes in water by the electric-arc technique *Chem. Phys. Lett.* **366** 664–9

[18] Alexandrou I, Wang H and Sano N *et al* 2004 Structure of carbon onions and nanotubes formed by arc in liquids *J. Chem. Phys.* **120** 1055–8

[19] Journet C, Maser W and Bernier P *et al* 1997 Large-scale production of single-walled carbon nanotubes by the electric-arc technique *Nature* **388** 756–8

[20] Liu C, Cong H and Li F *et al* 1999 Semi-continuous synthesis of single-walled carbon nanotubes by a hydrogen arc discharge method *Carbon (New York, NY)* **37** 1865–8

[21] Kroto H W, Heath J R and O'Brien S C *et al* 1985 C60: buckminsterfullerene *Nature* **318** 162–3

[22] Guo T, Nikolaev P and Thess A *et al* 1995 Catalytic growth of single-walled manotubes by laser vaporization *Chem. Phys. Lett.* **243** 49–54

[23] Guo T, Nikolaev P and Rinzler A G *et al* 1995 Self-assembly of tubular fullerenes *J. Phys. Chem.* **99** 10694–7

[24] Thess A, Lee R and Nikolaev P *et al* 1996 Crystalline ropes of metallic carbon nanotubes *Science* **273** 483

[25] Li Y, Mann D and Rolandi M *et al* 2004 Preferential growth of semiconducting single-walled carbon nanotubes by a plasma enhanced CVD method *Nano Lett.* **4** 317–21

[26] Lu C and Liu J 2006 Controlling the diameter of carbon nanotubes in chemical vapor deposition method by carbon feeding *J. Phys. Chem.* B **110** 20254–7

[27] Sinnott S B, Andrews R and Qian D *et al* 1999 Model of carbon nanotube growth through chemical vapor deposition *Chem. Phys. Lett.* **315** 25–30

[28] Li Y, Liu J and Wang Y *et al* 2001 Preparation of monodispersed Fe–Mo nanoparticles as the catalyst for CVD synthesis of carbon nanotubes *Chem. Mater.* **13** 1008–14

[29] Li Y, Kim W and Zhang Y *et al* 2001 Growth of single-walled carbon nanotubes from discrete catalytic nanoparticles of various sizes *J. Phys. Chem.* B **105** 11424–31

[30] Cheung C L, Kurtz A and Park H *et al* 2002 Diameter-controlled synthesis of carbon nanotubes *The J. Phys. Chem.* B **106** 2429–33

[31] Zhang R, Zhang Y and Zhang Q *et al* 2013 Growth of half-meter long carbon nanotubes based on Schulz–Flory distribution *ACS Nano* **7** 6156–61

[32] Kunadian I, Andrews R and Qian D *et al* 2009 Growth kinetics of MWCNTs synthesized by a continuous-feed CVD method *Carbon* **47** 384–95

[33] Zhang S, Zhang N and Zhang J 2020 Controlled synthesis of carbon nanotubes: past, present and future *Acta Phys. Chim. Sin.* **36** 1907021

[34] Hong G, Zhang B and Peng B *et al* 2009 Direct growth of semiconducting single-walled carbon nanotube array *J. Am. Chem. Soc.* **131** 14642–3

[35] Li P and Zhang J 2011 Sorting out semiconducting single-walled carbon nanotube arrays by preferential destruction of metallic tubes using water *J. Mater. Chem.* **21** 11815–21

[36] Yu B, Liu C and Hou P-X *et al* 2011 Bulk synthesis of large diameter semiconducting single-walled carbon nanotubes by oxygen-assisted floating catalyst chemical vapor deposition *J. Am. Chem. Soc.* **133** 5232–5

[37] LI W-S, HOU P-X and LIU C *et al* 2013 High-quality, highly concentrated semiconducting single-wall carbon nanotubes for use in field effect transistors and biosensors *ACS Nano* **7** 6831–9

[38] Wang Z, Zhao Q and Tong L *et al* 2017 Investigation of etching behavior of single-walled carbon nanotubes using different etchants *J. Phys. Chem.* C **121** 27655–63

[39] Li X, Tu X and Zaric S *et al* 2007 Selective synthesis combined with chemical separation of single-walled carbon nanotubes for chirality selection *J. Am. Chem. Soc.* **129** 15770–1

[40] Miyauchi Y, Chiashi S and Murakami Y *et al* 2004 Fluorescence spectroscopy of single-walled carbon nanotubes synthesized from alcohol *Chem. Phys. Lett.* **387** 198–203

[41] Wang B, Poa C H P and Wei L *et al* 2007 (*n,m*) Selectivity of single-walled carbon nanotubes by different carbon precursors on Co–Mo catalysts *J. Am. Chem. Soc.* **129** 9014–9

[42] Yang F, Wang X and Zhang D *et al* 2014 Chirality-specific growth of single-walled carbon nanotubes on solid alloy catalysts *Nature* **510** 522–4

[43] Zhang S, Kang L and Wang X *et al* 2017 Arrays of horizontal carbon nanotubes of controlled chirality grown using designed catalysts *Nature* **543** 234–8

[44] Nikolaev P, Bronikowski M J and Bradley R K *et al* 1999 Gas-phase catalytic growth of single-walled carbon nanotubes from carbon monoxide *Chem. Phys. Lett.* **313** 91–7

[45] Chen G Z, Fan X and Luget A *et al* 1998 Electrolytic conversion of graphite to carbon nanotubes in fused salts *J. Electroanal. Chem.* **446** 1–6

[46] Hsu W K, Terrones M and Hare J P *et al* 1996 Electrolytic formation of carbon nanostructures *Chem. Phys. Lett.* **262** 161–6

[47] Hsu W K, Hare J P and Terrones M *et al* 1995 Condensed-phase nanotubes *Nature* **377** 687–7

[48] Huang J Y, Yasuda H and Mori H 1999 Highly curved carbon nanostructures produced by ball-milling *Chem. Phys. Lett.* **303** 130–4

[49] Yuan L, Saito K and Pan C *et al* 2001 Nanotubes from methane flames *Chem. Phys. Lett.* **340** 237–41

[50] Vander Wal R L 2000 Flame synthesis of substrate-supported metal-catalyzed carbon nanotubes *Chem. Phys. Lett.* **324** 217–23

[51] Lee G W, Jurng J and Hwang J 2004 Synthesis of carbon nanotubes on a catalytic metal substrate by using an ethylene inverse diffusion flame *Carbon* **42** 682–5

[52] Delzeit L, Mcaninch I and Cruden B A *et al* 2002 Growth of multiwall carbon nanotubes in an inductively coupled plasma reactor *J. Appl. Phys.* **91** 6027–33

[53] Neupane S, Lastres M and Chiarella M *et al* 2012 Synthesis and field emission properties of vertically aligned carbon nanotube arrays on copper *Carbon* **50** 2641–50

[54] Qu L, Du F and Dai L 2008 Preferential syntheses of semiconducting vertically aligned single-walled carbon nanotubes for direct use in FETs *Nano Lett.* **8** 2682–7

[55] Frank S, Poncharal P and Wang Z *et al* 1998 Carbon nanotube quantum resistors *Science* **280** 1744–6

[56] Liang W, Bockrath M and Bozovic D *et al* 2001 Fabry-perot interference in a nanotube electron waveguide *Nature* **411** 665–9

[57] Baierle R, Fagan S B and Mota R *et al* 2001 Electronic and structural properties of silicon-doped carbon nanotubes *Phys. Rev.* B **64** 085413

[58] Liu K, Avouris P and Martel R *et al* 2001 Electrical transport in doped multiwalled carbon nanotubes *Phys. Rev.* B **63** 161404

[59] Vavro J, Llaguno M C and Satishkumar B *et al* 2002 Electrical and thermal properties of C 60-filled single-wall carbon nanotubes *Appl. Phys. Lett.* **80** 1450–2

[60] Liu X, Pichler T and Knupfer M *et al* 2004 Electronic properties of FeCl 3-intercalated single-wall carbon nanotubes *Phys. Rev.* B **70** 205405

[61] Bai Y, Zhang R and Ye X *et al* 2018 Carbon nanotube bundles with tensile strength over 80 GPa *Nat. Nanotechnol.* **13** 589–95

[62] Zhao Q, Nardelli M B and Bernholc J 2002 Ultimate strength of carbon nanotubes: a theoretical study *Phys. Rev.* B **65** 144105

[63] Bai Y, Yue H and Wang *et al* 2020 Super-durable ultralong carbon nanotubes *Science* **369** 1104

[64] Kim H-I, Wang M and Lee S K *et al* 2017 Tensile properties of millimeter-long multi-walled carbon nanotubes *Sci. Rep.* **7** 9512

[65] Jensen K, Mickelson W and Kis A *et al* 2007 Buckling and kinking force measurements on individual multiwalled carbon nanotubes *Phys. Rev.* B **76** 195436

[66] Wang J and Wang J-S 2006 Carbon nanotube thermal transport: ballistic to diffusive *Appl. Phys. Lett.* **88** 111909

[67] Kim P, Shi L and Majumdar A *et al* 2001 Thermal transport measurements of individual multiwalled nanotubes *Phys. Rev. Lett.* **87** 215502

[68] Pop E, Mann D and Wang Q *et al* 2006 Thermal conductance of an individual single-wall carbon nanotube above room temperature *Nano Lett.* **6** 96–100

[69] Fischer J E, Zhou W and Vavro J *et al* 2003 Magnetically aligned single wall carbon nanotube films: preferred orientation and anisotropic transport properties *J. Appl. Phys.* **93** 2157–63

[70] Hamasaki H, Takimoto S and Hirahara K 2021 Visualization of thermal transport within and between carbon nanotubes *Nano Lett.* **21** 3134–8

[71] Begtrup G E, Ray K G and Kessler B M *et al* 2007 Probing nanoscale solids at thermal extremes *Phys. Rev. Lett.* **99** 155901

[72] Thostenson E T, Li C and Chou T-W 2005 Nanocomposites in context *Compos. Sci. Technol.* **65** 491–516

[73] Itkis M E, Perea D E and Jung R *et al* 2005 Comparison of analytical techniques for purity evaluation of single-walled carbon nanotubes *J. Am. Chem. Soc.* **127** 3439–48

[74] Theocharous E, Deshpande R and Dillon A C *et al* 2006 Evaluation of a pyroelectric detector with a carbon multiwalled nanotube black coating in the infrared *Appl. Opt.* **45** 1093–7

[75] Mizuno K, Ishii J and Kishida H *et al* 2009 A black body absorber from vertically aligned single-walled carbon nanotubes *Proc. Natl Acad. Sci.* **106** 6044

[76] Zhang M, Ban D and Xu C *et al* 2019 Large-area and broadband thermoelectric infrared detection in a carbon nanotube black-body absorber *ACS Nano* **13** 13285–92

[77] Yin Z, Wang H and Jian M *et al* 2017 Extremely black vertically aligned carbon nanotube arrays for solar steam generation *ACS Appl. Mater. Interfaces* **9** 28596–603

[78] Tange M, Okazaki T and Iijima S 2011 Selective extraction of large-diameter single-wall carbon nanotubes with specific chiral indices by poly(9,9-dioctylfluorene-alt-benzothiadia-zole) *J. Am. Chem. Soc.* **133** 11908–11

[79] Wang H, Wang B and Quek X-Y *et al* 2010 Selective synthesis of (9,8) single walled carbon nanotubes on cobalt incorporated TUD-1 catalysts *J. Am. Chem. Soc.* **132** 16747–9

[80] Tange M, Okazaki T and Iijima S 2012 Selective extraction of semiconducting single-wall carbon nanotubes by poly(9,9-dioctylfluorene-alt-pyridine) for 1.5 µm emission *ACS Appl. Mater. Interfaces* **4** 6458–62

[81] Rao A M, Richter E and Bandow S *et al* 1997 Diameter-selective Raman scattering from vibrational modes in carbon nanotubes *Science* **275** 187

[82] Zhang D, Yang J and Li M *et al* 2016 (n,m) assignments of metallic single-walled carbon nanotubes by raman spectroscopy: the importance of electronic raman scattering *ACS Nano* **10** 10789–97

[83] Basiuk E V, Monroy-peláez M and Puente-lee I *et al* 2004 Direct solvent-free amination of closed-cap carbon nanotubes: a link to fullerene chemistry *Nano Lett.* **4** 863–6

[84] Mylvaganam K and Zhang L C 2004 Nanotube functionalization and polymer grafting: an *ab initio* study *J. Phys. Chem.* B **108** 15009–12

[85] Mallakpour S and Soltanian S 2016 Surface functionalization of carbon nanotubes: fabrication and applications *RSC Adv.* **6** 109916–35

[86] Vizuete M, Barrejón M and Gómez-escalonilla M J *et al* 2012 Endohedral and exohedral hybrids involving fullerenes and carbon nanotubes *Nanoscale* **4** 4370–81

[87] Dai L, Xue Y and Qu L *et al* 2015 Metal-free catalysts for oxygen reduction reaction *Chem. Rev.* **115** 4823–92

[88] Zhao J, Park H and Han J *et al* 2004 Electronic properties of carbon nanotubes with covalent sidewall functionalization *J. Phys. Chem.* B **108** 4227–30

[89] Park H, Zhao J and Lu J P 2006 Effects of sidewall functionalization on conducting properties of single wall carbon nanotubes *Nano Lett.* **6** 916–9

[90] Zhao Y-L and Stoddart J F 2009 Noncovalent functionalization of single-walled carbon nanotubes *Acc. Chem. Res.* **42** 1161–71

[91] Thess A, Lee R and Nikolaev P *et al* 1996 Crystalline ropes of metallic carbon nanotubes *Science* **273** 483–7

[92] Ding F, Larsson P and Larsson J A *et al* 2008 The importance of strong carbon–metal adhesion for catalytic nucleation of single-walled carbon nanotubes *Nano Letters* **8** 463–8

[93] Rafique M M A and Iqbal J 2011 Production of carbon nanotubes by different routes-a review *J. Encapsulation Adsorption Sci.* **1** 29

[94] Tsang S, Chen Y and Harris P *et al* 1994 A simple chemical method of opening and filling carbon nanotubes *Nature* **372** 159–62

[95] Dujardin E, Ebbesen T W and Krishnan A *et al* 1998 Purification of single-shell nanotubes *Adv. Mater.* **10** 611–3

[96] Gromov A, Dittmer S and Svensson J *et al* 2005 Covalent amino-functionalisation of single-wall carbon nanotubes *J. Mater. Chem.* **15** 3334–9

[97] Xu Y and Lin X 2007 Selectively attaching Pt-nano-clusters to the open ends and defect sites on carbon nanotubes for electrochemical catalysis *Electrochim. Acta* **52** 5140–9

[98] Chai G-L, Hou Z and Shu D-J *et al* 2014 Active sites and mechanisms for oxygen reduction reaction on nitrogen-doped carbon alloy catalysts: Stone–Wales defect and curvature effect *J. Am. Chem. Soc.* **136** 13629–40

[99] He H-Y and Pan B-C 2009 Studies on structural defects in carbon nanotubes *Frontiers of Physics in China* **4** 297–306

[100] Balasubramanian K and Burghard M 2005 Chemically functionalized carbon nanotubes *Small* **1** 180–92

[101] Kabir M and Van Vliet K J 2016 Kinetics of topological Stone–Wales defect formation in single-walled carbon nanotubes *J. Phys. Chem.* C **120** 1989–93

[102] Bettinger H F 2005 The reactivity of defects at the sidewalls of single-walled carbon nanotubes: the Stone–Wales defect *J. Phys. Chem.* B **109** 6922–4

[103] Hirsch A 2002 Functionalization of single-walled carbon nanotubes *Angew. Chem. Int. Ed.* **41** 1853–9

[104] Gao C, Guo Z and Liu J-H *et al* 2012 The new age of carbon nanotubes: an updated review of functionalized carbon nanotubes in electrochemical sensors *Nanoscale* **4** 1948–3

[105] Tasis D, Tagmatarchis N and Bianco A *et al* 2006 Chemistry of carbon nanotubes *Chem. Rev.* **106** 1105–36

[106] Xu Q, Li W and Ding L *et al* 2019 Function-driven engineering of 1D carbon nanotubes and 0D carbon dots: mechanism, properties and applications *Nanoscale* **11** 1475–504

[107] Mickelson E, Huffman C and Rinzler A *et al* 1998 Fluorination of single-wall carbon nanotubes *Chem. Phys. Lett.* **296** 188–94

[108] Yan Y, Miao J and Yang Z *et al* 2015 Carbon nanotube catalysts: recent advances in synthesis, characterization and applications *Chem. Soc. Rev.* **44** 3295–346

[109] Hamwi A, Alvergnat H and Bonnamy S *et al* 1997 Fluorination of carbon nanotubes *Carbon* **35** 723–8

[110] Zhang W, Bonnet P and Dubois M *et al* 2012 Comparative study of SWCNT fluorination by atomic and molecular fluorine *Chem. Mater.* **24** 1744–51

[111] Khare B N, Wilhite P and Meyyappan M 2004 The fluorination of single wall carbon nanotubes using microwave plasma *Nanotechnology* **15** 1650

[112] Khabashesku V N, Billups W E and Margrave J L 2002 Fluorination of single-wall carbon nanotubes and subsequent derivatization reactions *Acc. Chem. Res.* **35** 1087–95

[113] Wang Y, Iqbal Z and Mitra S 2006 Rapidly functionalized, water-dispersed carbon nanotubes at high concentration *J. Am. Chem. Soc.* **128** 95–9

[114] Liang F, Beach J M and Rai P K *et al* 2006 Highly exfoliated water-soluble single-walled carbon nanotubes *Chem. Mater.* **18** 1520–4

[115] Wei Y, Ling X and Zou L *et al* 2015 A facile approach toward preparation of sulfonated multi-walled carbon nanotubes and their dispersibility in various solvents *Colloids Surfaces* A **482** 507–13

[116] Chen J, Hamon M A and Hu H *et al* 1998 Solution properties of single-walled carbon nanotubes *Science* **282** 95–8

[117] Holzinger M, Abraham J and Whelan P *et al* 2003 Functionalization of single-walled carbon nanotubes with (R-)oxycarbonyl nitrenes *J. Am. Chem. Soc.* **125** 8566–80

[118] Basheer B V, George J J and Siengchin S *et al* 2020 Polymer grafted carbon nanotubes— synthesis, properties, and applications: a review *Nano-Structures Nano-Objects* **22** 100429

[119] Chen R J, Zhang Y and Wang D *et al* 2001 Noncovalent sidewall functionalization of single-walled carbon nanotubes for protein immobilization *J. Am. Chem. Soc.* **123** 3838–9

[120] O'connell M J, Boul P and Ericson L M *et al* 2001 Reversible water-solubilization of single-walled carbon nanotubes by polymer wrapping *Chem. Phys. Lett.* **342** 265–71

[121] Kuang Z, Kim S N and Crookes-Goodson W J *et al* 2010 Biomimetic chemosensor: designing peptide recognition elements for surface functionalization of carbon nanotube field effect transistors *ACS nano* **4** 452–8

[122] Wu Y, Jiang Z and Lu X *et al* 2019 Domino electroreduction of CO_2 to methanol on a molecular catalyst *Nature* **575** 639–42

[123] Smith B W, Monthioux M and Luzzi D E 1998 Encapsulated C60 in carbon nanotubes *Nature* **396** 323–4

[124] Yudasaka M, Ajima K and Suenaga K *et al* 2003 Nano-extraction and nano-condensation for C60 incorporation into single-wall carbon nanotubes in liquid phases *Chem. Phys. Lett.* **380** 42–6

[125] Kodama T, Ohnishi M and Park W *et al* 2017 Modulation of thermal and thermoelectric transport in individual carbon nanotubes by fullerene encapsulation *Nat. Mater.* **16** 892–7

[126] Govindaraj A, Satishkumar B and Nath M *et al* 2003 Metal nanowires and intercalated metal layers in single-walled carbon nanotube bundles *Advances in Chemistry: A Selection of CNR Rao's Publications (1994–2003)* (Singapore: World Scientific) pp 334–7

[127] Okubo S, Okazaki T and Hirose-Takai K *et al* 2010 Electronic structures of single-walled carbon nanotubes encapsulating ellipsoidal C70 *Chem. Phys. Lett.* **132** 15252–8

[128] Stephan O, Ajayan P and Colliex C *et al* 1994 Doping graphitic and carbon nanotube structures with boron and nitrogen *Science* **266** 1683–5

[129] Wang S, Iyyamperumal E and Roy A *et al* 2011 Vertically aligned BCN nanotubes as efficient metal-free electrocatalysts for the oxygen reduction reaction: a synergetic effect by co-doping with boron and nitrogen *Angew. Chem. Int. Ed.* **50** 11756–60

[130] Kim N S, Lee Y T and Park J *et al* 2003 Vertically aligned carbon nanotubes grown by pyrolysis of iron, cobalt, and nickel phthalocyanines *J. Phys. Chem.* B **107** 9249–55

[131] Yu D, Zhang Q and Dai L 2010 Highly efficient metal-free growth of nitrogen-doped single-walled carbon nanotubes on plasma-etched substrates for oxygen reduction *J. Am. Chem. Soc.* **132** 15127–9

[132] Jiang L and Gao L 2003 Modified carbon nanotubes: an effective way to selective attachment of gold nanoparticles *Carbon* **41** 2923–9

[133] Qu K, Zheng Y and Jiao Y *et al* 2017 Polydopamine-inspired, dual heteroatom-doped carbon nanotubes for highly efficient overall water splitting *Adv. Energy Mater.* **7** 1602068

[134] Gong K, Du F and Xia Z *et al* 2009 Nitrogen-doped carbon nanotube arrays with high electrocatalytic activity for oxygen reduction *Science* **323** 760–4

[135] Zhao Y, Yang L and Chen S *et al* 2013 Can boron and nitrogen co-doping improve oxygen reduction reaction activity of carbon nanotubes? *J. Am. Chem. Soc.* **135** 1201–04

[136] Li J-C, Hou P-X and Cheng M *et al* 2018 Carbon nanotube encapsulated in nitrogen and phosphorus co-doped carbon as a bifunctional electrocatalyst for oxygen reduction and evolution reactions *Carbon* **139** 156–63

[137] Yang L, Jiang S and Zhao Y *et al* 2011 Boron-doped carbon nanotubes as metal-free electrocatalysts for the oxygen reduction reaction *Angew. Chem. Int. Ed.* **50** 7132–5

[138] Li W, Yang D and Chen H *et al* 2015 Sulfur-doped carbon nanotubes as catalysts for the oxygen reduction reaction in alkaline medium *Electrochim. Acta* **165** 191–7

[139] Wang X, Ouyang C and Dou S *et al* 2015 Oxidized carbon nanotubes as an efficient metal-free electrocatalyst for the oxygen reduction reaction *RSC Adv.* **5** 41901–4

[140] Sohn G-J, Choi H-J and Jeon I-Y *et al* 2012 Water-dispersible, sulfonated hyperbranched poly (ether-ketone) grafted multiwalled carbon nanotubes as oxygen reduction catalysts *ACS nano* **6** 6345–55

[141] Suryanto B H R, Chen S and Duan J *et al* 2016 Hydrothermally driven transformation of oxygen functional groups at multiwall carbon nanotubes for improved electrocatalytic applications *ACS Appl. Mater. Interfaces* **8** 35513–22

[142] Wang S, Yu D and Dai L 2011 Polyelectrolyte functionalized carbon nanotubes as efficient metal-free electrocatalysts for oxygen reduction *J. Am. Chem. Soc.* **133** 5182–5

[143] Carini M, Shi L and Chamberlain T W *et al* 2019 Wall- and hybridisation-selective synthesis of nitrogen-doped double-walled carbon nanotubes *Angew. Chem. Int. Ed.* **58** 10276–80

[144] Lu X, Yim W-L and Suryanto B H R *et al* 2015 Electrocatalytic oxygen evolution at surface-oxidized multiwall carbon nanotubes *J. Am. Chem. Soc.* **137** 2901–7

[145] El-Sawy A M, Mosa I M and Su D *et al* 2016 Controlling the active sites of sulfur-doped carbon nanotube–graphene nanolobes for highly efficient oxygen evolution and reduction catalysis *Adv. Energy Mater.* **6** 1501966

[146] Zhang Y, Fan X and Jian J *et al* 2017 A general polymer-assisted strategy enables unexpected efficient metal-free oxygen-evolution catalysis on pure carbon nanotubes *Energy Environ. Sci.* **10** 2312–7

[147] Mo C, Jian J and Li J *et al* 2018 Boosting water oxidation on metal-free carbon nanotubes via directional interfacial charge-transfer induced by an adsorbed polyelectrolyte *Energy Environ. Sci.* **11** 3334–41

[148] Zhang S, Kang P and Ubnoske S *et al* 2014 Polyethylenimine-enhanced electrocatalytic reduction of CO_2 to formate at nitrogen-doped carbon nanomaterials *J. Am. Chem. Soc.* **136** 7845–8

[149] Wu J, Yadav R M and Liu M *et al* 2015 Achieving highly efficient, selective, and stable CO_2 reduction on nitrogen-doped carbon nanotubes *ACS Nano* **9** 5364–71

[150] Ventrapragada L K, Zhu J and Creager S E *et al* 2018 A versatile carbon nanotube-based scalable approach for improving interfaces in Li-ion battery electrodes *ACS Omega* **3** 4502–8

[151] Xie J and Lu Y-C 2020 A retrospective on lithium-ion batteries *Nat. Commun.* **11** 2499

[152] Yan Y, Li C and Liu C *et al* 2019 Bundled and dispersed carbon nanotube assemblies on graphite superstructures as free-standing lithium-ion battery anodes *Carbon* **142** 238–44

[153] Wang G, Chandrasekhar N and Biswal B P *et al* 2019 A crystalline, 2D polyarylimide cathode for ultrastable and ultrafast Li storage *Adv. Mater.* **31** 1901478

[154] Ning G, Zhang S and Xiao Z *et al* 2018 Efficient conducting networks constructed from ultra-low concentration carbon nanotube suspension for Li ion battery cathodes *Carbon* **132** 323–8

[155] Park S-H, King P J and Tian R *et al* 2019 High areal capacity battery electrodes enabled by segregated nanotube networks *Nat. Energy* **4** 560–7

[156] Yu D and Dai L 2010 Self-assembled graphene/carbon nanotube hybrid films for super-capacitors *The Journal of Physical Chemistry Letters* **1** 467–70

[157] Cao C, Zhou Y and Ubnoske S *et al* 2019 Highly stretchable supercapacitors via crumpled vertically aligned carbon nanotube forests *Adv. Energy Mater.* **9** 1900618

[158] Gong J, Li C and Wasielewski M R 2019 Advances in solar energy conversion *Chem. Soc. Rev.* **48** 1862–4

[159] Creutzig F, Agoston P and Goldschmidt J C *et al* 2017 The underestimated potential of solar energy to mitigate climate change *Nat. Energy* **2** 17140

[160] Habisreutinger S N, Leijtens T and Eperon G E *et al* 2014 Carbon nanotube/polymer composites as a highly stable hole collection layer in perovskite solar cells *Nano Lett.* **14** 5561–8

[161] Hu X-G, Hou P-X and Liu C *et al* 2018 Small-bundle single-wall carbon nanotubes for high-efficiency silicon heterojunction solar cells *Nano Energy* **50** 521–7

[162] Gong M, Shastry T A and Xie Y *et al* 2014 Polychiral semiconducting carbon nanotube–fullerene solar cells *Nano Lett.* **14** 5308–14

[163] Ramuz M P, Vosgueritchian M and Wei P *et al* 2012 Evaluation of solution-processable carbon-based electrodes for all-carbon solar cells *ACS Nano* **6** 10384–95

[164] Park S, Pitner G and Giri G *et al* 2015 Large-area assembly of densely aligned aingle-walled carbon nanotubes using solution shearing and their application to field-effect transistors *Adv. Mater.* **27** 2656–62

[165] Borghetti J, Derycke V and Lenfant S *et al* 2006 Optoelectronic switch and memory devices based on polymer-functionalized carbon nanotube transistors *Adv. Mater.* **18** 2535–40

[166] Bishop M D, Hills G and Srimani T *et al* 2020 Fabrication of carbon nanotube field-effect transistors in commercial silicon manufacturing facilities *Nat. Electron.* **3** 492–501

[167] Sireesha M, Jagadeesh Babu V and Kranthi Kiran A S *et al* 2018 A review on carbon nanotubes in biosensor devices and their applications in medicine *Nanocomposites* **4** 36–57

[168] Tîlmaciu C-M and Morris M C 2015 Carbon nanotube biosensors *Frontiers in Chemistry* **3** 59

[169] Merum S, Veluru J B and Seeram R 2017 Functionalized carbon nanotubes in bio-world: applications, limitations and future directions *Materials Science Engineering: B* **223** 43–63

[170] Gupta S, Murthy C and Prabha C R 2018 Recent advances in carbon nanotube based electrochemical biosensors *Int. J. Biol. Macromol.* **108** 687–703

[171] Yang N, Chen X and Ren T *et al* 2015 Carbon nanotube based biosensors *Sensors Actuators* B **207** 690–715

[172] He P, Li S and Dai L 2005 DNA-modified carbon nanotubes for self-assembling and biosensing applications *Synth. Met.* **154** 17–20

[173] Zheng T-T, Zhang R and Zou L *et al* 2012 A label-free cytosensor for the enhanced electrochemical detection of cancer cells using polydopamine-coated carbon nanotubes *Analyst* **137** 1316–8

[174] Feng J, Chen C and Sun X *et al* 2021 Implantable fiber biosensors based on carbon nanotubes *Accounts of Materials Research* **2** 138–46

[175] Wang L, Xie S and Wang Z *et al* 2020 Functionalized helical fibre bundles of carbon nanotubes as electrochemical sensors for long-term *in vivo* monitoring of multiple disease biomarkers *Nat. Biomed. Eng.* **4** 159–71

[176] Kim J-W, Galanzha E I and Shashkov E V *et al* 2009 Golden carbon nanotubes as multimodal photoacoustic and photothermal high-contrast molecular agents *Nat. Nanotechnol.* **4** 688–94

[177] Chen Z, Tabakman S M and Goodwin A P *et al* 2008 Protein microarrays with carbon nanotubes as multicolor Raman labels *Nat. Biotechnol.* **26** 1285–92

[178] Gong H, Peng R and Liu Z 2013 Carbon nanotubes for biomedical imaging: the recent advances *Adv. Drug Delivery Rev.* **65** 1951–63

[179] Tan C W, Tan K H and Ong Y T *et al* 2012 Energy and environmental applications of carbon nanotubes *Environ. Chem. Lett.* **10** 265–73

[180] Perreault F, De Faria A F and Elimelech M 2015 Environmental applications of graphene-based nanomaterials *Chem. Soc. Rev.* **44** 5861–96

[181] Gopinath K P, Vo D-V N and Prakash D G *et al* 2021 Environmental applications of carbon-based materials: a review *Environ. Chem. Lett.* **19** 557–82

[182] Das R, Ali M E and Abd Hamid S B *et al* 2014 Carbon nanotube membranes for water purification: a bright future in water desalination *Desalination* **336** 97–109

[183] Kar S, Bindal R and Tewari P 2012 Carbon nanotube membranes for desalination and water purification: challenges and opportunities *Nano Today* **7** 385–9

[184] Keller L, Ohs B and Abduly L *et al* 2019 Carbon nanotube silica composite hollow fibers impregnated with polyethylenimine for CO_2 capture *Chem. Eng. J.* **359** 476–84

[185] Keller L, Ohs B and Lenhart J *et al* 2018 High capacity polyethylenimine impregnated microtubes made of carbon nanotubes for CO_2 capture *Carbon* **126** 338–45

[186] Ahamad T, Naushad M and Eldesoky G E *et al* 2019 Effective and fast adsorptive removal of toxic cationic dye (MB) from aqueous medium using amino-functionalized magnetic multiwall carbon nanotubes *J. Mol. Liq.* **282** 154–61

[187] Hadavifar M, Bahramifar N and Younesi H *et al* 2014 Adsorption of mercury ions from synthetic and real wastewater aqueous solution by functionalized multi-walled carbon nanotube with both amino and thiolated groups *Chem. Eng. J.* **237** 217–28

IOP Publishing

Functional Carbon Materials

Jianmin Ma and Jiantie Xu

Chapter 5

Advanced functional graphene: characteristics, preparation, and applications

Weiyin Gao, Yonghua Chen and Chenxin Ran

Graphene has been under the spotlight in the science community during recent decades, and the unique characteristics of graphene and its derivatives have made this material a game-changer in various application fields, such as energy, electronics, biomedicine, materials science, and chemical engineering. This chapter is dedicated to giving the reader a full view of the development of graphene/graphene-based materials and their advanced applications. First, the intrinsic properties of graphene are reviewed, followed by an introduction to the synthesis methods of graphene and its derivatives. The state-of-the-art applications of functional graphene-based materials are then comprehensively summarized, including energy conversion and storage, transistors, bio-related applications, and environmental applications. Finally, the future prospects of this modern material are discussed.

5.1 Introduction

Graphene is defined as a monolayer of carbon atoms in the planar sp^2 hybridization configuration, which packs into a 2D honeycomb lattice. Two-dimensional graphene is the building block for all the other dimensional carbon materials, such as 0D fullerene, 1D nanotubes, and 3D graphite [1]. Due to its 2D single-crystal lattice and unique long-range π-conjugated electronic structure, graphene possesses impressive properties including high electron mobility, high specific surface area, high chemical stability, great mechanical strength, and superior thermal conductivity [2–6]. As a result, researchers have paid much attention to the development of graphene and graphene-based materials and their advanced applications during the last few decades.

As early as 1947, Wallace predicted the existence of atomic layers of graphite (i.e. graphene) and their extremely high carrier mobility [7]. However, it was believed that strict 2D crystals were thermodynamically unstable and could not exist [8], as it was confirmed by experimental observation that the melting temperature of thin films could decrease rapidly with decreasing thickness, and the film could easily segregate into islands or decompose at a thickness of dozens of atomic layers [9, 10]. Therefore, early in 1986, when an isolated graphene was assumed to be impossible, Boehm and co-workers defined 'graphene' as a hypothetical member of the polycyclic aromatic hydrocarbon (PAH) series (i.e. naphthalene, anthracene, phenanthrene, tetracene, coronene, ovalene, etc.) with infinite size [11]. Nevertheless, efforts were still made to produce thin layers of graphene, because graphene, with its tightly packed 2D sp^2-hybridized orthohexagonal lattice, was considered to be capable of stable existence. In the early stages of graphene research, a method involving the mechanical exfoliation of graphite was adopted to produce graphene, in which an atomic force microscope (AFM) tip was used to peel away highly oriented pyrolytic graphite (HOPG), but the thinnest layer obtained at that time had a thickness of 200 nm (i.e. ~600 layers of monolayer graphene) [12]. Later, Kim *et al* improved the method by using a tipless cantilever to successively reduce the thickness of the peeled graphite down to 10 nm (i.e. ~30 layers of monolayer graphene). However, actual monolayer graphene was not obtained until 2004, when Geim and co-workers at Manchester University first successfully isolated single-layer graphene from a graphite flake by simple cellophane-tape-assisted mechanical exfoliation, and the impressively strong ambipolar electric field effect of the graphene was observed [13]. Since then, the intrinsic characteristics of graphene have been extensively investigated, and graphene has been found to exhibit great superiority in a wide range of electronic applications [14]. Therefore, after just six years, the Nobel Prize in Physics in 2010 was granted to Geim and Novoselov for their pioneering studies of graphene, which started a 'graphene trend' in the materials science community. Figure 5.1(a) shows the numbers of graphene-related

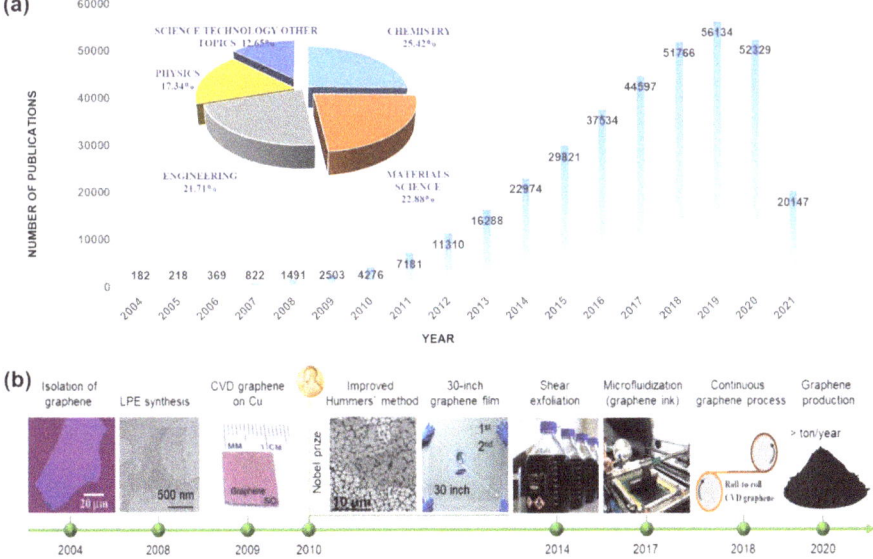

Figure 5.1. (a) Annual publications of studies with the keyword 'graphene' in the past 17 years (2004–21) from the *Web of Science*. (b) Timeline of representative developments in graphene preparation [15], reproduced with permission © Elsevier.

publications reported since 2004; explosive growth in the number of publications can be observed after 2010. In addition, multidisciplinary research areas are involved in graphene (see the inset of figure 5.1(a)), demonstrating the enormous potential and bright future of graphene-based materials.

The development of a material is highly dependent on the feasibility of its large-scale production and universality in various applications. However, those early mechanical exfoliation methods made it difficult to produce monolayer graphene at scale, which created a great obstacle for its practical application [16]. Thus, some solution-based exfoliation approaches have been developed as alternative methods for the large-scale production of high-quality graphene, including the exfoliation of graphite assisted by surfactants [17], ultrasonication [18], shear mixing [19], and electrical fields [20]. Meanwhile, chemical approaches, either top-down (functional-ization into graphene oxide (GO) followed by chemical reduction) [21] or bottom-up (assembled from various carbon sources) [22], have been developed. Figure 5.1(b) presents the timeline of representative developments in graphene preparation. Most importantly, these chemical routes allow the introduction of functional groups and defects on the graphene plate, which provides a large number of active sites for interactions between graphene and other molecules. It is also worth noting that some unique properties of graphene can be induced by tuning the dimensionality of graphene, including 0D graphene quantum dots (GQDs) [23], 1D graphene nano-ribbons (GNRs) [24], and 3D graphene [6]. These merits greatly expand the functionality of graphene and promote its widespread application, for example, in catalysis [25], detection [26], sensing [27], energy conversion and storage [28], smart composite materials [29], and bio-related applications [30].

In this chapter, we will comprehensively summarize the current understanding and recent developments of graphene and its derivatives in terms of the fundamental characteristics of graphene, the preparation methods of graphene and its derivatives, and the applications of graphene-based materials.

5.2 Fundamental characteristics of graphene

The ideal graphene structure is composed of tightly connected carbon atoms with planar sp^2-hybridization bonds, resulting in atomically thin 2D geometry (figure 5.2(a)) and a π-conjugated electronic configuration (figure 5.2(b)). These unique geometric and electric features of graphene endow it with various unique intrinsic characteristics, including optical, electrical, mechanical, thermal, chemical, and superconductive properties.

5.2.1 Optical properties

Due to its 2D atomic structure, graphene has been shown to possess mirrored valence and conduction bands with a single zero-state point at the Dirac point (figures 5.2(b) and (c)) based on Bloch wavefunction analysis and tight-binding models [34]. In this case, the optical absorption (i.e. optical conductivity) of graphene from the infrared through the visible range of the spectrum originates from the high-frequency conductivity of Dirac fermions, which is a constant equal to

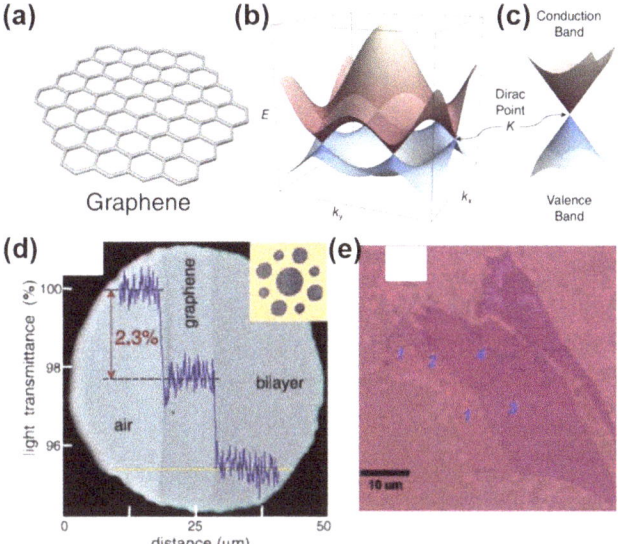

Figure 5.2. Illustration of the (a) crystal structure and (b) the band structure of 2D graphene. (c) The vertically mirrored Dirac cones of the conductance band and the valence band in graphene, which do not have a bandgap at the Dirac points [31], reproduced with permission © John Wiley & Sons. (d) Photograph of graphene and its bilayer; the line scan profile shows the intensity of light transmittance at different distances. The inset shows a photograph of the sample [32], reproduced with permission © American Association for the Advancement of Science. (e) Optical image of graphene flakes with one, two, three, and four layers on a SiO_2/Si substrate [33], reproduced with permission © American Chemical Society.

$\pi e^2/2h$, where e is the electron charge and h is Planck's constant [35]. For normal incident light, the optical transmittance T and reflectance R of graphene are $T = (1 + 1/2\pi\alpha)^{-2}$ and $R = 1/4\pi^2\alpha^2 T$, where $\alpha = 2\pi e^2/h.c. \approx 1/137$ and c is light speed. Thus, the opacity of graphene is $(1 - T) \approx \pi\alpha \approx 2.3\%$, that is, $T \approx 97.7\%$ [36]. It should be noted that the calculation of the T value of graphene is a constant that does not involve material parameters, which is proposed to be due to the structure and electronic properties of graphene [37]. Experimental results support the calculated transmittance value (i.e. 97.7%) for graphene in the visible range, in which the transmittance linearly decreases with increasing numbers of graphene layers (figure 5.2(d)) [32]. It should be noted that for incident photons with energies lower than 0.5 eV, a deviation from this transmittance value has been observed, which is attributed to the finite temperature and a doping-induced chemical potential shift in the charge neutrality (Dirac) point of graphene [38].

The constant transmittance value of single graphene can be useful for distinguishing the number of layers in multilayer graphene. For example, the optical contrast of graphene on various substrates, such as SiO_2/Si [39], Si_3N_4/Si [40], SiC [41], and Al_2O_3/Si [42], has been investigated, and the contrast of graphene on these substrates is found to depend on the wavelength and the angle of the incident light [43]. Ni *et al* reported that the number of graphene layers was identified as being less than ten on a 285 nm SiO_2/Si substrate using white light illumination (figure 5.2(e)) [33].

The refractive index n of monolayer graphene was determined to be $n = 2.0 - 1.1i$ in the visible range, which is different from that of bulk graphite with $n = 2.6 - 1.3i$ [44]. By fitting the experimental spectra as a function of wavelength, Bruna *et al* proposed that the complex refractive index of graphene and graphite could be generally expressed as $n = 3 - iC/(3\lambda)$ (where $C = 5.446$ µm^{-1} and λ is wavelength) [45].

Due to the low density of states near the Dirac point in graphene, the optical transition of graphene shows a dependence on the gate-variable effect. Wang *et al* reported that the optical transitions of graphene can be substantially modified by electrical gating; they used infrared spectroscopy to probe the interband optical transitions in graphene [46]. A change of electrical gating causes a shift in the Fermi level, which leads to a significant variation of charge density that in turn results in a significant change in the optical transmission of graphene.

5.2.2 Electric properties

The band structure of graphene stems from the symmetry in its unit cell, which contains two inequivalent lattice points, with a carbon–carbon bond length of 1.42 Å and a lattice parameter of 2.46 Å (figure 5.3(a)) [47]. Thus, the first Brillouin zone of a graphene unit has two inequivalent points K and K′ (figure 5.3(b)), and a crossing of

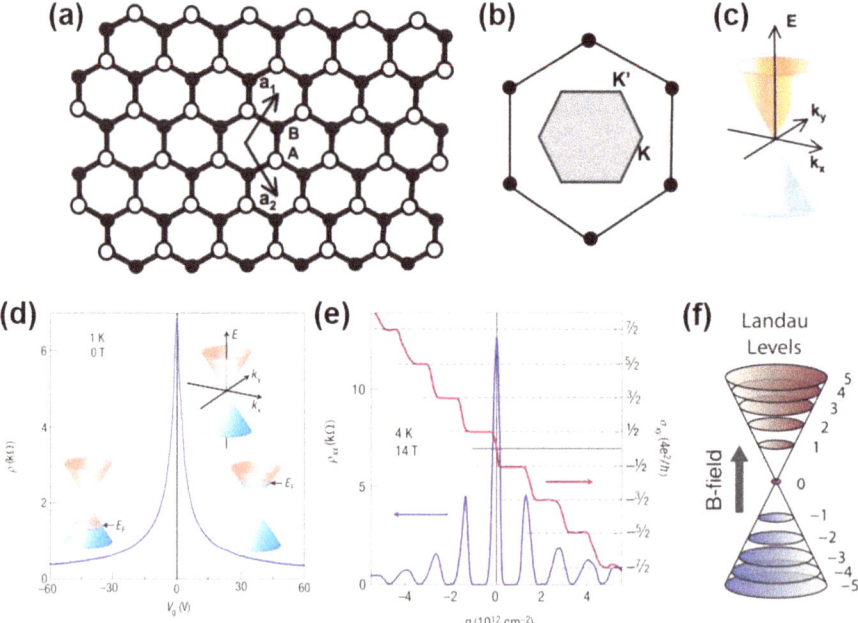

Figure 5.3. Schematics of the (a) crystal structure, (b) Brillouin zone, and (c) dispersion spectrum of graphene [36], reproduced with permission © John Wiley & Sons. (d) Ambipolar electric field effect in graphene; the insets show the changes in the position of the Fermi energy for changes in the gate voltage. (e) The hallmark of massless Dirac fermions in graphene [1], reproduced with permission © Springer Nature. (f) Landau levels for the anomalous and fractional quantum Hall effect (QHE) in graphene [31]. Source: [31], reproduced with permission © John Wiley & Sons.

the mirrored valence and conduction bands occurs, which intersects at a single point of zero states, i.e. the so-called Dirac point (figure 5.3(c)). This band intersection at the intrinsic Fermi level is responsible for the zero-gap semiconducting nature and semimetallic properties of graphene. Moreover, this crystal structure leads to the result that the bonding (π) and anti-bonding ($\pi*$) orbitals touch in each energy valley at K and K', considering that electron hopping only occurs between nearest-neighbor atomic sites [48].

This unique electronic structure of graphene leads to a number of extraordinary properties not seen in conventional materials. For example, the carriers' speed in graphene, the so-called Fermi velocity, is as high as 10^6 m s^{-1} ($\sim$1/300 of the velocity of light), and thus carriers in graphene act as photon-like relativistic Dirac fermions, leading to remarkable electrical transport properties [49]. In the band of graphene, the density of states (DOS) shows linear dispersion and converges to zero at the Dirac point, resulting in an effective mass of zero. Therefore, electrons in graphene essentially behave as massless Dirac fermions, leading to tremendously high carrier mobility which affects the way in which charge travels in a unit electrical field. Experimentally, the existence of massless Dirac quasi-particles in graphene has been shown to depend on the square root of the electronic density in graphene measured by the cyclotron mass [34, 50]. This property of rapid carrier transport under a weak electrical field has enabled the development of graphene-based field-effect transistors (FETs) that have ultrafast (up to 40 GHz) and efficient (6%–16% internal quantum efficiency) photo-responses [51], demonstrating the great potential of graphene to be used in high-speed optoelectronic devices for communications, detection, sensing, and so on.

As a zero bandgap semiconductor, graphene also exhibits an expressive ambipolar electric field effect, in which the charge carriers can be continuously tuned between electrons and holes with concentrations as high as 10^{13} cm^{-2} and room-temperature mobilities of up to 1.5×10^4 cm^2 V^{-1}s^{-1} (figure 5.3(d)) [4, 13, 34, 50]. Experimentally, an exceptional mobility of 2×10^5 cm^2 V^{-1} s^{-1} has been reported in suspended graphene by minimizing impurity scattering effects [3]. Most importantly, even at high carrier densities, carrier mobility in graphene remains high [52], evidencing the ballistic transport of carriers on the sub-micrometer scale in graphene [53].

Another measure of the electronic quality of graphene is whether the quantum Hall effect (QHE) can be observed at room temperature [54]. By strictly confining electrons to two dimensions, the QHE and Berry's phase of graphene can be activated [55]. In a magnetic field, the conventional QHE forms quantized orbital energies, called Landau levels. However, due to the additional complexities of the electronic structure of graphene [31], graphene has been found to exhibit an anomalous QHE with a Landau level residing at zero, as well as half-integer, fractional, spin, and valley QHEs (figures 5.3(e) and (f)). This magnetically tunable conductivity of graphene explains the experimentally observed physical complexities of graphene [56].

5.2.3 Mechanical properties

Graphene exhibits impressive mechanical properties, which have brought graphene to the fore both as an individual mechanical material and as a reinforcing agent in composites [57]. Note that graphite is not mechanically strong because it is easy to shear between layers, similarly to smooth pencil tips. The reason for the exceptional mechanical properties of graphene lies in the stability of the C–C sp^2 bonds that form the hexagonal lattice, which greatly resists different forms of in-plane deformation of the structure. Lee *et al* studied the elastic properties and intrinsic breaking strength of freestanding monolayer graphene by nanoindentation using an AFM tip (figures 5.4(a) and (b)) [2]. The elastic property of graphene was measured by force-displacement curves (figure 5.4(c)), from which the second-order elastic stiffness (E^{2D}) of graphene was calculated to be = 340 ± 50 N m^{-1}, corresponding to a Young's modulus of 1.0 ± 0.1 TPa (figure 5.4(d)). In addition, the intrinsic strength of the monolayer graphene was measured to be 42 N m^{-1}, which equates to an intrinsic strength of 130 GPa.

Another important mechanical property of graphene is its fracture toughness, which is a crucial property that is very relevant to engineering applications. Zhang *et al* developed an *in situ* micromechanical testing device and a nanoindenter within an SEM to determine the fracture toughness of chemical vapor deposition (CVD)-synthesized graphene [58]. To observe the brittle fracture of graphene, a central crack was introduced into a graphene membrane by a focused ion beam (FIB) and then a load was applied. It was found that the fracture stress of graphene decreases with increasing crack length. The critical strain energy release rate of graphene was found to be 15.9 J m^{-2}, while measurement of the fracture toughness determined that the critical stress intensity factor was 4.0 ± 0.6 MPa. It is worth noting that

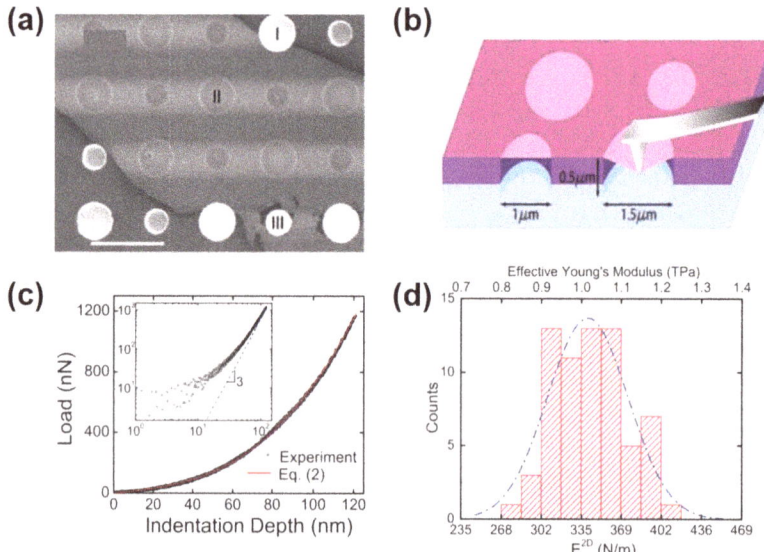

Figure 5.4. (a) Scanning electron microscopy (SEM) image of a graphene flake containing an array of circular holes 1 μm and 1.5 μm in diameter. (b) Schematic of nanoindentations in suspended graphene produced by an AFM tip. (c) Loading/unloading test with increasing AFM tip indentation depth. (d) Histogram of the elastic stiffness of graphene. Source: [2], reproduced with permission © American Association for the Advancement of Science.

wrinkling in graphene, which is an inevitable consequence of the transfer process of CVD-produced graphene, has a great impact on the mechanical properties of graphene [59]. Min *et al* performed molecular dynamic simulations to study the effect of wrinkles on the fracture stress of zigzag graphene [60]. They found that the fracture stress of flat graphene is 97.5 GPa, while that of wrinkled graphene is reduced to ~60 GPa at room temperature; this occurs because the presence of wrinkles softens the material to some extent.

5.2.4 Thermal properties

As mentioned above, the carrier density of pristine graphene is relatively low, and thus the electronic contribution to its thermal conductivity (Wiedemann–Franz law) is negligible. Therefore, the thermal conductivity (κ) of graphene is dominated by phonon transport, that is, diffusive conduction at high temperatures and ballistic conduction at sufficiently low temperatures [61]. Molecular dynamics simulations based on the Green–Kubo approach have shown that κ is directly proportional to $1/T$ for defect-free graphene as the temperature T increases beyond ~100 K [62]. Accordingly, a room-temperature thermal conductivity of ~6000 W m^{-1} K^{-1} has been predicted for suspended monolayer graphene, which is proposed to be much higher than that of other graphitic carbon materials [62].

Experiments have also been carried out to measure the thermal conductivity of graphene. Balandin *et al* studied the thermal conductivity of suspended monolayer graphene by measuring the peak position shift of the G band in the Raman spectra at room temperature [5]. Figure 5.5(a) demonstrates the setup used for the measurement, in which a trench is used to suspend the graphene and a laser beam at an excitation wavelength of 488 nm is focused on the center of the graphene. When the laser beam irradiates the graphene, the heat flows radially from the center of the graphene to the support, and the heat loss via air is considered to be negligible compared to heat conduction in the graphene [64]. The increased temperature of the heated graphene causes a redshift of the Raman G peak due to bond softening, and the redshift of the G peak is linearly correlated with the sample temperature at low laser powers [65]. Figure 5.5(b) shows the positional shift of the Raman G peak as a function of the excitation power applied to the graphene; thermal conductivity as high as ~5000 W m^{-1} K^{-1} can be determined from the slope of the trend line. Among various carbon materials, suspended graphene exhibits superior in-plane thermal conductivity, especially at room temperature (figure 5.5(c)) [63].

Although high thermal conductivity has been observed in freely suspended graphene, the in-plane thermal conductivity of graphene has been found to decrease significantly if the graphene contacts a substrate or is confined to a limited width (e.g. GNRs). This effect is to be expected, because the propagation of phonons in an atomically thin graphene sheet should be highly sensitive to heterogeneous surfaces or edge effects [63]. For example, Cai *et al* measured the thermal conductivity of CVD-grown graphene deposited on a thin silicon nitride membrane with an array of through holes, and a thermal conductivity of ~2500 W m^{-1} K^{-1} (at 350 K) was observed [66]. However, in the case of graphene that was micromechanically exfoliated and deposited on a SiO$_2$ substrate, the

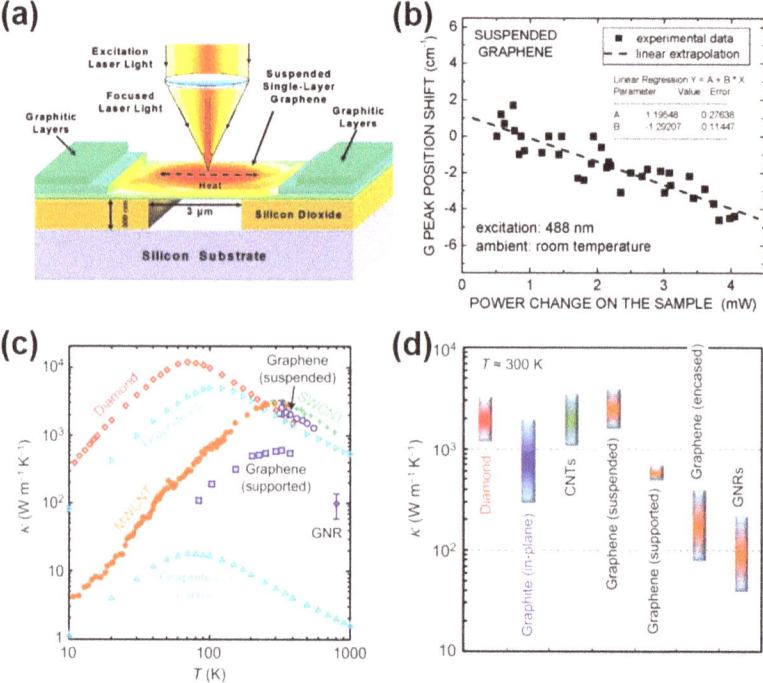

Figure 5.5. Schematic of the experimental setup used to measure the thermal conductivity of single-layer graphene suspended across a trench by laser light excitation. (b) Shift of the G peak in the Raman spectra of the function of power change on the sample [5], reproduced with permission © American Chemical Society. Thermal conductivity κ of different carbon materials (c) as a function of temperature and (d) at room temperature [63], reproduced with permission © Springer Nature.

thermal conductivity was only ~600 W m^{-1} K^{-1}, which still exceeds those of metals such as Cu [67]. Meanwhile, the thermal conductivity of SiO$_2$-encased graphene was measured at ~160 W m^{-1} K^{-1} [68], and that of supported GNRs with a width of ~20 nm was estimated to be ~80 W m^{-1} K^{-1} (figure 5.5(d)) [69]. The decreased thermal conductivity was attributed to the coupling and scattering of phonons in graphene with the vibrational modes of the substrate [70]. In addition, the reduced thermal conductivity may also be attributed to the edge effect of graphene, and molecular dynamics simulations have revealed that the heat flows of graphene can be greatly influenced by introducing atomistic alterations into the pristine graphene structural lattice, such as defects, strains, chemical groups, edge roughness, and folding [63].

5.2.5 Chemical properties

The surfaces of defect-free graphene with an ideal lattice appear to be chemically inert, and these surfaces usually interact with other molecules via physical adsorption (π–π interactions). However, the carbon atoms at the edges of graphene contain hydrogen atoms, which have unique chemical reactivity, and thus several chemical groups (e.g. hydroxyls, carboxyls, hydrogenated compounds, and amines) can be anchored at these edges [73]. In addition, the reactivity of graphene edges is sensitive to the termination type of the edges (i.e. armchair or zigzag), which have different features in terms of the energy per atom and their density [74]. Boukhvalov *et al* studied the effect of the edge on the reactivity of the graphene structure (figure 5.6(a)) [71]. The computational results (figure 5.6(b)) show that the chemisorption of hydrogen is energetically most favorable near the edge and is less favorable in the middle of the graphene structure, where the formation energy is similar to that of bulk graphite (1.44 eV) [75]. In addition, the chemisorption energy is also sensitive to the distortions created by the adsorbed hydrogen atoms, and the resultant geometric frustration of the graphene structure leads to the zigzag dependence of the chemisorption energy on atomic position shown in figure 5.6(b). In practical terms, ripples and wrinkles are generally present in produced graphene, and the sites at ripples and wrinkles are accordingly reactive. In corrugated graphene with high degrees of curvature (figure 5.6(c)), the chemical reactivity of the graphene surface is notably enhanced, especially when the ratio of height h to radius R of the ripple is greater than 0.07 (figure 5.6(d)) [72].

In addition, the chemical reactivity of graphene can be also enhanced by introducing the structural defects associated with dangling bonds, where hydroxyl, carboxyl, and other groups can be easily attached to the defects [71]. Notably, reconstructed defects without dangling bonds can still show increased local

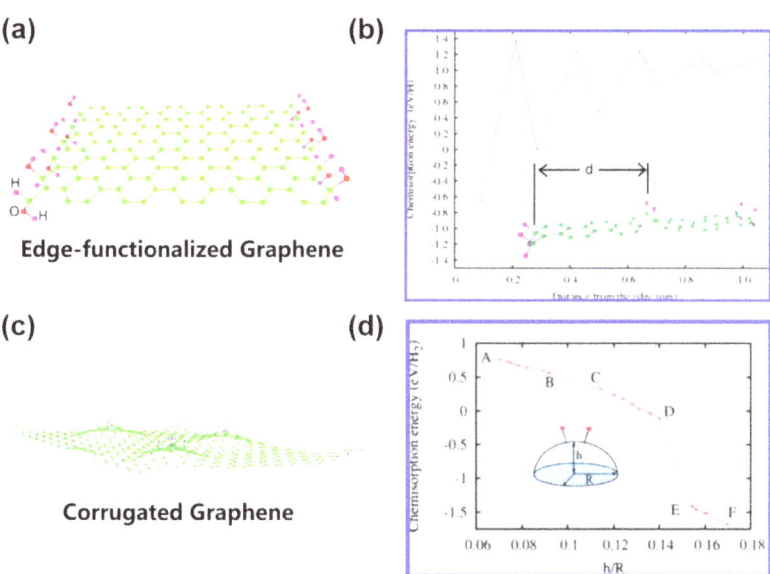

Figure 5.6. (a) Illustration of a graphene structure whose edges are functionalized by hydrogen atoms and hydroxyl groups. (b) Chemisorption energy of hydrogen in edge-functionalized graphene as a function of the distance from the edge [71], reproduced with permission © American Chemical Society. (c) Illustration of rippled graphene structure with corrugated morphology produced by the chemisorption of hydrogen atoms. (d) Dependence of the chemisorption energy on the curvature of the ripple, which is defined by the ratio of the height h to the radius R of the ripple (inset) [72], reproduced with permission © American Chemical Society.

reactivity due to the locally changed density of the π electrons [76]. Moreover, the reactivity of graphene can also be enhanced by introducing other heteroatoms with high reactivity. The replacement of carbon atoms in the graphitic lattice by non-carbon atoms, such as nitrogen and boron dopants that possess similar valence electrons to those of carbon, can significantly increase the surface reactivity [77]. Typically, nitrogen-doped graphene is a classic electrocatalyst for efficient chemical reduction reactions [78]. In addition, the introduction of halogen atoms (Cl and F) [79, 80] and oxygenated groups has also been shown to enhance the reactivity and hydrophilicity of graphene to some extent [81].

5.2.6 Superconductive properties

Another interesting feature of graphene is its strong interaction and correlation effects, which can create many exotic states in which superconductivity may be possible. The chemical potential of graphene can be tuned through an electric field effect, which paves the way for realizing superconductivity [50]. Various theoretical works have discussed the onset of superconductivity and the pairing symmetry [83]. However, a constrained-path quantum Monte Carlo study suggests that intrinsic superconductivity cannot exist in single-layer graphene, because the electron correlation in lightly doped graphene is not strong enough to produce intrinsic superconductivity due to the low density of states (DOS) [*Phys. Rev. B* 2011, 84. 121410]. Although there have been many theoretical predictions, reliable experimental studies are still rare.

In 2018, Cao *et al* reported breakthrough progress in the discovery of novel electronic ground states in twisted bilayer graphene (TBG) [82]. In this work, typical fully encapsulated TBG devices based on a four-probe measurement structure were used (figure 5.7(a)), and the two sheets of graphene were prepared from the same exfoliated flake by sequential graphene and hexagonal boron-nitride (hBN) flake pick-up steps using a hemispherical handle substrate to produce TBG with a relative twist angle that was precisely controlled to within ~0.1°–0.2° [84]. Figure 5.7(b) shows the four-probe resistance of devices with a TBG angle of 1.05° as a function of density n and temperature T; the device exhibits two pronounced superconducting domes on each side of the half-filling correlated insulating state, which is similar to those associated with high-temperature superconductivity in cuprate materials [82].

It is proposed that rotating the top layer of TBG away from Bernal stacking to the so-called 'magic angle' of ~1.1° results in a moiré superlattice in TBG (figure 5.7(c)). Figure 5.7(d) shows the typical ordered moiré pattern observed in TBG. The interplay between the moiré superlattice and hybridization leads to the formation of an isolated flat band at the charge neutrality point, near which it forms novel states (known as Mott insulators) at half band filling [83]. In this case, the introduction of charge carriers into the TBG drives the 'Mott insulator' away from half band filling, and a superconducting phase appears. This discovery has inspired a huge number of physicists to study TBG. Because of the electron–electron interactions, magic-angle TBG displays unusual spectral characteristics, which have been observed by high-resolution scanning tunneling microscopy (STM) [85]. By further controlling the interlayer coupling via controlling the hydrostatic

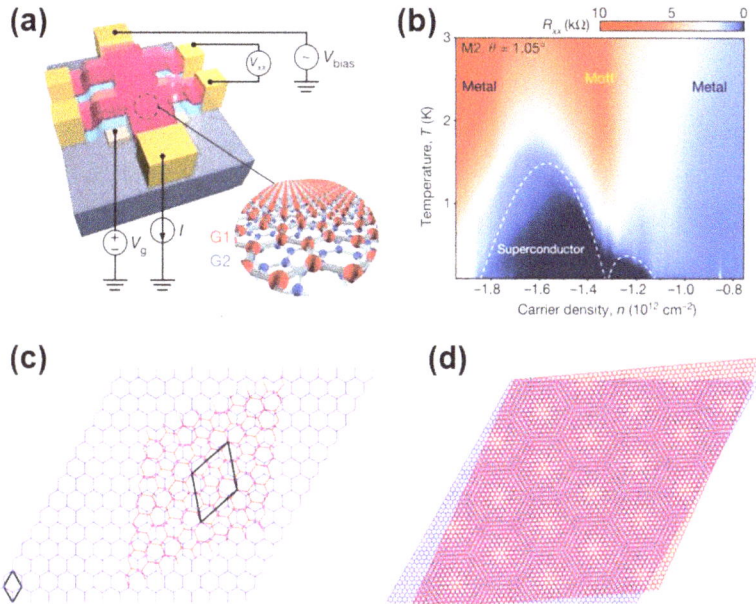

Figure 5.7. (a) Schematic of a typical four-probe measurement device used for the characterization of the superconductive properties of TBG. (b) Phase diagram of a TBG with a twist angle of 1.05° at a critical temperature of 1.7 K [82], reproduced with permission © Springer Nature. (c) Illustration of an enlarged unit cell of twisted bilayer graphene (TBG), in which the top layer is rotated out of alignment with the lower layer. (d) Schematic of the formation of a typical moiré pattern when the rotation angle is small [83], reproduced with permission © Institute of Physics.

pressure, M. Yankowitz *et al* observed superconductivity at a twist angle larger than 1.1° [86]. In addition, adding an insulating tungsten diselenide (WSe2) monolayer between the hBN and the TBG has been found to stabilize superconductivity at twist angles much smaller than the magic angle [87].

In addition to TBG, trilayer graphene (TLG) system also provides a platform for the study of superconductivity in graphene. TLG has two types of naturally stable stacking structure, that is, ABA and ABC. In the ABC stacking structure, the atoms of different layers are staggered, leading to the potential superconductivity of ABC-TLG [88]. In 2019, Chen *et al* fabricated a gate-tunable Mott insulator and observed signatures of superconductivity in ABC-TLG/hBN heterostructures, which is rather similar to that in high T_c doped cuprates [89]. The study of multilayer twisted graphene is quite an active field, and more exciting results are emerging in this trending field.

5.3 Preparation methods of graphene and its derivatives

Graphene can be prepared either directly from natural graphite (i.e. the so-called top-down strategy) or by chemical synthesis from other carbon sources (i.e. the so-called bottom-up strategy). Figure 5.8 summarizes the state-of-the-art methods used for the production of graphene. In this section, the currently available methods for

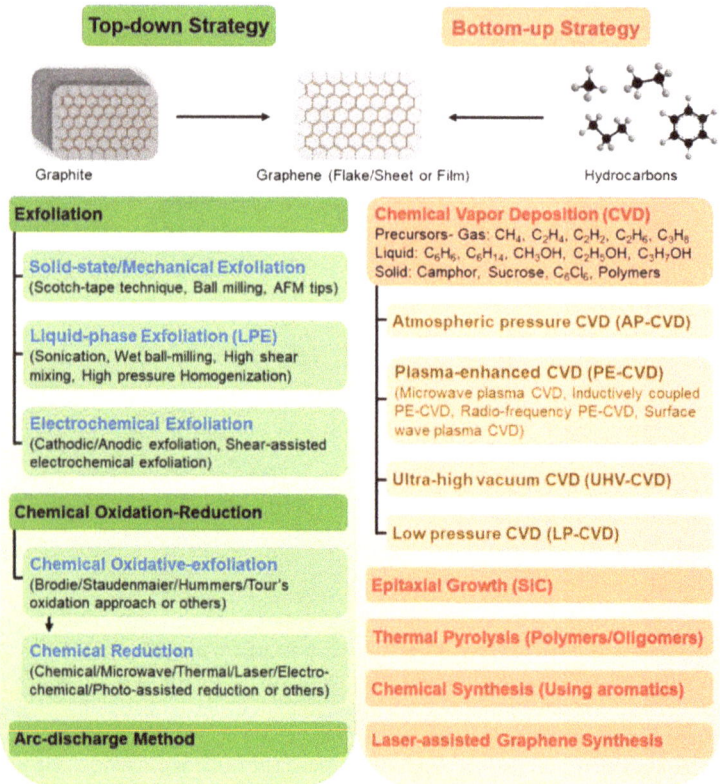

Figure 5.8. Schematic of graphene production using the top-down and bottom-up methods [15], reproduced with permission © Elsevier.

graphene production will be introduced in the two following main categories: (i) the top-down strategy and (ii) the bottom-up strategy.

5.3.1 Top-down strategy

Because graphite is composed of graphene monolayers stacked via van der Waals forces, the direct preparation of high-quality graphene from readily available graphite is naturally feasible. The use of graphite as the precursor to prepare graphene is the so-called 'top-down strategy.' In this section, various top-down methods are introduced, including the exfoliation method, the chemical oxidation–reduction method, and the arc-discharge method.

5.3.1.1 Exfoliation method

5.3.1.1.1 Mechanical exfoliation

The principle of mechanical exfoliation is to peel graphene from graphite by overcoming the van der Waals forces between the graphite layers. During this process, both a normal force and a lateral force may be involved. Figures 5.9(a)–(c)

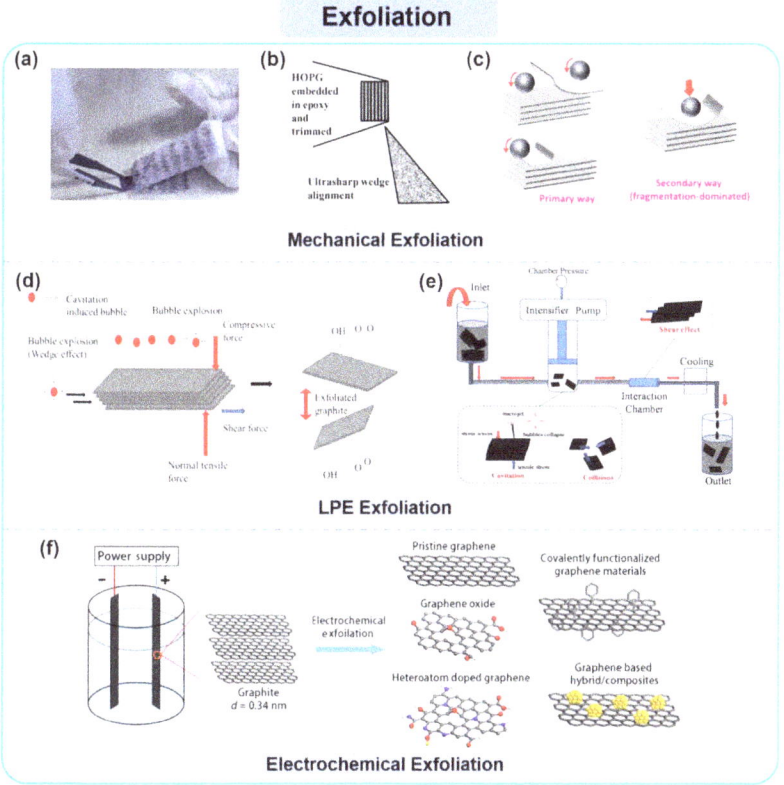

Figure 5.9. Schematics of different exfoliation methods. (a) The Scotch tape method [90], reproduced with permission © Springer Nature. (b) The ablation method [91], reproduced with permission © Springer. (c) The ball-milling method [92], reproduced with permission © The Royal Society of Chemistry. (d) The sonication-force-assisted liquid-phase exfoliation (LPE) method [93], reproduced with permission © MDPI. (e) The high-pressure-assisted LPE method [94], reproduced with permission © American Scientific Publishers. (f) The electrochemical exfoliation method [95], reproduced with permission © Elsevier.

illustrate the typical mechanical exfoliation methods reported in the literature, including the Scotch tape method, the ablation method, the ball-milling method, and the roll milling method.

Scotch tape method. The first attempt to mechanically exfoliate graphite was completed using an AFM tip, but it only produced graphite slabs composed of ~600 layers instead of actual graphene [12]. In 2004, the Nobel prizewinner demonstrated the first preparation of monolayer graphene with a lateral size of micrometers using Scotch tape to mechanically peel graphene from HOPG (figure 5.9(a)) [13]. Although the Scotch-tape method is simple and cost-effective and creates high-quality graphene monolayers quickly, it suffers from low yields and only produces small pieces of graphene. Therefore, this method is limited to lab-scale scientific research and is not suitable for large-scale production.

Ablation method. Since the Scotch tape method is effective but inefficient, other, more advanced techniques have been developed. Jayasena *et al* reported the use of an ultrasharp single-crystal diamond wedge to mechanically cleave HOPG, in which a high-frequency oscillation system was aligned to the surface of the HOPG (figure 5.9(b)) [91]. In this setup, the diamond wedge remained still while the HOPG pyramid moved slowly toward the wedge. As the wedge and the HOPG came into contact, the top layer of the HOPG pyramid was cleaved off and collected by a water bath. However, although this method is able to produce graphene at a thickness of tens of nanometers, it cannot be used for large-scale production. Janowska *et al* [96] and Pirzado *et al* [97] cleaved different carbon sources (i.e. pencil lead and an HOPG disc) using a rotational quartz disk; the carbon sources rotated in the opposite direction to that of the quartz disk. The ablation process starts as the rotating carbon source touches the rotating quartz disk, and the ablated fragment is subsequently sonicated in a solvent (e.g. toluene or ethanol) to extract the graphene layers. This method could produce graphene with a thickness of five to eight layers and a yield of 60%–70%.

Ball-milling method. Another effective method for peeling graphene layers from bulk graphite is mechanical ball milling. The ball-milling method primarily involves the normal and shear forces on the surface of graphite caused by the collision of balls (figure 5.9(c)). Generally, the normal force is stronger than the shear force in overcoming the van der Waals force. However, the application of a strong normal force to the graphite can destroy the crystalline structure of the graphene sheets, leading to fragmented instead of exfoliated graphene layers [98]. Therefore, it is important to balance the two forces in the ball-milling process: the normal force should be reduced, while the shear force should be dominant and large enough to overcome the van der Waals force. To achieve this, Zhao *et al* reduced the rotational speed to 300 rpm to make shear exfoliation dominant during the ball-milling process [99]. After milling for 30 h and dispersing the product in a DMF solvent, graphene sheets with an average thickness of three layers were obtained after high-speed centrifugation. Lv *et al* developed a Na_2SO_4 salt-assisted route to enable the production of graphene with two layers [100]. Knieke *et al* demonstrated that by optimizing the stir milling conditions, such as the delamination time and the initial concentration of the suspension, a higher yield of graphene with a reduced thickness of up to 3 nm could be obtained [101]. Jeon *et al* used halogenated compounds as a milling agent to assist in the exfoliation of graphite during a wet ball-milling process [102]. The presence of halogenated compounds promotes an edge-halogenation reaction, which not only peels off few-layer graphene but also functionalizes the edge of the graphene. Such halogenated graphene with reactive sites is promising for oxygen reduction reactions. Lin *et al* used oxalic acid as a milling agent, which could adhere to the defect sites of graphite and exfoliate graphene sheets via the shear force. The adhered carboxyl groups could facilely be removed by an annealing treatment at 600 °C. In the ball-milling method, the yield and quality of graphene are determined by the processing parameters, such as the initial graphite concentration, the ratio of graphite to ball, rotational speed, the type of milling agent, and treatment time.

Roll milling method. In addition to balls, milling can also be conducted with rolls. Chen *et al* developed a three-roll milling method which was inspired by the

Scotch tape method [103]. In this method, an adhesive mixture of polyvinyl chloride (PVC) and dioctyl phthalate (DOP) is prepared and loaded into the three-roll milling equipment along with the graphite. When the rolls start to rotate, the graphite mixes with, and spreads on, the adhesive. During the intensive mixing of the graphite and PVC/DOP adhesive between the rolls, the graphite is exfoliated and disperses within the adhesive. To extract the few-layered graphene from the adhesive, the mixture is soaked in alcohol followed by a high-temperature (500 °C) treatment. However, this process is complex, which makes it difficult for this method to be scaled up. Nevertheless, the roll milling process is a continuous process that uses readily available equipment, which could be useful for the direct preparation of a polymer–graphene nanocomposite.

5.3.1.1.2 Liquid-phase exfoliation (LPE)

The LPE method was developed to exfoliate graphite under mild conditions. The LPE method uses liquid media to assist the exfoliation of graphite; the liquid media can be an organic solvent with or without the addition of a surfactant agent, which can stabilize the dispersed graphene flakes by minimizing the interfacial tension between the liquid and the graphene [104]. Further, to promote the complete exfoliation of graphite in the LPE method, external forces are introduced, including sonication force, high shear force, and high-pressure force. In this section, various LPE methods will be introduced according to the three external forces.

Sonication-force-assisted LPE. The force most commonly used for LPE is sonication, specifically, bath sonication and probe/tip sonication, which can be applied either individually or simultaneously [105]. As shown in figure 5.9(d), the mechanism of sonication-force-assisted LPE is liquid cavitation, in which the sonication generates cavitation bubbles that can create high-speed microjets and shock waves. These waves produce normal and shear forces on the graphite plane, leading to the exfoliation of graphite to obtain graphene [93]. The effectiveness of sonication-force-assisted LPE depends on the type of liquid medium, the power and duration of sonication, the rate of centrifugation, and the position of the sonication source [106]. The liquid medium should stabilize graphene dispersion during the LPE process by balancing the intersheet attractive forces of graphene via the solvent–graphene interactions. In this sense, solvents that can minimize the contact surface area between the graphene and the solvent molecules are ideal for graphene dispersion. So far, various solvents have been investigated, including N-methyl-2-pyrrolidone (NMP) [107], orthodichlorobenzene (O-DCB) [108], H_2O/Triton X-100 [109], H_2O/ammonia solution [110], H_2O/polyvinylpyrrolidone (PVP) [111], H_2O/sodium cholate (NaC) [112], NMP/NaOH [113], NMP/azobenzene [114], and ionic liquids [115]. Because the LPE method is conducted in air, oxygen-based functional groups, such as ethers, carboxylic acids, and epoxides, are introduced and are independent of the solvent or the surfactant [116]. In addition, defects are induced in the graphene lattice during sonication, which are dependent on the sonication time [117]. These defects are located at the edges of the graphene in the early stages of sonication, while they grow in the basal plane after more than 2 h of sonication. Different sizes of graphene sheet can be separated by controlling the centrifugation rate [118]. At high centrifugation rates, small graphene sheets remain dispersed in solution, but larger sheets sediment out. Thus, graphene sheets at the desired size can be obtained by consecutively centrifuging the re-dispersed sample and separating it at a specific centrifugation speed. Moreover, the effectiveness of sonica- tion-assisted LPE has also been found to be influenced by the dispersion volume and vessel geometry; in other words, the position of the sonication source [119]. For example, in the probe sonication method, when the position of the sonication source is fixed, a considerable amount of graphite settles at the bottom of the sonication vessel and remains un-exfoliated because of the static cavitation effect. Although the use of a sonication bath could resolve this problem, the bath weakens the sonication intensity, greatly reducing the effectiveness of this method. Therefore, it is highly desirable to develop new routes that can directly apply mobile mechanical energy to the graphite solution.

High-shear-assisted LPE. To improve the exfoliation efficiency of the LPE method, extra shear force can be introduced into the LPE solution by adopting a high-shear generator. The high-shear force can be generated by a rotor–stator mixer [19], a Taylor–Couette flow reactor [120], a rotating-blade mixer (e.g. a kitchen blender) [121], or a micro-fluidizer [122]. The key role of the shear generator is to mix the immiscible phases (solid/liquid/gas) into one continuous liquid phase using shear force or shear stress, which shears the graphite layer when the high-velocity solution passes through the perforations in the stator and the main body of the mixer. To obtain a high shear rate over the entire volume of the mixer, the use of a mixer with rotating blades is proposed. This can be simply realized by a kitchen blender [121], in which the high shear rate (generated by strong turbulence at a high rotation speed) is present throughout the container instead of being localized at a single point. There are four fluid dynamics stresses that are responsible for the exfoliation of graphite in a rotating-blade mixer: (i) viscous shear stress induced by the velocity gradient; (ii) Reynolds shear stress created by severe velocity fluctua- tions in the turbulence; (iii) high inertial stress induced by high turbulence; and (iv) normal stress induced by the pressure gradient due to the turbulence pressure difference. In summary, high-shear-assisted LPE is a potential technique for the high-volume production of high-quality graphene.

High-pressure-assisted LPE. The force applied for graphite exfoliation can also be reinforced by increasing the pressure of the LPE system. A high-pressure system can create a strong jet cavitation force, which is more effective, more uniform, and more reproducible in exfoliating graphite than the cavitation force of the sonication methods [123]. Generally, the high pressure in the LPE system is achieved by injecting the LPE solution into narrow channels at a high speed, which generates two significant forces for the exfoliation of graphene; these are the normal force induced by the pressure difference and the shear force induced by the velocity gradient and turbulence [124]. Liang *et al* reported that by increasing the jet pressure to 20 MPa, an increased graphene yield at a thickness of >2 nm could be acquired after treatment for 0.5 h and that further prolonging the treatment time to 8 h resulted in graphene with a thickness of 1.3 nm [125]. In addition to jet pressure, the pressure of the LPE solution can also be enhanced to 207 MPa by the design of the channel. Wang *et al* reported the design of a 75 μm thick Y-shaped microchannel to conduct the high-pressure-assisted LPE of graphite [94]. The homogenization and exfoliation mechanisms in this channel are based on the impact of two opposing high-pressure jets. In addition to the Y-shaped microchannel, researchers have also developed other channels, such as the Z-shape [126], the narrow valve slit [127], the long tube [128], bi-cylinders [129], supercritical fluids [130], and vortex fluidics [131]. Due to the high pressures involved, undesirable fragmentation instead of graphene sheets can take place as a by-product, which might fortunately be useful for the synthesis of GQDs [132].

5.3.1.1.3 Electrochemical exfoliation

The electrochemical exfoliation of graphite has great potential for the commercialization of graphene owing to its scalability. Compared to other methods, the electrochemical exfoliation method works under milder conditions and is less time-consuming, more eco-friendly, and more accessible. In the electrochemical exfoliation method, various graphite sources (e.g. powder, rods, flakes, foils, or plates) can be employed as the working electrodes in electrolytes. Aqueous or solid electrolytes are generally used as the media, in which an electrical current is applied to the electrode to promote the structural exfoliation or expansion of graphite by charged ions and intercalating species [95]. In addition, according to the bias direction applied to the graphite electrodes, both anodic (using a positive bias) and cathodic (using negative bias) exfoliation have been investigated [133–135]. Using this electrochemical exfoliation method, various graphene materials (e.g. pristine graphene, GO, functionalized graphene, heteroatom-doped graphene, and graphene-based composites/hybrid materials) can be obtained by choosing different precursors, electrolytes, intercalating agents, and functionalizing agents (figure 5.9(f)) [136]. In addition, the properties of exfoliated graphene can easily be adjusted by changing the above electrochemical parameters [137]. It is worth mentioning that introducing an extra shear force into the electrochemical exfoliation process could further increase the quality of the produced graphene. For example, Shinde et al designed a microfluidic reactor to conduct the electrochemical exfoliation of graphite using graphite as the anode [138]. In this approach, the exfoliation of graphite is done under a shear field (400–74 400 S^{-1}) induced by a flowing electrolyte. However, it is still challenging to scale up electrochemical methods because designing a large electrochemical cell with large electrodes is expensive and complicated. Nevertheless, continuous-flow electrochemical exfoliation systems show great potential for scalability in the future.

5.3.1.2 Chemical oxidation–reduction method

Although the direct exfoliation of graphite discussed above could produce graphene with high quality, the functionality and yield of those methods were limited. Therefore, chemical routes were developed to exfoliate graphite and produce functionalized graphene, which involve an oxidation process followed by a reduction process (figure 5.10) [15]. In the oxidation process, strong oxidation agents are used to prepare graphite oxide by introducing oxygen groups, which increase the interlayer spacing between the graphite sheets. As a result, the intersheet van der

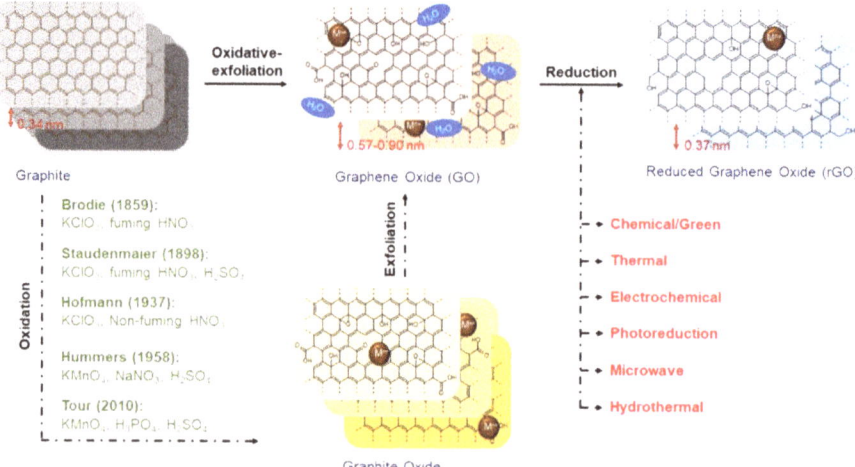

Figure 5.10. Schematic of the chemical oxidation–reduction method used for the preparation of GO and rGO from graphite [15], reproduced with permission © Elsevier.

Waals force can easily be overcome by dispersing graphite oxide in solvents (e.g. organic solvents or water) to produce well-dispersed single-, bi-, and few-layer GO [139]. After the oxidation process, GO can be chemically transformed into reduced GO (rGO) to regain the properties of graphene using various reduction methods, such as thermal, chemical, electrochemical, hydrothermal, microwave, and photoreduction methods. Due to the presence of functional groups in GO and rGO, this chemical oxidation–reduction method is suitable for the large-scale production of various graphene-based materials.

5.3.1.2.1 Oxidation methods for the synthesis of GO

To date, various methods have been developed to oxidize graphite into GO. There are five main established methods for the oxidation of graphite into GO: Brodie's [140], Staudenmaier's [141], Hofmann's [142, 143], Hummers' [144], and Tour's (improved Hummers') [145] methods. These methods are based on the chemical reaction between a common strong acid (e.g. H_2SO_4, H_3PO_4, or concentrated HNO_3) and graphite in the presence of an oxidizer (such as $KClO_3$, $KMnO_4$, or $NaNO_3$) for the oxidation of the basal and edge carbons in graphite. The oxidation process introduces large numbers of functional groups on the out-plane of the 2D graphite layer, which can expand the space between the graphite layers. In addition to the above conventional oxidation methods, numerous improved and modified strategies have also been reported that used alternative oxidizers, milder chemical and thermal conditions, and shorter reaction times; the functional groups, defects, and lateral size of the produced GO could be controlled by changing the reaction parameters of the oxidation process [146]. A particular oxidation method can be adopted according to the required GO product.

Hummers' method and Tour's method (an improved version of Hummers' method) are the most commonly used methods in current studies. In the classic Hummers' method, graphite powder and $NaNO_3$ are stirred in 0 °C H_2SO_4 in an ice bath, and then $KMnO_4$ powder is slowly added to the above suspension under stirring, followed by gradually increasing the temperature of the system to 35 °C. After that, H_2O is slowly added, and the temperature of the system is increased to 98 °C and maintained there for a short period of time. After completion of the reaction, the suspension is treated with H_2O_2 and water to remove the residual permanganate and manganese dioxide. Finally, the diluted mixture is purified by filtration (centrifugation and washing several times in H_2O until the pH of the supernatant is seven) to remove the dissolved mellitic acid salt. However, this method suffers from the unavoidable emission of toxic nitrogen-based gases (NO_2 and N_2O_4), because $NaNO_3$ is converted into HNO_3 upon acidification. To solve this problem, in 2010, Tour and colleagues improved the method in order to synthesize GO with a high oxygen content [145]. A small amount of phosphoric acid H_3PO_4 (an etching and dispersive agent) was used to replace $NaNO_3$ in the presence of large amounts of $KMnO_4$ and H_2SO_4. The GO produced by Tour's method exhibits a high oxidation degree, a more ordered structure, higher yield, and fewer defects compared to other methods.

5.3.1.2.2 *Reduction methods for the synthesis of rGO*
GO is broadly utilized as a raw material for the synthesis of graphene via chemical reduction assisted by various reductants. The reduction process is intended to minimize the functional groups and oxygen content of the GO as well as repair and restore defects in the GO. It has been found that after reduction, rGO can partially restore the lattice structure of graphene and shows similar characteristics to pristine graphene (e.g. mechanical, thermal, and electrical properties) to varying degrees, which is related to its reduction extent [146]. Due to the functionality of GO and rGO, both of them show great superiority in various applications, which will be introduced in section 5.4. This section briefly summarize various reduction methods used for the preparation of rGO from GO.

Thermal reduction. The reduction of GO via thermal annealing is one of the most widely used methods for the synthesis of rGO. When GO is rapidly heated to a high temperature under either air, an inert atmosphere, a reducing atmosphere, or vacuum conditions, it is exfoliated into fewer graphene sheets [147]. The rapid heating process decomposes the oxygen-containing functional groups bonded on the graphene plane in the GO into CO_2 gas, which produces defect-rich graphene sheets with a small and wrinkled morphology [148]. In the thermal reduction method, the annealing temperature has a considerable impact on the oxygen content, size, and defects of rGO. For example, low-temperature annealing of GO produces rGO with low electrical conductivity due to the high content of residual oxygen-containing groups [149], while annealing temperatures >1000 °C can yield highly crystalline rGO with a largely restored graphitic structure, which is essential for producing large-area rGO sheets with good intrinsic electrical properties [150]. However, the thermal reduction method has many disadvantages that limit its large-scale application, including high energy consumption (due to the requirement for high temperatures), long processing duration (due to the requirement for a slow heating rate to prevent structural explosion), and low compatibility (incompatibility with applications that require low-temperature processing).

Microwave irradiation (MWI) reduction. MWI replaces the thermal annealing process by supplying external microwave energy for the reduction of GO either in a solid state, in solution, or by directly heating intercalated graphite compounds. During MWI, high-density energy is directly transferred to GO and rapidly creates an extremely high local temperature, which is sufficient for the rapid reduction of GO [151]. Zhu *et al* used a microwave oven to simultaneously realize the exfoliation and reduction of GO in a very short time interval (~1 min) [152]. Compared to hydrazine-reduced rGO, the MWI-reduced rGO exhibits an improved C/O ratio of 2.75 and a conductivity of 274 S m^{-1}. Chen *et al* reported the fast and mild MWI reduction of GO in a mixed solution of N,N-dimethylacetamide (DMAc) and water [153]. The electrical conductivities of GO and MWI-reduced rGO were measured to be 0.015 S m^{-1} and 200 S m^{-1}, respectively, while their C/O ratios were 2.09 and 5.46, respectively. In addition, it was also found that under reducing/inert conditions or by pretreating GO with strong reducing agents (i.e. NaBH$_4$), the reduction degree of MWI-reduced GO was considerably improved [154, 155].

Photoreduction. The photoreduction of GO can be realized by a photochemical route, in which a photocatalyst is typically used to convert photon energy into active electrons and thermal energy, which result in the reduction of GO [156, 157]. Recently, the photocatalyst-free photoreduction of GO has been reported. Matsumoto *et al* demonstrated a simple GO reduction process using relatively weak UV irradiation in H$_2$/N$_2$ at room temperature without a photocatalyst, and the conductivity of the produced rGO increased by a factor of 10^5–10^6 [158]. Li *et al* photoreduced a sealed GO suspension by irradiation with a xenon lamp for 4 h, and the obtained rGO aqueous dispersion was stable without the addition of any surfactant [159]. Konkena *et al* found that the optimized control of UV exposure times and pH could produce rGO with conductive sp^2-graphitic domains and more acidic carboxylic groups, which resulted in the both conductive and water-dispersible the rGO sheets [160]. In sum, photoreduction is a simple and promising approach for producing rGO because no chemicals, reductants, or stabilizers are involved, and thus no further purification is required.

Hydrothermal reduction. The hydrothermal method is a classic synthesis strategy for the fabrication of all kinds of nanostructure with controlled morphology under high-pressure and high-temperature conditions [161]. It has been demonstrated that the hydrothermal method can also be applied for the reduction of aqueous GO. In the hydrothermal reduction, an aqueous solution of GO is heated to a temperature higher than the boiling point of water (~200 °C), which produces an overheated/subcritical/supercritical phase of water that can easily overcome the energy threshold for GO reduction [162]. GO reduction in H$_2$O is possible because the diffusivity of supercritical H$_2$O is much superior to that of normal H$_2$O—it can chemically dissociate O groups from the edges (carboxyl and carbonyl) and the basal plane (epoxy and hydroxy) of graphene sheets [162]. In addition, the hydrothermal treatment of GO is pH dependent: a basic pH solution produces better rGO with fewer defects, sufficient reinstatement of sp^2 domains, and a greatly improved C/O ratio [163]. This is the case because the reduction of GO in acidic media is associated with the intermolecular dehydration effect, leading to the aggregation of rGO sheets, while the intramolecular dehydration effect is dominant for GO reduction in alkaline media, which is responsible for higher-quality rGO [164]. Because the hydrothermal reduction of GO is cost-effective and easily controlled and functionalized, it can be adopted to produce various graphene-based composite materials and used in various applications.

Electrochemical reduction. GO can also be reduced by an electrochemical reduction method, which is performed at room temperature using an electrochemical cell with a non-toxic aqueous buffer solution or supporting electrolyte. The GO is coated or deposited either on conducting substrates or insulating substrates [165]. An external bias is used to perform the reduction, and the degree of GO reduction can be modified by changing the electrolyte, buffer solution, and electrolysis parameters [166]. Currently, various electrolytes, such as organic solvents [167], alkali and alkaline earth metal chlorides [168], propylene carbonate and acetonitrile [169], phosphate buffer solutions [170], and $NaSO_4$ [171], have been explored for the reduction of GO. Electrochemical reduction is a facile, rapid, economical, and eco-friendly method for producing high-quality rGO. However, this method may not be scalable, because depositing GO onto large-scale electrodes is greatly challenging.

Chemical reduction. The chemical reduction of GO is one of the most promising techniques for obtaining graphene-based composite materials in massive quantities. The chemical reduction of GO can be realized under mild conditions with simple equipment, thus it is regarded as an economical and facile technique for the scalable production of rGO [172]. The chemical reduction method requires chemical reducing agents to eliminate the oxygen-containing groups in GO sheets and produce rGO [173]. So far, various reducing agents have been developed for GO reduction, such as hydrazine, lithium aluminum hydride, iron/aluminum powder, hydrochloric acid with acetic acid, sodium borohydride, sodium/potassium hydroxide, hexylamine, hydroxylamine hydrochloride, ammonia, urea, sulfur compounds, hydrohalic acids, poly(norepinephrine), alcohol, thiourea, $Na–NH_3$, and NMP [15, 174]. However, some of the abovementioned reducing agents are hazardous/toxic chemicals, which require safety precautions and the removal of unwanted hazardous by-products [175]. With this in mind, the development of a green chemical reduction agent is highly anticipated.

Green chemical reduction. Over the last few years, researchers have paid much attention to exploring alternative green reductants for GO reduction, such as alcohols/ phenols, metals with acids/bases, organic acids, sugars, microorganisms, plant extracts, amino acids, antioxidants, vitamins, proteins, and hormones [176–180]. In some cases, a single green reducing agent is insufficient for the complete reduction of GO, so the process might need an assistant agent for reduction or a stabilizing agent to avoid aggregation. However, the use of additional agents leads to additional energy costs for the filtration or centrifugation needed to remove excess reducing agent and side products. However, the green chemical reduction method still has disadvantages, such as the long time required for the process, the extreme difficulty in removing all the extracts after reduction, and the considerably low conductivity of the produced rGO. Nevertheless, it is still meaningful to continue the search for novel efficient green reductants to achieve the large-scale and eco-friendly reduction of GO.

5.3.1.3 Arc-discharge method

In 1990, Krastchmer *et al* demonstrated the first example of the preparation of fullerenes (C_{60}) using an arc discharge [181]. Inspired by this work, Subrahmanyam *et al* reported the use of the arc-discharge method to produce graphene flakes from graphite; graphene with two to four layers could be obtained from the inner wall region of the arc chamber [182]. The typical setup of the arc-discharge chamber used a sacrificial graphite rod anode and a cathode, which were placed inside a steel vacuum chamber cooled by water. To date, various gas atmospheres have been studied for the discharge evaporation of graphite rods, such as He, H_2, Ar, NH_3, and CO_2 as well as their mixtures [182–185], and the different gas atmospheres have been found to influence the type of graphene produced (pure or doped). For example, nitrogen-doped graphene can be produced by the arc-discharge method in a NH_3–He atmosphere [186], while a CO_2–He atmosphere can produce graphene with high dispersivity and electrical conductivity [187]. In the arc-discharge method, a constant direct current (DC) or alternating current (AC) is applied to the electrodes, and the contact between the current and the graphite anode leads to the formation of plasma and vaporization of the anode. As a result, graphene forms in the plasma and deposits on the side of the chamber. The arc-discharge method is a facile strategy for the production of heteroatom (N, B, and F)-doped graphene [187]. Recently, Kim *et al* modified the previous DC arc-discharge method using anodic carbon fillers (made of graphene and polyaniline) as the anode in a mixed gas atmosphere (H_2, He, and NH_3), and it was found that the N-doping content in the graphene increased from 1% to ~3.5% while maintaining high crystallinity [188]. Moreover, the AC arc-discharge method produced not only high-quality graphene but also a higher yield compared to the DC method. In addition to its scalability, the AC arc-discharge method is easier, cheaper, and compatible, and can prepare graphene based on the supercooling effect [189], catalysts [185], aqueous environments [190], and gaseous environments [191]. Nevertheless, the main shortcomings of the arc-discharge method are the uncontrolled formation of graphene, undesired doping, low doping content, and the difficulty of preparing large-scale electrodes.

5.3.2 Bottom-up strategy

The bottom-up strategy is based on the use of hydrocarbon compounds, which serve as the carbon source and the precursor, to chemically synthesize graphene. Some of the most widely used bottom-up approaches include CVD, epitaxial growth, thermal pyrolysis, chemical synthesis, and laser-assisted synthesis. These bottom-up methods are introduced in this section.

5.3.2.1 Chemical vapor deposition

The CVD method is the classic bottom-up synthesis route for producing high-quality, large-area, and single-crystal graphene, which is attractive from the viewpoint of industry [197]. CVD is the deposition of gaseous carbon sources as a reactant onto a catalytic substrate to form graphene. CVD typically requires a vacuum environment to ensure that the reactants reach the substrate without interference. In the CVD method, the graphene growth process is a heterogeneous chemical reaction on a metal catalytic substrate, which plays two different roles of substrate and catalyst. The catalytic substrate can promote the formation of the graphene structure, but the covered graphene reduces the catalytic activity due to the catalyst poisoning effect, leading to the end of the graphene growth. Thus, the balance between catalysis and the graphene growth rate is crucial. If the overall process is performed on the surface of a substrate with low solubility, the growth process is governed by the adsorption, decomposition, and diffusion of molecules, and in this case, monolayer graphene is preferentially grown (figure 5.11(a)). In this case, carbon atoms derived from hydrocarbon precursors can directly diffuse onto the metal surface and build up thermodynamically stable graphene. This process easily results in the formation of large-area monolayer graphene. Notably, over 95% of monolayer graphene films have been achieved using binary Ni–Cu alloys as the catalytic substrates [198]. On Ni and other transition metals (Co, Ru, Ir, etc.) that exhibit high carbon solubility and segregation, the growth of graphene via CVD is governed by carbon bulk diffusion during the cooling step. In this case, a solid solution of a mixture of elements is formed near the surface, and the growth of the graphene depends on the kinetic parameters selected for the synthesis. To suppress the formation of multilayer graphene, a fast cooling rate is required [199, 200]. Note that the resulting graphene quality is mainly determined by the processing parameters, such as the catalysts, precursors, gas flow rate, temperature, pressure, and time [201]. This model for CVD-produced graphene can effectively be applied to various CVD systems based on different carbon precursors, catalytic substrates, and technique parameters [202]. It is worth noting that the CVD method could be a scalable, high-yield, and facile strategy for preparing 3D graphene materials (e.g. foams, shells, and hierarchical structures) with relatively high crystallinity and controllable layer numbers [203].

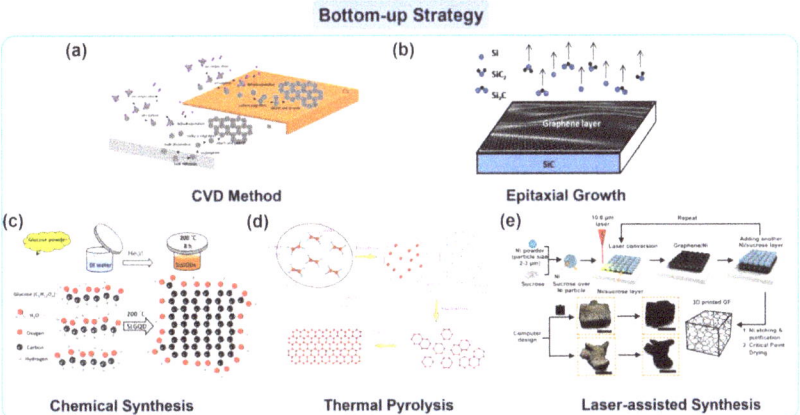

Figure 5.11. Schematics of different bottom-up methods. (a) CVD method [192], reproduced with permission © John Wiley & Sons. (b) Epitaxial growth [193], reproduced with permission © John Wiley & Sons. (c) Chemical synthesis [194], reproduced with permission © Elsevier. (d) Thermal pyrolysis [195], reproduced with permission © MDPI. (e) Laser-assisted synthesis [196], reproduced with permission © American Chemical Society.

To facilitate the production of graphene via CVD for various practical applications, the prepared graphene should be perfectly removed from the catalytic metal substrate and transferred onto the target substrate [204]. Typically, the CVD-grown graphene is first coated with a thin layer of polymethyl methacrylate (PMMA), followed by annealing at 120 °C to remove the solvent. The catalytic metal substrate is then etched by a Ni or Cu etchant, leaving the freestanding PMMA/graphene film. The PMMA/graphene film is further cleaned using deionized (DI) water and then transferred onto the target substrate. After evaporation of the residual water, the PMMA is removed using acetone, leaving a graphene film on top of the target substrate. Using this transfer technique, graphene at full-wafer size can readily be transferred from Ni film and Cu foil onto different substrates, including glass, patterned Si/SiO2 substrates, and polyethylene terephthalate (PET) film [204].

5.3.2.2 Epitaxial growth

The word epitaxy comes from the Greek roots 'epi,' which means 'above,' while 'taxis' means 'an ordered manner.' So the epitaxial growth method consists of growing a crystalline film on a crystalline substrate; the structure of the film follows that of the substrate [205], and the deposited layer is called the epitaxial layer. For graphene preparation, the epitaxial growth of graphene typically takes place on SiC substrates, which is realized by the depletion of the surface Si via high-temperature thermal annealing. In a typical process, a commercially available SiC sample is annealed at high temperature (>1400 °C) in vacuum or under atmospheric pressure conditions. Since the vapor pressure of carbon is negligible compared to that of silicon, at high temperatures, Si atoms sublimate and leave C atoms behind on the surface of SiC, which subsequently re-construct into graphene layers (figure 5.11(b)) [193]. The main disadvantage of this method is that even without a transfer process, the epitaxial graphene is not perfectly homogeneous due to the presence of defects or grain boundaries. Nevertheless, graphene/SiC-based electronic devices are believed to have inspiring potential for future high-frequency applications [206]. Although this technique produces high-quality graphene, the production cost is high due to the expensive SiC substrate and the low yield; thus, this method might not be suitable for industrial manufacturing.

Epitaxial growth is a promising method for the production of large-area graphene films with high uniformity, thickness, and quality, which are ideal candidates for electronics applications [207]. The main advantage of epitaxial growth compared to other techniques is that the graphene layers can be directly produced on a commercially available semiconducting or semi-insulating substrate, thus no transfer process is needed and the target electronic devices can be directly fabricated on the substrate [208].

5.3.2.3 Chemical synthesis

The chemical synthesis of graphene is another bottom-up approach that is based on the assembly of atomic or molecular carbon sources, such as aromatic molecules, the oxidative cyclodehydrogenation of polyphenylene, and oligophenylene sources [209]. This process mostly results in small graphene molecules (e.g. GNRs and GQDs) with zigzag-edged structures, helically coiled structures, or armchair structures [210–212]. Recently, Bayat et al reported the chemical synthesis of low-cost and high-yield single-layer GQDs using only deionized water and glucose as precursors via a hydrothermal method (figure 5.11(c)) [194]. During the hydrothermal process, the hydrogen atoms of a glucose molecule interact with the hydroxyl groups of an adjacent glucose molecule, promoting the formation of H_2O and the dehydration of glucose molecules to form C=C bonds, which are the elementary units of the graphene structure. Consequently, carbon atoms covalently interact with each other and finally form a graphene structure. In sum, as this method produces graphene with atomically precise and uniform nanostructures, it is suitable for the fabrication of controllable and reproducible components for optoelectronics, nano-electronics, and spintronics as well as biomedical applications [213].

5.3.2.4 Thermal pyrolysis

Thermal pyrolysis is another bottom-up method for the synthesis of graphene, which has the potential to produce graphene on a massive scale. This method involves the pyrolysis of a carbon precursor (such as a polymer, an oligomer, or a prepolymer) under solvothermal conditions followed by the formation of graphene (figure 5.11(d)). Carbon atoms in the carbon source normally link with other atoms in sp^3-hybridization bonds which are not in the same 2D plane. Thus, in order to produce graphene, individual carbon atoms are first released via thermal energy, and then the carbon atoms nucleate with each other to form benzene ring structures through sp^2-hybridization bonds, followed by the growth of graphene. Choucair *et al* reported the gram-scale production of graphene using the thermal pyrolysis method [214]. In this method, a 1:1 molar ratio of sodium (2 g) and ethanol (5 mL) is used as the carbon source and heated in a sealed vessel at 220 °C for 72 h to obtain a graphene precursor, which is then rapidly pyrolysed. The resulting graphene can be produced with a yield of ~0.1 g mL^{-1}. Although the thermal pyrolysis method is cost-effective and could be scaled up for production, the yield and purity of the product are low.

5.3.2.5 Laser-assisted synthesis

The laser-assisted synthesis of graphene is a fast-growing graphene production method for specialized electronics applications, because it can produce graphene with a specific shape in a single step. Typically, the laser-assisted synthesis approach is derived from the CVD method, with the difference that the applied energy originates from laser irradiation instead of high temperatures. Specifically, a carbon precursor is activated using laser irradiation to promote its dissolution in a metal catalyst-substrate, and graphene is precipitated on the surface of the substrate upon cooling. Normally, Ni is used as the substrate to produce graphene [215], and various low-cost carbon precursors have been developed to synthesize laser-assisted graphene, such as cotton, paper, wood, food, lignin, and commercial polymers [216–218]. The laser-assisted method is good at producing flexible, defective, and porous graphene films, which are useful for widespread applications in which higher mechanical flexibility and surface area are desired [219]. Moreover, the laser-assisted method can be performed under ambient conditions without the precise control of pressure and ultra-pure gases, so it can be compatible with large-scale techniques, such as roll-to-roll manufacturing [220] and 3D printing [196]. Notably, Sha *et al* reported the preparation of 3D graphene foam using a laser-assisted 3D printing method (figure 5.11(e)) [196]. Briefly, a 10.6 μm laser was applied to the Ni/sucrose mixture layer, where Ni acted as the catalyst to convert sucrose into graphene. The 3D graphene structure was constructed by repeating this process consecutively. After the etching of the Ni and the removal of the solvent, the 3D printed graphene foam was obtained, which exhibited high porosity (~99.3%), high conductivity (~8.7 S cm^{-1}), and a remarkable storage modulus (~11 kPa). Most importantly, the shape of the 3D graphene foam could facilely be regulated by computer design.

5.4 Applications of graphene

The fascinating properties of graphene have attracted great attention from researchers in the field of materials science, in which graphene and graphene-based materials have been extensively developed and shown to play disruptive roles in various applications. In this section, the advanced applications of graphene-based materials are discussed comprehensively.

5.4.1 Energy conversion

5.4.1.1 Solar cells
Owing to the optical transparency, high carrier mobility, flexibility, and large-scale processibility of graphene, the strength of graphene in photovoltaic applications has been extensively exploited to serve various functions in different parts of solar cell devices, such as the cathode, the anode, the charge-transport layer, and the electron acceptor in the photoactive layer (figure 5.12(a)). For multi-junction solar cell devices, graphene has been shown to be an ideal candidate for the intermediate layer in both series-connected and parallel-connected tandem devices (figure 5.12(b)).

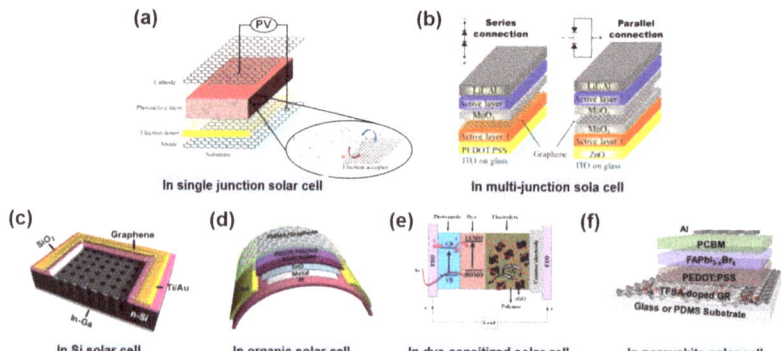

Figure 5.12. Illustration of the application of graphene-based materials in different components of (a) single-junction solar cells [221], reproduced with permission © American Chemical Society; and (b) multi-junction solar cells [222], reproduced with permission © John Wiley & Sons. Schematics of the application of graphene-based materials in different solar cells: (c) Si solar cell [223], reproduced with permission © The Royal Society of Chemistry; (d) organic solar cell [224], reproduced with permission © John Wiley & Sons; (e) dye-sensitized solar cell [225], reproduced with permission © Elsevier; (f) perovskite solar cell [226], reproduced with permission © The Royal Society of Chemistry.

In fact, different features are requested for the different components of different solar cell devices, and this calls for different processing methods for the preparation of graphene-based materials. In this section, the application of graphene-based materials in mainstream solar cells will be briefly introduced, including Si solar cells, polymer solar cells, dye-sensitized solar cells, and perovskite solar cells.

Si solar cells. Although graphene was first discovered in 2004, the first example of the use of graphene on a Si solar cell was only reported in 2010, when Li *et al* reported the transfer of CVD-grown graphene onto a pre-patterned n-Si/SiO$_2$ substrate to fabricate an n-type Si solar cell [227]. The graphene sheet formed a conformal coating on the exposed n-Si substrate, where electrons in the n-Si transferred to graphene due to the work function difference to form a Schottky junction and a built-in electric field at the interface. This first graphene–Si solar cell exhibited an overall power conversion efficiency (PCE) of 1%–1.7%. After this first study, graphene-based materials were widely applied in different parts of Si solar cells. Because the sheet resistance of CVD-grown graphene lies in the range of 125–1000 Ω/sq [228], a graphene doping strategy was developed to facilitate the use of graphene as a transparent conducting electrode. Xie *et al* reported the fabrication of a graphene/Si hole array (figure 5.12(c)), which used four layers of graphene doped with AuCl$_3$ and a Si hole depth of 12.8 mm [223]. AuCl$_3$ doping reduces the sheet resistance of graphene from 300 Ω/sq to 80 Ω/sq, which enhances the PCE of the device from 6.02% to 10.40%. Using p-type or n-type graphene doping, the work function, band structure, and the type and concentration of the graphene carriers could be facilely controlled [229], and the barrier height at the graphene/silicon junction could be improved, thereby improving the separation and extraction of photogenerated carriers [230]. Therefore, this feature of graphene enables it to be used as an anode, a cathode, or a junction layer [231–233]. It has been reported that after appropriate graphene doping, the Schottky barriers at the graphene/n-Si junctions could be 0.52–0.67 eV, while those at the graphene/p-Si junctions could be 0.61–0.73 eV at 300 K [234].

Organic solar cells. Organic solar cells employing polymer or organic small molecules as the active layer have been considered a promising low-cost photo-voltaic technology [235]. Graphene-based materials have been applied in organic solar cells as electrodes, charge-transport layers, and one of the ternary components in the active layer. Graphene doping can help in matching its work function with that of the adjacent charge-transport layer. If the work function of doped graphene matches the highest occupied molecular orbital (HOMO) of the hole transport layer, the graphene works as an anode [236], while matching the lowest unoccupied molecular orbital (LUMO) of the electron transport layer could make graphene serve as the cathode [237]. Notably, researchers have also investigated the use of doped graphene with different work functions to serve as both anode and cathode in the same device [238]. Benefiting from the excellent flexibility of graphene, the used of graphene-based electrodes could enable fabricated organic solar cell devices to have promising flexibility. Liu *et al* reported package-free flexible organic solar cells using a graphene composite as the top electrode (figure 5.12(d)). A PCE of 3.2% was achieved, and excellent bending stability was observed, as the PCE decreased by 8%

after about 1000 bending cycles. Moreover, the sheet resistance of graphene and the series resistance of the device were found to degrade by only 7% and 16%, respectively, after 1000 bending cycles [224]. Due to the ambipolar behavior of graphene after the work function is adjusted, graphene and its derivatives are suitable for use as both the electron transfer layer (ETL) and the hole transfer layer (HTL). It has been demonstrated that if the contents and the types of functional groups on graphene derivatives (e.g. GO) are controlled, functionalized GO can serve as both the ETL and the HTL because of its aligned energy levels as well as the good compatibility of its functional groups with the active layer [239]. For example, replacing poly(3,4-ethylenedioxythiophene) (PEDOT):poly(styrenesulfonate) (PSS) with sulfonic or fluorinated GO as the HTL was found to produce polymer solar cells with PCEs of 7.18% [240] and 8.6% [241], respectively. Graphene and its derivatives have also been used as the ternary component within the photoactive layer of solution-processable ternary organic solar cells, where graphene can act as a bridge structure that helps to prevent charging due to its high electrical conductivity. In addition, the doping of graphene (e.g. nitrogen-doped graphene) [242] or the incorporation of inorganic nanocrystal on graphene (e.g. an $rGO–Sb_2S_3$ composite) [243] were found to effectively enhance the PCEs (4.5% for nitrogen-doped graphene and 8.6% for the $rGO–Sb_2S_3$ composite) of polymer solar cells by improving the charge selectivity or energy-level alignment in the active layer.

Dye-sensitized solar cells (DSSCs). The merits of graphene, such as its large specific surface area, absorption over a wide spectral range, light weight, flexibility, and high mechanical, thermal, and chemical stability, are all vital for DSSC fabrication [244]. As a result, modifying the various components of DSSCs with graphene-based materials has proven to be an attractive strategy, and the PCEs of such devices have shown a rapid increase from ~0.13% [245] to more than 12% [246]. In DSSCs, graphene-based materials can play a role in modifying transparent conductive electrodes, semiconducting layers, photosensitizers, electrolytes, and counter electrodes. Roh *et al* employed rGO-modified fluorine-doped tin oxide (FTO) substrates as transparent conducting electrodes in DSSCs. The rGO sheets were found to possess few defects and to be firmly attached to the FTO surface, which helped to decease the charge transfer resistance and suppress charge recombination at the $TiO_2–FTO$ interface, leading to an impressive PCE of 8.44% [247]. Krishnamoorthy *et al* reported the use of a NiS_2–graphene composite that served as a semiconducting layer with a large surface area, a porous structure, and a continuous interpenetrating network, which showed synergistic effects of efficient photon harvesting, photocurrent generation, and electron extraction, leading to DSSCs with record PCEs as high as 12.56% [246]. Yang *et al* reported the incorporation of oxygen-functionalized GQDs into N719 dye to serve as a photosensitizer. They showed that the GQDs–N719 hybrid composite could enable effective light harvesting, better electron transport, and suppressed carrier recombination in the photosensitizer, and the corresponding DSSC could yield an enhanced PCE of 8.9% [248]. Prabakaran *et al* incorporated rGO into polymer electrolytes made from poly(ethylene oxide) (PEO)/poly(vinyliden fluoride) (PVDF) hexafluoro propylene (HFP), which was found to increase the ionic conductivity, charge carrier concentration, diffusion coefficient, and stability of the composite electrolytes (figure 5.12(e)). As a result, the corresponding DSSC achieved an improved PCE of 4.58% [225]. Oh *et al* reported the development of Pt-free counter electrodes based on graphene-based $Cu_2ZnNiSe_4$ with incorporated tungsten trioxide (WO_3) nano-rods (G-CZNS@W) [249]. The DSSC device could obtain an exceptional PCE of 12.16%, which was attributed to the synergistic effect between the highly catalytic CZNS@W nanoparticles and the electrically conductive and electrochemically stable rGO sheets.

Perovskite solar cells (PSCs). Graphene and graphene-based materials have been applied in each layer of perovskite solar cells, including the transparent conductive substrate (TCS), the charge-transport layers (CTLs), the perovskite active layer, the counter electrode, and the interfacial buffer layer [250]. Heo *et al* developed a graphene-based TCS, in which bis(trifluoromethanesulfonyl)-amide (TFSA) was introduced onto the surface of a graphene layer to simultaneously realize high transmittance, superior conductivity, and a reduced energy barrier (figure 5.12(f)) [226]. The optimized flexible PSC that used TFSA-doped graphene as the TCS yielded superb PCEs of 17.8% and 18.2% as measured by forward and reverse scans, respectively. Moreover, the device also exhibited superior bending stability at various curvature radii, suggesting the robustness of the graphene-based material. In the case of CTLs, graphene-based materials can either serve as a modification agent or directly as the CTL. Xie *et al* prepared SnO_2:GQD composite ETLs by introducing GQDs into an SnO_2 precursor, and the optimized PSC displayed an outstanding PCE of 20.31% [251]. Li *et al* developed CNT@graphene (CNT@G)-incorporated Spiro-OMeTAD to serve as the HTL in PSC, and a PCE of 19.56% was achieved [252]. Jokar *et al* directly used 4-hydrazino benzene sulfonic-acid-reduced GO as the HTL to prepare inverted planar PSC; the hole extraction was shown to be effective, and the device produced a PCE of 16.4% [253]. Li *et al* directly incorporated an ultrathin oxo-functionalized graphene/dodecylamine (oxo-G/DA) composite into a triple-cation perovskite layer, in which oxo-G/DA was proposed to block the pathways of ionic migration at grain boundaries and encapsulate the perovskite crystals [254]. The Oxo-G/DA-based PSC exhibited a remarkable PCE of 21.1% with enhanced PCE stability in air. Zhang *et al* designed an innovative modular PSC architecture using carbon sources (e.g. graphene, graphite, and carbon black) as the counter electrodes. Among these carbon sources, the graphene-based modular PSCs exhibited an outstanding PCE of 18.65% [255]. Graphene-based materials can also serve as the interfacial buffer layer at CTL/perovskite and CTL/electrode interfaces, where they can passivate the superficial defects in these layers and promote charge transfer [256–259].

5.4.1.2 Photodetectors

As a result of the exceptional optical and electrical properties of graphene and graphene-based materials, they exhibit great promise for novel applications in photodetection [263]. There are four mechanisms for graphene-based photodetectors, as shown in figures 5.13(a)–(d). Photocurrent generation via the photovoltaic effect (figure 5.13(a)) is based on the separation of photogenerated electron–hole pairs by

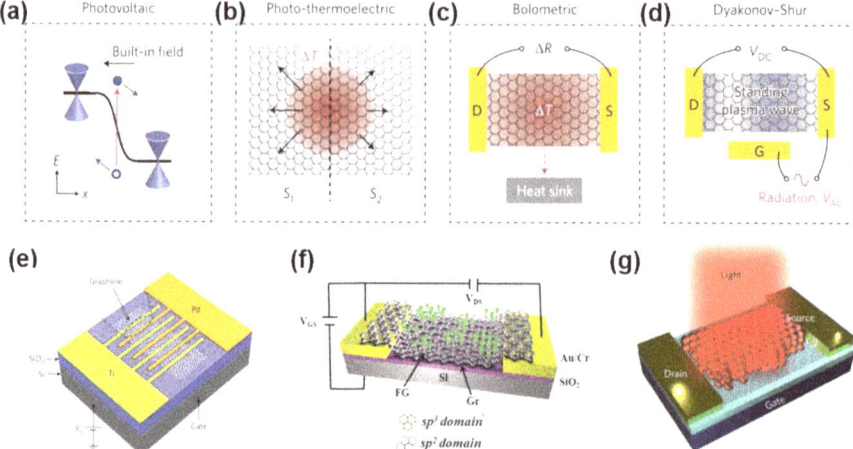

Figure 5.13. Schematic representation of the four photocurrent generation mechanisms in graphene-based photodetectors: (a) the photovoltaic effect, (b) the photothermoelectric effect, (c) the bolometric effect, and (d) the plasma-wave-assisted resonant response (Dyakonov–Shur effect) [26], reproduced with permission © Springer Nature. Device schematics of photodetectors based on different composites: (e) metal–graphene–metal [260], reproduced with permission © Springer Nature. (f) Functional graphene (FG)/graphene (Gr) [261], reproduced with permission © John Wiley & Sons. (g) Quantum dots (QDs)/graphene [262], reproduced with permission © Springer Nature.

built-in electric fields, which can be created using positive (p-type) and negative (n-type) doping of graphene [264]. Photothermoelectric (PTE)-effect-induced photo-current is based on hot-carrier-assisted transport (figure 5.13(b)) [265]. Upon light irradiation, a photoexcited electron–hole pair can promote the ultrafast (~10–50 fs) heating of the carriers in graphene due to strong electron–electron interactions [266]. Due to the high optical phonon energy in graphene (~200 meV) [267], hot carriers created by the radiation field can remain at a temperature higher than that of the lattice for many picoseconds, and they can produce a photovoltage via the PTE effect (the Seebeck effect), which generates the electronic response. In addition to hot carriers, light irradiation can also induce the bolometric effect, which is associated with a light-induced change in the conductance of graphene, including a change in the carrier mobility or the number of carriers. Because direct photocurrent generation is not involved, an externally applied bias is required in order for this effect to operate on homogeneous graphene (figure 5.13(c)) [268]. Dyakonov and Shur proposed a photodetector based on the so-called resonant regime of plasma-wave photodetection (figure 5.13(d)) [269]. This is based on the fact that a graphene-based FET that hosts a 2D electron gas can act as a cavity for plasma waves (collective density oscillations) [26]. When these plasma waves are weakly damped (that is, when a plasma wave launched at the source can reach the drain in a time shorter than the momentum relaxation time), constructive interference between the plasma waves in the cavity can enable the light irradiation to be detected.

Mueller *et al* first constructed a metal–graphene–metal photodetector based on Pd and Ti metal electrodes with an interdigitated finger structure (figure 5.13(e)), which increased the effective photodetection area of the device [260]. In this structure, photocurrent is generated by local illumination of the metal/graphene interfaces of a back-gated graphene FET, which is attributed to the photovoltaic effect. In addition, the interdigitated structure employs an asymmetric metallization effect, which enables zero-bias/dark current operation, resulting in responsivities between 1.5 and 6.1 mA W^{-1} in the NIR. Moreover, the high carrier mobility and short carrier lifetime in graphene allow this device to operate at high data rates, which allows the error-free recovery of a 10 Gbit s^{-1} data stream in an optical link. This work addresses the great potential of graphene-based materials in photo-detector applications.

However, there are severe challenges for metal–graphene–metal photodetectors, including the limited linear dynamic range (LDR) of graphene photodetectors, the lack of efficient generation and extraction of photoexcited charges, the smearing of photoactive junctions due to hot-carrier effects, large-scale fabrication, and environmental stability [263]. To overcome these issues, various graphene-based materials with tunable properties have been developed, in which the performance, stability, and versatility of the resultant photodetectors have been improved by the modification of graphene via chemical functionalization, composite hybridization, or heterostructure construction. Du *et al* investigated a photogating-based photodetector device based on van der Waals heterostructures of fluorine-doped graphene/graphene, in which the graphene served as a charge-transport layer and the fluorine-doped graphene served as the charge-trapping layer (figure 5.13(f)) [261]. The fluorination of graphene has been found to modify the C–C bond hybridization from sp^2 to sp^3, where different confined areas have different charge-trapping times. This photodetector has been found to work over a broad range of wavelengths, from UV ($\lambda = 255$ nm) to the mid-infrared (MIR) ($\lambda = 4.29$ μm). The broadband photoresponse is attributed to the synergistic effects of the spatially non-uniform collective quantum confinement of sp^2 domains and the trapping of photoexcited charge carriers in the localized states in sp^3 domains. In addition, an LDR of 4 dB and an operating bandwidth of 3 Hz were obtained, which is in line with the high responsivity values and the slow response due to the trap states in the fluorine-doped graphene. Moreover, the performance of the photo-detector exhibited a dependence on the degree of fluorination (i.e. the C/F ratio) in the fluorine-doped graphene, and the optimal C/F ratio of 3.5–3.75 was observed to deliver the best device performance. In another study, Konstanatos *et al* reported the first graphene–quantum dot (QD) hybrid photodetector, in which a 80 nm thick film of lead sulfide (PbS) QDs with the first exciton peak at 950 nm or 1450 nm was spin-coated onto a two-terminal graphene FET (figure 5.13(g)) [262]. In the graphene/QDs hybrid, photogenerated holes in the photosensitive PbS QDs transferred to the graphene, whilst electrons remained in the QDs. The electron QDs could be trapped for a timescale τ due to the built-in field at the graphene/QD interface and charge traps within the PbS, while the transferred holes changed the conductivity of the graphene channel, leading to a shift in the Dirac point voltage of graphene. As a result of the high mobility of graphene, the holes could be recirculated multiple times, and the photodetector achieved an ultrahigh gain of more than 10^8 electrons/photon and a responsivity of ~10^7 A W^{-1}.

5.4.1.3 Photocatalysis

For photocatalysis applications, graphene is generally used to modify an existing photocatalyst, but it can play various roles, such as conductive support, adsorbent, photosensitizer, photostabilizer, primary photocatalyst, and co-catalyst in composites (figure 5.14(a)) [271]. It is known that photocatalytic redox reactions are driven by photogenerated electrons and holes on the surface active sites, and thus it is essential to increase the surface active sites in graphene-based photocatalysts to drive different photocatalytic reactions. Generally, there are two strategies for increasing the active sites on graphene: one is to dope the graphene with substitutional heteroatoms, while the other is to load new active sites onto the surface of graphene (figure 5.14(b)). To date, various semiconductor/graphene composites based on Schottky junctions and Type II heterojunctions have been constructed for photocatalytic applications, taking advantage of the semimetallic and semiconducting properties of graphene [272]. Importantly, graphene-based materials have been found to be able to enhance photocatalysts' performance in various photocatalytic reactions, such as water splitting [273–275], pollution degradation [276–278], CO_2 reduction [271, 279, 280], H_2O_2 production [281, 282], and organic synthesis [283].

Water splitting. Water splitting by the photocatalytic route can be divided into the H_2 evolution reaction (HER) and the O_2 evolution reaction (OER). To achieve overall water splitting, the conduction band (CB) (valence band (VB)) levels of the photocatalysis must be more negative (positive) than the reduction (oxidation) potentials of H_2O [270]. HER is the typical half-reaction of the water splitting process, which is driven by photogenerated electrons to promote the reaction of H^+/H_2. The HER efficiency of conventional photocatalysts is quite low due to their low e^-/h^+ conversion and separation efficiencies [284], which have shown to be greatly improved by the incorporation of graphene-based materials [285]. For example, the HER performance of conventional photocatalysts such as TiO_2 [286], MoS_2 [287], g-C_3N_4 [288], CdS [289], and $BiVO_4$ [290] have been shown to be substantially enhanced in the presence of graphene and/or rGO. OER is the other typical half-reaction of the water splitting process, which is driven by

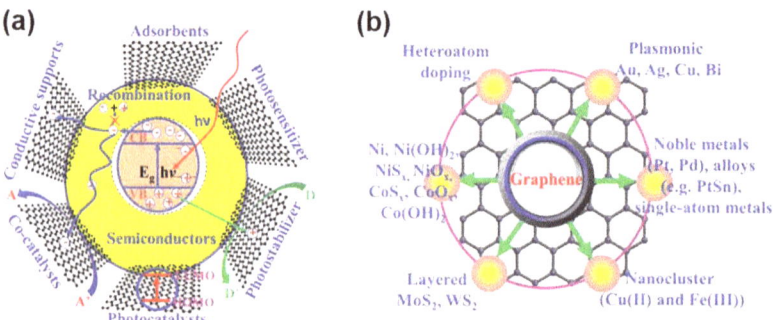

Figure 5.14. (a) Illustration of the different roles of graphene in graphene-based composite photocatalysts. (b) Illustration of strategies for modifying graphene with other co-catalysts [270], reproduced with permission © John Wiley & Sons.

photogenerated holes to promote the reaction of H_2O/O_2. However, holes generally have much slower mobility than electrons for migration from the interiors to the surfaces of photocatalytic particles, so electron-driven reactions are dominant in the system [291]. In addition, from a thermodynamic viewpoint, four units of photo-generated holes are required to generate one unit of O_2, so the protection of photogenerated holes is of great importance for the OER. Therefore, to enhance the OER process, many studies have focused on using sacrificial reagents to rapidly remove photogenerated electrons, thus protecting photogenerated holes from recombination [292]. In these systems, graphene-based materials have been shown to prolong the lifetime of photogenerated holes during the OER process [293]. For example, GO [294], heteroatom-doped GO [295], graphene-supported $Ag_3PO/Ag/AgBr$ [296], an α-Fe_2O_3/N-graphene composite [297], and nitrogen-doped GQDs (N-GQDs) [295] have been shown to exhibit better photocatalytic OER activity compared to traditional photocatalysts.

Pollution degradation. Photocatalytic pollutant degradation is usually performed within water, where photogenerated $O_2\cdot^-$ and OH· radicals are responsible for the initiation of the cascade reactions that convert pollutants into non-toxic substances and remove them from the environment [298]. Graphene-based materials are used as supports for photocatalysts to enhance their degradation performance, and graphene can play the roles of electron acceptor, reactive site, or conductor bridge. The oxidation potential of rGO/rGO^- was reported to be about -0.08 eV (vs. a standard hydrogen electrode (SHE)), which is less negative than the conduction band of TiO_2 (-0.24 V), enabling facile electron transfer from TiO_2 to graphene [299]. Note that the degree of GO reduction was found to greatly influence the degradation performance of the photocatalyst due to the changed work function [300]. In addition, the doping of heteroatoms into graphene has been found to introduce extra reactive sites into the material. Huang *et al* reported that sulfur-doped GQDs (S-GQDs) could efficiently remove 81% of basic fuchsine (BF) after 2 h of irradiation, while S-free GQDs only removed 18% under equivalent conditions [301]. In a graphene-based ternary photocatalyst system, graphene can act as a conductor bridge to integrate different photocatalyst components. For example, CdS–ZnO–GO has been reported to photocatalytically degrade methyl orange (MO), in which the photogenerated electrons in the CdS are first transferred to the conduction band of ZnO and finally accumulate in the graphene. This process inhibits the electrons from migrating back to the CdS, which prolongs the lifetime of the electrons and the efficiency of MO degradation [302].

CO2 reduction. The use of photocatalysts to convert CO_2 into hydrocarbon compounds and fuels has great potential to simultaneously solve the global warming problem and promote the fuel industry [303]. Because the formation of CO_2 is thermodynamically favorable if electrons and protons are transferred to CO_2 at same time, the key to enhancing the efficiency of CO_2 reduction is to effectively separate electrons and protons [304]. In addition, the photoreduction of CO_2 by traditional photocatalysts is also hampered by low energy conversion efficiency, low selectivity, and poor control in suppressing the competing hydrogen evolution reaction, especially in the presence of water [305]. Fortunately, graphene-based materials have been found to promote the separation of photogenerated electrons and substantially improve the efficiency of CO_2 reduction in different photocatalyst systems, including TiO_2/MoS_2/3D graphene aerogel [306], TiO_2/N-rGO [307], CeO_2/N-graphene [308], CdS/graphene [309], Cu nanoparticles/GO [310], WSe_2/graphene [311], and g-C_3N_4/graphene [312].

H_2O_2 production. Theoretically, H_2O_2 can be produced from the reaction of H_2O and O_2 at a certain voltage, and the photocatalytic process is as follows: photo-generated electrons in the CB of the photocatalyst can reduce O_2 into the superoxide radical $O_2\cdot^-$, which then disproportionates with another $O_2\cdot^-$ to form H_2O_2. This facilitates solar-to-H_2O_2 energy conversion with a relatively high free-energy gain ($\Delta G^\circ = 117$ kJ mol^{-1}) [313]. The H_2O_2 production reaction is highly dependent on the pH of the system; a low pH is required for increased production because the formation of H_2O_2 is governed by the proton [314]. However, graphene-based photocatalysts can break this limit; for instance, a TiO_2/rGO composite exhibited much higher activity than TiO_2 at a pH of seven [315], while WO_3/graphene hybrids showed outstanding performances at higher pH values [316]. In addition, the incorporation of rGO into a carbon nitride aromatic diimide (g-C_3N_4/PDI) photo-catalyst produced a twofold enhancement in the production efficiency of O_2 from pure water under visible light irradiation ($\lambda > 420$ nm) [317].

Organic synthesis. Organic synthesis with high selectivity can be achieved by using well-designed photocatalysts under well-controlled reaction conditions, while graphene-based materials can play a key role in improving the photocatalytic efficiency and selectivity of photocatalysts [318]. Zhang *et al* reported a 20% enhancement in the organic synthesis of aldehydes from alcohols after incorporating graphene into CdS nanoparticles [319]. Zhang *et al* further developed a ternary hybrid composed of CdS and TiO_2 on GO sheets, and this hybrid not only enhanced the photo-catalytic activities of alcohol oxidation but also increased the selectivity to aldehyde products [320]. Reduced GO–silver vanadate nanocomposite (rGO–Ag_3VO_4) photo-catalysts were designed for the organic synthesis of catechol from phenol, and the selectivity toward catechol at room temperature under visible light irradiation was 80.23% within 2 h when the graphene content in the photocatalyst was 4 wt% [321]. Tan *et al* reported the oxidative C–H functionalization of tertiary amines by introducing GO into Rose Bengal (RB), in which the addition of certain amounts of GO into the reaction system enhanced the reactivity and product yield of tertiary amines [322]. Interestingly, it was found that natural graphite and activated carbon showed no activities in this reaction, and thus it was proposed that the slightly acidic GO with its high surface area can promote the acidic stabilization of the iminium intermediate, which is attributed to the enhancement of the reaction rate and yield.

5.4.1.4 Fuel cells

Fuel cells are devices that are used for the direct conversion of the chemical energy of a fuel into electricity, and thus fuel cells possess significantly high energy conversion efficiency along with low pollutant emissions [323]. A typical fuel cell contains an electrolyte layer that is sandwiched between two electrodes (figure 5.15(a)) [323]. Under working conditions, the fuel is oxidized on the anode surface, and the released electrons

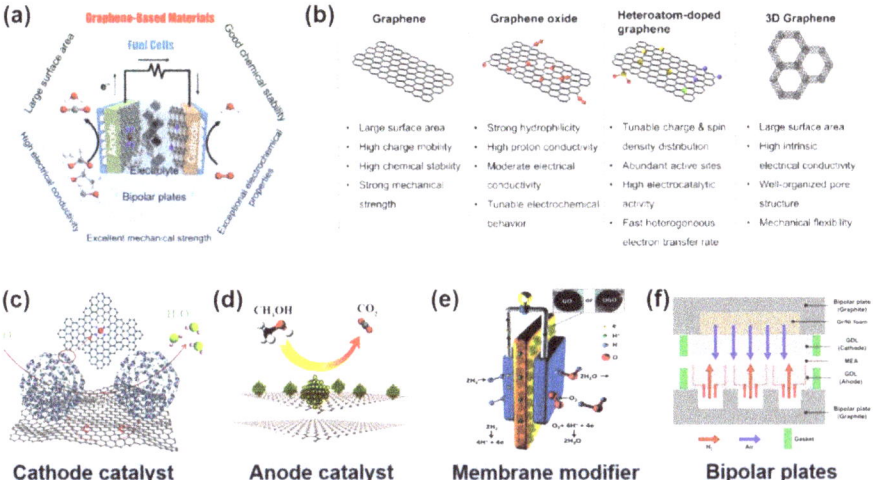

Cathode catalyst   **Anode catalyst**   **Membrane modifier**   **Bipolar plates**

Figure 5.15. Schematic of the (a) superiorities and (b) main properties of graphene-based materials used in a fuel cell [323], reproduced with permission © John Wiley & Sons. The different roles of graphene-based materials in fuel cells: (c) cathode catalyst [324], reproduced with permission © Elsevier; (d) anode catalyst [325], reproduced with permission © The Royal Society of Chemistry; (e) membrane modifier [326], reproduced with permission © John Wiley & Sons; (f) bipolar plates [327], reproduced with permission © The Royal Society of Chemistry.

flow via an external circuit to promote the reduction of O_2 at the cathode. To complete the circuit, the mobile charge carriers (H^+, OH^-, CO_2^{-3}, or O^{2-}) are simultaneously transferred from the cathode to the anode through the electrolyte. So far, various electrolytes have been developed, and according to the type of electrolyte, fuel cells are classified into phosphoric acid fuel cells, polymer electrolyte membrane fuel cells, alkaline fuel cells, molten carbonate fuel cells, and solid-oxide fuel cells. The traditional materials used for fuel cell components face great challenges in terms of electrochemical performance, efficiency, and durability, while the excellent chemical, physical, electronic, and mechanical properties of graphene and its derivatives enable them to be ideal candidates for the construction of efficient fuel cells. Figure 5.15(b) demonstrates the main properties of different graphene-based materials that are related to fuel cell applications, in which graphene-based materials are found to play vital roles in every component of fuel cells.

Graphene-based materials are ideal supports for electrocatalysts, and can increase the number of active sites and facilitate electron transport for both the fuel oxidation and ORR reactions on the electrocatalysts. Chen *et al* developed a highly efficient electrocatalyst with single Fe atoms anchored by N-doped short-range ordered carbon loading on 2D rGO (figure 5.15(c)), and a fuel cell that used this electrocatalyst as the cathode exhibited a half-wave potential of 0.84 V versus the reversible hydrogen electrode and a <5 mV loss of half-wave potential during 15 000 potential cycles [324]. In addition, metal-free graphene materials have been demonstrated to be attractive candidates for direct oxygen reduction due to their high electrocatalytic activity, high tolerance to poisoning, and low cost as well as the designable nature of the graphene materials [328].

Due to the good dispersion of metal nanoparticles (e.g. Pt, Ni) on graphene as well as the superior conductivity of graphene, the activity of anode catalysts can also be significantly enhanced by the introduction of graphene [329, 330], and this high activity results in a decrease in the loading of the metal catalysts as well as an improvement in the stability of the catalyst in the anode of the fuel cells. Sun *et al* reported the construction of pure Ni nanoparticles at ultrafine sizes embedded on rGO, which exhibited anode reactivity with ultrahigh catalytic activity (1600 mA mg^{-1}) and excellent stability (1020 mA mg^{-1} retained after 1000 cycles) (figure 5.15(d)) [325].

Furthermore, the incorporation of graphene-based materials into polymer electrolyte membranes has been shown to effectively improve the ionic conductivity and minimize the crossover of fuel cells [331]. In addition, graphene-based materials are promising as proton-exchange membranes due to their high proton conductivity and impermeability to water, H$_2$, and methanol. Gao *et al* reported the preparation of freestanding ozonated GO film, which was used as an efficient polymer electrolyte fuel cell membrane (figure 5.15(e)) [326]. The ozonation of GO leads to a higher content of oxygenated functional groups in the basal planes and edges of GO, which further increases the protonic conductivity of GO due to enhanced proton hopping.

In addition to electrolytes and electrodes, graphene-based materials can also improve the current collection, fuel/air distribution, and stability of bipolar plates in fuel cells. Sim *et al* developed a 3D graphene-coated Ni foam to act as a bipolar plate with long-term operating stability (figure 5.15(f)) [327]. The synthesized graphene layers were shown to have a low defect density and could completely cover the 3D-structured surface of the Ni foam, thus dramatically enhancing its corrosion resistance, interfacial contact resistance, and hydrophobicity. Specifically, a fuel cell device based on the 3D graphene-coated Ni foam maintained its outstanding interfacial contact resistance of 9.3 mΩ·cm^2 at 10.1 kg f cm^{-2}.

5.4.1.5 *Piezoelectric nanogenerators*

In 1880, the brothers Pierre and Jacques Curie first discovered the direct piezoelectric effect, which is attributed to the asymmetric shift of charges or ions of piezoelectric materials when they are exposed to mechanical strain. This effect can be reversed if an electric potential is applied to the same materials, leading to their mechanical deformation [332]. Typically, some inorganic crystalline ceramic materials are found to exhibit piezoelectricity, generating electricity under the effect of applied pressure [332]. However, the inherent brittleness of these inorganic materials greatly limits their application in flexible devices. To solve this problem, piezoelectric composites made by dispersing nanosized ceramics in a piezoelectric polymer matrix have been developed [333]. For example, a lead zirconate titanate (PZT)/polyvinylidene fluoride polymer (PVDF) composite was found to exhibit a greater piezoelectric coefficient than pristine PVDF without compromising the flexibility of pristine PVDF [334]. The energy conversion capacity of piezoelectric nanogenerators (PENGs) based on PZT/PVDF is low, but it can be enhanced by the incorporation of graphene-based materials. In PENGs, graphene-based materials enhance the mechanical-to-electrical energy conversion capacity of the device via two paths: (i) serving as an interfacial layer and (ii) being incorporated into piezoelectric composites.

Bhavanasi *et al* reported an enhancement of the piezoelectric energy-harvesting performance of poled P(VDF-TrFE) using GO as the interfacial layer [335]. The enhanced performance was attributed to multiple effects, including the electrostatic contribution of GO, the residual tensile stress and enhanced Young's modulus of the bilayer films, and the presence of a space charge at the interface of the bilayer films. The developed bilayer film exhibited a superior voltage output of 4 V and a power output of 4.41 μW cm^{-2}. Yaqoob *et al* reported the fabrication and characterization of a trilayer PENG assembled by stacking a PVDF–BaTiO$_3$ layer on an N-graphene layer; the device generated a maximum output voltage of 10 V and a current of 2.5 μA at an applied force of 2 N [336]. N-graphene has been found to play a role in enhancing the energy-harvesting performance by aligning the dipoles in one direction. However, as an interfacial layer, graphene has been found to have a limited ability to enhance the piezoelectricity of PENGs.

Graphene-based materials are proposed to serve as nucleation and crystallization sites that could improve β-phase formation (the piezoelectricity phase of PVDF), crystal orientation, and thus the piezoelectric coefficient of PVDF [338]. Karan *et al* synthesized an Fe-doped rGO/PVDF (Fe-rGO/PVDF) nanocomposite via the solution casting method, and this piezoelectric nanocomposite material could yield a high output voltage without the use of the electric poling process [339]. Specifically, when excited by the touch of a human finger, the nanocomposite film exhibited a maximum output voltage of 5.1 V and a short circuit current of 0.254 μA, which are 12 times and 10^5 times greater than those of pure PVDF film, respectively. Shi *et al* demonstrated that the co-incorporation of BaTiO$_3$ nanoparticles and graphene nanosheets into PVDF fibers using the electrospinning process could lead to a synergistic contribution to the enhanced piezoelectric performance of the device (figure 5.16(a)) [337]. Graphene, acting as a nucleating agent and a substrate, could promote the formation of small PVDF fibers with a favorable β-phase. The composite fibers were adopted to construct a flexible PENG device, which could be easily wrapped around a finger (figure 5.16(b)). As the polarization of the fibers was parallel to the in-plane direction of the PENG, small strains and slow strain rates could also be detected during subtle body movements, such as wrist bending, finger tapping, and treading (figure 5.16(c)). Impressively, the generator produced a peak voltage of 112 V during a finger pressing–releasing process, which was sufficient to light 15 LEDs and drive an electric watch.

5.4.1.6 *Thermal energy conversion devices*

The use of integrated circuits for thermal energy conversion is a very important research topic. Graphene-based materials have been emerging in the field of thermal devices, providing new opportunities as well as challenges for the conversion and utilization of thermal energy [63]. To promote efficient thermal energy conversion, suitable materials and structures should be used to extract heat energy, which further need to be effectively coupled with other physical quantities to maximize the conversion of thermal energy [340]. Therefore, the key factors in the performance of these thermal energy conversion devices are the thermal sensitivity, the thermal conductivity and thermal conversion rate of the thermal active layer, where

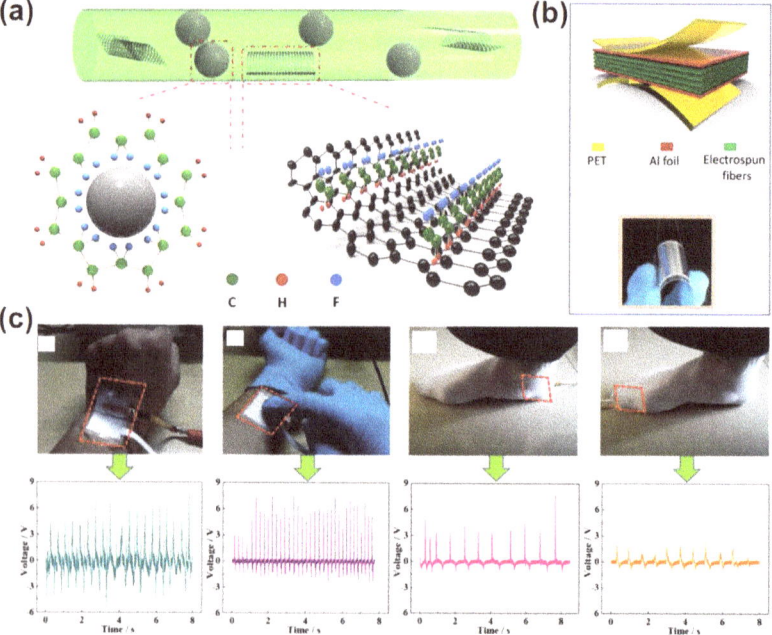

Figure 5.16. (a) Illustration of the mechanism of β-phase formation of PVDF on BaTiO₃ nanoparticles and graphene nanosheets in the nanocomposite fiber. (b) Schematic and photograph of the device structure of a flexible PENG. (c) Photographs and output voltages generated by human motions of different body parts, from left to right: wrist bending, finger taping, heel stepping, and toe stepping [337], reproduced with permission © Elsevier.

graphene-based materials can play a vital role due to their high thermal conductivity and thermal rectification. Using different coupling models of graphene with different physical quantities (i.e. thermo self-coupling, thermoacoustic coupling, thermo-electric coupling, and thermo-optical coupling), several thermal energy conversion devices have been developed, such as thermoacoustic devices, the artificial throat, thermal rectifiers, thermal radiation light sources, photo-to-thermal conversion devices, and electrothermal actuators (figure 5.17).

Graphene is an ideal thermoacoustic material because of its excellent conductivity, low-cost preparation, and thinness. After graphene has been incorporated into a thermoacoustic device, it can produce a high sound pressure level (SPL) over a wide audio range. The merits of graphene-based thermoacoustic devices are mainly reflected by their high performance, high reliability, and low drive voltage [341]. Most importantly, graphene-based materials can provide thermoacoustic devices with advantages that cannot be found in conventional materials, such as bendability, stretchability, and transparency [342]. Tian *et al* prepared a single-layer graphene thermoacoustic device on an anodic aluminum oxide (AAO) substrate, and the effect of the graphene thickness on the device performance was studied [342]. It was found that thinner graphene produces a better-sounding performance, and that the SPL value of the single-layer graphene device could reach up to 95 dB. Moreover, a

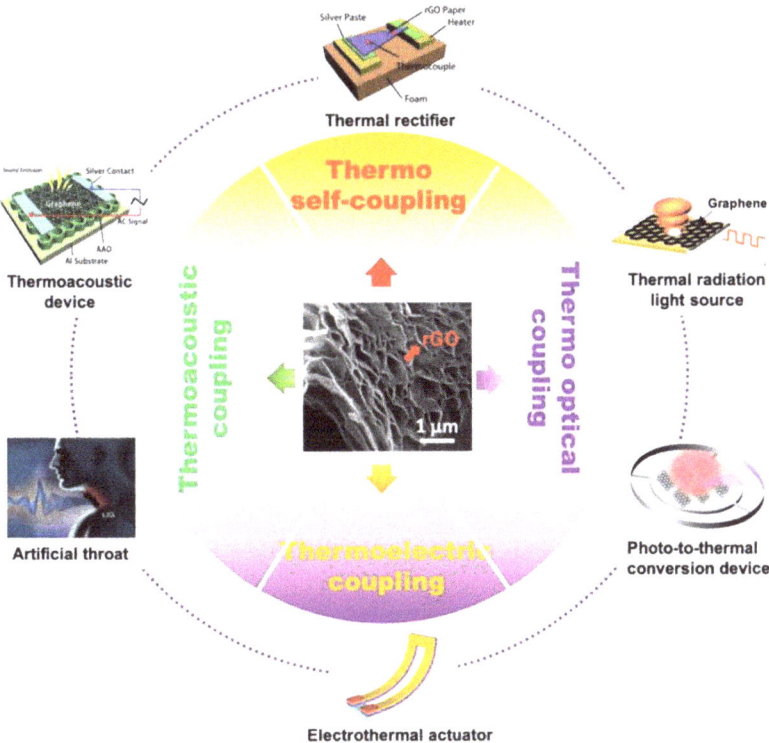

Figure 5.17. Four coupling models of graphene with different physical quantities and the thermal energy conversion devices derived from them [340], reproduced with permission © John Wiley & Sons.

thermal infrared imager was used to confirm that the source of the sound was heat dissipated by the device. Tao *et al* prepared an artificial throat based on laser-induced graphene; sound emission and detection were both achieved on an integrated single device [343]. It was found that the SPL value of the artificial throat increased with a decrease in the thickness of the graphene.

Tian *et al* developed a thermal rectifier based on rGO using both triangular and two-rectangle structures [344]. In the device with a 60° triangular structure, the rectification ratio could reach up to 1.28 and increased with increasing angles. The two-rectangle-shaped device only exhibited a rectification ratio of 1.10, which could also be enhanced by increasing the width ratio. These results demonstrate the great potential of large-area graphene-based materials in thermal circuits and thermal management. In addition to the shape of the graphene, Wang *et al* showed that the inherent strain of suspended graphene can also induce good thermal rectification characteristics; using suspended graphene, thermal rectification was experimentally demonstrated in various asymmetric monolayer graphene nanostructures [345]. A large thermal rectification factor of 26% was achieved in defect-engineered monolayer graphene with nanopores on one side, while a thermal rectification factor of 10% was achieved in pristine monolayer graphene with nanoparticles deposited on one side or with a tapered width. These results suggest the potential application of monolayer graphene in the design of high-performance thermal rectifiers for heat flow control and energy harvesting.

Miyoshi *et al* reported integrated graphene-based on-silicon-chip blackbody emitters in the NIR [346]. They demonstrated that the graphene represented blackbody radiation generated by Joule heating, and that the number of graphene layers had a great impact on the emission responses. This work shows that graphene-based emitters can be directly integrated with silicon platforms, which opens new paths for highly integrated optoelectronics. Manjavacas *et al* demonstrated an optical-to-thermal converter that adopted graphene-based plasmonic structures, which could release over 90% of its emission through individual infrared lines with 1% bandwidth [347]. They proposed that radiative emission was the main cooling mechanism of this system. This concept of optical thermal conversion demonstrates a novel approach toward the efficient generation of infrared light.

Graphene has been widely used as the heater or the heating layer of electrothermal actuators, as its temperature distribution is more uniform and its electrothermal response is faster. Zhang *et al* utilized laser-reduced GO to fabricate a super-flexible and shape-designable graphene heater [348]. Due to the high performance of the graphene-based heater, it is expected to be used in three scenarios, including medical infusion apparatus, flexible custom-shaped heaters for special requirements, and displays. In addition, graphene-based materials also possess high flexibility that is suitable for electrothermal actuators that are exposed to bending deformation. Sang *et al* designed and fabricated an electrothermal bilayer actuator by spin-coating rGO onto a polymer substrate [349]. The bilayer actuator exhibited a fast and large bending response to a DC voltage applied to the graphene layer, which extends the potential use of the device into the fields of soft electronics, sensors, and energy conversion.

5.4.2 Energy storage

Graphene-based materials have been extensively investigated in the field of energy storage applications, as they provide better means of storing electricity. Specifically, graphene could endow energy storage devices with several new features, resulting in miniaturized, flexible, rollable, high-capacity, and transparent devices as well as fast-charging devices [350]. In this section, we will briefly describe the advanced development of graphene-based materials in energy storage applications, including ion batteries and supercapacitors.

5.4.2.1 Ion batteries

Graphene-based materials have been shown to substantially enable large specific capacity, high rate capability, and long cycle life in various ion batteries, including those made using Li^+, Na^+, Mg^{2+}, and Al^{3+} [351]. Among these ion batteries, lithium-ion batteries (LIBs) are the most representative, and the application of graphene-based materials in LIBs has been extensively studied. All those ion batteries basically share the same device architecture, in which graphene-based materials play similar roles. Therefore, this section will take LIBs as an example to introduce the application of graphene-based materials in ion batteries.

The charge/discharge mechanism of rechargeable LIBs is based on the shuttling of Li^+ between a negative electrode (commonly graphite) and a positive electrode (commonly a layered lithium metal oxide) via an electrolyte (figure 5.18(a)). Due to their advantages of high energy density, high voltage, absence of a memory effect, long cycle life, and environmental friendliness, LIBs have revolutionized modern portable electronic devices. However, current commercial LIBs still face the challenges of low specific energy (per weight) and low energy density (per volume), which greatly hinder their application in diverse applications in expanded fields [352]. The main cause of the low performance of LIBs lies in the poor lithium storage capability of most electrode materials due to their intrinsically low electronic conductivities, volume-change-induced structural and interfacial instabilities, inter-particle connection issues, and low-tapped-density problems [351].

Benefiting from graphene's unique crystal structure and extraordinary physical and chemical properties, graphene-based composite materials have provided inspiring solutions to the limitations and challenges faced by today's electrode materials for LIBs. For graphene-based composite electrodes, there are six different structural models, as shown in figure 5.18(b) [352]. It is worth noting that many graphene-based composites may be based on more than one type of model, and can thus be described by several of the above models. In the different interaction models of graphene-based composites, graphene plays different roles that provide fundamental performance enhancements over conventional materials/electrodes, including (i) as a material support to improve electronic and ionic conductivity [353], (ii) as a material mediator to manage

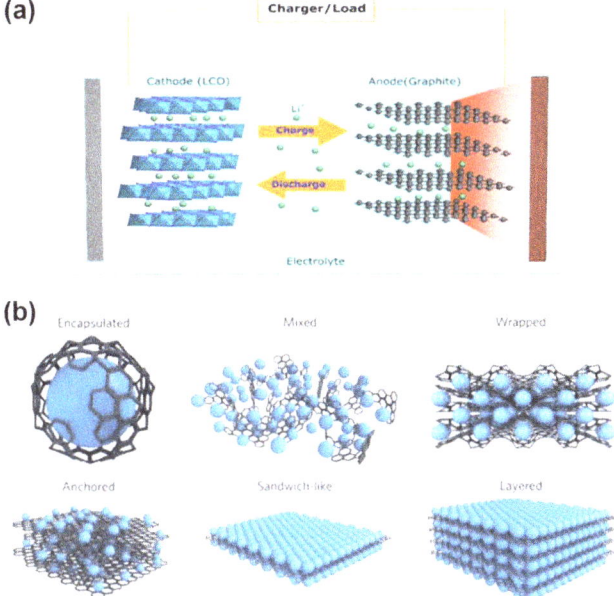

Figure 5.18. (a) Schematic diagram of a typical LIB device. (b) Schematic of the different structures of graphene-based composites as electrode materials for LIBs [352], reproduced with permission © The Royal Society of Chemistry.

the material component interface and the material/electrolyte interface [354], (iii) as an electrode framework to offer 3D continuous conductive channels [355], (iv) as an electrode modifier to stabilize the electrode/electrolyte interface [356], and (v) in combined media to synergistically modulate material structures and electrode architectures [357].

To examine its role as a material support, Paek *et al* developed a simple approach to *ex situ* prepare SnO_2/graphene hybrids SnO_2 nanoparticles homogeneously anchored on graphene sheets, in which the graphene was found to dimensionally confine the volume change of SnO_2, induce nanopores as buffered spaces, and create efficient conductive channels, resulting in superior cyclic performances and higher reversible capacities [353]. To examine graphene's role as a material mediator, Yang *et al* fabricated a graphene encapsulated metal oxide Co_3O_4 composite, and the resulting hybrids possessed flexible and ultrathin graphene shells that effectively enwrapped the oxide nanoparticles [354]. This classical core–shell structured hybrid design suppressed the aggregation of oxide nanoparticles, accommodated the volume change that occurred during cycling, and maintained high electrical conductivity in the active material. As a result, LIBs based on the graphene/Co_3O_4 anode exhibited a very high reversible capacity of 1100 mA h g^{-1} in the first ten cycles, and over 1000 mA h g^{-1} after 130 cycles, with excellent cyclic performance. To examine graphene's role as an electrode framework, Wang *et al* prepared paper-like layered hybrid electrodes containing stable, ordered, alternating layers of nanocrystalline SnO_2 metal oxide and graphene sheets, in which the graphene sheets served as framework. They proposed that the well-controlled electrode architecture with the locally ordered alternating layers of graphene sheets and SnO_2 nanocrystals improved the conductivity and the structural integrity of the composite [355]. To examine graphene's role as an electrode modifier, Wang *et al* developed a hybrid array electrode configuration in which well-ordered perforated silicon nanowires were conformally coated with graphene sheets, thus forming mechanically robust, free-standing textured hybrid nanowire arrays [356]. The deposited graphene coating was found to completely mimic the modality of Si, thus improving the silicon interface with the electrolyte, and at the same time afforded efficient highways, thus promoting the lateral transport of both electrons and lithium ions to/from each encapsulated nanowire. To examine graphene's role in combined media, Wang *et al* developed an adaptable, self-supporting, flexible, and binder-free paper-like silicon-based anode, in which silicon nanowires were sandwiched between rGO sheets [357]. Within this architecture, the overlapping graphene sheets functioned as sealed mediators, which prevented direct contact between the silicon and the electrolyte and enabled the structural and interfacial stabilization of the encapsulated silicon nanowire materials during repeated cycling. Benefiting from the graphene hybridization, the anode electrodes exhibited a great enhancement in lithium storage performance with reversible specific capacity, robust stability, and superior rate capability.

5.4.2.2 *Supercapacitors*

Supercapacitor devices are used for short load cycle applications due to their static electrical energy storage with optimum power density and rapid charging/discharging abilities [361]. Supercapacitor devices mainly consist of two current collectors followed by two electrodes, which are separated by an ion-permeable separator in an electrolyte solution to avoid short-circuits (figure 5.19(a)) [350]. The performance of a supercapacitor is largely determined by the equivalent series resistance, cell operating voltage, and capacitance of the device, which are closely dependent on the properties of the electrode's matrix material. Effective electrode materials for supercapacitors should possess a high surface area with suitable pore sizes, high electrical conductivity with good stability, and high power density to attain highest volumetric energy densities [362]. Graphene has been also reported to be an ideal candidate for supercapacitor electrodes because of its chemical stability, low cost, ultrahigh specific surface area, outstanding electrical conductivity, good mechanical flexibility, and high intrinsic capacitance [363]. However, the main drawback of graphene as an electrode material in supercapacitors is its agglomeration and restacking between adjacent sheets via the van der Waals force [363]. To address this issue, various structures for assembled graphene materials have been developed, including 1D fibers or yarns, 2D graphene in the form of thin films or sheets, and 3D graphene in the form of foams or composite networks [364].

1D graphene fibers or yarns. 1D graphene materials are considered to be the most promising candidates for next-generation supercapacitors for portable and wearable electronics, as they exhibit the advantages of high flexibility, light weight, tiny volume, and high electrical conductivity. Chen *et al* developed textile carbon nanotube/graphene (CNT/G) hybrid fibers, which had a large surface area and high electrical conductivity and were adopted as electrodes in flexible supercapacitors (figure 5.19(b)) [358]. The performance of the flexible device included an areal capacitance of 1.2–1.3 mF cm^{-2} and a high tolerance to repeated bending cycles. To improve the electric performance of 1D graphene fiber, it can also be combined with

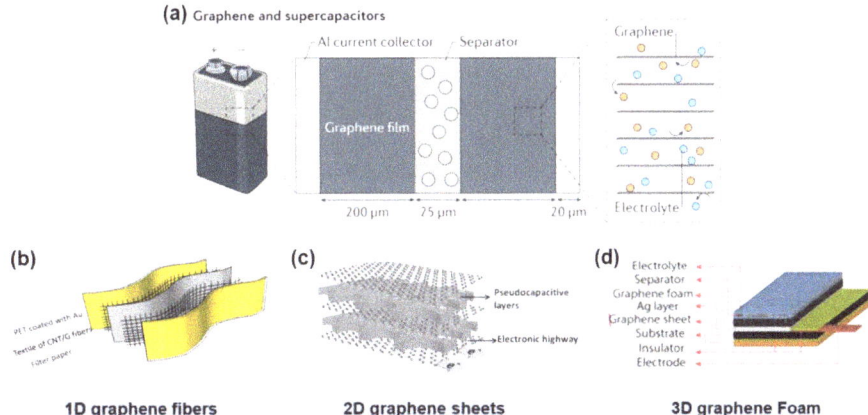

Figure 5.19. (a) Schematic representation of a two-electrode supercapacitor, in which graphene-based materials typically serve as electrode materials [350], reproduced with permission © Springer Nature. Illustrations of supercapacitors based on electrodes with different graphene structures: (b) 1D graphene fibers [358], reproduced with permission © The Royal Society of Chemistry. (c) 2D graphene sheets [359], reproduced with permission © John Wiley & Sons. (d) 3D graphene foam as an electrode [360], reproduced with permission © Elsevier.

selected electrically active materials with faradaic pseudocapacitance, such as metal oxides, hydroxides, and conductive polymers. Kou *et al* designed a coaxial wet-spinning assembly process to continuously fabricate polyelectrolyte-wrapped graphene-CNT core–sheath fibers; the sheath layer of the polyelectrolyte effectively avoided the risk of the short circuits [365]. The maximum capacitances achieved were C_l (5.3 mF cm^{-1}), CA (177 mF cm^{-2}), and CV (158 F cm^{-3}), which were ascribed to the large surface area and efficient ion transportation between electrodes.

2D graphene thin films or sheets. 2D graphene thin films are frequently used as an electrolyte transport platform in flexible supercapacitors due to their light weight, tunable thickness, high electrical conductivity, high power density, and high surface area [366]. However, the restacking of graphene sheets due to van der Waals attractive forces and inter-planar π–π interactions leads to a reduction in the surface area of graphene films and limited electrolyte ion diffusion in between them. Thus, appropriate spacers are created by inserting other materials to overcome the stacking of graphene films, such as conductive polymers, transition metal oxides/hydroxides, metals (Au or Pt)/metal oxides (SnO$_2$) materials, or carbonaceous materials (CNT or carbon powders) [367]. For instance, Li *et al* used Ni(OH)$_2$ nanoplates to intercalate in between densely stacked graphene sheets (figure 5.19(c)) [359]. A supercapacitor based on this electrode structure exhibited superior supercapacitive performance, including high gravimetric capacitance (≈573 F g^{-1}), high volumetric capacitance (≈655 F cm^{-3}), excellent rate capability, and superior cycling stability.

3D graphene foam or composite. 3D graphene can take the form of either foam or a composite with macro/micro/meso-interconnected pores, which are extremely appropriate for supercapacitors with outstanding performance and optimum power and energy density due to the large surface area available for rapid electron/ion transport channels [368]. For instance, Manjakkal *et al* reported the construction of a flexible supercapacitor based on 3D porous graphene foam (figure 5.19(d)) [360]. The device exhibited excellent electrochemical and supercapacitive performance with a capacitance of 38 mF cm^{-2}, an energy density of 3.4 µWh cm^{-2}, and a power density of 0.27 mW cm^{-2}. Other well-known 3D graphene materials are the porous and ultralight graphene aerogels, which show high ratios of surface area to volume and strength to weight [208]. The unique 3D structure of graphene aerogels provides multi-dimensional transport pathways for ions/electrons, resulting in a reduction of the transport distances between the electrolyte and the bulk electrode material. Thus, 3D graphene aerogels are used to fabricate binder-free electrodes for numerous electrochemical applications. Fuertes *et al* reported an all-solid-state flexible supercapacitor based on high-density and free-standing graphene hydrogel on a flexible current collector, which resulted in an efficient device with a specific capacitance of 298 F g^{-1} and a well-retained capacitance of ~87% after 10 000 charge–discharge cycles [369].

5.4.3 Transistors

5.4.3.1 Light-emitting diodes (LEDs)

Light-emitting diode (LEDs), including organic (OLEDs) and inorganic LEDs, are very mature and have already been applied in several fields, including signage, display backlights, general illumination, and communications [371]. As a result of their superior properties, graphene-based materials have been successfully applied in LEDs to enhance their device performance. For example, graphene and doped graphene have been shown to improve LED performance in terms of current-spreading enhancement [372], ohmic contact formation, and reduced growth temperatures [373]. Theoretical studies have been conducted to measure the light emitted by graphene and related structures, such as single and multilayer graphene [374], rGO [375], GNRs [376], and QDs [377], in which the corresponding emission was explained using the mechanisms of electroluminescence [378], thermal emission radiation [379], and plasmon-assisted emission (figure 5.20) [380].

Graphene can be used as an active material for LED devices, in which the electroluminescent (EL) effect can be observed in graphene and graphene-related structures. By applying an electrical bias to graphene supported on a substrate, phonon-assisted EL emission in the visible region can be obtained from the device [381]. Visible emission from graphene can also be generated by the excitation of an electron tunneling current in a scanning tunneling microscope (STM) using a voltage-biased tip, which is attributed to the hot electroluminescence effect [382]. In addition, the EL emission spectrum of graphene-based LEDs can be tuned in the range of the entire visible spectrum by applying a tunable gate voltage, which is quite challenging in the traditional solid-state LED industry [383].

Emission in the spectral region from the infrared (IR) to the visible range can also be created by thermal stress applied to graphene-based LEDs [384]. Thermal emissions from the graphene layer were attributed to local heating over almost

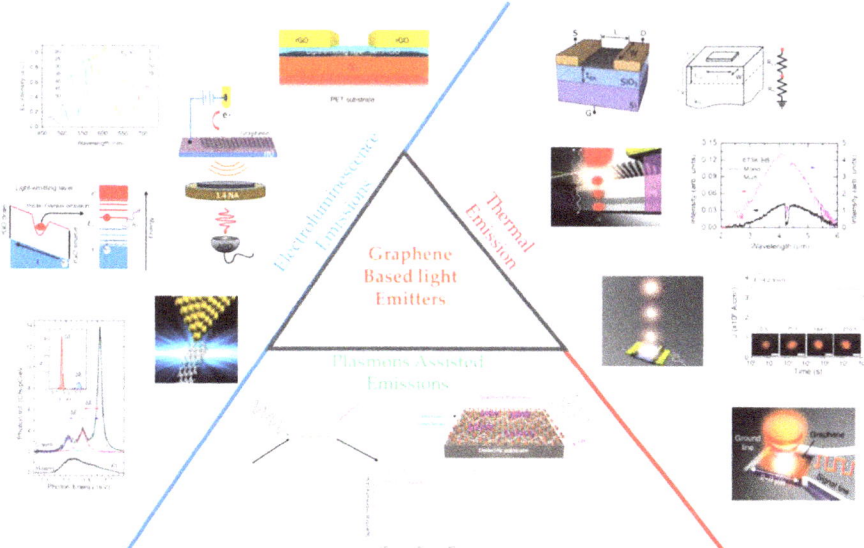

Figure 5.20. Functional light-emitting transistors that have adopted graphene-based materials. The emissions radiated by graphene have been categorized into thermal, electroluminescent, and plasmon-assisted emissions [370], reproduced with permission © MDPI.

the entire spectrum of gray-body radiation, in which a small fraction of the energy was converted into light emission [385]. Notably, LEDs based on micron-sized CVD graphene exhibit a sustained high current density (10^7 A cm^{-2}) compared to those based on conventional tungsten filaments (~100 A cm^{-2}), which indicates their suitability for high-frequency operation. Moreover, the impressive thermal emission from a large-area graphene layer, together with its extraordinary thermal conductivity, suggests a potential high-efficiency IR light source [386].

Plasmon-assisted emission from graphene can be realized by applying an electron beam to the optically excited surface plasmons of graphene, and in this way, unidirectional, chromatic, and tunable emissions from the IR region to the x-ray region can be obtained [387]. Theoretical investigations and experimental results have confirmed the existence of plasmons in graphene at visible and IR wavelengths [388], and the plasmon-assisted light emission from graphene at visible and even shorter wavelengths is attributed to the interaction between surface plasmons and charged particles [389]. Significantly, due to graphene's unique properties of high field confinement, surface plasmons, and low phase velocity, the 2D quantum Cerenkov effect can also be achieved, which is related to the emissions that occur when shockwave plasmons are excited by hot carriers [390].

5.4.3.2 Thin-film transistors

Thin-film transistors (TFTs) are the central building blocks of all modern electronic systems. However, the rigid substrate used for traditional TFTs greatly limits their application scenarios, and thus the integration of TFTs with flexible substrates is of great significance for a new generation of flexible, wearable, and disposable electronics, such as electronic paper, wearable displays, and artificial skin [391]. Currently, the major challenges for the realization of high-performance flexible TFTs are the difficulties in producing TFTs with high carrier mobility, a high on/off ratio, and a low operating voltage using a low-temperature-compatible fabrication process. Due to its high carrier mobility, excellent optical transparency, and high mechanical strength, atom-thick graphene is one of the most ideal candidates for realizing efficient flexible FETs. In TFTs, graphene-based materials can be used as the channel material and the electrode material.

As a channel material. Graphene-based thin films with high transparency and conductivity have been usually used as the conducting channel layer in TFTs, and the atomic thinness achievable in graphene-based devices leads to an almost perfect control of the channel potential by controlling the graphene structure [396]. Wang *et al* successfully synthesized graphene nanoribbons with smooth edges and widths smaller than 10 nm, which could open up a bandgap that was large enough for room-temperature transistor operation (figure 5.21(a)) [392]. The graphene nanoribbon-based TFTs exhibited an on/off ratio of ~1×10^6 and a carrier mobility of up to ~200 cm^2 V$^{-1} \cdot$s^{-1} at room temperature. However, graphene nanoribbon-based devices have been shown to suffer from low driving currents or transconductances. To solve this, the graphene nanomesh, which can be viewed as orderly and periodically arranged graphene nanoribbons, was developed for use as a conductive channel material in TFTs (figure 5.21(b)). The nanomesh construct not only

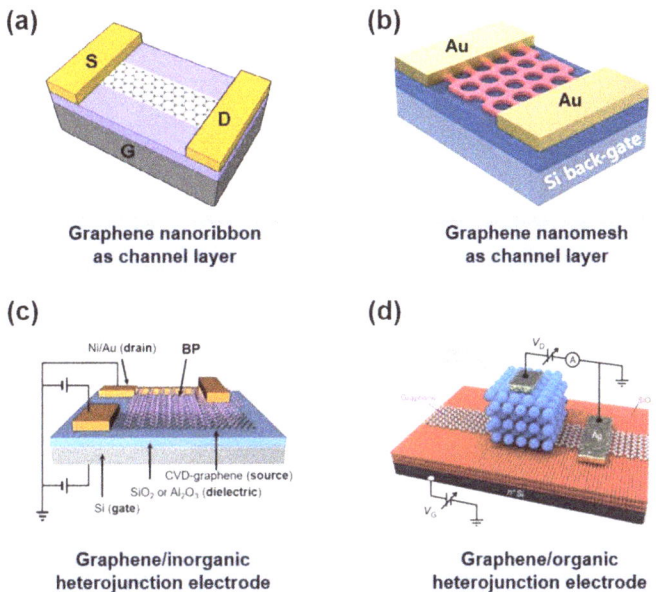

Figure 5.21. Illustrations of TFTs constructed using graphene-based materials. (a) Graphene nanoribbon as the channel layer [392], reproduced with permission © American Physical Society. (b) Graphene nanomesh as the channel layer [393], reproduced with permission © The Royal Society of Chemistry. (c) Graphene/inorganic heterojunction electrode [394], reproduced with permission © American Chemical Society. (d) Graphene/inorganic heterojunction electrode [397], reproduced with permission © American Chemical Society.

introduced a bandgap for graphene, but also dramatically increased the driving current of the devices [393]. Bai *et al* successfully synthesized a graphene nanomesh with a width of 5 nm, and the corresponding TFTs exhibited an on/off ratio of ~100 with a current density ~100 times higher than that of a device based on graphene nanoribbons [397].

As an electrode material. Graphene has been found to effectively overcome the disadvantages of the traditional metal/metal oxide electrodes for TFTs, including insensitivity, opacity, non-flexibility, and a large contact resistance between the metal electrode and the active layer [396]. Generally, graphene films serve as an efficient electrode in TFT by forming a heterojunction structure between the graphene electrode and the inorganic and/or organic semiconductor active layer. Kang *et al* presented a vertical TFT device based on a graphene/black phosphorus (graphene/BP) van der Waals heterostructure (figure 5.21(c)), and the resulting device exhibited high on-state current densities (>1600 A cm^{-2}) and current on/off ratios of more than 800 at low temperatures [394]. Two distinct charge-transport mechanisms were proposed: the Schottky barrier between graphene and BP determines charge transport at high temperatures and positive gate voltages, whereas the tunneling effect dominates at low temperatures and negative gate voltages. Shih *et al* synthesized a series of vertical TFTs based on a graphene/organic semiconductor heterostructure, including graphene/pentacene and graphene/C$_{60}$, and the devices exhibited good optoelectronic properties (figure 5.21(d)) [395]. Interestingly, the results experimentally evidenced that the underlying mechanism of the electronic transport at the heterointerface is dominated by the partially screened field effect and selective carrier injection through graphene.

5.4.3.3 Sensors based on FETs

Due to their distinct advantages of robustness, high flexibility, and electrical conductivity, graphene-based materials have been developed for use in sensors [401]. Graphene-based sensing materials exhibit less electrical noise and crystal defects than conventional sensing materials, and sensor devices constructed using graphene-based materials have been shown to operate with simpler techniques than conventional methods [402]. Figure 5.22(a) shows an optical micrograph of a typical graphene-based FET sensor device. Depending on the sensing mechanism, these sensors could be classified as electrochemical sensors, strain sensors, or electrical sensors.

Electrochemical sensors. Graphene-based materials are useful as electrochemical sensors because of their advantages of a wide electrochemical potential range, a fast electron transfer rate, and a high electrocatalytic capability [401]. Electrodeposition, polymerization, and electrochemical doping methods have been adopted to synthesize different graphene-based composite materials for electrochemical sensor use. The mechanism of electrochemical sensing stems from the chemical interaction between the graphene-based material and the molecules of interest (figure 5.22(b)) [398], and this interaction can alter the conductivity of graphene. Graphene-based electrochemical sensors can be utilized to effectively detect various molecules, including heavy metal ions [403], proteins [404], biomolecules [405], and organic molecules [406].

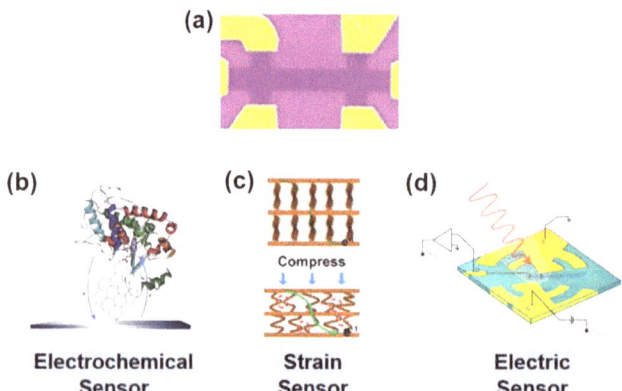

Figure 5.22. (a) Optical micrograph of a typical graphene-based sensor device [52], reproduced with permission © Springer Nature. Different graphene-based sensors classified by sensing mechanism: (b) electrochemical sensor [398], reproduced with permission © Elsevier. (c) Strain sensor [399], reproduced with permission © American Chemical Society. (d) Electric sensor [400], reproduced with permission © Springer Nature.

Strain sensors. Graphene-based sensors have proved to be an excellent candidate for strain-sensing applications. The advantage of using graphene over other conductive materials for strain sensing is attributed to the generation of a pseudo-magnetic field owing to the shift in the Dirac cones and reduction of the Fermi velocity of graphene when strains are applied [401]. This change in magnetic field, which leads to a change in the electronic structure of graphene, is the origin of the activity of the strain sensor device (figure 5.22(c)) [399]. Moreover, it has been found that strain that applied parallel to the C–C bonds can induce a larger electronic structural change in graphene compared to a strain applied perpendicular to the C–C bonds, which is due to a higher increase in the bandgap for the parallel case [407]. To promote the desired strain detection, strain sensors have been developed using graphene on various polymer substrates, such as polydimethylsiloxane (PDMS), polyethylene terephthalate (PET), and polyimide (PI) [399, 408, 409].

Electrical sensors. Graphene-based electrical sensors are based on typical transistor architecture (figure 5.22(d)) [400], and graphene has been utilized in the form of nanoribbons, nanowires, and other forms of nanoparticle array. Graphene-based electrical sensors can be applied in temperature sensing, physical and chemical sensing, inverters, and radio-frequency (RF) applications [410]. The merits of graphene-based electrical sensors lie in their design simplicity, volume production, and high signal sensitivity.

5.4.4 Bio-related applications

5.4.4.1 Bioimaging

Bioimaging plays critical roles in both research and clinical practice, as it allows the observation and study of biological processes from the cellular and sub-cellular levels to small animals [412]. Graphene-based materials have been shown to be easily functionalized by small molecular dyes, polymers, nanoparticles, drugs, or biomolecules to produce graphene-based biocomposites for different bioimaging applications [413]. Therefore, the superiority of graphene-based materials in bioimaging application has been thoroughly investigated due to the versatile surface functionalization and ultrahigh surface area of graphene and its derivatives. In this section, we will briefly introduce the utilization of graphene-based materials for different bioimaging applications, including optical imaging, magnetic resonance imaging (MRI), photoacoustic imaging (PAI), and computed tomography (CT).

Optical imaging. Optical imaging is a noninvasive technique that uses visible light to obtain detailed images of organs and tissues [414]. Graphene-based materials have been widely explored for optical imaging; the main applications include fluorescence imaging, two-photon fluorescence imaging (TPFI), and Raman imaging. GO/rGO functionalized by dyes, photosensitizers, QDs, and gold nanoclusters have been widely investigated for FL imaging [415]. GQDs with intrinsic photoluminescence can be directly used as photoluminescent probes for FL imaging. Nitrogen-doped GQDs, which have a strong two-photon absorption cross-section, have been demonstrated to be an efficient two-photon fluorescent probe for TPFI [416]. Interestingly, graphene and its derivatives also exhibit an efficient photoluminescence quenching effect, which can be applied in graphene-based nanosensors for many fluorescent species, including small molecule dyes, QDs, and conjugated polymers via photoluminescence resonance energy transfer or charge transfer [417]. Yan *et al* reported a photo-theranostic agent based on a GO–sinoporphyrin sodium (DVDMS) composite for enhanced optical imaging guided photodynamic therapy (figure 5.23) [411]. The fluorescence of DVDMS is drastically enhanced via the intramolecular charge transfer of GO. For Raman imaging, the intrinsic Raman signals of graphene-based materials can be further enhanced by decorating them with metal nanoparticles (Au or Ag) to support surface-enhanced Raman scattering (SERS) applications in Raman imaging [418].

Magnetic resonance imaging (MRI). MRI has been widely used to image the anatomy and function of tissues in a quantitative manner, because MRI is a noninvasive technique that uses iron oxide nanoparticles as a magnetic sensitizer without ionizing radiation and has excellent spatial resolution [414]. Nevertheless, the ions of paramagnetic metals used in MRI, such as gadolinium (Gd) and manganese (Mn), are generally toxic owing to their non-selective coordination with biomolecules, which increases the risks of MRI [419]. However, GO possesses many oxygenated functional groups and cavities, and thus it can be readily coordinated with these ions by chelation or by sequestering them between graphene layers, which limits these ions and reduces the toxicity of MRI [415]. Moreover, GO has also been found to enhance the MRI contrast of iron oxide nanoparticles, as they form clusters or aggregates on GO [420]. In addition, fluorinated graphene oxide (FGO) has been also explored as non-magnetic-nanoparticle carbon-based MRI contrast agent by creating dipolar C–F bonds as the paramagnetic centers [421].

Photoacoustic imaging (PAI). Photoacoustic imaging (PAI) is a hybrid imaging method based on the PA effect, in which the absorbed short-pulsed electromagnetic

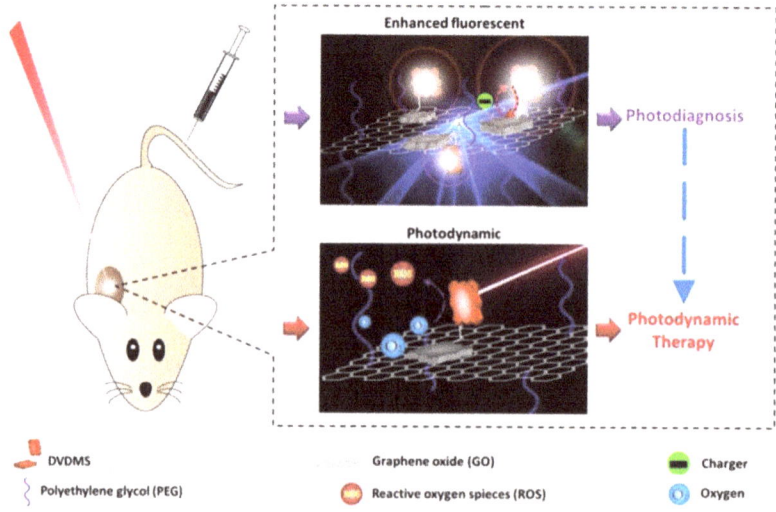

Figure 5.23. Illustration of the GO–sinoporphyrin sodium (DVDMS) composite used in enhanced fluorescence bioimaging to guide photodynamic therapy [411], reproduced with permission © Elsevier.

energy (from non-ionizing laser pulses) is converted into heat, resulting in acoustic emission due to transient thermoelastic expansion [422]. Among graphene-based materials, rGO shows best performance as a PA contrast agent, which is due to the more efficient NIR light absorption of rGO with larger sp^2 domains. However, the hydrophilicity and water solubility of rGO is poor due to its reduced functional groups, and the absorption coefficient of rGO in the NIR region is limited. Many methods have been explored to solve these problems, such as the production of less oxygenated nanosized graphene sheets [423], loading the rGO with indocyanine green (ICG) (which has strong absorbance in the NIR region) [424], and stabilizing nanosized rGO using bovine serum albumin (BSA) [425].

Computed tomography (CT). CT is is carried out by measuring the absorption signals of x-rays when they pass through targets; CT can provide complementary anatomical information [426]. The mechanism by which CT distinguishes tissues is based on the fact that different tissues provide distinct degrees of x-ray attenuation, in which the attenuation coefficient depends on the atomic number and the electron density of the tissues. Therefore, materials containing electron-dense elements with high atomic numbers such as iodine, bismuth, or gold, have been proposed for use as CT contrast agents [427]. To enhance the concentration of these elements, GO is used to serve as a growth substrate for those CT contrast agents to form composites, such as GO@Ag [428] and GO@Au [429], which have been found to substantially improve the efficiency of CT.

5.4.4.2 Biomedical

Owing to the functionality of GO, its potential in biomedical applications has been investigated (figure 5.24) [430]. The principal advantage of using GO over other

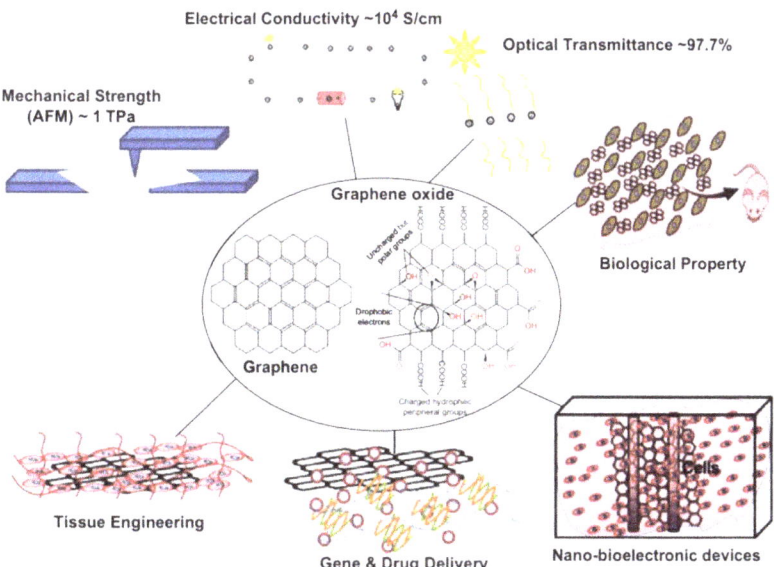

Figure 5.24. Schematic of the properties of graphene-based materials and their biomedical applications [430]. Source: [430], reproduced with permission © Elsevier.

carbon-based materials is due to its aqueous and colloidal stability, the formation of hydrogen bonds between polar functional groups on the GO surface and water molecules forms a stable GO colloidal suspension for potential biomedical applications of GO [431]. However, for actual biomedical application in biology and medicine, the concern about the biocompatibility of the material has the highest priority. Although graphene and its derivatives consist of solely carbon, it is a matter of serious concern to understand how graphene derivatives like GO and rGO behave in a biological system and how long it takes to excrete from the human body [36]. Fortunately, non-covalent functionalization of GO could improve its dispersibility, biocompatibility, reactivity, binding capacity [432]. In this section, biomedical application of GO will be briefly introduced including tissue engineering, gene/drug delivery, and biosensor.

Tissue engineering. Tissue engineering is an emerging research area in the life sciences that aims to develop biological substitutes, made of biodegradable materials, to modify the function of a tissue in order to repair and maintain its properties [433]. However, different tissues possess different mechanical, electrical, and physical properties, and a single material might not meet the requirements for the physical and biological properties of the native tissue; thus, hybrid bioactive materials are widely used to fabricate artificial tissues. Because of the functional groups on their surfaces, GO and rGO can easily be chemically modified to interact with various biological molecules, such as DNA, proteins, peptides, and enzymes, which allows GO/rGO-based composites to be used in tissue engineering to induce specific cellular functions, to direct cell differentiation, and to modulate cell–cell interactions. In addition, GO/rGO-based composites are also applied in cardiac, neural, bone, cartilage, skeletal muscle, and skin/adipose tissue engineering [434]. Moreover, the antimicrobial activity of GO/rGO-based composites may enable their application in reducing microbial infections and thereby advance human health [435].

Gene/drug delivery. The surface area of graphene is extremely high, which allows graphene-based materials to exhibit high drug loading capacity and makes them suitable for drug delivery exploration [436]. The two prominent modifications for drug delivery using graphene-based materials are (i) chemical modification via electrostatic interaction and (ii) binding to the aromatic molecule via the p–p stacking interaction [413]. Another advantage of graphene-based materials in drug delivery is their controllable rate of sustainable drug release [437]. Thus, GO has become quite a competitive drug delivery system and has the potential to be applied for systemic targeting and local effective drug delivery [438]. In addition to drugs, graphene-based materials can also interact with gene biomolecules, such as nucleic acids, DNA, and RNA, and serve as carriers due to their large sp^2-hybridized carbon area [439]. The basic requirements for a gene delivery vector include protecting DNA from degradation and ensuring high transfection efficiency, and GO has also been demonstrated to adsorb nucleobases by p–p interaction and efficiently protect nucleotides from enzymatic cleavage. Effective gene delivery can be applied in gene therapy applications, and GO materials complexed with different genes were recently shown to be efficient deliverers of myocardial therapy [440] and non-viral therapy [441].

Biosensors. Biosensors are useful in the detection of biomolecules and chemical analytes, as they can characterize the spectrochemical, electrochemical, or magneto-chemical behaviors of specific types of analyte [442]. The capacity of graphene-based materials to adsorb a variety of aromatic molecules via the p–p stacking interaction makes them ideal materials for the fabrication of biosensors [443]. Graphene-based biosensors were developed to detect small molecules such as glucose, nicotinamide, dinucleotide adenine, adenine triphosphate, hydrogen peroxide, and estrogen [444]. GO-based materials are capable of low detection limits, fast response times, high sensitivities, and increased signal-to-noise ratios when used in the fabrication of biosensors [445]. Moreover, graphene-based biosensors have been found to work efficiently in combination with metal nanoparticles, auxiliary biomolecules (chitosan), and bioenzymes (horseradish peroxidase) due to their enhanced electronic and synergistic compositions that catalyze glucose enzymatic reactions for electrochemical sensing [446].

5.4.5 Environmental applications

Graphene-based materials exhibit unique laminate and porous structures, which enable a wide range of nanomaterials to be combined, resulting in novel properties for environmental applications [448]. For example, ultrathin and non-porous laminate GO-based membranes can act as barriers for molecular separation (e.g. gaseous separation and filtration, small molecule separation, and water filtration), while GO with porous structures can be applied as a building block for next-generation water treatment. There are two types of molecular separation mechanism in graphene-based membranes: (i) nanoporous single-layer graphene can achieve the molecular selectivity by size exclusion and electrostatic repulsion between charged species and the graphene pores (figure 5.25(a)); (ii) stacked GO sheets can achieve

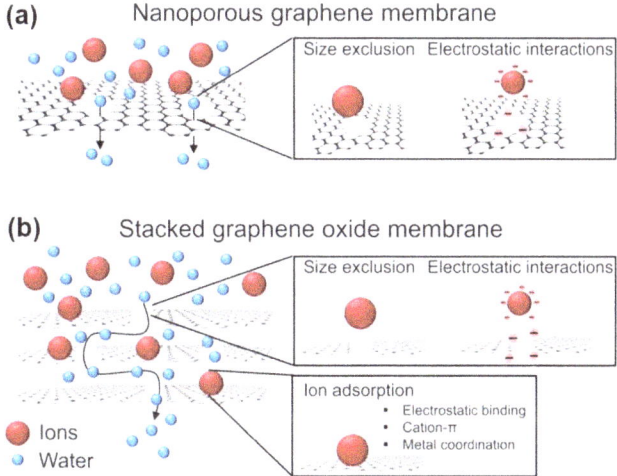

Figure 5.25. Schematic representation of the two types of molecular separation mechanism of graphene-based membranes [447], reproduced with permission © The Royal Society of Chemistry.

molecular selectivity not only by size exclusion and electrostatic repulsion, but also via the interlayer distance between two sheets (figure 5.25(b)) [447]. These unique properties of graphene-based materials permit them to be superior in environmental-related applications, including gaseous, water, and antibacterial treatments.

Gaseous treatment. Due to the hydrophilic features and polar natures of hydroxyl, carboxyl, and epoxide groups in GO, GO is found to favor interactions with the individual polar C–O bonds of CO_2, which provides preferential sites for selective CO_2 adsorption and removal. Notably, owing to their infinitesimal thickness, desirable mechanical strength, chemical robustness, and favorable permeation and mass transport, GO-based membranes can be ideal candidates for the separation of gas mixtures, such as CO_2/N_2 [449], H_2/CH_4 [366], H_2/CO_2 [450], and H_2/C_3H_8 [450]. In addition, adsorption-based applications of GO-based membranes have been developed for gaseous contaminant removal (e.g. sulfur dioxide [451] and hydrogen sulfide [452]). In addition to the oxygen functional groups in GO, the surface chemistry of GO has been found to play a vital role in its adsorption behaviors. Nevertheless, there are still challenges for GO membranes in this application, such as the membranes' structural robustness under the very harsh conditions of industrial gas treatment. Due to the principles of gas–solid reactions, promoting the contact area and providing loose but strongly interacting GO laminates are of interest for fundamental research [448].

Water treatment. Because water can freely pass through GO-based membranes by pressure filtration, GO-based membranes are suitable for water filtration and purification. GO-based membranes have been investigated for different filtration processes, such as ultrafiltration, nanofiltration, forward osmosis, and reverse osmosis [453]. The size of the interlayer space between the GO sheets in the membrane has been found to determine the rejection of the permeating solutes, such as large molecular weight dyes (several dozens of nanometers) [454] and inorganic salts (less than 5 nm) with different hydrated ionic radii [455]. This merit of GO enables its application in seawater desalination, in which GO can serve as an ideal barrier for the separation of small molecular species from seawater, such as Na^+ and Cl^- [456]. Due to the high content of functionalized oxygen groups available in GO for interaction with metal ions, GO is a suitable substrate for aqueous contaminant removal application through the adsorption of metal ions [457] and organic pollutants [458]. Compared with its counterparts, GO offers other superior properties for fast water transport, such as a smooth and frictionless surface, ultralow thickness, and superior mechanical strength, all of which increase its water permeability via a special phase transition in the molecular water channels.

Antibacterial. The antibacterial activity of a material with a sheet-like structure can mainly be attributed to membrane stress induced by sharp edges, which may result in the physical destruction of cell bodies via a piercing effect, leading to the loss of bacterial membrane integrity and the leakage of living matter. Several studies have shown that the antimicrobial activity of graphene-based materials is related to not only physical but also chemical routes [459]. Physically, when bacteria directly contact defective or wrinkled graphene, intensive physical interactions cause physical damage to cell membranes and result in the release of intracellular matter [460].

From a chemical perspective, GO-based membranes with functional groups or nanoparticles are found to increase the cellular oxidative stress and the number of active radicals, which can disrupt specific microbial processes [461].

5.5 Conclusions and perspective

Graphene, as a rising star in the research community, has an interesting history as well as a bright future. The sudden discovery of graphene in 2004 led to an explosion of interest in the study of graphene, and the development of graphene opens up a new research area for condensed-matter physics and materials science. The unique physical, chemical, thermal, and mechanical properties of graphene enable its application in wide-ranging and diversified applications, which demonstrates its great industrial and commercial potential. Inspiringly, the unique properties of graphene are still being continuously explored today, for instance, the super-conductivity of bilayer graphene arising from the 'magic angle.' Therefore, graphene and graphene-based materials are expected to bring us more surprises in the future.

Currently, the superiority of graphene-based materials in various applications have been evidenced in the science community, and the next step in the advancement of graphene-based materials should lie in their large-scale production for industrial manufacture. Therefore, for the realization of practical applications, it is highly desirable to develop cost-effective methods to synthesize various functional graphene-based materials at scale with an acceptable degree of reproducibility. Moreover, considering the unique properties of graphene-based materials, incorporating them into the existing smart systems and developing graphene-based multi-functional systems from an interdisciplinary point of view should be highly valued.

Acknowledgements

This work was financially supported by the Natural Science Foundation of China (Grants 51802253, 51972172, 61705102, and 91833304), the China Postdoctoral Science Foundation (Grant 2021M692630), the Natural Science Basic Research Plan in Shaanxi Province of China (2019JM-326), the Joint Research Funds of the Department of Science and Technology of Shaanxi Province and Northwestern Polytechnical University (No. 2020GXLH-Z-007), the Natural Science Foundation of Jiangsu Province for Distinguished Young Scholars, China (Grant BK20200034), the Young 1000 Talents Global Recruitment Program of China, the Jiangsu Specially Appointed Professor program, 'Six talent peaks' Project in Jiangsu Province, China, and the Fundamental Research Funds for the Central Universities.

References

[1] Geim A K and Novoselov K S 2007 *Nat. Mater.* **6** 183–91
[2] Lee C, Wei X D, Kysar J W and Hone J 2008 *Science* **321** 385–88
[3] Bolotin K I, Sikes K J, Jiang Z, Klima M, Fudenberg G, Hone J, Kim P and Stormer H L 2008 *Solid State Commun.* **146** 351–55
[4] Morozov S V, Novoselov K S, Katsnelson M I, Schedin F, Elias D C, Jaszczak J A and Geim A K 2008 *Phys. Rev. Lett.* **100** 016602

[5]   Balandin A A, Ghosh S, Bao W Z, Calizo I, Teweldebrhan D, Miao F and Lau C N 2008 *Nano Lett.* **8** 902–7

[6]   Sun Z X, Fang S Y and Hu Y H 2020 *Chem. Rev.* **120** 10336–453

[7]   Wallace P R 1947 *Phys. Rev.* **71** 622–34

[8]   Peierls R E 1935 *Ann. I. H. Poincare* **5** 177–222

[9]   Venables J A, Spiller G D T and Hanbucken M 1984 *Rep. Prog. Phys.* **47** 399–459

[10] Mermin N D 1968 *Phys. Rev.* **176** 250–54

[11] Boehm H P, Setton R and Stumpp E 1986 *Carbon* **24** 241–45

[12] Lu X, Yu M, Huang H and Ruoff R S 1999 *Nanotechnology* **10** 269

[13] Novoselov K S, Geim A K, Morozov S V, Jiang D, Zhang Y, Dubonos S V, Grigorieva I V and Firsov A A 2004 *Science* **306** 666–69

[14] Novoselov K S, Fal'ko V I, Colombo L, Gellert P R, Schwab M G and Kim K 2012 *Nature* **490** 192–200

[15] Kumar N, Salehiyan R, Chauke V, Botlhoko O J, Setshedi K, Scriba M, Masukume M and Ray S S 2021 *Flatchem* **27** 100224

[16] Li Z, Liu Z, Sun H Y and Gao C 2015 *Chem. Rev.* **115** 7046–117

[17] Lotya M, Hernandez Y, King P J, Smith R J, Nicolosi V, Karlsson L S, Blighe F M and De S *et al* 2009 *J. Am. Chem. Soc.* **131** 3611–20

[18] Hernandez Y, Nicolosi V, Lotya M, Blighe F M, Sun Z Y, De S, McGovern I T and Holland B *et al* 2008 *Nat. Nanotechnol.* **3** 563–68

[19] Paton K R, Varrla E, Backes C, Smith R J, Khan U, O'Neill A, Boland C and Lotya M *et al* 2014 *Nat. Mater.* **13** 624–30

[20] Achee T C, Sun W M, Hope J T, Quitzau S G, Sweeney C B, Shah S A, Habib T and Green M J 2018 *Sci. Rep.* **8** 14525

[21] Park S and Ruoff R S 2010 *Nat. Nanotechnol.* **5** 309

[22] Luong D X, Bets K V, Algozeeb W A, Stanford M G, Kittrell C, Chen W Y, Salvatierra R V and Ren M Q *et al* 2020 *Nature* **577** 647–51

[23] Tian P, Tang L, Teng K S and Lau S P 2018 *Mater. Today Chem.* **10** 221–58

[24] Chen Z P, Narita A and Mullen K 2020 *Adv. Mater.* **32** 2001893

[25] Haag D and Kung H H 2014 *Top. Catal.* **57** 762–73

[26] Koppens F H L, Mueller T, Avouris P, Ferrari A C, Vitiello M S and Polini M 2014 *Nat. Nanotechnol.* **9** 780–93

[27] Coros M, Pruneanu S and Stefan-van Staden R I 2019 *J. Electrochem. Soc.* **167** 037528

[28] Bai L Q, Zhang Y H, Tong W S, Sun L, Huang H W, An Q, Tian N and Chu P K 2020 *Electrochem Energy* R **3** 395–430

[29] Mohan V B, Lau K T, Hui D and Bhattacharyya D 2018 *Compos Part B-Eng* **142** 200–20

[30] Catania F, Marras E, Giorcelli M, Jagdale P, Lavagna L, Tagliaferro A and Bartoli M 2021 *Appl Sci-Basel* **11** 614

[31] Weiss N O, Zhou H L, Liao L, Liu Y, Jiang S, Huang Y and Duan X F 2012 *Adv. Mater.* **24** 5782–825

[32] Nair R R, Blake P, Grigorenko A N, Novoselov K S, Booth T J, Stauber T, Peres N M R and Geim A K 2008 *Science* **320** 1308

[33] Ni Z H, Wang H M, Kasim J, Fan H M, Yu T, Wu Y H, Feng Y P and Shen Z X 2007 *Nano Lett.* **7** 2758–63

[34] Zhang Y B, Tan Y W, Stormer H L and Kim P 2005 *Nature* **438** 201–4

[35] Peres N M R, Guinea F and Castro Neto A H 2006 *Phys. Rev.* B **73** 125411

[36] Zhu Y W, Murali S, Cai W W, Li X S, Suk J W, Potts J R and Ruoff R S 2010 *Adv. Mater.* **22** 3906–24

[37] Peres N M R 2009 *J Phys-Condens Mat* **21** 095501

[38] Mak K F, Sfeir M Y, Wu Y, Lui C H, Misewich J A and Heinz T F 2008 *Phys. Rev. Lett.* **101** 196405

[39] Blake P, Hill E W, Castro Neto A H, Novoselov K S, Jiang D, Yang R, Booth T J and Geim A K 2007 *Appl. Phys. Lett.* **91** 063124

[40] Jung I, Pelton M, Piner R, Dikin D A, Stankovich S, Watcharotone S, Hausner M and Ruoff R S 2007 *Nano Lett.* **7** 3569–75

[41] Abergel D S L, Russell A and Fal'ko V I 2007 *Appl. Phys. Lett.* **91** 063125

[42] Gao L B, Ren W C, Li F and Cheng H M 2008 *ACS Nano* **2** 1625–33

[43] Yu V and Hilke M 2009 *Appl. Phys. Lett.* **95** 151904

[44] Palik E D (ed) 1991 *Handbook of Optical Constant of Solids* vol 2 (New York: Academic) https://www.sciencedirect.com/book/9780125444224/handbook-of-optical-constants-of-solids

[45] Bruna M and Borini S 2009 *Appl. Phys. Lett.* **94** 031901

[46] Wang F, Zhang Y B, Tian C S, Girit C, Zettl A, Crommie M and Shen Y R 2008 *Science* **320** 206–9

[47] Slonczewski J C and Weiss P R 1958 *Phys. Rev.* **109** 272–79

[48] Castro Neto A H, Guinea F, Peres N M R, Novoselov K S and Geim A K 2009 *Rev. Mod. Phys.* **81** 109–62

[49] Zhan D, Yan J X, Lai L F, Ni Z H, Liu L and Shen Z X 2012 *Adv. Mater.* **24** 4055–69

[50] Novoselov K S, Geim A K, Morozov S V, Jiang D, Katsnelson M I, Grigorieva I V, Dubonos S V and Firsov A A 2005 *Nature* **438** 197–200

[51] Xia F N, Mueller T, Lin Y M, Valdes-Garcia A and Avouris P 2009 *Nat. Nanotechnol.* **4** 839–43

[52] Schedin F, Geim A K, Morozov S V, Hill E W, Blake P, Katsnelson M I and Novoselov K S 2007 *Nat. Mater.* **6** 652–55

[53] Du X, Skachko I, Barker A and Andrei E Y 2008 *Nat. Nanotechnol.* **3** 491–95

[54] Novoselov K S, Jiang Z, Zhang Y, Morozov S V, Stormer H L, Zeitler U, Maan J C and Boebinger G S *et al* 2007 *Science* **315** 1379

[55] Jiang Z, Zhang Y, Tan Y W, Stormer H L and Kim P 2007 *Solid State Commun.* **143** 14–9

[56] Li G and Andrei E Y 2007 *Nat. Phys.* **3** 623–27

[57] Papageorgiou D G, Kinloch I A and Young R J 2017 *Prog. Mater Sci.* **90** 75–127

[58] Zhang P, Ma L L, Fan F F, Zeng Z, Peng C, Loya P E, Liu Z and Gong Y J *et al* 2014 *Nat. Commun.* **5** 3782

[59] Bao W Z, Miao F, Chen Z, Zhang H, Jang W Y, Dames C and Lau C N 2009 *Nat. Nanotechnol.* **4** 562–66

[60] Min K and Aluru N R 2011 *Appl. Phys. Lett.* **98** 013113

[61] Yu C H, Shi L, Yao Z, Li D Y and Majumdar A 2005 *Nano Lett.* **5** 1842–46

[62] Berber S, Kwon Y K and Tomanek D 2000 *Phys. Rev. Lett.* **84** 4613–16

[63] Pop E, Varshney V and Roy A K 2012 *MRS Bull.* **37** 1273–81

[64] Hsu I K, Pettes M T, Bushmaker A, Aykol M, Shi L and Cronin S B 2009 *Nano Lett.* **9** 590–94

[65] Calizo I, Balandin A A, Bao W, Miao F and Lau C N 2007 *Nano Lett.* **7** 2645–49

[66] Cai W W, Moore A L, Zhu Y W, Li X S, Chen S S, Shi L and Ruoff R S 2010 *Nano Lett.* **10** 1645–51

[67] Seol J H, Jo I, Moore A L, Lindsay L, Aitken Z H, Pettes M T, Li X S and Yao Z *et al* 2010 *Science* **328** 213–16

[68] Jang W Y, Chen Z, Bao W Z, Lau C N and Dames C 2010 *Nano Lett.* **10** 3909–13

[69] Liao A D, Wu J Z, Wang X R, Tahy K, Jena D, Dai H J and Pop E 2011 *Phys. Rev. Lett.* **106** 256801

[70] Qiu B and Ruan X L 2012 *Appl. Phys. Lett.* **100** 193101

[71] Boukhvalov D W and Katsnelson M I 2008 *Nano Lett.* **8** 4373–79

[72] Boukhvalov D W and Katsnelson M I 2009 *J. Phys. Chem. C* **113** 14176–78

[73] Yang G, Li L H, Lee W B and Ng M C 2018 *Sci. Technol. Adv. Mat.* **19** 613–48

[74] Liu Y Y, Dobrinsky A and Yakobson B I 2010 *Phys. Rev. Lett.* **105** 235502

[75] Boukhvalov D W, Katsnelson M I and Lichtenstein A I 2008 *Phys. Rev. B* **77** 035427

[76] Peng X Y and Ahuja R 2008 *Nano Lett.* **8** 4464–68

[77] Maldonado S, Morin S and Stevenson K J 2006 *Carbon* **44** 1429–37

[78] Qu L T, Liu Y, Baek J B and Dai L M 2010 *ACS Nano* **4** 1321–26

[79] Robinson J T, Burgess J S, Junkermeier C E, Badescu S C, Reinecke T L, Perkins F K, Zalalutdniov M K and Baldwin J W *et al* 2010 *Nano Lett.* **10** 3001–5

[80] Bousa D, Luxa J, Mazanek V, Jankovsky O, Sedmidubsky D, Klimova K, Pumera M and Sofer Z 2016 *RSC Adv.* **6** 66884–92

[81] Dikin D A, Stankovich S, Zimney E J, Piner R D, Dommett G H B, Evmenenko G, Nguyen S T and Ruoff R S 2007 *Nature* **448** 457–60

[82] Cao Y, Fatemi V, Fang S, Watanabe K, Taniguchi T, Kaxiras E and Jarillo-Herrero P 2018 *Nature* **556** 43–50

[83] Chu Y H, Zhu F D, Wen L Z, Chen W Y, Chen Q N and Ma T X 2020 *Chinese Phys B* **29** 117401

[84] Kim K, Yankowitz M, Fallahazad B, Kang S, Movva H C P, Huang S Q, Larentis S and Corbet C M *et al* 2016 *Nano Lett.* **16** 1989–95

[85] Xie Y L, Lian B, Jack B, Liu X M, Chiu C L, Watanabe K, Taniguchi T and Bernevig B A *et al* 2019 *Nature* **572** 101–5

[86] Padhi B and Phillips P W 2019 *Phys. Rev. B* **99** 205141

[87] Arora H S, Polski R, Zhang Y R, Thomson A, Choi Y, Kim H, Lin Z and Wilson I Z *et al* 2020 *Nature* **583** 379–84

[88] Bao W, Jing L, Velasco J, Lee Y, Liu G, Tran D, Standley B and Aykol M *et al* 2011 *Nat. Phys.* **7** 948–52

[89] Chen G R, Sharpe A L, Gallagher P, Rosen I T, Fox E J, Jiang L L, Lyu B S and Li H Y *et al* 2019 *Nature* **572** 215–19

[90] Van Noorden R 2012 *Nature* **483** S32–3

[91] Jayasena B and Subbiah S 2011 *Nanoscale Res. Lett.* **6** 95

[92] Wei J C, Vo T and Inam F 2015 *RSC Adv.* **5** 73510–24

[93] Lin Z, Karthik P S, Hada M, Nishikawa T and Hayashi Y 2017 *Nanomaterials-Basel* **7** 125

[94] Wang Y Z, Chen T, Liu H H, Wang X C and Zhang X X 2019 *J. Nanosci. Nanotechno.* **19** 2078–86

[95] Yu P, Lowe S E, Simon G P and Zhong Y L 2015 *Curr. Opin. Colloid Interface Sci.* **20** 329–38

[96] Janowska I, Vigneron F, Begin D, Ersen O, Bernhardt P, Romero T, Ledoux M J and Pham-Huu C 2012 *Carbon* **50** 3106–10

[97] Pirzado A A, Normand F L, Romero T, Paszkiewicz S, Papaefthimiou V, Ihiawakrim D and Janowska I 2019 *ChemEngineering* **3** 37

[98] Yi M and Shen Z G 2015 *J. Mater. Chem. A* **3** 11700–15

[99] Zhao W F, Fang M, Wu F R, Wu H, Wang L W and Chen G H 2010 *J. Mater. Chem.* **20** 5817–19

[100] Lv Y Y, Yu L S, Jiang C M, Chen S M and Nie Z X 2014 *RSC Adv.* **4** 13350–54

[101] Knieke C, Berger A, Voigt M, Taylor R N K, Rohrl J and Peukert W 2010 *Carbon* **48** 3196–204

[102] Jeon I Y, Choi H J, Choi M, Seo J M, Jung S M, Kim M J, Zhang S and Zhang L P *et al* 2013 *Sci. Rep.* **3** 1810

[103] Chen J F, Duan M and Chen G H 2012 *J. Mater. Chem.* **22** 19625–28

[104] Ciesielski A and Samori P 2014 *Chem. Soc. Rev.* **43** 381–98

[105] Xu Y Y, Cao H Z, Xue Y Q, Li B and Cai W H 2018 *Nanomaterials-Basel* **8** 942

[106] Pavlova A S, Obraztsova E A, Belkin A V, Monat C, Rojo-Romeo P and Obraztsova E D 2016 *J. Nanophotonics* **10** 012525

[107] Liu W, Tanna V A, Yavitt B M, Dimitrakopoulos C and Winter H H 2015 *ACS Appl. Mater. Inter.* **7** 27027–30

[108] Gayathri S, Jayabal P, Kottaisamy M and Ramakrishnan V 2014 *AIP Adv.* **4** 027116

[109] Buzaglo M, Shtein M, Kober S, Lovrincic R, Vilan A and Regev O 2013 *Phys. Chem. Chem. Phys.* **15** 4428–35

[110] Ma H, Shen Z G, Yi M, Ben S, Liang S S, Liu L, Zhang Y X and Zhang X J *et al* 2017 *J. Colloid. Interf. Sci.* **503** 68–75

[111] Wajid A S, Das S, Irin F, Ahmed H S T, Shelburne J L, Parviz D, Fullerton R J and Jankowski A F *et al* 2012 *Carbon* **50** 526–34

[112] Lotya M, King P J, Khan U, De S and Coleman J N 2010 *ACS Nano* **4** 3155–62

[113] Liu W W and Wang J N 2011 *Chem. Commun.* **47** 6888–90

[114] Dobbelin M, Ciesielski A, Haar S, Osella S, Bruna M, Minoia A, Grisanti L and Mosciatti T *et al* 2016 *Nat. Commun.* **7** 11090

[115] Nuvoli D, Valentini L, Alzari V, Scognamillo S, Bon S B, Piccinini M, Illescas J and Mariani A 2011 *J. Mater. Chem.* **21** 3428–31

[116] Skaltsas T, Ke X X, Bittencourt C and Tagmatarchis N 2013 *J. Phys. Chem.* C **117** 23272–78

[117] Bracamonte M V, Lacconi G I, Urreta S E and Torres L E F F 2014 *J. Phys. Chem.* C **118** 15455–59

[118] Khan U, O'Neill A, Porwal H, May P, Nawaz K and Coleman J N 2012 *Carbon* **50** 470–75

[119] Khan U, O'Neill A, Lotya M, De S and Coleman J N 2010 *Small* **6** 864–71

[120] Park W K, Yoon Y, Song Y H, Choi S Y, Kim S, Do Y, Lee J and Park H *et al* 2017 *Sci. Rep.* **7** 16414

[121] Varrla E, Paton K R, Backes C, Harvey A, Smith R J, McCauley J and Coleman J N 2014 *Nanoscale* **6** 11810–19

[122] Wang Y Z, Zhang X Y, Liu H H and Zhang X X 2019 *Nanomaterials-Basel* **9** 1653

[123] Liu L, Shen Z G, Liang S S, Yi M, Zhang X J and Ma S L 2014 *J. Mater. Sci.* **49** 321–28

[124] Yi M, Shen Z G and Zhu J Y 2014 *Chinese Sci. Bull.* **59** 1794–99

[125] Liang S S, Shen Z G, Yi M, Liu L, Zhang X J, Cai C J and Ma S L 2015 *J. Nanosci. Nanotechno.* **15** 2686–94

[126] Karagiannidis P G, Hodge S A, Lombardi L, Tomarchio F, Decorde N, Milana S, Goykhman I and Su Y *et al* 2017 *ACS Nano* **11** 2742–55

[127] Qi X, Zhang H B, Xu J T, Wu X Y, Yang D Z, Qu J and Yu Z Z 2017 *ACS Appl. Mater. Inter.* **9** 11025–34

[128] Zhang K, Tang J, Yuan J S, Li J, Sun Y G, Matsuba Y, Zhu D M and Qin L C 2018 *ACS Appl. Nano Mater.* **1** 2877–84

[129] Tran T S, Park S J, Yoo S S, Lee T R and Kim T 2016 *RSC Adv.* **6** 12003–8

[130] Rangappa D, Sone K, Wang M S, Gautam U K, Golberg D, Itoh H, Ichihara M and Honma I 2010 *Chem.-Eur. J.* **16** 6488–94

[131] Chen X J, Dobson J F and Raston C L 2012 *Chem. Commun.* **48** 3703–5

[132] Buzaglo M, Shtein M and Regev O 2016 *Chem. Mater.* **28** 21–4

[133] Parvez K, Li R J, Puniredd S R, Hernandez Y, Hinkel F, Wang S H, Feng X L and Mullen K 2013 *ACS Nano* **7** 3598–606

[134] Wang J Z, Manga K K, Bao Q L and Loh K P 2011 *J. Am. Chem. Soc.* **133** 8888–91

[135] Paredes J I and Munuera J M 2017 *J. Mater. Chem.* A **5** 7228–42

[136] Liu F, W. C, Sui X, Riaz M A, Xu M, Wei L and Chen Y 2019 *Carbon Energy* **1** 173–99

[137] Yang Y C, Hou H S, Zou G Q, Shi W, Shuai H L, Li J Y and Ji X B 2019 *Nanoscale* **11** 16–33

[138] Shinde D B, Brenker J, Easton C D, Tabor R F, Neild A and Majunider M 2016 *Langmuir* **32** 3552–59

[139] Lee X J, Hiew B Y Z, Lai K C, Lee L Y, Gan S Y, Thangalazhy-Gopakumar S and Rigby S 2019 *J. Taiwan Inst. Chem. E* **98** 163–80

[140] Brodie B C 1859 *Philos. Trans. R. Soc. Lond.* **149** 249–59

[141] Staudenmaier L 1899 *Ber. Dtsch. Chem. Ges.* **32** 1394–99

[142] Hofmann U and K. E 1937 *Z. Anorg. Allg. Chem.* **234** 311–36

[143] Hofmann U and H. R 1939 *Ber. Dtsch. Chem. Ges. (A and B Series)* **72** 754–71

[144] Hummers W S and Offeman R E 1958 *J. Am. Chem. Soc.* **80** 1339

[145] Marcano D C, Kosynkin D V, Berlin J M, Sinitskii A, Sun Z Z, Slesarev A, Alemany L B and Lu W *et al* 2010 *ACS Nano* **4** 4806–14

[146] Dreyer D R, Park S, Bielawski C W and Ruoff R S 2010 *Chem. Soc. Rev.* **39** 228–40

[147] De Silva K K H, Huang H H, Joshi R and Yoshimura M 2020 *Carbon* **166** 74–90

[148] Schniepp H C, Li J L, McAllister M J, Sai H, Herrera-Alonso M, Adamson D H, Prud'homme R K and Car R *et al* 2006 *J. Phys. Chem.* B **110** 8535–39

[149] Lopez V, Sundaram R S, Gomez-Navarro C, Olea D, Burghard M, Gomez-Herrero J, Zamora F and Kern K 2009 *Adv. Mater.* **21** 4683–86

[150] Negishi R, Akabori M, Ito T, Watanabe Y and Kobayashi Y 2016 *Sci. Rep.* **6** 28936

[151] Singh R K, Kumar R and Singh D P 2016 *RSC Adv.* **6** 64993–5011

[152] Zhu Y W, Murali S, Stoller M D, Velamakanni A, Piner R D and Ruoff R S 2010 *Carbon* **48** 2118–22

[153] Chen W F, Yan L F and Bangal P R 2010 *Carbon* **48** 1146–52

[154] Xie X X, Zhou Y P and Huang K M 2019 *Front Chem.* **7** 355

[155] Wen C Y, Zhao N, Zhang D W, Wu D P, Zhang Z B and Zhang S L 2014 *Synthetic Met.* **194** 71–6

[156] Li H L and Bubeck C 2013 *Macromol. Res.* **21** 290–97

[157] Rahimi K and Yazdani A 2020 *Mater. Lett.* **262** 127078

[158] Matsumoto Y, Koinuma M, Kim S Y, Watanabe Y, Taniguchi T, Hatakeyama K, Tateishi H and Ida S 2010 *ACS Appl. Mater. Inter.* **2** 3461–66

[159] Li X H, Chen J S, Wang X C, Schuster M E, Schlogl R and Antonietti M 2012 *ChemSusChem* **5** 642–46

[160] Konkena B and Vasudevan S 2015 *J. Phys. Chem.* C **119** 6356–62

[161] Darr J A, Zhang J Y, Makwana N M and Weng X L 2017 *Chem. Rev.* **117** 11125–238
[162] Diez N, Sliwak A, Gryglewicz S, Grzyb B and Gryglewicz G 2015 *RSC Adv.* **5** 81831–37
[163] Hu K W, Xie X Y, Szkopek T and Cerruti M 2016 *Chem. Mater.* **28** 1756–68
[164] Zheng X L, Peng Y S, Yang Y, Chen J L, Tian H W, Cui X Q and Zheng W T 2017 *J. Raman Spectrosc.* **48** 97–103
[165] Zhou M, Wang Y L, Zhai Y M, Zhai J F, Ren W, Wang F A and Dong S J 2009 *Chem.-Eur. J.* **15** 6116–20
[166] Toh S Y, Loh K S, Kamarudin S K and Daud W R W 2014 *Chem. Eng. J.* **251** 422–34
[167] Harima Y, Setodoi S, Imae I, Komaguchi K, Ooyama Y, Ohshita J, Mizota H and Yano J 2011 *Electrochim. Acta* **56** 5363–68
[168] Karacic D, Korac S, Dobrota A S, Pasti I A, Skorodumova N V and Gutic S J 2019 *Electrochim. Acta* **297** 112–17
[169] Kauppila J, Kunnas P, Damlin P, Viinikanoja A and Kvarnstrom C 2013 *Electrochim. Acta* **89** 84–9
[170] Marrani A G, Motta A, Schrebler R, Zanoni R and Dalchiele E A 2019 *Electrochim. Acta* **304** 231–38
[171] Shao Y Y, Wang J, Engelhard M, Wang C M and Lin Y H 2010 *J. Mater. Chem.* **20** 743–48
[172] Pei S F and Cheng H M 2012 *Carbon* **50** 3210–28
[173] Chua C K and Pumera M 2014 *Chem. Soc. Rev.* **43** 291–312
[174] Mao S, Pu H H and Chen J H 2012 *RSC Adv.* **2** 2643–62
[175] Zhu C Z, Guo S J, Fang Y X and Dong S J 2010 *ACS Nano* **4** 2429–37
[176] De Silva K K H, Huang H H, Joshi R K and Yoshimura M 2017 *Carbon* **119** 190–99
[177] Ismail Z 2019 *Ceram. Int.* **45** 23857–68
[178] Yang S, Yue W B, Huang D Z, Chen C F, Lin H and Yang X J 2012 *RSC Adv.* **2** 8827–32
[179] Dreyer D R, Murali S, Zhu Y W, Ruoff R S and Bielawski C W 2011 *J. Mater. Chem.* **21** 3443–47
[180] Luo F B, Wu K, Shi J, Du X X, Li X Y, Yang L and Lu M G 2017 *J. Mater. Chem.* A **5** 18542–50
[181] Kratschmer W, Lamb L D, Fostiropoulos K and Huffman D R 1990 *Nature* **347** 354–58
[182] Subrahmanyam K S, Panchakarla L S, Govindaraj A and Rao C N R 2009 *J. Phys. Chem. C* **113** 4257–59
[183] Dallas P, Meysami S S, Grobert N and Porfyrakis K 2016 *RSC Adv.* **6** 24912–20
[184] Su Y J and Zhang Y F 2015 *Carbon* **83** 90–9
[185] Huang L P, Wu B, Chen J Y, Xue Y Z, Geng D C, Guo Y L, Yu G and Liu Y Q 2013 *Small* **9** 1330–35
[186] Li N, Wang Z Y, Zhao K K, Shi Z J, Gu Z N and Xu S K 2010 *Carbon* **48** 255–59
[187] Agudosi E S, Abdullah E C, Numan A, Mubarak N M, Khalid M and Omar N 2020 *Crit. Rev. Solid State* **45** 339–77
[188] Pham T V, Kim J G, Jung J Y, Kim J H, Cho H, Seo T H, Lee H and Kim N D *et al* 2019 *Adv. Funct. Mater.* **29** 1905511
[189] Li B, Song X L and Zhang P 2014 *Carbon* **66** 426–35
[190] Kim S, Song Y J, Wright J and Heller M J 2016 *Carbon* **102** 339–45
[191] Wu X H, Liu Y, Yang H and Shi Z J 2016 *RSC Adv.* **6** 93119–24
[192] Munoz R and Gomez-Aleixandre C 2013 *Chem. Vapor. Depos.* **19** 297–322
[193] Mishra N, Boeckl J, Motta N and Iacopi F 2016 *Phys. Status Solidi a* **213** 2277–89
[194] Bayat A and Saievar-Iranizad E 2017 *J. Lumin.* **192** 180–83

[195] Tan H, Wang D G and Guo Y B 2018 *Coatings* **8** 40

[196] Sha J W, Li Y L, Salvatierra R V, Wang T, Dong P, Ji Y S, Lee S K and Zhang C H *et al* 2017 *ACS Nano* **11** 6860–67

[197] Ani M H, Kamarudin M A, Ramlan A H, Ismail E, Sirat M S, Mohamed M A and Azam M A 2018 *J. Mater. Sci.* **53** 7095–111

[198] Li Y L Z, Sun L Z, Liu H Y, Wang Y C and Liu Z F 2020 *Catalysts* **10** 1305

[199] Reina A, Jia X T, Ho J, Nezich D, Son H B, Bulovic V, Dresselhaus M S and Kong J 2009 *Nano Lett.* **9** 30–5

[200] Kim K S, Zhao Y, Jang H, Lee S Y, Kim J M, Kim K S, Ahn J H and Kim P *et al* 2009 *Nature* **457** 706–10

[201] Cabrero-Vilatela A, Weatherup R S, Braeuninger-Weimer P, Caneva S and Hofmann S 2016 *Nanoscale* **8** 2149–58

[202] Hussain A, Mehdi S M, Abbas N, Hussain M and Naqvi R A 2020 *Mater. Chem. Phys.* **248** 122924

[203] Chen K, Shi L R, Zhang Y F and Liu Z F 2018 *Chem. Soc. Rev.* **47** 3018–36

[204] Zhang Y, Zhang L Y and Zhou C W 2013 *Acc. Chem. Res.* **46** 2329–39

[205] Yazdi G R, Iakimov T and Yakimova R 2016 *Crystals* **6** 53

[206] Lin Y M, Dimitrakopoulos C, Jenkins K A, Farmer D B, Chiu H Y, Grill A and Avouris P 2010 *Science* **327** 662

[207] de Heer W A, Berger C, Ruan M, Sprinkle M, Li X B, Hu Y K, Zhang B Q and Hankinson J *et al* 2011 *Proc. Natl Acad. Sci. USA* **108** 16900–5

[208] Emtsev K V, Bostwick A, Horn K, Jobst J, Kellogg G L, Ley L, McChesney J L and Ohta T *et al* 2009 *Nat. Mater.* **8** 203–7

[209] Wang X Y, Narita A and Mullen K 2018 *Nat. Rev. Chem.* **2** 0100

[210] Paterno G M, Chen Q, Wang X Y, Liu J Z, Motti S G, Petrozza A, Feng X L and Lanzani G *et al* 2017 *Angew. Chem. Int. Ed.* **56** 6753–57

[211] Yang W L, Lucotti A, Tommasini M and Chalifoux W A 2016 *J. Am. Chem. Soc.* **138** 9137–44

[212] Daigle M, Miao D, Lucotti A, Tommasini M and Morin J F 2017 *Angew. Chem. Int. Ed.* **56** 6213–17

[213] Haque E, Kim J, Malgras V, Reddy K R, Ward A C, You J, Bando Y and Hossain M S A *et al* 2018 *Small Methods* **2** 1800050

[214] Choucair M, Thordarson P and Stride J A 2009 *Nat. Nanotechnol.* **4** 30–3

[215] Wan Z F, Streed E W, Lobino M, Wang S J, Sang R T, Cole I S, Thiel D V and Li Q 2018 *Adv. Mater. Technol.* **3** 1700315

[216] Chyan Y, Ye R Q, Li Y L, Singh S P, Arnusch C J and Tour J M 2018 *ACS Nano* **12** 2176–83

[217] Ye R Q, Chyan Y, Zhang J B, Li Y L, Han X, Kittrell C and Tour J M 2017 *Adv. Mater.* **29** 1702211

[218] Lin J, Peng Z W, Liu Y Y, Ruiz-Zepeda F, Ye R Q, Samuel E L G, Yacaman M J and Yakobson B I *et al* 2014 *Nat. Commun.* **5** 5714

[219] Ye R Q, James D K and Tour J M 2019 *Adv. Mater.* **31** 1803621

[220] Ye R Q, James D K and Tour J M 2018 *Accounts Chem. Res.* **51** 1609–20

[221] Loh K P, Tong S W and Wu J S 2016 *J. Am. Chem. Soc.* **138** 1095–102

[222] Tong S W, Wang Y, Zheng Y, Ng M F and Loh K P 2011 *Adv. Funct. Mater.* **21** 4430–35

[223] Xie C, Zhang X J, Ruan K Q, Shao Z B, Dhaliwal S S, Wang L, Zhang Q and Zhang X W *et al* 2013 *J. Mater. Chem.* A **1** 15348–54

[224] Liu Z K, Li J H and Yan F 2013 *Adv. Mater.* **25** 4296–301

[225] Prabakaran K, Jandas P J, Mohanty S and Nayak S K 2018 *Sol. Energy* **170** 442–53

[226] Heo J H, Shin D H, Song D H, Kim D H, Lee S J and Im S H 2018 *J. Mater. Chem.* A **6** 8251–58

[227] Li X M, Zhu H W, Wang K L, Cao A Y, Wei J Q, Li C Y, Jia Y and Li Z *et al* 2010 *Adv. Mater.* **22** 2743–48

[228] Bae S, Kim H, Lee Y, Xu X F, Park J S, Zheng Y, Balakrishnan J and Lei T *et al* 2010 *Nat. Nanotechnol.* **5** 574–78

[229] Garg R, Dutta N K and Choudhury N R 2014 *Nanomaterials-Basel* **4** 267–300

[230] Lee H, Paeng K and Kim I S 2018 *Synthetic Met.* **244** 36–47

[231] Xia F N, Yan H G and Avouris P 2013 *Proc. IEEE* **101** 1717–31

[232] Pykal M, Jurecka P, Karlicky F and Otyepka M 2016 *Phys. Chem. Chem. Phys.* **18** 6351–72

[233] Park C S, Zhao Y, Lee J H, Whang D, Shon Y, Song Y H and Lee C J 2013 *Appl. Phys. Lett.* **102** 032106

[234] Mohammed M, Li Z R, Cui J B and Chen T P 2012 *Nanoscale Res. Lett.* **7** 1–6

[235] Rwenyagila E R 2017 *Int. J. Photoenergy* **2017** 1656512

[236] Kim H, Bae S H, Han T H, Lim K G, Ahn J H and Lee T W 2014 *Nanotechnology* **25** 014012

[237] Yi Y, Choi W M, Kim Y H, Kim J W and Kang S J 2011 *Appl. Phys. Lett.* **98** 013505

[238] Park H, Chang S, Zhou X, Kong J, Palacios T and Gradecak S 2014 *Nano Lett.* **14** 5148–54

[239] Liu J, Durstock M and Dai L M 2014 *Energy Environ Sci.* **7** 1297–306

[240] Yeo J S, Yun J M, Jung Y S, Kim D Y, Noh Y J, Kim S S and Na S I 2014 *J. Mater. Chem.* A **2** 292–98

[241] Cheng X F, Long J, Wu R, Huang L Q, Tan L C, Chen L and Chen Y W 2017 *ACS Omega* **2** 2010–16

[242] Jun G H, Jin S H, Lee B, Kim B H, Chae W S, Hong S H and Jeon S 2013 *Energy Environ Sci.* **6** 3000–6

[243] Balis N, Konios D, Stratakis E and Kymakis E 2015 *Chemnanomat.* **1** 346–52

[244] Muchuweni E, Martincigh B S and Nyamori V O 2020 *RSC Adv.* **10** 44453–69

[245] Majumder T and Mondal S P 2019 *B Mater. Sci.* **42** 65

[246] Krishnamoorthy D and Prakasam A 2020 *Inorg. Chem. Commun.* **119** 108063

[247] Roh K M, Jo E H, Chang H, Han T H and Jang H D 2015 *J. Solid State Chem.* **224** 71–5

[248] Yang W, Park I W, Lee J M and Choi H 2020 *J. Nanosci. Nanotechno.* **20** 3432–36

[249] Oh W C, Cho K Y, Jung C H and Areerob Y 2020 *Sci. Rep.* **10** 4738

[250] Zhang J J, Fan J J, Cheng B, Yu J G and Ho W K 2020 *Sol. RRL* **4** 2000502

[251] Xie J S, Huang K, Yu X G, Yang Z R, Xiao K, Qiang Y P, Zhu X D and Xu L B *et al* 2017 *ACS Nano* **11** 9176–82

[252] Li X H, Tong T T, Wu Q J, Guo S H, Song Q, Han J and Huang Z X 2018 *Adv. Funct. Mater.* **28** 1800475

[253] Jokar E, Huang Z Y, Narra S, Wang C Y, Kattoor V, Chung C C and Diau E W G 2018 *Adv. Energy Mater.* **8** 1701640

[254] Li M, Zuo W W, Wang Q, Wang K L, Zhuo M P, Kobler H, Halbig C E and Eigler S *et al* 2020 *Adv. Energy Mater.* **10** 1902653

[255] Zhang C Y, Wang S, Zhang H, Feng Y L, Tian W M, Yan Y, Bian J M and Wang Y C *et al* 2019 *Energy Environ. Sci.* **12** 3585–94

[256] Zhou Z M, Li X, Cai M L, Xie F X, Wu Y Z, Lan Z, Yang X D and Qiang Y H *et al* 2017 *Adv. Energy Mater.* **7** 1700763

[257] Tu B, Shao Y F, Chen W, Wu Y H, Li X, He Y L, Li J X and Liu F Z *et al* 2019 *Adv. Mater.* **31** 1805944

[258] Feng S L, Yang Y G, Li M, Wang J M, Cheng Z D, Li J H, Ji G W and Yin G Z *et al* 2016 *ACS Appl. Mater. Inter.* **8** 14503–12

[259] Wang Y B, Wu T H, Barbaud J, Kong W Y, Cui D Y, Chen H, Yang X D and Han L Y 2019 *Science* **365** 687–91

[260] Mueller T, Xia F N A and Avouris P 2010 *Nat. Photonics* **4** 297–301

[261] Du S C, Lu W, Ali A, Zhao P, Shehzad K, Guo H W, Ma L L and Liu X M *et al* 2017 *Adv. Mater.* **29** 1700463

[262] Konstantatos G, Badioli M, Gaudreau L, Osmond J, Bernechea M, de Arquer F P G, Gatti F and Koppens F H L 2012 *Nat. Nanotechnol.* **7** 363–68

[263] De Sanctis A, Mehew J D, Craciun M F and Russo S 2018 *Materials* **11** 1762

[264] Peters E C, Lee E J H, Burghard M and Kern K 2010 *Appl. Phys. Lett.* **97** 193102

[265] Gabor N M, Song J C W, Ma Q, Nair N L, Taychatanapat T, Watanabe K, Taniguchi T and Levitov L S *et al* 2011 *Science* **334** 648–52

[266] Kotov V N, Uchoa B, Pereira V M, Guinea F and Castro Neto A H 2012 *Rev. Mod. Phys.* **84** 1067–125

[267] Piscanec S, Lazzeri M, Mauri F, Ferrari A C and Robertson J 2004 *Phys. Rev. Lett.* **93** 185503

[268] Bock J J, Chen D, Mauskopf P D and Lange A E 1995 *Space Sci. Rev.* **74** 229–35

[269] Dyakonov M and Shur M 1996 *IEEE Trans. Electron. Dev.* **43** 380–87

[270] Li X, Yu J G, Wageh S, Al-Ghamdi A A and Xie J 2016 *Small* **12** 6640–96

[271] Low J X, Yu J G and Ho W K 2015 *J. Phys. Chem. Lett.* **6** 4244–51

[272] Singh S, Faraz M and Khare N 2020 *ACS Omega* **5** 11874–82

[273] Wang P F, Zhan S H, Xia Y G, Ma S L, Zhou Q X and Li Y 2017 *Appl. Catal.* B **207** 335–46

[274] Yeh T F, Cihlar J, Chang C Y, Cheng C and Teng H S 2013 *Mater. Today* **16** 78–84

[275] Albero J, Mateo D and Garcia H 2019 *Molecules* **24** 906

[276] Al Kausor M and Chakrabortty D 2021 *Inorg. Chem. Commun.* **129** 108630

[277] Gao W Y, Wang M Q, Ran C X and Li L 2015 *Chem. Commun.* **51** 1709–12

[278] Gao W Y, Wang M Q, Ran C X, Yao X, Yang H H, Liu J, He D L and Bai J B 2014 *Nanoscale* **6** 5498–508

[279] Yang M Q and Xu Y J 2016 *Nanoscale Horiz.* **1** 185–200

[280] Ge J, Zhang Y and Park S J 2019 *Materials* **12** 1916

[281] Hou W C and Wang Y S 2017 *ACS Sustain. Chem. Eng.* **5** 2994–3001

[282] Yu T and Breslin C B 2020 *J. Electrochem. Soc.* **167** 126502

[283] Radhika N P, Selvin R, Kakkar R and Umar A 2019 *Arab. J. Chem.* **12** 4550–78

[284] Kudo A and Miseki Y 2009 *Chem. Soc. Rev.* **38** 253–78

[285] Carminati S A, Rodriguez-Gutierrez I, de Morais A, da Silva B L, Melo M A, Souza F L and Nogueira A F 2021 *RSC Adv.* **11** 14374–98

[286] Xu D F, Li L L, He R A, Qi L F, Zhang L Y and Cheng B 2018 *Appl. Surf. Sci.* **434** 620–25

[287] Yuan Y J, Yang Y, Li Z J, Chen D Q, Wu S T, Fang G L, Bai W F and Ding M Y *et al* 2018 *ACS Appl. Energy Mater.* **1** 1400–7

[288] Liao G Z, Chen S, Quan X, Yu H T and Zhao H M 2012 *J. Mater. Chem.* **22** 2721–26

[289] Ben Ali M, Jo W K, Elhouichet H and Boukherroub R 2017 *Int. J. Hydrogen Energy* **42** 16449–58

[290] Ng Y H, Iwase A, Kudo A and Amal R 2010 *J. Phys. Chem. Lett.* **1** 2607–12

[291] Liu X Y, Zheng H W, Zhang J W, Xiao Y and Wang Z Y 2013 *J. Mater. Chem.* A **1** 10703–12

[292] Xiang Q J, Yu J G and Jaroniec M 2012 *J. Am. Chem. Soc.* **134** 6575–78

[293] Pan H, Zhu S, Lou X, Mao L, Lin J, Tian F and Zhang D 2015 *RSC Adv.* **5** 6543–52

[294] Yeh T F, Chan F F, Hsieh C T and Teng H S 2011 *J. Phys. Chem.* C **115** 22587–97

[295] Yeh T F, Teng C Y, Chen S J and Teng H S 2014 *Adv. Mater.* **26** 3297–303

[296] Hou Y, Zuo F, Ma Q, Wang C, Bartels L and Feng P Y 2012 *J. Phys. Chem.* C **116** 20132–39

[297] He L M, Jing L Q, Luan Y B, Wang L and Fu H G 2014 *ACS Catal.* **4** 990–98

[298] Wang C H, Zhang X T and Liu Y C 2015 *Appl. Surf. Sci.* **358** 28–45

[299] Adan-Mas A and Wei D 2013 *Nanomaterials-Basel* **3** 325–56

[300] Gao W Y, Ran C X, Wang M Q, Li L, Sun Z W and Yao X 2016 *Phys. Chem. Chem. Phys.* **18** 18219–26

[301] Huang B T, He J B, Bian S Y, Zhou C J, Li Z Y, Xi F N, Liu J Y and Dong X P 2018 *Chinese Chem. Lett.* **29** 1698–701

[302] Khan Z, Chetia T R, Vardhaman A K, Barpuzary D, Sastri C V and Qureshi M 2012 *RSC Adv.* **2** 12122–28

[303] White J L, Baruch M F, Pander J E, Hu Y, Fortmeyer I C, Park J E, Zhang T and Liao K *et al* 2015 *Chem. Rev.* **115** 12888–935

[304] Habisreutinger S N, Schmidt-Mende L and Stolarczyk J K 2013 *Angew. Chem. Int. Ed.* **52** 7372–408

[305] Sun Z Y, Talreja N, Tao H C, Texter J, Muhler M, Strunk J and Chen J F 2018 *Angew. Chem. Int. Ed.* **57** 7610–27

[306] Jung H, Cho K M, Kim K H, Yoo H W, Al-Saggaf A, Gereige I and Jung H T 2018 *ACS Sustain. Chem. Eng.* **6** 5718–24

[307] Lin L Y, Nie Y, Kavadiya S, Soundappan T and Biswas P 2017 *Chem. Eng. J.* **316** 449–60

[308] Zhou S S and Liu S Q 2017 *Photoch. Photobio. Sci.* **16** 1563–69

[309] Cho K M, Kim K H, Park K, Kim C, Kim S, Al-Saggaf A, Gereige I and Jung H T 2017 *ACS Catal.* **7** 7064–69

[310] Shown I, Hsu H C, Chang Y C, Lin C H, Roy P K, Ganguly A, Wang C H and Chang J K *et al* 2014 *Nano Lett.* **14** 6097–103

[311] Ali A and Oh W C 2017 *Sci. Rep.* **7** 1867

[312] Liang Y T, Vijayan B K, Gray K A and Hersam M C 2011 *Nano Lett.* **11** 2865–70

[313] Hoffman A J, Carraway E R and Hoffmann M R 1994 *Environ. Sci. Technol.* **28** 776–85

[314] Qiang Z M, Chang J H and Huang C P 2002 *Water Res.* **36** 85–94

[315] Moon G H, Kim W, Bokare A D, Sung N E and Choi W 2014 *Energy Environ. Sci.* **7** 4023–28

[316] Weng B, Wu J, Zhang N and Xu Y J 2014 *Langmuir* **30** 5574–84

[317] Kofuji Y, Isobe Y, Shiraishi Y, Sakamoto H, Tanaka S, Ichikawa S and Hirai T 2016 *J. Am. Chem. Soc.* **138** 10019–25

[318] Koehler F M and Stark W J 2013 *Accounts Chem. Res.* **46** 2297–306

[319] Zhang N, Yang M Q, Tang Z R and Xu Y J 2013 *J. Catal.* **303** 60–9

[320] Zhang N, Zhang Y H, Pan X Y, Yang M Q and Xu Y J 2012 *J. Phys. Chem.* C **116** 18023–31

[321] Das D P, Barik R K, Das J, Mohapatra P and Parida K M 2012 *RSC Adv.* **2** 7377–79

[322] Pan Y H, Wang S, Kee C W, Dubuisson E, Yang Y Y, Loh K P and Tan C H 2011 *Green Chem.* **13** 3341–44

[323] Su H R and Hu Y H 2021 *Energy Sci. Eng.* **9** 958–83

[324] Chen S Q, Zhang N J, Villarrubia C W N, Huang X, Xie L, Wang X Y, Kong X D and Xu H *et al* 2019 *Nano Energy* **66** 104164

[325] Sun H M, Ye Y X, Liu J, Tian Z F, Cai Y Y, Li P F and Liang C H 2018 *Chem. Commun.* **54** 1563–66

[326] Gao W, Wu G, Janicke M T, Cullen D A, Mukundan R, Baldwin J K, Brosha E L and Galande C *et al* 2014 *Angew. Chem. Int. Ed.* **53** 3588–93

[327] Sim Y, Kwak J, Kim S Y, Jo Y, Kim S, Kim S Y, Kim J H and Lee C S *et al* 2018 *J. Mater. Chem.* A **6** 1504–12

[328] Yang L J, Shui J L, Du L, Shao Y Y, Liu J, Dai L M and Hu Z 2019 *Adv. Mater.* **31** 1804799

[329] Zhao J, Li H Q, Liu Z S, Hu W B, Zhao C Z and Shi D L 2015 *Carbon* **87** 116–27

[330] Barakat N A M, Moustafa H M, Nassar M M, Abdelkareem M A, Mahmoud M S, Almajid A A and Khalil K A 2015 *Electrochim. Acta* **182** 143–55

[331] Perez-Page M, Sahoo M and Holmes S M 2019 *Adv. Mater. Interfaces* **6** 1801838

[332] Sezer N and Koc M 2021 *Nano Energy* **80** 105567

[333] Fu J, Hou Y D, Gao X, Zheng M P and Zhu M K 2018 *Nano Energy* **52** 391–401

[334] Kapat K, Shubhra Q T H, Zhou M and Leeuwenburgh S 2020 *Adv. Funct. Mater.* **30** 1909045

[335] Bhavanasi V, Kumar V, Parida K, Wang J X and Lee P S 2016 *ACS Appl. Mater. Inter.* **8** 521–29

[336] Yaqoob U, Uddin A S M I and Chung G S 2017 *Appl. Surf. Sci.* **405** 420–26

[337] Shi K M, Sun B, Huang X Y and Jiang P K 2018 *Nano Energy* **52** 153–62

[338] Ghosh S K, Sinha T K, Mahanty B and Mandal D 2015 *Energy Technol-Ger* **3** 1190–97

[339] Karan S K, Mandal D and Khatua B B 2015 *Nanoscale* **7** 10655–66

[340] Li Y T, Tian Y, Sun M X, Tu T, Ju Z Y, Gou G Y, Zhao Y F and Yan Z Y *et al* 2020 *Adv. Funct. Mater.* **30** 1903888

[341] Tian H, Yang Y, Xie D, Ge J and Ren T L 2013 *RSC Adv.* **3** 17672–76

[342] Tian H, Xie D, Yang Y, Ren T L, Wang Y F, Zhou C J, Peng P G and Wang L G *et al* 2012 *Nanoscale* **4** 2272–77

[343] Tao L Q, Tian H, Liu Y, Ju Z Y, Pang Y, Chen Y Q, Wang D Y and Tian X G *et al* 2017 *Nat. Commun.* **8** 14579

[344] Tian H, Xie D, Yang Y, Ren T L, Zhang G, Wang Y F, Zhou C J and Peng P G *et al* 2012 *Sci Rep.* **2** 523

[345] Wang H D, Hu S Q, Takahashi K, Zhang X, Takamatsu H and Chen J 2017 *Nat. Commun.* **8** 15843

[346] Miyoshi Y, Fukazawa Y, Amasaka Y, Reckmann R, Yokoi T, Ishida K, Kawahara K and Ago H *et al* 2018 *Nat. Commun.* **9** 1279

[347] Manjavacas A, Thongrattanasiri S, Greffet J J and de Abajo F J G 2014 *Appl. Phys. Lett.* **105** 211102

[348] Zhang T Y, Zhao H M, Wang D Y, Wang Q, Pang Y, Deng N Q, Cao H W and Yang Y *et al* 2017 *Nanoscale* **9** 14357–63

[349] Sang W, Zhao L M, Tang R, Wu Y P, Zhu C H and Liu J 2017 *Macromol. Mater. Eng.* **302** 1700239

[350] El-Kady M F, Shao Y L and Kaner R B 2016 *Nat. Rev. Mater.* **1** 16033

[351] Li X L and Zhi L J 2018 *Chem. Soc. Rev.* **47** 3189–216

[352] Cai X Y, Lai L F, Shen Z X and Lin J Y 2017 *J. Mater. Chem.* A **5** 15423–46

[353] Paek S M, Yoo E and Honma I 2009 *Nano Lett.* **9** 72–5

[354] Yang S B, Feng X L, Ivanovici S and Mullen K 2010 *Angew. Chem. Int. Ed.* **49** 8408–11

[355] Wang D H, Kou R, Choi D, Yang Z G, Nie Z M, Li J, Saraf L V and Hu D H *et al* 2010 *ACS Nano* **4** 1587–95

[356] Wang B, Li X L, Qiu T F, Luo B, Ning J, Li J, Zhang X F and Liang M H *et al* 2013 *Nano Lett.* **13** 5578–84

[357] Wang B, Li X L, Zhang X F, Luo B, Jin M H, Liang M H, Dayeh S A and Picraux S T *et al* 2013 *ACS Nano* **7** 1437–45

[358] Cheng H H, Dong Z L, Hu C G, Zhao Y, Hu Y, Qu L T, Chen N and Dai L M 2013 *Nanoscale* **5** 3428–34

[359] Li M, Tang Z, Leng M and Xue J M 2014 *Adv. Funct. Mater.* **24** 7495–502

[360] Manjakkal L, Nunez C G, Dang W T and Dahiya R 2018 *Nano Energy* **51** 604–12

[361] Yan J, Fan Z J, Wei T, Qian W Z, Zhang M L and Wei F 2010 *Carbon* **48** 3825–33

[362] Vickery J L, Patil A J and Mann S 2009 *Adv. Mater.* **21** 2180–84

[363] Lehtimaki S, Suominen M, Damlin P, Tuukkanen S, Kvarnstrom C and Lupo D 2015 *ACS Appl Mater Inter* **7** 22137–47

[364] Tung V C, Kim J, Cote L J and Huang J X 2011 *J. Am. Chem. Soc.* **133** 9262–65

[365] Kou L, Huang T Q, Zheng B N, Han Y, Zhao X L, Gopalsamy K, Sun H Y and Gao C 2014 *Nat. Commun.* **5** 3754

[366] Jiang D E, Cooper V R and Dai S 2009 *Nano Lett.* **9** 4019–24

[367] Xia J L, Chen F, Li J H and Tao N J 2009 *Nat. Nanotechnol.* **4** 505–9

[368] Ritter K A and Lyding J W 2008 *Nanotechnology* **19** 015704

[369] Fuertes A B and Alvarez S 2004 *Carbon* **42** 3049–55

[370] Junaid M, Khir M H M, Witjaksono G, Ullah Z, Tansu N, Saheed M S M, Kumar P and Wah L H *et al* 2020 *Molecules* **25** 4217

[371] Steranka F M, Bhat J, Collins D, Cook L, Craford M G, Fletcher R, Gardner N and Grillot P *et al* 2002 *Phys. Status Solidi a* **194** 380–8

[372] Min J H, Son M, Bae S Y, Lee J Y, Yun J, Maeng M J, Kwon D G and Park Y *et al* 2014 *Opt. Express* **22** A1040–50

[373] Park P S, Reddy K M, Nath D N, Yang Z C, Padture N P and Rajan S 2013 *Appl. Phys. Lett.* **102** 153501

[374] Kim Y D, Kim H, Cho Y, Ryoo J H, Park C H, Kim P, Kim Y S and Lee S *et al* 2015 *Nat. Nanotechnol.* **10** 676–81

[375] Ghosh T and Prasad E 2015 *J. Phys. Chem.* C **119** 2733–42

[376] Soavi G, Dal Conte S, Manzoni C, Viola D, Narita A, Hu Y B, Feng X L and Hohenester U *et al* 2016 *Nat. Commun.* **7** 11010

[377] Kim D H and Kim T W 2018 *Nano Energy* **51** 199–205

[378] Wang Z G, Chen Y F, Li P J, Hao X, Liu J B, Huang R and Li Y R 2011 *ACS Nano* **5** 7149–54

[379] Liu Z W, Bushmaker A, Aykol M and Cronin S B 2011 *ACS Nano* **5** 4634–40

[380] Fares H and Almokhtar M 2019 *Phys. Lett.* A **383** 1005–10

[381] Essig S, Marquardt C W, Vijayaraghavan A, Ganzhorn M, Dehm S, Hennrich F, Ou F and Green A A *et al* 2010 *Nano Lett.* **10** 1589–94

[382] Beams R, Bharadwaj P and Novotny L 2014 *Nanotechnology* **25** 055206

[383] Fei Z, Rodin A S, Andreev G O, Bao W, McLeod A S, Wagner M, Zhang L M and Zhao Z *et al* 2012 *Nature* **487** 82–5

[384] Barnard H R, Zossimova E, Mahlmeister N H, Lawton L M, Luxmoore I J and Nash G R 2016 *Appl. Phys. Lett.* **108** 131110

[385] Freitag M, Chiu H Y, Steiner M, Perebeinos V and Avouris P 2010 *Nat. Nanotechnol.* **5** 497–501

[386] Buss I J, Nash G R, Rarity J G and Cryan M J 2008 *J. Opt. Soc. Am.* B **25** 810–17

[387] Miskovic Z L, Segui S, Gervasoni J L and Arista N R 2016 *Phys. Rev.* B **94** 125414

[388] Jablan M, Buljan H and Soljacic M 2009 *Phys. Rev.* B **80** 245435

[389] Xie K X, Cao S H, Wang Z C, Weng Y H, Huo S X, Zhai Y Y, Chen M and Pan X H *et al* 2017 *Sensor Actuat B-Chem* **253** 804–8

[390] Kaminer I, Katan Y T, Buljan H, Shen Y C, Ilic O, Lopez J J, Wong L J and Joannopoulos J D *et al* 2016 *Nat. Commun.* **7** 11880

[391] Wang X W, Liu Z and Zhang T 2017 *Small* **13** 1602790

[392] Wang X R, Ouyang Y J, Li X L, Wang H L, Guo J and Dai H J 2008 *Phys. Rev. Lett.* **100** 206803

[393] Jung I, Jang H Y, Moon J and Park S 2014 *Nanoscale* **6** 6482–86

[394] Kang J, Jariwala D, Ryder C R, Wells S A, Choi Y, Hwang E, Cho J H and Marks T J *et al* 2016 *Nano Lett.* **16** 2580–85

[395] Shih C J, Pfattner R, Chiu Y C, Liu N, Lei T, Kong D S, Kim Y and Chou H H *et al* 2015 *Nano Lett.* **15** 7587–95

[396] Zhu Z C, Murtaza I, Meng H and Huang W 2017 *RSC Adv.* **7** 17387–97

[397] Bai J W, Zhong X, Jiang S, Huang Y and Duan X F 2010 *Nat. Nanotechnol.* **5** 190–94

[398] Lawal A T 2015 *Talanta* **131** 424–43

[399] Qin Y Y, Peng Q Y, Ding Y J, Lin Z S, Wang C H, Li Y, Li J J and Yuan Y *et al* 2015 *ACS Nano* **9** 8933–41

[400] Vicarelli L, Vitiello M S, Coquillat D, Lombardo A, Ferrari A C, Knap W, Polini M and Pellegrini V *et al* 2012 *Nat. Mater.* **11** 865–71

[401] Nag A, Mitra A and Mukhopadhyay S C 2018 *Sensor Actuat a-Phys* **270** 177–94

[402] Yavari F and Koratkar N 2012 *J. Phys. Chem. Lett.* **3** 1746–53

[403] Liu F M, Zhang Y, Yin W, Hou C J, Huo D Q, He B, Qian L L and Fa H B 2017 *Sensor Actuat B-Chem* **242** 889–96

[404] Kang X H, Wang J, Wu H, Aksay I A, Liu J and Lin Y H 2009 *Biosens. Bioelectron.* **25** 901–5

[405] Matsumoto K, Maehashi K, Ohno Y and Inoue K 2014 *J. Phys. D: Appl. Phys.* **47** 094005

[406] Gao F, Cai X L, Wang X, Gao C, Liu S L, Gao F and Wang Q X 2013 *Sensor Actuat B-Chem* **186** 380–87

[407] Boland C S, Khan U, Backes C, O'Neill A, McCauley J, Duane S, Shanker R and Liu Y *et al* 2014 *ACS Nano* **8** 8819–30

[408] Bae S H, Lee Y, Sharma B K, Lee H J, Kim J H and Ahn J H 2013 *Carbon* **51** 236–42

[409] Tian H, Shu Y, Cui Y L, Mi W T, Yang Y, Xie D and Ren T L 2014 *Nanoscale* **6** 699–705

[410] Zhan B B, Li C, Yang J, Jenkins G, Huang W and Dong X C 2014 *Small* **10** 4042–65

[411] Yan X F, Niu G, Lin J, Jin A J, Hu H, Tang Y X, Zhang Y J and Wu A G *et al* 2015 *Biomaterials* **42** 94–102

[412] Satpathy M, Wang L Y, Zielinski R J, Qian W P, Wang Y A, Mohs A M, Kairdolf B A and Ji X *et al* 2019 *Theranostics* **9** 778–95

[413] Yang K, Feng L Z, Shi X Z and Liu Z 2013 *Chem. Soc. Rev.* **42** 530–47

[414] Janib S M, Moses A S and MacKay J A 2010 *Adv. Drug Deliver Rev.* **62** 1052–63

[415] Yoo J M, Kang J H and Hong B H 2015 *Chem. Soc. Rev.* **44** 4835–52

[416] Liu Q, Guo B D, Rao Z Y, Zhang B H and Gong J R 2013 *Nano Lett.* **13** 2436–41

[417] Ran C X, Wang M Q, Gao W Y, Ding J J, Shi Y H, Song X H, Chen H W and Ren Z Y 2012 *J. Phys. Chem.* C **116** 23053–60

[418] Wang X S, Huang P, Feng L L, He M, Guo S W, Shen G X and Cui D X 2012 *RSC Adv.* **2** 3816–22

[419] Caravan P, Ellison J J, McMurry T J and Lauffer R B 1999 *Chem. Rev.* **99** 2293–352

[420] Chen W H, Yi P W, Zhang Y, Zhang L M, Deng Z W and Zhang Z J 2011 *ACS Appl. Mater. Inter.* **3** 4085–91

[421] Hu Y H 2014 *Small* **10** 1451–52

[422] Huang P, Rong P F, Lin J, Li W W, Yan X F, Zhang M G, Nie L M and Niu G *et al* 2014 *J. Am. Chem. Soc.* **136** 8307–13

[423] Patel M A, Yang H, Chiu P L, Mastrogiovanni D D T, Flach C R, Savaram K, Gomez L and Hemnarine A *et al* 2013 *ACS Nano* **7** 8147–57

[424] Wang Y W, Fu Y Y, Peng Q L, Guo S S, Liu G, Li J, Yang H H and Chen G N 2013 *J. Mater. Chem.* B **1** 5762–67

[425] Sheng Z H, Song L, Zheng J X, Hu D H, He M, Zheng M B, Gao G H and Gong P *et al* 2013 *Biomaterials* **34** 5236–43

[426] Lusic H and Grinstaff M W 2013 *Chem. Rev.* **113** 1641–66

[427] Lee N, Choi S H and Hyeon T 2013 *Adv. Mater.* **25** 2641–60

[428] Shi J J, Wang L, Zhang J, Ma R, Gao J, Liu Y, Zhang C F and Zhang Z Z 2014 *Biomaterials* **35** 5847–61

[429] Shi X Z, Gong H, Li Y J, Wang C, Cheng L and Liu Z 2013 *Biomaterials* **34** 4786–93

[430] Goenka S, Sant V and Sant S 2014 *J. Control. Release* **173** 75–88

[431] Boukhvalov D W and Katsnelson M I 2008 *J. Am. Chem. Soc.* **130** 10697–701

[432] Georgakilas V, Tiwari J N, Kemp K C, Perman J A, Bourlinos A B, Kim K S and Zboril R 2016 *Chem. Rev.* **116** 5464–519

[433] Song P, Zhang X Y, Sun M X, Cui X L and Lin Y H 2012 *Nanoscale* **4** 1800–4

[434] Nayak T R, Andersen H, Makam V S, Khaw C, Bae S, Xu X F, Ee P L R and Ahn J H *et al* 2011 *ACS Nano* **5** 4670–78

[435] Li D P, Liu T J, Yu X Q, Wu D and Su Z Q 2017 *Polym. Chem.* **8** 4309–21

[436] Kuila T, Bose S, Mishra A K, Khanra P, Kim N H and Lee J H 2012 *Prog. Mater. Sci.* **57** 1061–105

[437] Moore T L, Podilakrishna R, Rao A and Alexis F 2014 *Part. Part. Syst. Char.* **31** 886–94

[438] Sun X M, Liu Z, Welsher K, Robinson J T, Goodwin A, Zaric S and Dai H J 2008 *Nano Res.* **1** 203–12

[439] Li J B, Tan S B, Kooger R, Zhang C Y and Zhang Y 2014 *Chem. Soc. Rev.* **43** 506–17

[440] Paul A, Hasan A, Al Kindi H, Gaharwar A K, Rao V T S, Nikkhah M, Shin S R and Krafft D *et al* 2014 *ACS Nano* **8** 8050–62

[441] Niidome T and Huang L 2002 *Gene Ther.* **9** 1647–52

[442] Nanda S S, Papaefthymiou G C and Yi D K 2015 *Crit. Rev. Solid State* **40** 291–315

[443] Yang K, Feng L Z, Hong H, Cai W B and Liu Z 2013 *Nat. Protoc.* **8** 2392–403

[444] Lin J, Chen X Y and Huang P 2016 *Adv. Drug Deliver. Rev.* **105** 242–54

[445] Chen D, Feng H B and Li J H 2012 *Chem. Rev.* **112** 6027–53

[446] Lin D J, Wu J, Ju H X and Yan F 2014 *Biosens. Bioelectron.* **52** 153–58

[447] Perreault F, de Faria A F and Elimelech M 2015 *Chem. Soc. Rev.* **44** 5861–96

[448] Sun M and Li J H 2018 *Nano Today* **20** 121–37

[449] Li H, Song Z N, Zhang X J, Huang Y, Li S G, Mao Y T, Ploehn H J and Bao Y *et al* 2013 *Science* **342** 95–8

[450] Huang A S, Liu Q, Wang N Y, Zhu Y Q and Caro J 2014 *J. Am. Chem. Soc.* **136** 14686–89

[451] Seredych M and Bandosz T J 2010 *J. Phys. Chem.* C **114** 14552–60

[452] Chen C, Xu K, Ji X, Miao L and Jiang J J 2014 *Phys. Chem. Chem. Phys.* **16** 11031–36

[453] Mahmoud K A, Mansoor B, Mansour A and Khraisheh M 2015 *Desalination* **356** 208–25

[454] Liu F, Chung S, Oh G and Seo T S 2012 *ACS Appl. Mater. Inter.* **4** 922–27

[455] Sun P Z, Chen Q, Li X D, Liu H, Wang K L, Zhong M L, Wei J Q and Wu D H *et al* 2015 *NPG Asia Mater.* **7** e162

[456] Nair R R, Wu H A, Jayaram P N, Grigorieva I V and Geim A K 2012 *Science* **335** 442–44

[457] Zhao G X, Li J X, Ren X M, Chen C L and Wang X K 2011 *Environ. Sci. Technol.* **45** 10454–62

[458] Chowdhury I, Duch M C, Mansukhani N D, Hersam M C and Bouchard D 2013 *Environ. Sci. Technol.* **47** 6288–96

[459] Chong Y, Ge C C, Fang G, Wu R F, Zhang H, Chai Z F, Chen C Y and Yin J J 2017 *Environ. Sci. Technol.* **51** 10154–61

[460] Zou F M, Zhou H J, Jeong D Y, Kwon J, Eom S U, Park T J, Hong S W and Lee J 2017 *ACS Appl. Mater. Inter.* **9** 1343–51

[461] Kim T I, Kwon B, Yoon J, Park I J, Bang G S, Park Y, Seo Y S and Choi S Y 2017 *ACS Appl. Mater. Inter.* **9** 7908–17

IOP Publishing

Functional Carbon Materials

Jianmin Ma and Jiantie Xu

Chapter 6

Graphite and its main applications

Haiying Lu, Xianghong Chen, Jiakui Zhang, Yu Lei, Wenlu Min and Jiantie Xu

As a result of its unique physical and chemical properties, graphite has been widely used in a variety of technical fields, including batteries, graphene synthesis, refractory materials, nuclear materials, water purification, lubricants, and pencils. At present, the demand for graphite resources is growing rapidly. Graphite has been regarded as a key material for the industrial and national security of many countries or organizations. As expected, it has become critically important to realize the purification and rational utilization of the limited natural graphite resources. In this chapter, we briefly summarize the occurrence, classification, and purification methods of natural graphite and its main applications. Finally, a perspective on, and challenges for, the development of graphite are also presented.

6.1 Introduction

Graphite is a mineral form of the element carbon with layered structures in which each layer consists of carbon atoms linked by strong covalent bonds with a bond length of 0.142 nm in honeycomb-like hexagonal rings [1]. The graphitic layers are bound by weak Van der Waals forces and have an interlayer distance of 0.335 nm. Graphite has high strength, chemical inertness, and corrosion resistance, high electrical conductivity and thermal conductivity, and high lubricity [2]. These unique physical and mechanical properties make it versatile for a wide variety of industrial applications. Graphite can also form several allotropes, together with carbon nanotubes, fullerene, and diamond. In general, graphite can be classified into natural and synthetic graphite. Natural graphite is further divided into flake graphite, vein or lump graphite, and microcrystalline/amorphous graphite [3, 4]. Among these, the flake graphite has attracted great attention for industrial applications due to its merits (e.g., its degree of crystallinity and production cost).

Due to its unique physical and chemical properties, the application of graphite is highly developed in several fields [5]. For example, graphite is the most commonly used anode of lithium-ion batteries (LIBs), as well as a common raw material for the synthesis of graphene and its derivatives. In addition, graphite is commonly used as a conductive materials in modern industries, for example, in metallurgy, casting, and the production of steel. In addition, graphite can also be applied in nuclear reactors, water purification, and aerospace materials. Therefore, graphite has been regarded as a strategic material worldwide. However, with the growing use of graphite in many industries, balancing the supply against the demand for graphite has become critically urgent. According to the United States Geological Survey, there are 230 million tons of graphite reserves worldwide and their distribution is extremely nonuniform, as shown in figure 6.1. It should be expected that the contradiction between supply constraints and demand growth will strengthen the competition among graphite importers [6].

Proven graphite reserves are concentrated in a few countries, so that most countries acquire graphite resources only through the international trade market. China is the largest exporter of natural graphite in the world. According to the UN Comtrade Database, China has exported ~72% of the global natural graphite in the international market since 2001. Figure 6.2 shows the natural graphite trade network in China in 2018. As can be seen, the graphite trade partners of China mainly include Japan, Malaysia, North Korea, the EU, and South Korea [7]. According to the 2018 production data, global demand for graphite for energy storage applications is expected to increase by nearly 500% by 2050, indicating that graphite is a key and indispensable energy mineral for the green energy transition [7, 8]. Moreover, global demand for graphite in other fields is also expected to increase considerably in the near future. A sustainable society relies heavily on critical minerals due to their importance to economic prosperity and national security. This review summarizes the occurrence, classification, and purification methods of graphite and its main applications.

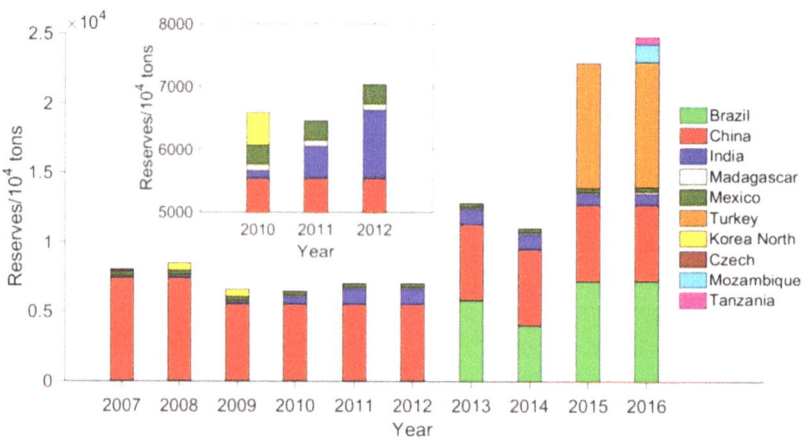

Figure 6.1. Graphite reserves in representative countries [6], reproduced with permission © Elsevier.

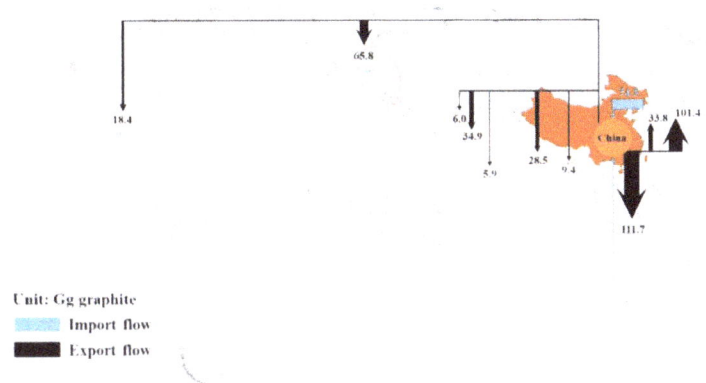

Figure 6.2. The trade flow of natural graphite between China and its top trade partners in 2018 [7], reproduced with permission © Elsevier.

6.2 The occurrence and classification of graphite

The occurrence of natural graphite is commonly believed to be generated by the conversion of carbonaceous material during metamorphism [9–12]. Natural graphite in the global market has been classified into three types: (1) crystalline flake, (2) microcrystalline or amorphous, and (3) crystalline vein or lump. All these types of graphite have their own characteristics and corresponding applications in various fields. In addition to natural graphite, a variety of carbon-based precursors have been also adopted to prepare synthetic graphite for multiple applications.

6.2.1 Vein or lump graphite

Vein or lump graphite is graphite in its most natural form, which is formed from the direct deposition of solid, graphitic carbon by subterranean, high-temperature fluids. Typically, this type of graphite has a needle-like macromorphology and a flake-like micromorphology. Due to the natural fluid-to-solid deposition process, vein graphite generally has a purity of up to 99.5% and is highly crystalline [13]. Moreover, this type of graphite exists in the form of large crystals and condenses into lumps and chips so that it possesses unique physical properties (e.g. high thermal and electrical conductivity, lubricity, and oxidation resistance). However, the scarcity and high price of vein or lump graphite tend to make it uncommon in electrochemical systems [14].

6.2.2 Microcrystalline or amorphous graphite

Microcrystalline graphite is an aggregate of randomly orientated graphite microcrystallites. It is widely known as amorphous graphite and is usually formed by the thermal metamorphism of coal. Microcrystalline graphite is abundantly found in natural graphite, but its grade is limited to the range of 20–40 wt%. Even after being processed, the carbon content is typically less than 85% [1]. Moreover, because of its lower crystallinity, smaller crystal particle size, poorer enrichment efficiency and lower grade than those of flake and vein graphite [15, 16] microcrystalline graphite has moderate electrical conductivity and lubrication properties. As a result, micro-crystalline graphite is commonly used as a raw material in pencils and crucibles and as a carburant for steelmaking [17].

6.2.3 Flake graphite

Flake graphite consists of isolated flat and plate-like particles in which each flake of graphite has a hexagonal structure with an irregular fragmented edge. Most flake graphite is generated by the heat and pressure metamorphism of dispersed organic material at high temperature and pressure under metamorphic geologic conditions [13, 18]. The flake size and product purity of flake graphite largely determine its price. Depending on the flake size and graphitic carbon content, flake graphite has two grades: coarse flake (150–850 μm in diameter) and fine flake (45–150 μm in diameter) [19, 20]. Among all sizes of flake graphite, flakes in the range of 250–1000 μm in diameter have the highest price [21].

6.2.4 Synthetic graphite

Synthetic graphite is a type of graphite prepared from carbon-based materials as precursors. Many efforts have been devoted to preparing graphite through the use of various carbon precursors, such as meso-carbon, microbeads, sponge coke, needle coke, petroleum coke, pitch, fly ash, and anthracite. In these precursors, petroleum coke is commonly used for the preparation of synthetic graphite [22]. In addition, coal is also an attractive precursor candidate. Among the different classes of coal, anthracites can be converted to graphitize by heating them to a temperature of >2000 °C [23]. Graphitized carbon materials, such as natural graphite, have been widely used as electrode materials and heat-resistant materials, as these carbons have high chemical stability and electrical conductivity [24].

6.3 The purification of graphite

Graphite ore is often associated with quartz, kaolinite, alusite, and sericite and small amounts of pyrite, limonite, tourmaline, and calcite. Before use, graphite ore needs to be purified to remove impurities [25]. Graphite purification is a technological process which depends upon the nature of the associated gangue minerals in the ore [26]. The graphite purification process mainly includes physical purification methods (e.g. comminution, flotation, and thermal purification) and chemical purification methods (e.g. the roasting method and acid leaching) [27].

6.3.1 Comminution

Comminution is the initial step in the treatment of mineral ore to get the desired graphite with reduced impurities. Normally, large flake graphite has greater industrial value and wider application compared to small flake graphite. During the mechanical process, the ore or flake is always easily destroyed [20]. Therefore, it is highly desirable to choose a suitable comminution method. For example, the type

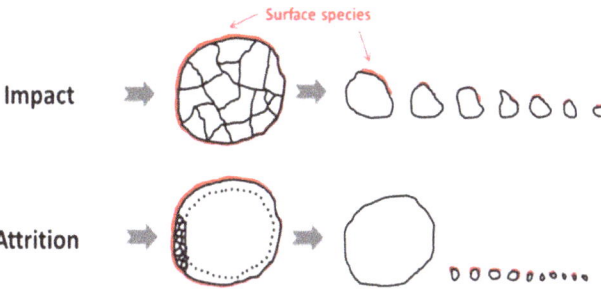

Figure 6.3. Schematic of the distribution of surface species during particle breaking achieved via the impact and attrition mechanisms [30], reproduced with permission © Elsevier.

of grinding mode (e.g. shear- and shock-type grinding) is a key factor that influences the mechanical grade of the products. The graphite obtained from 'shear-type' grinding has less damaged parts than that obtained using the 'shock-type' because of its weak mechanical strain [28]. In addition, rod milling with 'line-contact' grinding was found to be more favorable for retaining the flakes than the use of 'point-contact' mills [29, 30]. To reduce the breakage of graphite flakes, vibration milling and stirred milling are more energy efficient than the traditional tumbling milling [20, 29, 31]. The breakage mechanism can significantly influence the particle-size distribution, particle shape, and mineral liberation, which plays important roles in mineral flotation. As shown in figure 6.3, the surface species can be distributed on the surfaces of particles of all sizes particles when the breakage occurs on impact. If the breakage is due to attrition, the surface species are transferred from the large particles to ultrafine particles [30]. Apart from the effect of grinding type on the grade of the product, the mechanical environment (e.g. oxygen, air, nitrogen, and argon) is also an important factor [32]. For example, when natural graphite is milled in the presence of oxygen, the fracture rate of the graphite can be suppressed by the formation of oxides during the milling process.

6.3.2 Flotation

Due to the natural hydrophobicity and floatability of graphite [33–35] as well as the different surface chemistries and physical properties of minerals [36] flotation is one of the purification processes most commonly used to flow out solid minerals from a water suspension. Compared to vein and amorphous graphite, crystalline flake graphite has the best floatability, so it is easily purified by the flotation process [20]. In order to improve the floatability of graphite, several reagents (e.g. frother and collector, depressant and regulator) are often employed. The functions of these reagents can be summarized as follows: (1) to reduce the surface tension of the liquid–gas interface to enable froth formation; (2) hinder coalescence in order to stabilize bubble size and (3) facilitate hydrophobic particle adhesion to air bubbles [37]. The agents most widely used for flotation are methyl isobutyl carbinol (MIBC), pine oil, and polyglycolic ethers. Among these, MIBC is recognized as an effective frother

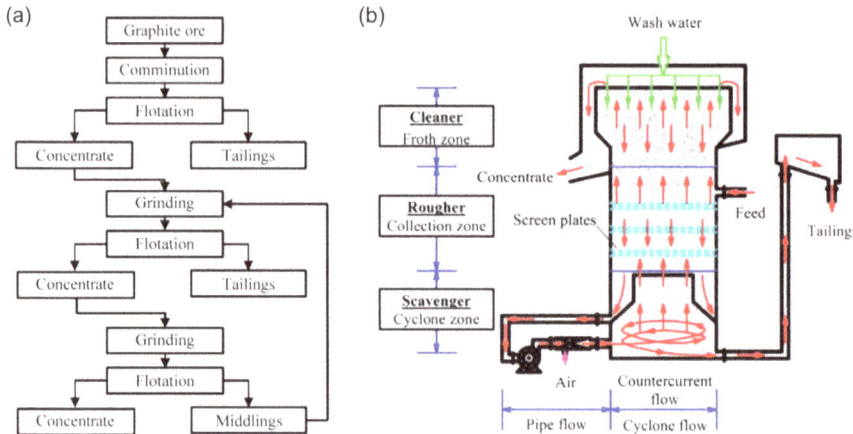

Figure 6.4. (a) Flowchart for laboratory graphite processing [43], reproduced with permission © Elsevier. (b) Schematic illustration of the flotation column [35], reproduced with permission © MDPI.

[21, 37–39]. However, the use of an MIBC frother can cause environmental issues due to its evaporation in high-temperature conditions.

During the graphite flotation process, the collector also plays an essential role. Non-ionic hydrocarbons (e.g. kerosene, fuel oil, paraffin, and diesel oil) and the biodegradable single reagent ether–alcohol have been widely used as collectors [27, 40]. In addition, sodium silicate, quebracho, gelatin, tannic acid, and starch are commonly used as depressants to prevent the collector from adsorbing on a particular mineral flotation pulp [41]. Typically, the flotation of graphite is a multistage process in a fixed-flow configuration, as shown in figure 6.4(a). However, mechanical flotation process stages are too long and have poor separation efficiency. In contrast, column flotation with reduced operational stages and an additional centrifugal force field, as shown in figure 6.4(b), could be favorable for improving the grade and recovery of graphite [35]. Overall, graphite purification by the flotation method can reduce the energy consumption and the production cost, while improving the grade to as much as ~96%; however, the further upgrading of purified graphite is difficult [33, 42]. Therefore, flotation is usually used as a preliminary process for the purification of graphite. To obtain higher-purity graphite, secondary purification by other physical or chemical methods (e.g. acid leaching or alkaline roasting) is required.

6.3.3 Roasting

Roasting purification methods mainly include alkali roasting, water washing, and acid leaching. For example, alkali roasting can effectively remove the impurities in graphite ores (e.g. quartz, mica, hematite, and silicate) at high temperature [44]. In this process, the impurities in the graphite react with sodium hydroxide to form a solution of sodium silicate and sodium aluminate. The possible reaction equations are described as follows: [45, 46]

$$SiO_2 + 2NaOH \rightarrow Na_2SiO_3 + H_2O$$

$$Al_2O_3 + 2NaOH \rightarrow 2NaAlO_2 + H_2O$$

$$P_2O_5 + 6NaOH \rightarrow 2Na_3PO_4 + 3H_2O$$

$$TiO_2 + 2NaOH \rightarrow Na_2TiO_3 + H_2O$$

$$V_2O_5 + 6NaOH \rightarrow 2Na_3VO_4 + 3H_2O$$

$$FeS + 2NaOH + 2O_2 \rightarrow FeO + Na_2SO_4 + H_2O$$

Normally, most of the impurities in graphite can be removed by initial alkali roasting to form water-soluble alkali silicates. The residual part of the slightly dissolved or insoluble products in the graphite (e.g. hydroxides of Fe and Al and oxides of Fe, Mg, and Ca) can be neutralized by an acid (e.g. HCl, H_2SO_4) to form water-soluble substances, which are further purified by washing [27]. The detailed reactions are as follows:[45]

$$CaO + 2HCl \rightarrow CaCl_2 + H_2O$$

$$MgO + 2HCl \rightarrow MgCl_2 + H_2O$$

$$Fe_2O_3 + 6HCl \rightarrow 2FeCl_2 + 3H_2O$$

$$Fe(OH)_3 + 3HCl \rightarrow FeCl_3 + 3H_2O$$

$$Al(OH)_3 + 3HCl \rightarrow AlCl_3 + 3H_2O$$

For the acid washing, some studies have shown that the carbon content of the graphite products obtained is directly related to the acid leaching time, the temperature, the liquid–solid ratio, and the concentration. As shown in figure 6.5, the purification process includes three steps: paths A, B, and C. The results indicate that different acids (e.g. HCl, HNO_3, H_2SO_4, $H_2SO_4 + H_2O_2$, and HF) can lead to a maximum carbon content of 98.4%. When the acid–alkali–acid (H_2SO_4/H_2O_2–NaOH –HCl) was employed, the carbon grade of graphite was significantly improved to 99.68% [47]. Although the roasting purification method can increase the carbon grade, the use of acid and alkali agents is highly corrosive to the equipment and the waste water that remains after purification is also dangerous to the environment. Moreover, the roasting purification process also requires a large amount of energy [48].

6.3.4 Acid leaching

Acid leaching purification is the most common technique for producing high-purity graphite. Different acids, such as HCl, HF, H_2SO_4, and HNO_3, or a mixture of these, are often used depending on the different mineral compounds present. Both HCl and H_2SO_4 are equally good at removing impurities from fine graphite ore;

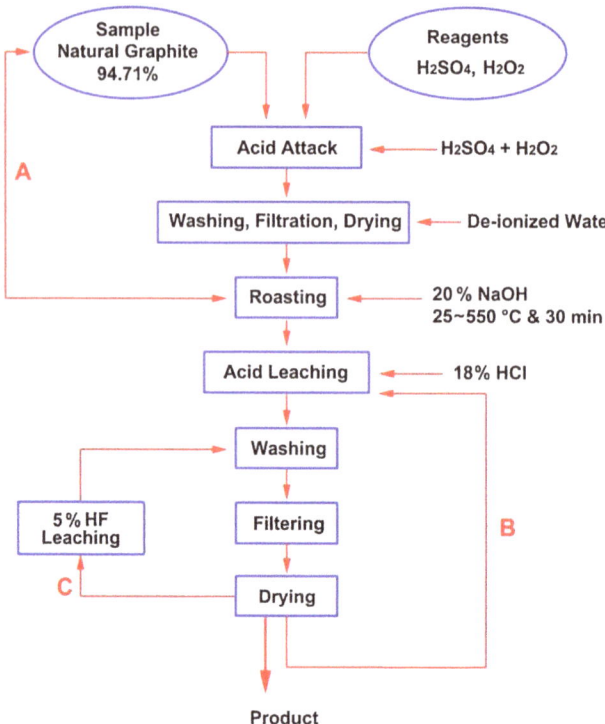

Figure 6.5. Flowchart of the integrated acid–alkali–acid treatments used for the purification of graphite [47], reproduced with permission © Elsevier.

however, HCl is volatile and more expensive [1]. HF is more effective than HCl and H_2SO_4 in leaching silicate impurities [46]. However, it is not efficient at removing pyrite and leaves insoluble fluoride compounds. Compared to a single acid, a mixture of acids such as $HNO_3 + H_2SO_4$, and $HF + HCl + H_2SO_4$ has been found to be better at obtaining high-purity graphite products [49]. In addition, the effects of acid type, acid concentration, leaching temperature, and leaching time have key impacts on the dissolution of graphite ore [50]. For example, optimized experimental parameters that obtain high-purity graphite are: 20% HCl + 4% HF at 85 °C for 4 h. The possible chemical equations for acid leaching are as follows:[50]

$$FeS + 2H^+ \rightarrow Fe^{2+} + H_2S$$

$$CaSO_4 \cdot 2H_2O + 2HCl \rightarrow CaCl_2 + H_2SO_4 + 2H_2O$$

$$CaCO_3 + 2HCl \rightarrow CaCl_2 + CO_2 + H_2O$$

$$SiO_2 + 4HF \rightarrow SiF_4 + 2H_2O$$

$$SiF_4 + 2HF \rightarrow H_2SiF_6$$

$$Al_2Si_2O_5(OH)_4 + 6H^+ \leftrightarrow 2Al^{3+} + 2H_4SiO_4 + H_2O$$

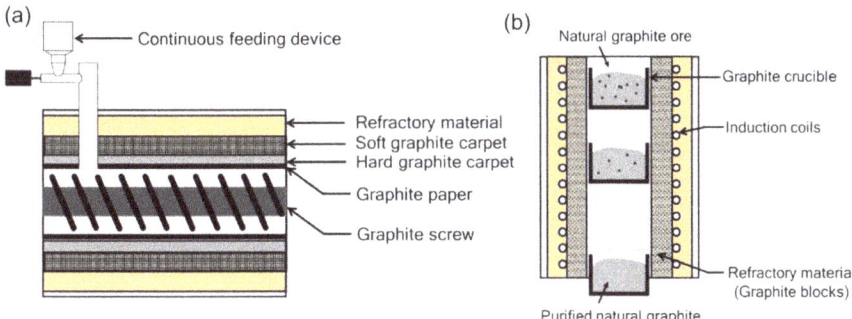

Figure 6.6. Schematic diagram of the furnaces used for thermal purification with (a) a horizontal arrangement and (b) a vertical arrangement [51], reproduced with permission © Elsevier.

Regardless of the pure graphite obtained by acid purification, the volatile acid solution and the waste water that remains after purification can result in serious environmental problems; in particular, HF is highly toxic and causes environmental pollution [27].

6.3.5 Thermal purification

Since the silicate impurities of graphite can be decomposed into metal oxides (e.g. SiO_2, Al_2O_3, Fe_3O_4, MgO, CaO) at high temperatures, thermal purification is also an efficient technique for graphite purification [51]. The purification equipment used for the high-temperature treatment is critically important. As can be seen, the arrangement can be designed to be either horizontal (figure 6.6(a)) or vertical (figure 6.6(b)) [52]. It has been reported that the approximately 80 wt% of carbon in graphite ore can be further increased to 99.9 wt% [17].

In particular, the purity of natural flake graphite and microcrystalline graphite can be upgraded to nuclear-grade graphite by thermal and gas purification [51]. It can be seen that flake graphite (figure 6.7(a)) and microcrystalline graphite (figure 6.7(b)) containing at least 99.9% carbon were produced by thermal purification at 3000 °C. After thermal purification, the metallic impurities were mainly changed to elements with extraordinarily high melting/boiling points, such as B, Ti, Ta, V, W, and Mo. Nevertheless, thermal processing is much more expensive than chemical processing.

6.4 The application of graphite for advanced technologies

6.4.1 Graphite intercalation compounds

As a unique layered material, graphite is composed of graphene layers stacked together by the Van der Waals force. Since its interlayer space are highly tunable, graphite can accommodate a variety of ions and molecules to form graphite intercalation compounds (GICs) [53]. Owing to their unique physicochemical properties, GICs have been widely used in a variety of applications, such as the synthesis of expanded graphite (EG) and electrode materials for alkali-ion batteries (AIBs) [54].

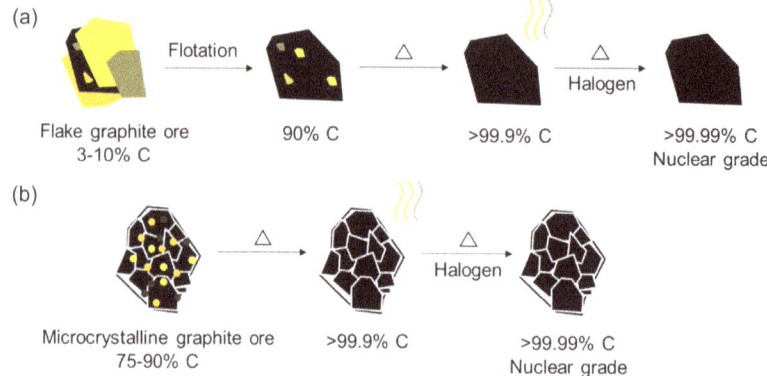

Figure 6.7. Scheme of (a) thermal (without halogen) and gas (with halogen) purification of flake graphite ore and (b) thermal (without halogen) and gas (with halogen) purification of microcrystalline graphite ore [51], reproduced with permission © Elsevier.

6.4.1.1 Expanded graphite

Expanded graphite (EG) is commonly prepared by the rapid heat treatment of GICs, followed by a washing process [55, 56]. EG retains a layered structure which is similar to that of natural graphite flakes and produces a tremendous variety of different-sized pores and nanosheets. Due to its remarkable physical and chemical properties (e.g. large specific surface area, high surface energy, strong adsorption, high temperature resistance, high pressure resistance, good sealing, and corrosion resistance to a variety of media), EG has been widely used in various fields, including gaskets, seals, thermal management, and cell applications [57]. For example, phase change materials (PCMs) with higher thermal storage densities and nearly isothermal processes, have been widely applied in aerospace, military applications, medical treatments, textiles, and building materials. EG, which has a well-defined mesoporous structure, can significantly improve the thermal conductivity and stability of stearic acid/EG composites [58]. In addition, polyethylene glycol/expanded graphite (PEG/EG), a PCM with good thermal stability, was also developed for indoor energy saving [59]. The synthetic route of the PEG/EG PCM is shown in figure 6.8. Furthermore, the EG can enhance the electrical and mechanical properties of high-density polyethylene [60]. Only 3 wt% filler content was required to reach the percolation threshold (ϕ_c).

In addition, EG is also an attractive anode candidate for LIBs [61, 62] and sodium-ion batteries (SIBs) [63]. For example, Chen *et al* reported an efficient approach for the synthesis of EG and expanded holey graphite (EhG) and studied them as anodes for LIBs, as shown in figure 6.9(a) [61]. To obtain holey graphite (hG), pristine graphite (PG, figure 6.9(b)) powder was annealed in an Ar/H$_2$O gas mixture. In contrast to the conventional routes that use water-based electrolytes and subsequent excessive washing water, GICs were formed during the charge process by the intercalation of PF$_6^-$ anions into the graphite layers. Benefiting from plenty of 'holey' and cross-linked 'worm-like' structures (figures 6.10(c)–(e)), the EhG

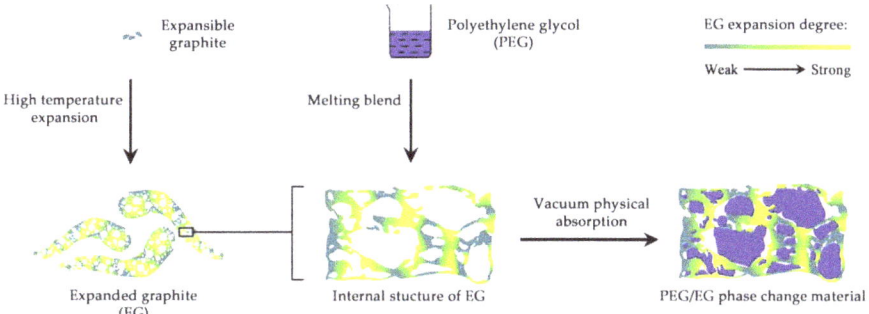

Figure 6.8. The synthetic route of PEG/EG PCM [59], reproduced with permission © Elsevier.

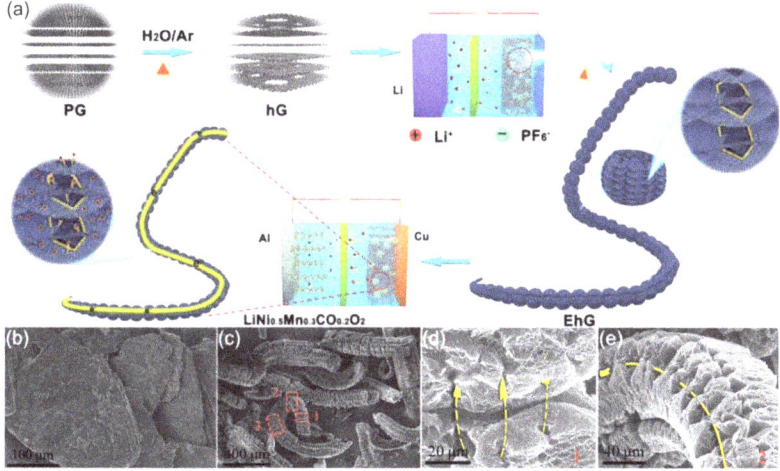

Figure 6.9. (a) Schematic of the fabrication and lithiation processes of EhG. Scanning electron microscopy (SEM) images of (b) pristine graphite (PG) and (c) expanded holey graphite (EhG). (d–e) are magnified SEM images of EhG (1, 2 in figure 6.9(c), respectively) [61], reproduced with permission © Elsevier.

displayed excellent rate capability in fabricated half-cell LIBs, as well as outstanding lithium storage properties in full-cell LIBs when used with a commercial $LiNi_{0.5}Mn_{0.3}Co_{0.2}O_2$ (NMC) cathode. Wang *et al* synthesized EG through a two-step oxidation-reduction process. The EG preserved the long-range-ordered layered structure of graphite, leading to an enlarged interlayer distance (>0.34 nm) [63]. When used in the anode of an SIB, the EG delivered a high reversible capacity of 284 mAh g^{-1} at 20 mA g^{-1}, maintained a capacity of 184 mAh g^{-1} at 100 mA g^{-1}, and retained 73.92% of its capacity after 2000 cycles.

6.4.1.2 Anodes for lithium-ion batteries

LIBs have been widely used in portable and smart devices because of their high energy densities, long cycle life, and environmental friendliness. As shown in figure 6.10, the electrode materials (i.e. the cathode and anode) used as core components of LIBs have largely determined the development of LIBs. As the most commonly used commercial

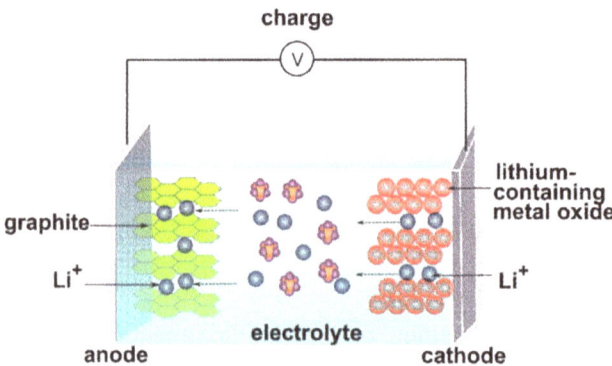

Figure 6.10. Schematic of the configuration of an LIB [71], reproduced with permission © John Wiley & Sons.

anode, graphite suffers from a limited theoretical capacity of 372 mAh g^{-1} and poor rate performance [61, 64, 65]. This is due to the intercalation of Li$^+$ into graphite (which has a maximum stoichiometry of LiC$_6$) and the sluggish charge-transfer kinetics of graphite [66]. Moreover, the sluggish kinetics and lithiation/delithiation of graphite at a relatively low potential (close to 0 V) versus Li$^+$/Li tend to form undesired lithium-metal plating on the graphite surface during deep cycles. The formation of lithium metal can increase the risk of internal shorting and create serious safety hazards in LIBs [67, 68]. To counter this problem, the hybridization or coating of graphite with conductive agents has been proven to be an effective way to improve graphite's kinetics [69]. This is the case because graphite suffers from the following drawbacks: (1) graphite is very sensitive to electrolytes and can easily be exfoliated in PC (propylene carbonate)-based electrolytes; (2) graphite is highly anisotropic, so it is not conducive on the collector substrate (copper foil) [1, 70].

To protect the surface of natural graphite from direct contact with the electrolyte, the surface of graphite can be coated by carbon through the thermal vapor decomposition (TVD) technique [72–74]. In addition, the electrochemical properties of graphite can also be significantly improved by rolling the graphite flakes into a spherical shape and then coating it with TVD carbon (figure 6.11(a)) [70, 75]. Figures 6.11(b) and (c) show the SEM images of graphite with different particle shapes before carbon coating, indicating that the spherical particles can be randomly orientated and easily spread uniformly and thinly on copper foil. Unfortunately, the yield of spherical graphite production is only 30%, leading to a lower energy density than that obtained using flake graphite [1]. In order to obtain high-performance graphite, many other coating materials have been also adopted [76], such as mild oxidation [77], metal modification [78, 79], polymer coating [80], pyrolytic carbon coating [73], as well as electrically conductive additives (e.g. traditional carbon black [81], novel carbon nanotubes [82], and bamboo-like carbon nanotubes [83]). Nevertheless, excessive and large-surface-area conductive agents can significantly decrease the amount of graphite available as an active material, inevitably leading to the majority of solid-state electrolyte interfaces (SEI), which have reduced reversible capacity and poor cycling stability. Alternatively, the structural modification of

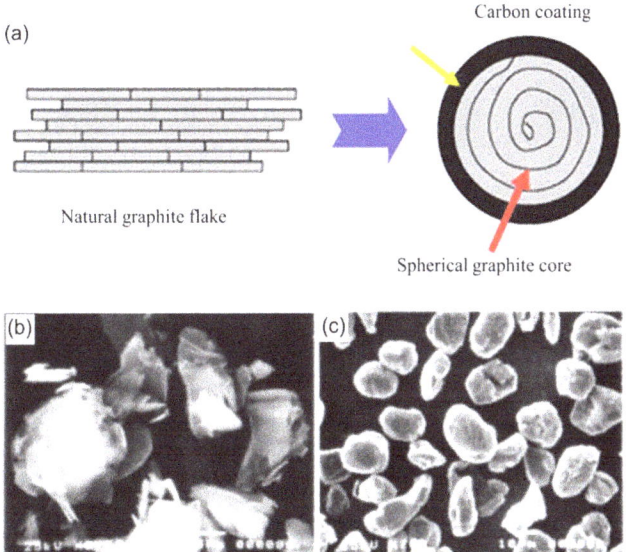

Figure 6.11. (a) Schematic diagram of the spherification of natural flake graphite. SEM images of the natural graphite before TVD-carbon coating: (b) flakes, and (c) spherical [70], reproduced with permission © The Royal Society of Chemistry.

graphite (e.g. hG) is also a promising route for improving electrode kinetics. Xiao *et al* proposed a feasible and eco-friendly approach to the large-scale production of hG with a well-defined hexagonal-hole structure, armchair edges, and a high-crystallinity basal plane [64]. Their synthesis approach employed low-cost common H_2O steam as a mild oxidative agent to directly anneal graphite at a high temperature. Benefiting from its unique structural features, the use of hG as the anode in LIBs leads to outstanding rate capability (e.g. higher average capacities of 225.0 and 95.7 mAh g^{-1} at 1 and 2 C, respectively, than those of pristine graphite, e.g. 120.7 and 48.2 mAh g^{-1}) and excellent cycling stability (e.g. 279.2 mAh g^{-1} after 500 cycles with an initial capacity retention of 93.2%).

In addition to flake graphite, synthetic graphite with its predictable morphology and higher graphitization has also proven to be an ideal anode candidate for LIBs [84]. Anthracite is widely recognized as the most promising raw material for the production of synthetic graphite. Anthracite-based graphite prepared by thermal treatment in the temperature range of 2400 °C–2800 °C delivered a reversible capacity of ~250 mAh g^{-1} [85]. In order to improve the graphitization degree of synthetic graphite, some additives (catalysts) have successfully been added [84, 86–90]. For example, Li *et al* studied the catalytic graphitization of coke carbon by adding iron [90]. As shown in figure 6.12, the lattice fringes of microcrystal graphite, which are better than those of commercial graphite, were clearly observed in coke carbon heated to >1200 °C.

With the rapid development of electric vehicles (EVs) and energy storage systems (ESSs), there is a strong demand for upgraded LIB technology. These developments

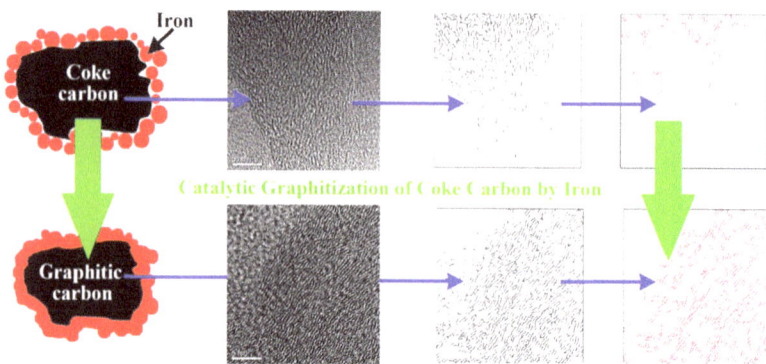

Figure 6.12. The lattice fringe changes of coke carbon and graphite carbon before and after catalytic graphitization at 1500 °C [90], reproduced with permission © Elsevier.

require LIBs with improved power and energy densities. In recent decades, a great deal of effort has been devoted to exploring alternative anode materials that can replace commercial graphite. So far, silicon (Si) has been widely regarded as a promising anode material for next-generation high-energy LIBs. This is due to its extremely high specific capacity of 4200 mAh g^{-1} (which is more than ten times that of commercial graphite), low working voltage (<0.4 V versus Li/Li$^+$), environmental compatibility, and natural abundance [91–94]. However, the utilization of Si alone suffers from its huge volume change (>300%) during the lithiation–delithiation process, leading to its limited poor electrochemical performance [95]. As a result, the combination of Si with graphite has been demonstrated to be one of the most appropriate approaches for realizing high-energy-density electrodes for LIBs [96, 97]. In graphite–Si electrodes, the graphite not only guarantees the high electrical conductivity of the electrode components [98], but also enhances the initial coulombic efficiency (CE) and cyclic stability [98–100]. So far, the practical utilization of Si in anodes for LIBs has been vigorously studied by blending it with graphite [101]. For example, Zhu *et al* designed a three-dimensional (3D) hierarchical structure in which Si nanoparticles were homogeneously dispersed on commercial graphite and then uniformly encapsulated in a hierarchical graphene oxide (GO) scaffold (figure 6.13(a)) [102]. As an anode for LIBs, the composite with 5 wt% Si exhibited a reversible capacity of 559 mAh g^{-1} at 75 mA g^{-1}. Sun *et al* designed a strategy that integrated boron doping and carbon nanotube wedging modified nano/microstructured silicon with graphite (B–Si/CNT@G) (figure 6.13(b)) [103]. The B–Si/CNT@G exhibited an areal capacity of 5.2 mAh cm^{-2} and good cycle retention of 83.4% over 100 cycles.

6.4.1.3 Cathode for dual-ion batteries

Dual-ion batteries (DIBs), which use anion-accepting host materials (e.g. graphite) as the cathode, cation-accepting host materials (e.g. graphite, lithium/sodium metal, or others) as the anode, and alkali-ion-containing electrolytes have recently attracted great interest [53, 105]. This is mainly due to their large voltage window. To date, the

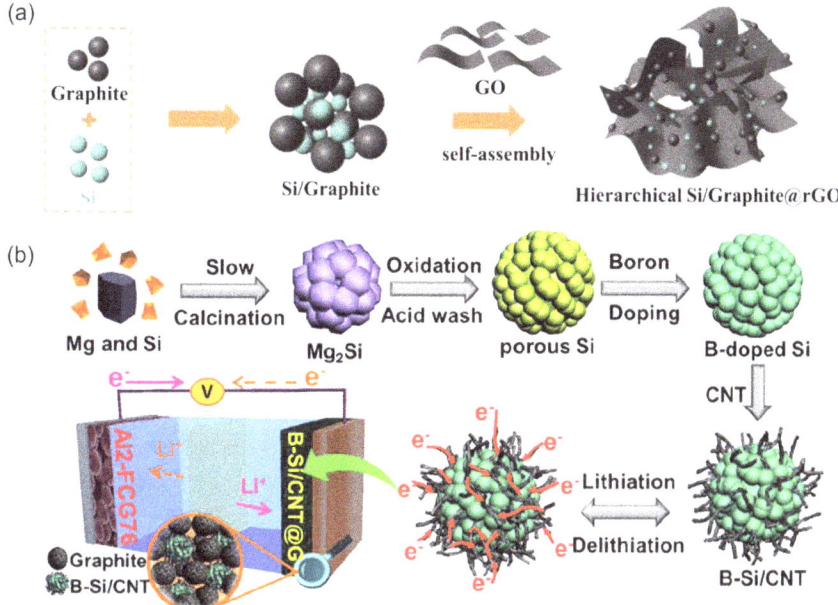

Figure 6.13. (a) Scheme of the synthesis process of the Si nanoparticle/graphite/graphene composite [102], reproduced with permission © John Wiley & Sons. (b) Scheme of the synthesis of B–Si/CNT@G and its application in LIBs [103], reproduced with permission © American Chemical Society.

use of graphite or modified carbon as the cathodes and anodes of DIBs has been intensively reported. Figure 6.14 is a schematic of the charge/discharge process of a DIB [106].

The reversible intercalation/de-intercalation of anions (e.g. PF_6^-, BF_4^-, ClO_4^-, $AlCl_4^-$, $CF_3SO_3^-$, and $C_6H_5COO^-$) into/from graphite has been studied extensively [104, 107]. Previous results indicated that the structure, morphologies, and size of graphite play a key role in affecting the electrochemistry of anion intercalation. It was found that an improved graphitization degree of graphite endows DIBs with higher charge/discharge capacity, as shown in figure 6.15 [106, 108]. Figures 6.16(a)–(f) compare the reaction mechanisms of non-graphitic carbon and graphitic nano-flakes with $AlCl_4^-$ anions during the charge–discharge process. As can be seen, the non-graphitic carbon with the turbostratic disordered structure mainly exhibits capacitive behavior, due to the adsorption of $AlCl_4^-$ anions in its micropores. In contrast, the graphite with the ordered graphitic structure demonstrates reversible battery behavior due to the intercalation of $AlCl_4^-$ anions between the graphene layers.

The particle size, specific surface area, and morphology of graphite also have direct impacts on the specific discharge capacity. A high specific surface area and a small particle size endow graphite with enhanced specific discharge capacity due to the enhanced kinetics during anion intercalation [106, 109]. As shown in figure 6.17, the pristine natural graphite flakes are broken into smaller particles with two patterns, namely, a retained flat morphology with unfolded edges and 'potato'

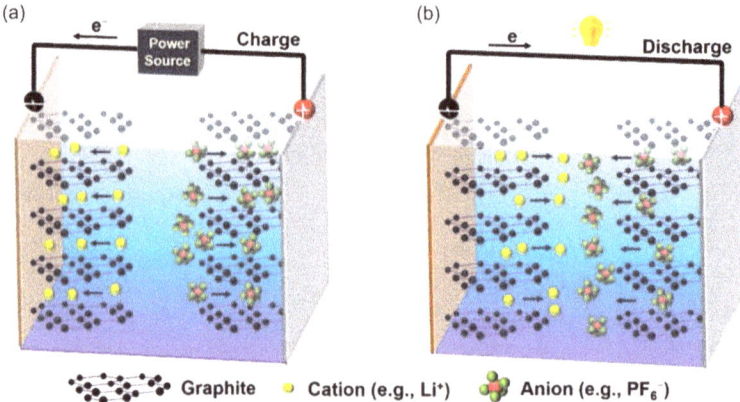

Figure 6.14. Schematic of the working principle of a DIB that uses graphite as both anode and cathode [104], reproduced with permission © John Wiley & Sons.

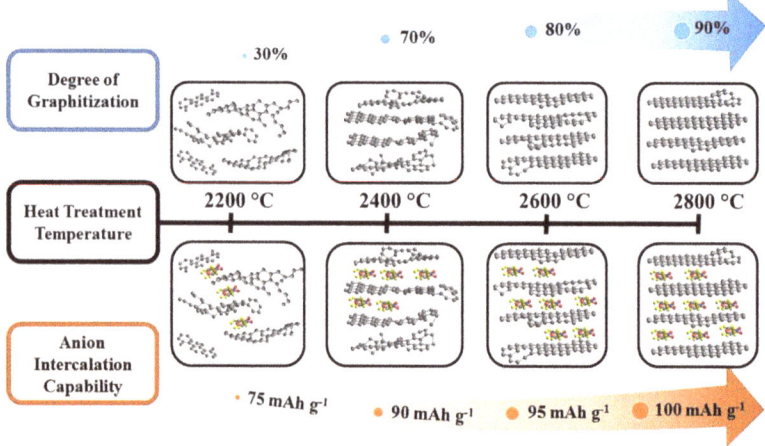

Figure 6.15. Schematic of the changes in graphitic structure and discharge capacity that occur with heat treatment at the temperatures shown [106], reproduced with permission © Elsevier.

graphite particles with imparted partial spheroidization. It was found that the 'smaller flaky' graphite exhibited higher capacities for $AlCl_4^-$ ions than the 'potato-shaped' graphite [110].

6.4.2 Water purification

Graphitic materials (in particular, EG) have also attracted great interest for use in water purification because they can serve as excellent platforms for the removal of aqueous pollutants (e.g. heavy oils and organic dyes) via adsorption routes [111, 112]. This is mainly attributed to the large specific area and low density of EG. EG is usually produced by heating GICs, which are obtained by inserting different chemical compounds such as alkali metals or acid molecules between graphitic layers [113–115].

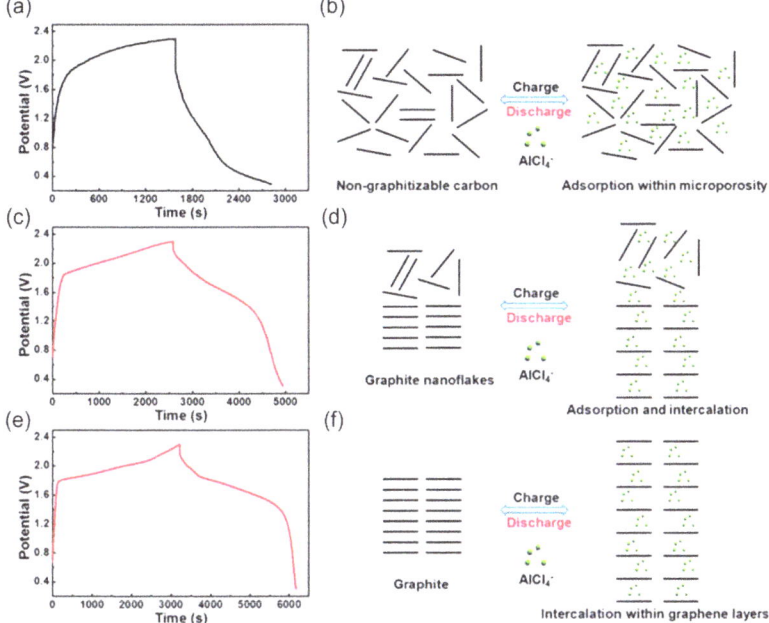

Figure 6.16. (a, c, and e) Typical potential vs. time profile and (b, d, and f) corresponding schematics of the reaction mechanisms of (a and b) non-graphitizable carbon, (c and d) graphite nanoflakes, and (e and f) commercial flake graphite in an aluminum-ion battery [108], reproduced with permission © The Royal Society of Chemistry.

Figure 6.17. SEM images of a pristine graphite flake (middle) and the results of mechanically processing it by sonication (left) and by knife milling (right) [110], reproduced with permission © American Chemical Society.

Pham *et al* employed conventional thermal heating and a microwave irradiation method to fabricate EGs from flake graphite. The prepared EGs presented a larger surface area and micropore volume, which created a higher adsorption capacity for heavy oils than that of the conventional thermal heating method [112]. Hou *et al* used a binary system of H_2SO_4 and H_2O_2 to prepare graphite-based sorbents and flexible materials in which EG is exfoliated at room temperature, as shown in figure 6.18(a) [116]. After several seconds, oil floating on the surface of water was completely adsorbed into an EG block (figure 6.18(b)). As shown in figure 6.18(c), the sorption capacities of the EG block for diesel, kerosene, and engine oil were 47 g/g, 49 g/g, and 63 g/g, respectively. Moreover, the sorption capacity for kerosene

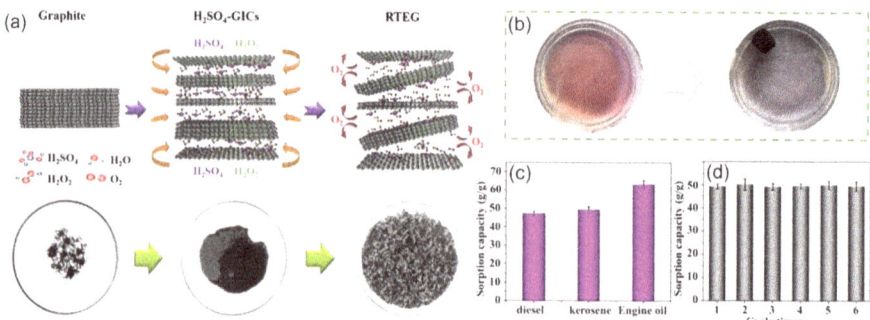

Figure 6.18. (a) Schematic illustration of the preparation process of exfoliated graphite. (b) The EG block adsorbs oil floating on water. (c) The sorption capacity of the EG block for diesel, kerosene, and engine oil. (d) The sorption capacity after several cycles [116], reproduced with permission © Elsevier.

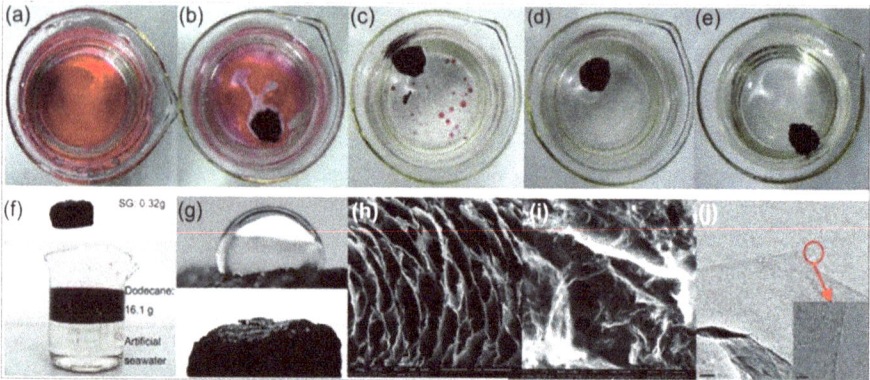

Figure 6.19. (a–e) Absorption of dodecane in SG at intervals of 20 s. (f) Efficiency of oil absorption. (g) Contact angles of SG surface (upper panel) and fast absorption of dodecane (lower panel). (h) SEM image of SG. The scale bar represents 1 mm. (i) SEM image of the graphene skeleton. Scale bar: 1 μm. (j) TEM image of the graphene skeleton. The scale bars of figure 6.20(j) and the inset of figure 6.20(j) are 50 nm and 5 nm, respectively [119], reproduced with permission © John Wiley & Sons.

was maintained at 50 g/g during at least 6 sorption/desorption cycles (figure 6.18(d)).

In addition to EG, the graphite-derived material graphene oxide (GO) is also considered to be a potential candidate for the removal of heavy metal ions from aqueous solutions [117, 118]. For example, spongy graphene (SG) was made by reducing a suspension of GO platelets followed by moulding [119]. The progress of absorption is shown in a series of photos (figures 6.19(a)–(e)); 0.32 g SG appears as a dark sponge (figure 6.19(f)) with a density of 12 ± 5 mg cm^{-3}, which can quickly adsorb dodecane (figure 6.19(g)). The outstanding results are attributed to the unique porous, wrinkled, and 'needle-like' structure of SG, as verified by SEM (figures 6.19(h) and (i)) and transmission electron microscopy (TEM) (figure 6.19(j)).

6.4.3 Synthesis of graphene

Since the discovery of graphene in 2004 [123], a large number of graphene-based materials have been developed for various applications. This is mainly due to the intrinsic extraordinary properties of graphene, such as its large theoretical surface area, high electrical conductivity, and excellent thermal/chemical stability [124–129]. The preparation of graphene can be realized by exfoliating it mechanically or chemically (or both), which needs to overcome the Van der Waals forces between the graphitic layers [130, 131]. The first successful method to produce single-layer or few-layer graphene was the mechanical exfoliation of graphite by Scotch tape [123]. However, the yield of graphene produced by this method is extremely low and the process is difficult to control. Ultrasound-assisted liquid-phase exfoliation is a common method. Due to the low interlayer pressure of graphite, however, it is difficult to exfoliate. Moreover, some solvents need to be added to assist with exfoliation. At present, the commonly used intercalation solvents are mainly organic solvents and surfactants [120]. Figures 6.20(a) and (b) show a schematic of the preparation of graphene via ultrasound-assisted Poly[9-(heptadecane-9-yl)-9H-carbazole] (PCz) exfoliation. It can be seen that graphene exhibits a typical colloidal dispersion in a PCz/toluene solution (figure 6.20(c)). Although this method can significantly improve the yield of graphene, it can also cause the destruction of graphene [132, 133]. Furthermore, ball milling can be used to exfoliate graphite platelets into graphene in a liquid medium (e.g. N,N-dimethylformamide) [134]. For example, Lin *et al* prepared few-layer graphene using a plasma-assisted ball-milling method with carbide, nitride, or oxides as the balling media (figure 6.20(d)) [121].

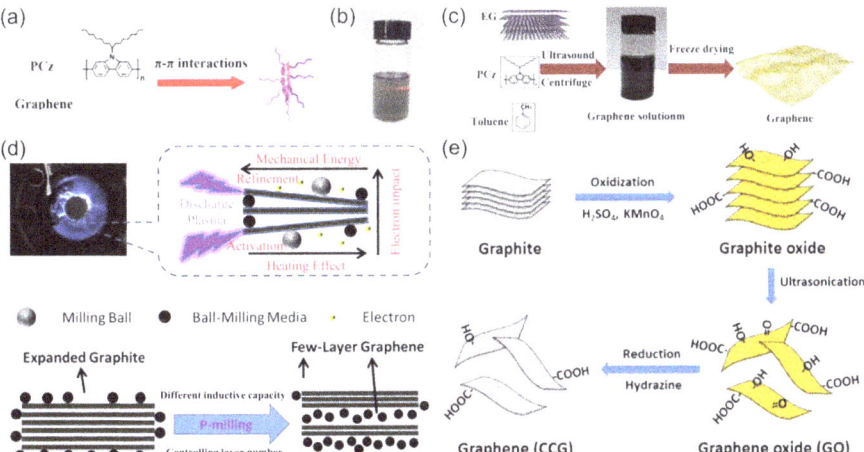

Figure 6.20. (a) Scheme of the principle of the interaction between Poly[9-(heptadecane-9-yl)–9H-carbazole] (PCz) and graphene. (b) The Tyndall effect of graphene dispersion. (c) Ultrasound-assisted PCz exfoliation process for graphene [120], reproduced with permission © Elsevier. (d) Scheme of the synthesis process of few-layer graphene by the plasma-assisted ball-milling process [121], reproduced with permission © Elsevier. (e) Scheme of Hummer's synthesis of GO and reduced GO [122], reproduced with permission © Elsevier.

To investigate the ball-milling technique, Xu *et al* prepared a series of edge-selective functionalized graphene nanoplatelets (XGnPs, X = F, Cl, Br, I, etc.) by ball milling graphite in the presence of various heteroatom-containing substances [124, 127, 135–139]. The performances of XGnPs as anode materials for LIBs have been studied. The XGnPs benefited from more electronegative X groups than carbon (C) and widened graphitic edges ($d_{X-X} > d_{C-C}$) and displayed better LIB performance than pristine graphite. Like many carbon-based materials, however, XGnPs still suffer in that a large proportion of their capacity is only available at high voltage (>1 V versus Li^+/Li), compared to pristine graphite (<0.2 V). This is because XGnPs possess sites that are different electronically and geometrically, and are non-equivalent to those in pristine graphite. As shown in figure 6.21(a), edge-thionic-acid (TA)-functionalized graphene nanoplatelets (GnPs) were also prepared using the

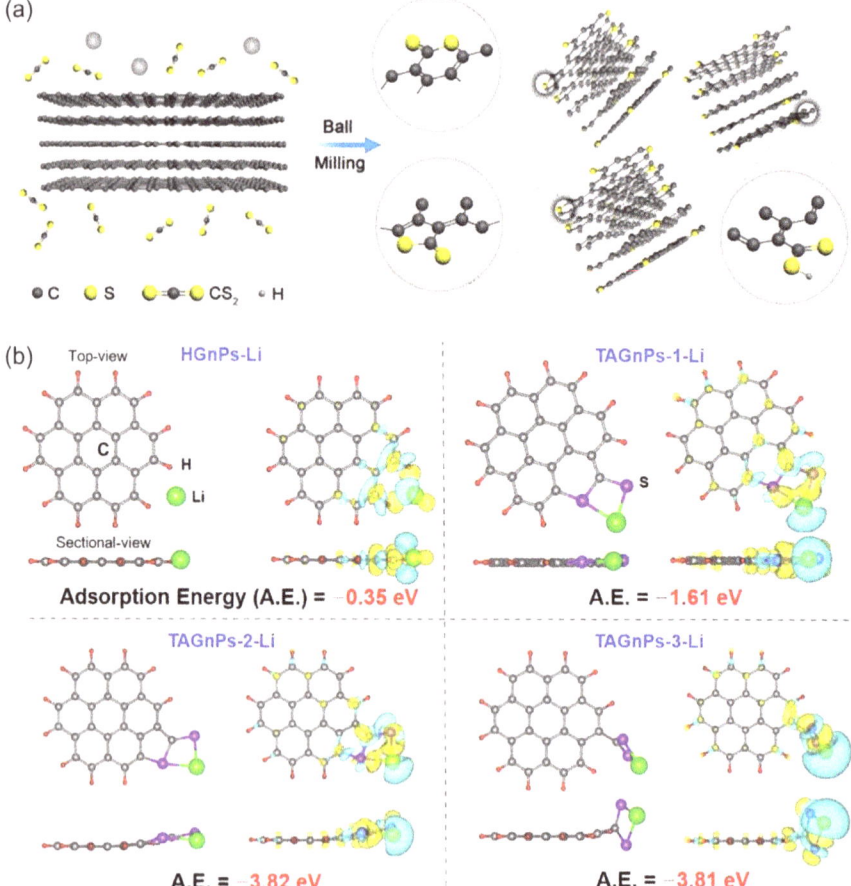

Figure 6.21. (a) Schematic showing the ball milling of TAGnPs. (b) Optimized configurations (left) and the corresponding charge density difference isosurfaces (right) of the active materials interacting with one Li atom. The active materials: (upper left) reference HGnPs, (upper right) TAGnPs-1, (lower left) TAGnPs-2, and (lower right) TAGnPs-3 [135], reproduced with permission © Elsevier.

similar ball-milling method, which not only exhibited electrochemical behavior similar to that of pristine graphite, with a low average working voltage (<0.5 V), but also a superior rate capability (>0.5 A g^{-1}) [135]. In particular, when TAGnPs were tested at 0.5, 1, 2, and 5 A g^{-1} in the voltage range of 0.02–1 V, they exhibited high average reversible capacities of 228.3, 208.1, 141.0, and 80.6 mAh g^{-1}, respectively. This result can be attributed to their relatively increased surface area, well-preserved 'graphite-like' layered structure, and edge functionalization (specifically, –C = S) with high lithium adsorption (figure 6.21(b)). These unique structural features of TAGnPs provide an efficient accessible surface area, increased electronic/ionic conductivity, strong adsorption, and effective storage of Li$^+$. Generally, the number of graphene layers can be controlled by varying the ball-milling media. To increase the supply of graphene sheets and facilitate their application and study, mechanical peel-off techniques still need to be further investigated [134].

Compared with physical synthesis, chemical synthesis seems to be much more attractive. The chemical synthesis of graphene mainly uses a modified version of Hummer's method to prepare GO as a precursor. This method includes three steps: typically, the graphite is first treated by oxidizing agents (e.g. sulfuric and nitric acid) to prepare GO, followed by exfoliation into individual GO sheets by sonication, and finally by reduction to form so-called chemically-converted graphene (CCG) [140, 141]. Figure 6.20(e) shows the typical procedures used [122]. As this graphene is obtained from graphite as a raw material, its wide application is discussed in detail in another chapter.

6.4.4 Nuclear materials

Flake graphite and microcrystalline graphite are important materials used in gas-cooled high-temperature reactors (HTRs) due to their high thermal stability, high compressive strength, easy fabricability, and low cost [142]. The HTR-10 is a pebble-bed-type high-temperature gas-cooled reactor. As shown in figure 6.22, spherical fuel elements are used in the pebble-bed core of the HTR-10 [143]. Natural graphite

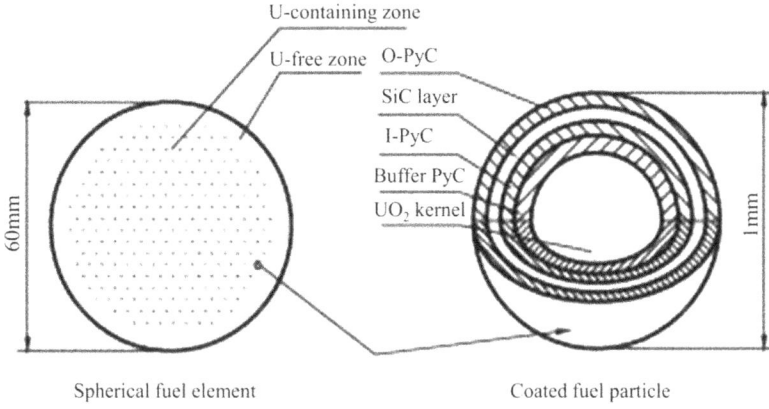

Figure 6.22. The HTR-10 spherical fuel element [143], reproduced with permission © Elsevier.

along with electrographite and resin are used to manufacture the graphite matrix, which serves as a neutron moderator as well as a heat conductor that transmits the power generated to the coolant. The graphite matrix of a spherical fuel element utilizes ~64 wt% natural flake graphite, 16 wt% artificial graphite, and 20 wt% phenol resin binder. With these fuel elements, the reactor has been successfully operated for 21 years [144, 145].

6.4.5 Refractory materials

Since the introduction of carbon/graphite into oxide-based castables with enhanced properties in the 1970s, graphite has been widely adopted in the production of refractory materials, including firebricks, crucibles, continuous casting powder, mold cores, mold detergents, and high-resistance materials [146]. This is mainly due to its unique physical properties of non-wettability by molten metal/slag, high thermal conductivity, low thermal expansion coefficient, low elastic modulus, and thermal shock resistance [147]. Refractories constructed using carbon composite/oxide–graphite bricks for the production of steel, glass, and ceramics can have long service lives with excellent thermal stability and remarkable slag resistance [148–150]. It should be noted that the high carbon contents of bricks made for refractories can also result in several drawbacks [151, 152]. Typically, magnesia–carbon (MgO–C) refractories containing ~12%–18% total carbon are widely used in basic oxygen furnaces, electric arc furnaces, and steel ladles [153]. Moreover, MgO–C refractories with low carbon contents have been developed [152, 154]. For example, Li *et al* developed MgO–C refractories containing 6% carbon and basic slag (CaO/SiO$_2$) for an induction furnace and a resistance furnace, respectively (figure 6.23) [155]. An electromagnetic field (EMF) can result the formation of spinel-type MgFe$_2$O$_4$ with a little Mn ions replacing Mg ions, which increases the anti-oxidative property of graphite and decreases the penetration of the slag. At the same time, the EMF

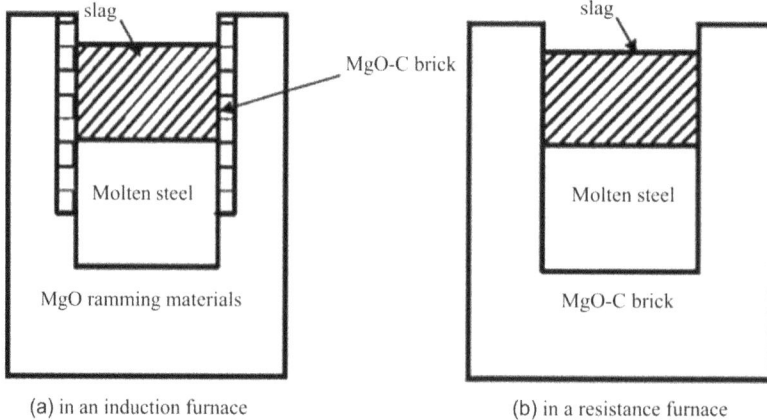

(a) in an induction furnace   (b) in a resistance furnace

Figure 6.23. Schematic diagram of a refractory for an induction furnace and (b) a refractory for a resistance furnace [155], reproduced with permission © Elsevier.

increases both the wettability and the collision frequency between the slag and MgO–C refractories, leading to more serious corrosion of the MgO–C refractories.

6.4.6 Aerospace materials

Owing to the remarkable features of graphite, such as its thermal, electrical, and mechanical features, polymer–graphite composites have been utilized in structural and aerospace applications [156]. For example, electrically conductive polymeric composites are employed as temperature-dependent resistors, heating elements, sensors, switching devices, self-limiting electrical heaters, and antistatic materials for the electromagnetic interference shielding of electronic devices [157]. The presence of graphite particles in composites can not only significantly influence their electrical and thermal conductivity, but also effectively change the mechanical properties of polymers [158–161]. Moreover, both the Young's modulus and thermal conductivity of polyethylene/graphite composites can be increased by increasing the graphite content [162].

6.4.7 Other applications

Graphite is anisotropic and possess excellent thermal and electrical conductivities. As a result of anisotropy, layers of carbon may slide easily with respect to one another, thus making graphite a good pencil material and lubricant [163]. In addition, highly conductive graphite composites can be also used for electromagnetic interference shielding [164, 165]. Graphitic materials are also helpful in fields such as paint production, grinding wheels, and coatings [1]. Moreover, graphite is also widely used as an engineering material for various applications, such as piston rings, thrust bearings, journal bearings, and vanes.

6.5 Conclusions

Graphite is a kind of non-metallic mineral with unique physical and chemical properties, including excellent thermal and electrical conductivities, resistance to a wide range of temperatures, high radiation resistance, high corrosion resistance, and excellent lubrication. As an indispensable strategic mineral resource in the national economy, graphite has been widely used in various fields, including batteries, graphene synthesis, refractory materials, nuclear applications, water purification, aerospace, and other fields. Graphite ore is mostly concentrated from crushed rocks using flotation separation techniques. The purification of graphite is the basis for the preparation of all graphitic materials. The purity of graphite largely determines the properties of further processed products. The higher the purity of graphite, the higher the application value. Moreover, natural graphite is not only an important strategic resource but also a non-renewable mineral resource. The key to realizing the high-value development and utilization of graphite is to make good use of graphite with desirable and controllable physicochemical properties according to the corresponding application, a particular example being applications in energy-related systems (e.g. LIBs).

Acknowledgements

The authors are grateful for financial support from the Pearl River Talent Recruitment Program (2019QN01L096), the Guangdong Innovative and Entrepreneurial Research Team Program (2019ZT08L075), the Guangdong Science and Technology Program (2020B121201003), and the 'Young Talent Fellowship' Program of South China University of Technology.

References

[1] Jara A D, Betemariam A, Woldetinsae G and Kim J Y 2019 *Int. J. Mining Sci. Tech.* **29** 671–89

[2] Kelly B T 1981 *Physics of Graphite* (London: Applied Science Publishers)

[3] Kavanagh A and Schlögl R 1988 *Carbon* **26** 23–32

[4] Sutphin D M and Bliss J D 1990 *CIM Bull.* **83** 85–9

[5] Wang X, Li H, Yao H, Zhu D and Liu N 2018 *Resour. Policy* **59** 200–9

[6] Wang X, Li H, Yao H, Chen Z and Guan Q 2019 *Resour. Policy* **60** 153–61

[7] Rui X, Geng Y, Sun X, Hao H and Xiao S 2021 *Resour. Conserv. Recycl.* **173** 105732

[8] Hund D L P K, Fabregas T P, Laing T and Drexhage J 2020 *Minerals for Climate Action: The Mineral Intensity of the Clean Energy Transition* (World Bank)

[9] Bonijoly M, Oberlin M and Oberlin A 1982 *Int. J. Coal Geol.* **1** 283-312

[10] Buseck P R and Huang B-J 1985 *Geochim. Cosmochim. Acta* **49** 2003–16

[11] Luque F J, Huizenga J M, Crespo-Feo E, Wada H, Ortega L and Barrenechea J F 2014 *Miner. Deposita* **49** 261–77

[12] Barma S D, Baskey P K, Rao D S and Sahu S N 2019 *Ultrason. Sonochem.* **56** 386–96

[13] Tamashausky A 1998 *Am. Ceram. Soc. Bull.* **77** 102–4

[14] Wissler M 2006 *J. Power Sources* **156** 142–50

[15] Peng W, Wang C, Hu Y and Song S 2017 *J. Dispersion Sci. Technol.* **38** 889–94

[16] Wang X, Bu X, Ni C, Zhou S, Yang X, Zhang J, Alheshibri M, Peng Y and Xie G 2021 *Miner. Eng.* **163** 106766

[17] Shen K, Huang Z-H, Hu K, Shen W, Yu S, Yang J, Yang G and Kang F 2015 *Carbon* **90** 197–206

[18] Rumble O B D 2014 *Elements* **10** 415–20

[19] Nakajima Y M T 1994 *Carbon* **32** 469–75

[20] Sun K, Qiu Y and Zhang L 2017 *Minerals* **7** 115

[21] Al-Ani T, Leinonen S, Ahtola T and Salvador D 2020 *Minerals* **10** 680

[22] Inagaki M 2001 *Graphite and Precursors* (Boca Raton, FL: CRC Press) pp 179–98

[23] Oberlin A and Terriere G 1975 *Carbon* **13** 367–76

[24] Ishii T, Kaburagi Y, Yoshida A, Hishiyama Y, Oka H, Setoyama N, Ozaki J-i and Kyotani T 2017 *Carbon* **125** 146–55

[25] Li H, Feng Q, Yang S, Ou L and Lu Y 2014 *Int. J. Miner. Process.* **127** 1–9

[26] Vasumathi N, Vijaya Kumar T V, Ratchambigai S, Subba S, Rao and Bhaskar Raju G 2015 *Int. J. Mining Sci. Tech.* **25** 415–20

[27] Chehreh Chelgani S, Rudolph M, Kratzsch R, Sandmann D and Gutzmer J 2016 *Miner. Process. Extr. Metall. Rev.* **37** 58–68

[28] Salver-Disma F, Tarascon J M, Clinard C and Rouzaud J N 1999 *Carbon* **37** 1941–59

[29] Yue C 2002 *Non-Met. Mines* **25** 36–7

[30] Chen X and Peng Y 2015 *Miner. Eng.* **83** 33–43

[31] Roufail R and Klein B 2010 *Can. Metall. Q.* **49** 419–28

[32] Ong T S and Yang H 2000 *Carbon* **38** 2077–85

[33] Lu X and Forssberg E 2001 *Miner. Eng.* **14** 1541–3

[34] Peng W, Qiu Y, Zhang L, Guan J and Song S 2017 *Minerals* **7** 208

[35] Bu X, Zhang T, Peng Y, Xie G and Wu E 2018 *Minerals* **8** 15

[36] Wakamatsu T and Numata Y 1991 *Miner. Eng.* **4** 975–82

[37] Veras M M, Baltar C A M, Paulo J B D A and Leite J Y P 2014 *Rem: Revista Escola de Minas* **67** 87–92

[38] Park J G and Dodd D S 1994 *Miner. Eng.* **7** 371–87

[39] Patil M R, Shivakumar K S, Rudramuniyappa M V and Bhima R R 2000 *Metall. Mater. Sci.* **42** 233–41

[40] Kaya O and Canbazoglu M 2007 *J. Ore Dressing* **9** 40–4

[41] Crozier R D 1992 *Flotation: Theory, Reagents and Ore Testing* (Oxford: Pergamon)

[42] Pugh R J 2000 *Miner. Eng.* **13** 151–62

[43] Jara A D, Woldetinsae G, Betemariam A and Kim J Y 2020 *Int. J. Mining Sci. Tech.* **30** 715–21

[44] Liu H Q, Xie Y Z, Li Y and Lin Y 2000 *Carbon Tech.* **1** 12–14

[45] Wang H, Feng Q, Tang X and Liu K 2016 *Sep. Sci. Technol.* **51** 2465–72

[46] Bhima Rao R and Patnaik N 2004 *Miner. Process. Extr. Metall. Rev.* **33** 257–60

[47] Jara A D and Kim J Y 2020 *Mater. Today Commun.* **25** 101437

[48] Wang H, Feng Q, Liu K, Zuo K and Tang X 2018 *Sep. Sci. Technol.* **53** 982–9

[49] Mustika D, Torowati S, Fisli A, Joni I M, Langenati R and Setiawan J 2019 *Int. J. Chem.* **11** 9–17

[50] Kaya Ö and Canbazoğlu M 2009 *Mining Metall. Explor.* **26** 158–62

[51] Shen K, Chen X, Shen W, Huang Z-H, Liu B and Kang F 2021 *Carbon* **173** 769–81

[52] Hu X-l, Tang X, Zhou Y-b, Dai Y and Huang Q-z 2017 *Carbon* **114** 753–4

[53] Xu J, Dou Y, Wei Z, Ma J, Deng Y, Li Y, Liu H and Dou S 2017 *Adv. Sci.* **4** 201700146

[54] Zhang M, Song X, Ou X and Tang Y 2019 *Energy Stor. Mater.* **16** 65–84

[55] Metrot A and Fischer J E 1981 *Synth. Met.* **3** 201–7

[56] Inagaki M, Iwashita N and Kouno E 1990 *Carbon* **28** 49–55

[57] Liu T, Zhang R, Zhang X, Liu K, Liu Y and Yan P 2017 *Carbon* **119** 544–7

[58] Yu H, Gao J, Chen Y and Zhao Y 2016 *J. Therm. Anal. Calorim.* **124** 87–92

[59] Yang Y, Pang Y, Liu Y and Guo H 2018 *Mater. Lett.* **216** 220–3

[60] Zheng W, Lu X and Wong S -C 2004 *Appl. Polym. Sci.* **91** 2781–8

[61] Chen X, Xiao F, Lei Y, Lu H, Zhang J, Yan M and Xu J 2021 *J. Energy Chem.* **59** 292–8

[62] Ma C-L, Hu Z-H, Song N-J, Zhao Y, Liu Y-Z and Wang H-Q 2021 *Rare Met.* **40** 837–47

[63] Wen Y, He K, Zhu Y, Han F, Xu Y, Matsuda I, Ishii Y, Cumings J and Wang C 2014 *Nat. Commun.* **5** 4033

[64] Xiao F, Chen X, Zhang J, Huang C, Hu T, Hong B and Xu J 2020 *J. Energy Chem.* **48** 122–7

[65] Yang X-X, Zhang G-J, Bai B-S, Li Y, Li Y-X, Yang Y, Jian X and Wang X-W 2021 *Rare Met.* **40** 1708–18

[66] Ma X, Zhang Z-J, Wang J-M, Sun S-H, Zhang S-F, Yuan S, Qiao Z-J, Yu Z-Y, Kang J-L and Li W-J 2021 *Rare Met.* **40** 2802–9

[67] Xu Q *et al* 2020 *Energy Stor. Mater.* **26** 73–82

[68] Huang H-F, Gui Y-N, Sun F, Liu Z-J, Ning H-L, Wu C and Chen L-B 2021 *Rare Met.* **40** 3494–500

[69] Fu L J, Liu H, Li C, Wu Y P, Rahm E, Holze R and Wu H Q 2006 *Solid State Sci.* **8** 113–28
[70] Yoshio M, Wang H, Fukuda K, Umeno T, Abe T and Ogumi Z 2004 *J. Mater. Chem.* **14** 1754–8
[71] Tarascon J M and Armand M 2001 *Nature* **414** 359–67
[72] Yoshio M, Wang H, Fukuda K, Hara Y and Adachi Y 2000 *J. Electrochem. Soc.* **147** 1245
[73] Wang H and Yoshio M 2001 *J. Power Sources* **93** 123–9
[74] Wang H, Yoshio M, Abe T and Ogumi Z 2002 *J. Electrochem. Soc.* **149** A499
[75] Yoshio M, Wang H and Fukuda K 2003 *Angew. Chem., Int. Ed. Engl.* **42** 4203–6
[76] Wu Y P, Rahm E and Holze R 2003 *J. Power Sources* **114** 228–36
[77] Wu Y P, Jiang C, Wan C and Holze R 2002 *J. Power Sources* **111** 329–34
[78] Kim S-S, Kadoma Y, Ikuta H, Uchimoto Y and Wakihara M 2001 *Electrochem. Solid-State Lett.* **4** A109
[79] Lee Y T, Yoon C S and Sun Y-K 2005 *J. Power Sources* **139** 230–4
[80] Veeraraghavan B, Paul J, Haran B and Popov B 2002 *J. Power Sources* **109** 377–87
[81] Wang H, Umeno T, Mizuma K and Yoshio M 2008 *J. Power Sources* **175** 886–90
[82] Zhang Y, Zhang X G, Zhang H L, Zhao Z G, Li F, Liu C and Cheng H M 2006 *Electrochim. Acta* **51** 4994–5000
[83] Zou L, Lv R, Kang F, Gan L and Shen W 2008 *J. Power Sources* **184** 566–9
[84] Shi M, Song C, Tai Z, Zou K, Duan Y, Dai X, Sun J, Chen Y and Liu Y 2021 *Fuel* **292** 120250
[85] Cameán I, Lavela P, Tirado J L and García A B 2010 *Fuel* **89** 986–91
[86] Ōya A and Ōtani S 1979 *Carbon* **17** 131–7
[87] Ōya A and Marsh H 1982 *J. Mater. Sci.* **17** 309–22
[88] Hu C, Yu C, Li M, Wang X, Dong Q, Wang G and Qiu J 2015 *Chem. Commun.* **51** 3419–22
[89] Zhang Y, Zhang K, Ren S, Jia K, Dang Y, Liu G, Li K, Long X and Qiu J 2019 *J. Alloys Compd.* **792** 828–34
[90] Li H, Zhang H, Li K, Zhang J, Sun M and Su B 2020 *Fuel* **279** 118531
[91] See-How N, Jiazhao W, David W, Konstantin K, Zai-Ping G and Hua-Kun L 2006 *Angew. Chem. Int. Ed.* **45** 6896–9
[92] Chan C K, Peng H, Liu G, McIlwrath K, Zhang X F, Huggins R A and Cui Y 2007 *Nat. Nanotechnol.* **3** 31
[93] Ke C-Z, Liu F, Zheng Z-M, Zhang H-H, Cai M-T, Li M, Yan Q-Z, Chen H-X and Zhang Q-B 2021 *Rare Met.* **40** 1347–56
[94] Zhang F-Z, Ma Y-Y, Jiang M-M, Luo W and Yang J-P 2021 *Rare Met.* **41** 1276–83
[95] McDowell M T, Lee S W, Harris J T, Korgel B A, Wang C, Nix W D and Cui Y 2013 *Nano Lett.* **13** 758–64
[96] Chae S, Choi S-H, Kim N, Sung J and Cho J 2020 *Angew. Chem., Int. Ed. Engl.* **9** 110–35
[97] Wang N, Liu Y-Y, Shi Z-X, Yu Z-L, Duan H-Y, Fang S, Yang J-Y and Wang X-M 2021 *Rare Met.* **41** 438–47
[98] Du Z, Dunlap R A and Obrovac M N 2014 *J. Electrochem. Soc.* **161** A1698–705
[99] Chevrier V L, Liu L, Le D B, Lund J, Molla B, Reimer K, Krause L J, Jensen L D, Figgemeier E and Eberman K W 2014 *J. Electrochem. Soc.* **161** A783–91
[100] Obrovac M N and Chevrier V L 2014 *Chem. Rev.* **114** 11444–502
[101] Chae S, Kim N, Ma J, Cho J and Ko M 2017 *Adv. Energy Mater.* **7** 1700071
[102] Zhu S, Zhou J, Guan Y, Cai W, Zhao Y, Zhu Y, Zhu L, Zhu Y and Qian Y 2018 *Small* **14** 1802457

[103] Li P, Hwang J-Y and Sun Y-K 2019 *ACS Nano* **13** 2624–33

[104] Zhang L, Wang H, Zhang X and Tang Y 2021 *Adv. Funct. Mater.* **31** 2010958

[105] Cui C, Wei Z, Xu J, Zhang Y, Liu S, Liu H, Mao M, Wang S, Ma J and Dou S 2018 *Energy Stor. Mater.* **15** 22–30

[106] Heckmann A, Fromm O, Rodehorst U, Münster P, Winter M and Placke T 2018 *Carbon* **131** 201–12

[107] Wang M and Tang Y 2018 *Adv. Energy Mater.* **8** 1703320

[108] Wang J, Tu J, Lei H and Zhu H 2019 *RSC Adv.* **9** 38990–7

[109] Placke T, Rothermel S, Fromm O, Meister P, Lux S F, Huesker J, Meyer H-W and Winter M 2013 *J. Electrochem. Soc.* **160** A1979–91

[110] Kravchyk K V, Wang S, Piveteau L and Kovalenko M V 2017 *Chem. Mater.* **29** 4484–92

[111] Zheng Y P, Wang H N, Kang F Y, Wang L N and Inagaki M 2004 *Carbon* **42** 2603–7

[112] Pham T V, Nguyen T T, Nguyen D T, Thuan T V, Bui P Q T, Viet V N D and Bach L G 2019 *J. Nanosci. Nanotechnol.* **19** 1122–5

[113] Chung D D L 1987 *J. Mater. Sci.* **22** 4190–8

[114] Tryba B, Przepiorski J and Morawski A W 2003 *Carbon* **41** 2013–6

[115] Wei X H, Liu L, Zhang J X, Shi J L and Guo Q G 2009 *Mater. Lett.* **63** 1618–20

[116] Hou S, He S, Zhu T, Li J, Ma L, Du H, Shen W, Kang F and Huang Z-H 2021 *J. Materiomics* **7** 136–45

[117] Cao Y and Li X B 2014 *Adsorption J. Int. Adsorption Soc.* **20** 713–27

[118] Fadlalla M I, Kumar P S, Selvam V and Babu S G 2020 *J. Mater. Sci.* **55** 7156–83

[119] Bi H, Xie X, Yin K, Zhou Y, Wan S, He L, Xu F, Banhart F, Sun L and Ruoff R S 2012 *Adv. Funct. Mater.* **22** 4421–5

[120] Gu X, Zhao Y, Sun K, Vieira C L Z, Jia Z, Cui C, Wang Z, Walsh A and Huang S 2019 *Ultrason. Sonochem.* **58** 104630

[121] Lin C, Yang L, Ouyang L, Liu J, Wang H and Zhu M 2017 *J. Alloys Compd.* **728** 578–84

[122] Liu Q, Shi J and Jiang G 2012 *Trends Anal. Chem.* **37** 1–11

[123] Novoselov K S 2004 *Science* **306** 666–9

[124] Xu J T, Jeon I Y, Seo J M, Dou S X, Dai L M and Baek J B 2014 *Adv. Mater.* **26** 7317–23

[125] Xu J T, Wang M, Wickramaratne N P, Jaroniec M, Dou S X and Dai L M 2015 *Adv. Mater.* **27** 2042–8

[126] Xu J T, Lin Y, Connell J W and Dai L M 2015 *Small* **11** 6179–85

[127] Jeon I Y *et al* 2015 *Adv. Funct. Mater.* **25** 1170–9

[128] Zhang J, Chen X, Lei Y, Lu H, Xu J, Wang S, Yan M, Xiao F and Xu J 2022 *Chem. Eng. J.* **428** 131025

[129] Xiao F *et al* 2021 *Energy Storage Mater.* **41** 61–8

[130] Niyogi S, Bekyarova E, Itkis M E, McWilliams J L, Hamon M A and Haddon R C 2006 *J. Am. Chem. Soc.* **128** 7720–1

[131] Yan Y, Meng Y, Zhao H, Lester E, Wu T and Pang C H 2021 *Bioresour. Technol.* **331** 124934

[132] Hernandez Y *et al* 2008 *Nat. Nanotechnol.* **3** 563–68

[133] Li X, Wang X, Zhang L, Lee S and Dai H 2008 *Science* **319** 1229

[134] Zhao W, Fang M, Wu F, Wu H, Wang L and Chen G 2010 *J. Mater. Chem.* **20** 5817–9

[135] Fan Q, Noh H-J, Wei Z, Zhang J, Lian X, Ma J, Jung S-M, Jeon I-Y, Xu J and Baek J-B 2019 *Nano Energy* **62** 419–25

[136] Xu J T, Shui J L, Wang J L, Wang M, Liu H K, Dou S X, Jeon I Y, Seo J M, Baek J B and Dai L M 2014 *ACS Nano* **8** 10920–30

[137] Xu J, Jeon I-Y, Choi H-J, Kim S-J, Shin S-H, Park N, Dai L and Baek J-B 2016 *2D Mater.* **4** 014002

[138] Xu J T, Jeon I Y, Ma J M, Dou Y H, Kim S J, Seo J M, Liu H K, Dou S X, Baek J B and Dai L M 2017 *Nano Res.* **10** 1268–81

[139] Noh H-J, Liu S, Yu S-Y, Fan Q, Xiao F, Xu J, Jeon I-Y and Baek J-B 2020 *Batteries Supercaps* **3** 928–35

[140] Hummers W and Offeman R 1958 *J. Am. Chem. Soc.* **80** 1339

[141] Park S and Ruoff R S 2009 *Nat. Nanotechnol.* **4** 217–24

[142] William E W, Timothy D B and Robert L B 2008 The next generation nuclear plant graphite program, United States https://inldigitallibrary.inl.gov/sites/sti/sti/4074980.pdf

[143] Tang C, Tang Y, Zhu J, Zou Y, Li J and Ni X 2002 *Nucl. Eng. Des.* **218** 91–102

[144] Bäumer V -G E R 1990 *AVR – Experimental High-Temperature Reactor. 21 Years of Successful Operation for a Future Energy Technology* (Düsseldorf: VDI-Verlag)

[145] Zhao H, Liang T, Zhang J, He J, Zou Y and Tang C 2006 *Nucl. Eng. Des.* **236** 643–7

[146] Zhang S and Lee W E 2003 *J. Eur. Ceram. Soc.* **23** 1215–21

[147] Ewais E M M 2004 *J. Ceram. Soc. Jpn.* **112** 517–32

[148] Zhang S and Lee W E 2002 *Br. Ceram. Trans.* **101** 1–8

[149] Gokce A S, Gurcan C, Ozgen S and Aydin S 2008 *Ceram. Int.* **34** 323–30

[150] Wang X, Chen Y, Yu C, Ding J, Guo D, Deng C and Zhu H 2019 *J. Alloys Compd.* **788** 739–47

[151] Bag M, Adak S and Sarkar R 2012 *Ceram. Int.* **38** 2339–46

[152] Peng X, Li L and Peng D 2003 *Refractories* **37** 355–7

[153] Ewais E M M 2004 *J. Ceram. Soc. Jpn.* **112** 517–32

[154] Tamura S, Ochiai T, Takanaga S, Kanai T and Nakamura H 2003 *Proc. UNITECR 2003 Congress* pp 517–20

[155] Li X, Zhu B and Wang T 2012 *Ceram. Int.* **38** 2105–9

[156] Ezquerra T A, Kulescza M and Baltá-Calleja F J 1991 *Synth. Met.* **41** 915–20

[157] Klason C, Mcqueen D H and Kubát J 1996 *Macro. Symp.* **108** 247–60

[158] Bigg D M 1995 *Thermal and Electrical Conductivity of Polymer Materials* (Berlin: Springer) pp 1–30

[159] Krupa I and Chodak I 2001 *Eur. Polym. J.* **37** 2159–68

[160] Thongruang W, Spontak R J and Balik C M 2002 *Polymer* **43** 2279–86

[161] Thongruang W, Spontak R J and Balik C M 2002 *Polymer* **43** 3717–25

[162] Krupa I, Novák I and Chodák I 2004 *Synth. Met.* **145** 245–52

[163] Dresselhaus M S and Dresselhaus G 1994 *Molecular Crystals and Liquid Crystals Science and technology. Section A. Mol. Cryst. Liq. Cryst.* **244** 1–12

[164] Harris G, Lennhoff J, Nassif J, Vinciguerra M, Rose P, Jaworski D and Gaier J 2000 *SAMPE J.* **36** 59–63

[165] Kausar A, Rafique I and Muhammad B 2017 *Polym.-Plast. Technol. Eng.* **56** 1438–56

IOP Publishing

Functional Carbon Materials

Jianmin Ma and Jiantie Xu

Chapter 7

The structures, synthesis, properties, and applications of diamond

Sheng-Yi Xie and Fuyang Liu

Diamond is an allotrope of carbon, and in common with other carbon materials, it has many extraordinary properties including superhardness for cutting materials and a wide bandgap for next-generation semiconductors; it is the ideal platform for quantum information. In this chapter, we summarize the recent developments in the field of diamonds and briefly introduce the structures, preparation, exotic properties, and the potential applications of diamond.

7.1 Introduction

As far back as ancient times, natural diamond was one of the most precious treasures because of its scarcity and dazzling beauty. As civilization has advanced and science has developed, people have come to know much about diamond and can synthesize it in the laboratory. We now know that diamond is an allotrope of the element carbon arranged in a cubic lattice with sp^3 hybrid bonds. In addition to natural diamonds, researchers have also developed different methods for producing diamonds in laboratory, such as growing them using the high-pressure high-temperature (HPHT) method, the chemical vapor deposition (CVD) method, and the hydrothermal reaction method. With improvements in these processes, the quality of produced diamond may approach that of natural diamond. Both natural diamonds and synthesized diamonds have many superior properties that support a wide range of applications in modern industrial techniques or fundamental science research. High-quality transparent natural diamonds as well as colorful synthesized diamonds doped with impurities are still among the most popular jewels for women worldwide. Diamond is considered to be an ideal material for cutting tools. Due to its wide bandgap and extraordinary thermal conductivity, diamond is very competitive for applications in high-power electronic devices. Mover, the diamond anvil cell (DAC) is the instrument most commonly used to research high-pressure physics,

doi:10.1088/978-0-7503-4972-7ch7

chemistry, and geoscience. Nitrogen-vacancy (N_V) centers in diamond provide an increasingly favored platform for quantum information, quantum sensing, and single-photon sources.

In this chapter, we introduce recent progress in the diamond field, as diamond is a significant carbon material. First, we summarize the atomic structures as well as the synthesis methods of diamond. We then also discuss novel properties and potential applications based on these properties of diamond. Finally, the outlook for diamond is also presented.

7.2 The structures of diamonds

Since diamond was first synthesized in the laboratory in 1955, research into this material has never stopped. Diamond is described as a solid form of the element carbon with its atoms arranged in a crystal structure called 'cubic diamond' in many dictionaries. The structure of diamond is not only limited to the cubic structure, but also extends to other structures, such as hexagonally structured diamond, nano-twinned diamond, and triclinic diamond, whose structure has not been determined to date [14, 44]. In addition, from the perspective of properties, researchers not only pay attention to the high hardness of diamond, but also are interested in its high thermal conductivity (TC), low coefficient of thermal expansion (CTE), high corrosion resistance, low density, and other properties [1–5]. Therefore, it is unwise to confine diamond to the framework of cubic structure and ultrahigh hardness. In this chapter, diamond is broadly defined as the carbon material formed wholly or partly by the sp^3 hybridization of the C–C bond. This broad definition includes diamonds in the traditional sense, as well as some diamond materials that have been prepared or are still at the stage of theoretical prediction.

7.2.1 Cubic diamond

Diamond is a typical atomic crystal, in which the basic structural particles are carbon atoms [6–9]. Each carbon atom forms a covalent single bond with four carbon atoms in an sp^3 hybrid orbital with a bond length of 1.55 Å and a bond angle of 109°28′; the carbon atoms together form a regular tetrahedron. The lattice of cubic diamond is a face-centered cubic (FCC) lattice with a lattice constant of $a = 3.5667$ Å and a space group of Fd-3m [9–11]. One carbon atom in diamond is located at the center of the tetrahedron formed by the surrounding four carbon atoms which are located at the four vertices. There are three main crystal planes (110), (100), (111) in an octahedral crystal of diamond, as shown in figure 7.1. The relative densities of the covalent bonds in these three planes are: (110) plane:(100) plane:(111) plane = 1.22:1.73:3, which means that the (110) plane is the most easily lapped, while the (111) plane is the hardest [12]. Depending on the direction, the stresses calculated by Whitlock and Ruoff in 1981 are 52, 53, and 98 GPa in the (111), (110), and (100) planes, which suggests that the (111) plane is the cleavage plane of cubic diamond [13].

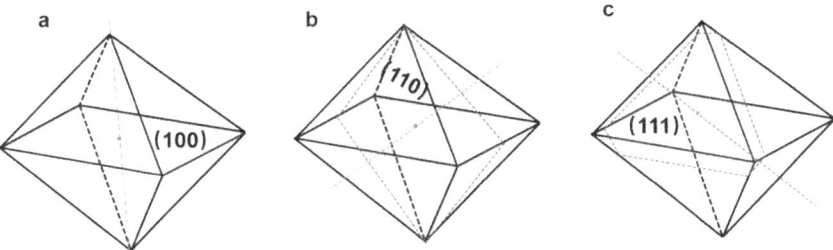

Figure 7.1. Schematic diagram of the three main crystal planes of diamond: (a) the (100), (b) (110), and (c) (111) planes.

7.2.2 Hexagonal diamond

Hexagonal diamond, also called lonsdaleite, was first found in nature in Canyon Diablo iron and Goalpara ureilite about 50 years ago [14, 15]. Lonsdaleite was reported to be a kind of carbon material with a wurtzite (ZnS)-type structure, a space group of $P6_3/mmc$, and the lattice parameters $a = 2.51$ Å and $c = 4.12$ Å [16, 17]. Figure 7.2 shows the structural schematics of hexagonal diamond and cubic diamond [16, 18]. The C–C bond length details can be found in figure 7.2(a). As a mineral, lonsdaleite is also expected to have extremely high hardness and rigidity [19–22]. But so far, it seems that when this mineral is discovered, it is accompanied by olivine, graphite, and cubic diamond [23–25]. Therefore, in the x-ray diffraction pattern, the characteristic diffraction peaks of hexagonal diamond appear in the form of cubic diamond shoulder peaks [16, 26]. A shoulder peak corresponding to a d-spacing of 2.18 Å indicates the (111) reflection of hexagonal diamond. Unfortunately, so far, we have not found any articles on the synthesis of high-quality hexagonal diamond. Pure hexagonal diamond in natural samples has not been reported [27–29]. At present, reports on the structure of hexagonal diamond are based on its powder x-ray diffraction (XRD) pattern [19, 30, 31]. Poor-resolution peaks at 0.218, 0.193, 0.151, and 0.116 nm were indexed as the hexagonal structure [32]. Based on the above experimental results, a new understanding of hexagonal diamond has been reported: hexagonal diamond is faulty and twisted cubic diamond. Scanning transmission electron microscope (STEM) images show that lonsdaleite is composed of a large number of twins and {111} stacked cubic diamond [11, 33]. Figure 7.3 shows a STEM image of a Canyon Diablo sample [32]. A structural model of the region marked with white corners is shown in figure 7.3(a). The schematic diagram in figure 7.3(d) shows that the crystal plane spacing of cubic diamond following dislocation is consistent with the reported characteristic crystal plane spacing of hexagonal diamond. Therefore, it is possible that the twins and dislocations of cubic diamond have led to the mistaken belief that hexagonal diamond exists, as shown in figures 7.3(c) and (d) [32]. Although the report on hexagonal diamond is still controversial, it seems to inspire people to anticipate this inconclusive diamond [28, 34, 35].

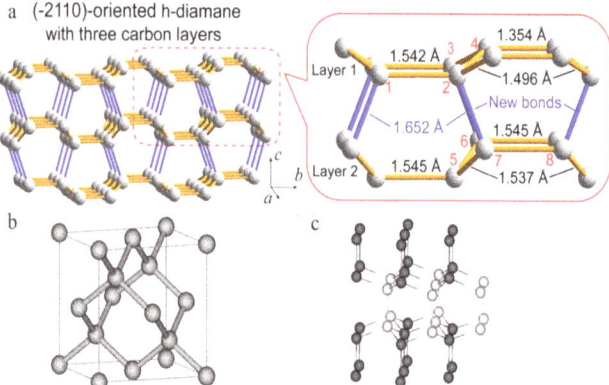

Figure 7.2. Schematic of the structure of hexagonal diamond (a and c) and cubic diamond (b). The bond length also is shown in the inset figure of (a). Source for (a) and (b): [16], reproduced with permission © American Chemical Society; source for (c): [18], reproduced with permission © IOP.

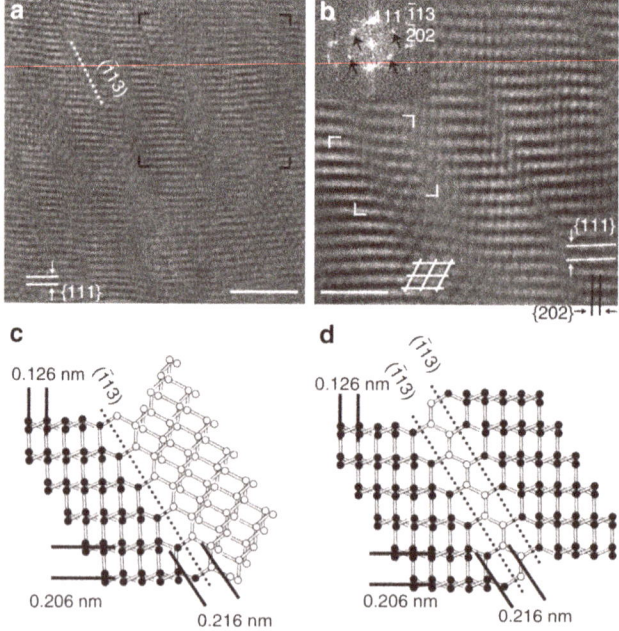

Figure 7.3. Twins provide an alternate explanation for the diffraction features of 'lonsdaleite.' (a) <121> STEM image from the natural sample (Canyon Diablo); {113} twins are indicated by the dotted white line. (b) Background-filtered image calculated from the region marked by black corners in (a). (c) Structural model of the {113} diamond twin. The structure across the twin consists of hexagonally arranged carbon atoms. Black and white atoms indicate twin domains. (d) Structural model of the region marked with white corners in (b). Source: [32], reproduced with permission © Springer Nature.

7.2.3 Doped diamond

Depending on the different contents of boron and nitrogen impurities in diamond, diamond is usually divided into four categories: Ia, Ib, IIa, and IIb [3, 36, 37]. In type Ia diamond, nitrogen exists in an aggregated state [38]. Usually, there is no absorption band in the visible region, and it is colorless and transparent. Most natural diamonds belong to the Ia type, and the percentage can reach 98%. Type Ib: the form of nitrogen present is a single substitute atom, the diamond has a blue absorption band and is yellow in the visible region. Synthetic diamond is mainly the Ib type. Type IIa: has no impurity nitrogen, or a small nitrogen content ($\leqslant$1 ppm); it is colorless and transparent. Type IIb has boron that is not fully compensated by nitrogen impurities [39]. It is blue and has the characteristics of a p-type semiconductor. Its acceptor center is dispersed boron atoms. The lattice parameters for various amounts of boron dopant in diamond are shown in figure 7.4.

7.2.4 Nanotwinned diamond

As we discussed before, diamond is partially or completely bonded by sp^3 hybridization between carbon atoms. Therefore, the connection mode between C atoms at the crystal boundary in nanocrystalline diamond is worth discussing [2, 40–42]. As early as the 1990s, the crystal structure of the grain boundary in natural nanodiamond was studied by high-resolution transmission electron microscopy [43]. The structures of nanodiamond separated from the acid-soluble residue of a carbonaceous meteorite and nanodiamond synthesized by shock wave were studied by T L Daulton *et al* [43]. The interface at the twin boundary represents a minimum energy lattice defect. Therefore, nanocrystals can form under relatively easy-to-accommodate double-structure growth constraints. In cubic diamond, twins along the {111} plane are unstable when the stacking order is {111}.

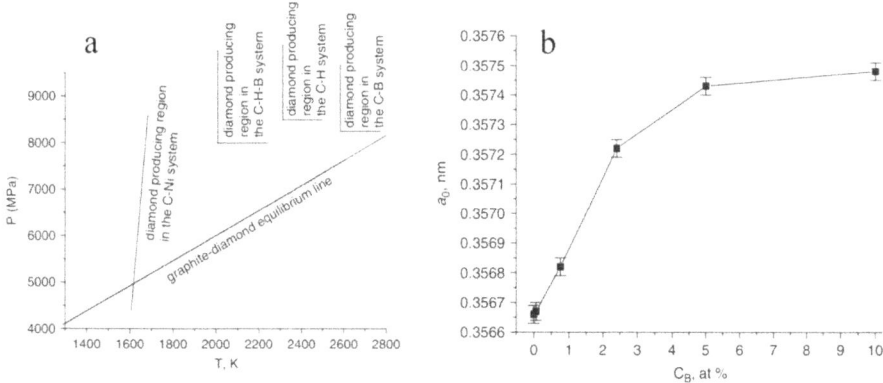

Figure 7.4. Cubic cell parameters of diamond versus boron atomic concentration. Reproduced with permission from [9] © Springer Nature.

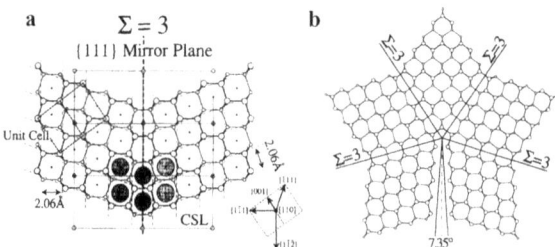

Figure 7.5. (a) Schematic illustration of the atomic model for a first-order $Z = 3$ twin structure. (b) Schematic illustration of an unrelaxed model for a star-twin microstructure; there is a 7.35° mismatch between the twin related lattices [43], reproduced with permission © Elsevier

A sudden reversal of the plane, such as {aabccbaa}, caused by the introduction of a continuous stacking fault is shown in figure 7.5(a). In the lattice representation of the same place, this type of twin structure is described is a first-order $y = 3$ {111} twin. In a perfectly rigid solid, it is impossible to have five undistorted and coherent $Z = 3$ twin boundaries radiating from a common central core because such a construction generates a 7.35° misalignment of the twin related lattices, as shown in figure 7.5(b) [43].

In recent years, Irifuni *et al* reported the synthesis of nanocrystalline diamond with a hardness higher than that of single-crystal diamond. Yongjun Tian *et al* also prepared nanocrystalline diamond with ultrahigh hardness and studied its grain boundary, dislocation, and other structural information in detail using a high-resolution transmission electron microscope (HRTEM) [44, 45]. HRTEM images of nanocrystalline diamond at different temperatures and pressures are shown in figure 7.6. It can be seen from the images that in addition to the twins of nanocrystalline cubic diamond, there are stacking faults and a new kind of diamond, M-diamond, which has a monoclinic structure. Figure 7.6 shows the evidence for monoclinic diamond. The nanocrystalline diamond produced by interweaving these fine structures at a nanometer scale exhibits greater hardness than single-crystal diamond [46–49].

7.2.5 Other kinds of diamond

When the ordered diamond units that combined with each other were further reduced, the existence of amorphous diamond was verified [50]. When the structure of diamond is only ordered over a short range, we get a very special carbon material obtained by carbon sp^3 hybridization. Glassy carbon samples were compressed to 50 GPa at room temperature in a diamond anvil cell, followed by laser heating at approximately 1800 K until the sample became transparent. At that point, a quenchable amorphous diamond was obtained [50]. The atoms of amorphous diamond are randomly arranged; HRTEM images and diffuse diffraction rings in the electron diffraction image both indicate that the recovered sample has no identifiable crystalline symmetry. The structure of amorphous diamond is shown in figure 7.7(b). The biggest difference between the structure of amorphous diamond

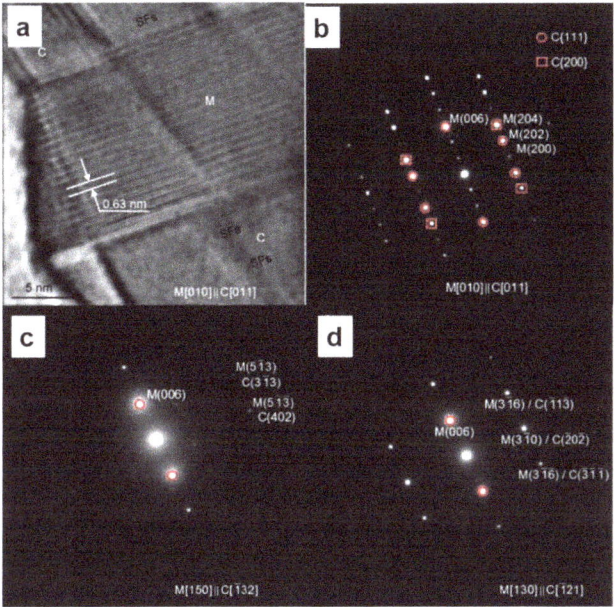

Figure 7.6. (a) Domains forming a {111} twin boundary (TB). Several M-diamond (M) domains are associated with cubic diamond twins that contain stacking faults (SFs). A monoclinic M-diamond (M) domain is observed between two cubic diamond (C) domains. (b)–(d) SAED patterns along the [010], [150], and [130] zone axes of M, recorded by rotating an M crystal. The (111) and (200) spots of the twinned C phase, overlapping with some spots of the M phase as a result of coherent growth, are marked by red circles and boxes, respectively. The determined orientation relations between M and C phases are M(001)//C(111) and M [010]//C[011] [44]. Source: [44], reproduced with permission © Springer Nature.

and the raw material, amorphous carbon, is the disappearance of the lamellar structure in amorphous diamond. Meanwhile, the electron energy loss spectroscopy (EELS) pattern of amorphous diamond implies that its atoms are fully sp^3-bonded, as shown in figure 7.7(a). This diamond shows ultrahigh incompressibility; therefore, this material is expected to exhibit great hardness.

Another special kind of diamond can be understood to be sp^2 carbon-doped diamond. Although all diamonds are composed of carbon elements, some of them are sp^2 hybrids and some are sp^3 hybrids. One of the more representative materials is called Q-carbon, which has a very high ratio (75%–85%) of sp^3 hybrid carbon (the rest is sp^2) and is expected to possess novel physical and chemical properties including great hardness, room-temperature ferromagnetism, and enhanced field emission [51]. Another classic example is pentadiamond, predicted by theoretical calculation, which can also be regarded as sp^2 hybrid carbon-doped diamond [52]. Its pentagonal covalent network is a metastable three-dimensional carbon allotrope with the *Fm−3m* space group which has a lattice parameter of 9.195 Å, as shown in figure 7.7(d) [52]. The network is composed of pentagons, and three of the five sides are shared by adjacent pentagons, as shown in figure 7.7. There are 22 carbon atoms in the unit cell, in which there are ten sp^3 hybrid carbon atoms and 12 sp^2 C atoms,

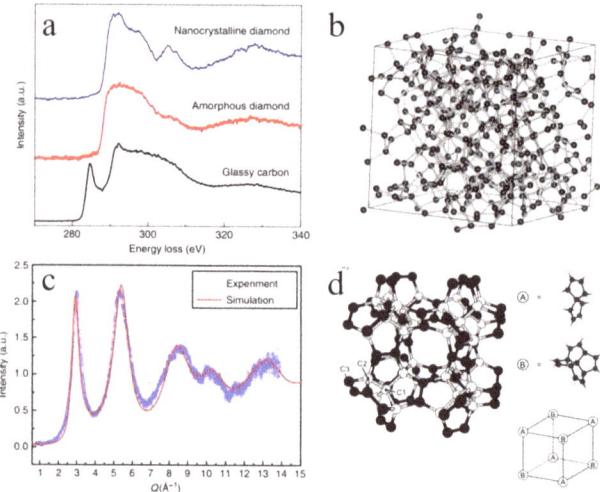

Figure 7.7. (a) Calculated electron energy loss spectroscopy (EELS) patterns for cubic diamond, amorphous diamond, and glassy carbon. (b) Structural model of amorphous diamond. (c) Calculated XRD pattern (structure factor) of amorphous diamond (solid red line) in comparison with experimental data (open blue circles). (d) An optimized geometry of pentadiamond for an optimum lattice constant of 9.195 Å and the *Fm-3m* space group. Pale and dark gray spheres denote sp^3 and sp^2 carbon atoms, respectively [50, 52]. Reproduced with permission from [50, 52] © 2017, The Author(s) and © American Physical Society.

respectively. It can be seen from the structural data that pentadiamond has a lower ratio of sp^3 carbon content than Q-carbon. Its calculated bulk modulus is 381 GPa, which is close to that of diamond and other ultrahard materials, indicating that pentadiamond is a potential candidate for the hard carbon allotropes.

7.3 The synthesis of diamonds

The purpose of this chapter is to give a brief review of the main methods and equipment used for diamond growth using the HPHT, CVD, and hydrothermal reaction methods [8, 53–56].

7.3.1 . The high-pressure, high-temperature method

The apparatuses most commonly used to generate high pressure and high temperature in order to produce diamonds are the belt, cubic, tetrahedral, and Walker/Kawai-type apparatuses [57, 58]. Among these apparatuses and systems, the cubic high-pressure apparatus is the most popular for growing single-crystal diamonds because its large chamber accepts more of the starting materials than other types of apparatus. In the cubic high-pressure apparatus, there are six cubic tungsten carbide (WC) anvils, each with a square anvil tip to generate high pressure. The six anvils are actuated by three pairs of hydraulic rams, forming a cubic cell. The cubic cell usually made of pyrophyllite, within which the sample assembly is compressed. The apparatuses and a high-pressure cell assembly are shown in the schematic diagrams figure 7.8(a) and (b) [57]. Diamond seed crystals are usually used to obtain

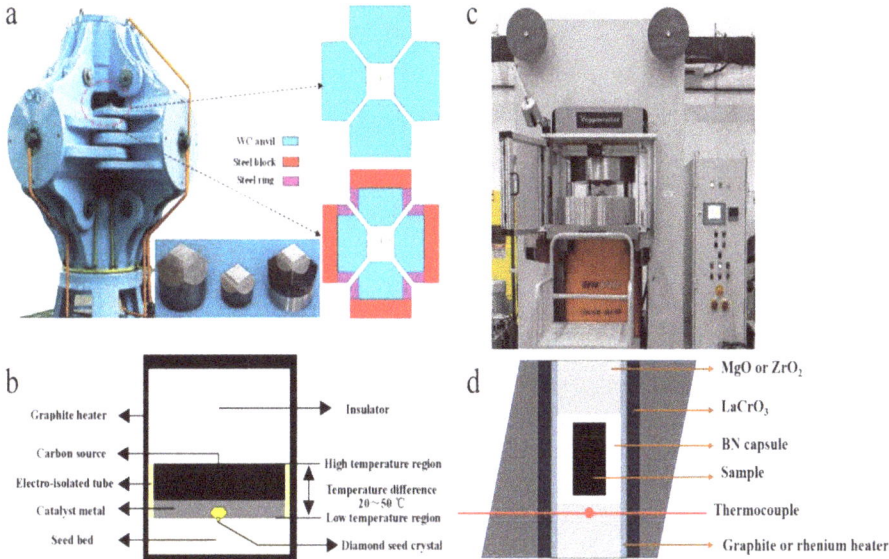

Figure 7.8. (a) Optical photo of the cubic high-pressure apparatus. (b) Schematic diagram of the high-pressure cell assembly used in the cubic high-pressure apparatus. (c) Optical photo of the Kawai-type apparatus. (b) Schematic diagram of the high-pressure cell assembly used in Kawai-type apparatuses [57, 58]. Source: [57, 58], reproduced with permission © 2022 American Chemical Society and © Elsevier.

gem-quality single-crystal diamonds. Highly pure graphite is used as the carbon source material. High-pressure synthesis with a metal solvent catalyst is the most mature process for manufacturing high-quality single-crystal diamond. The transition metals Fe, Co, Ni, Mn, Cr, Ta, and Nb help to reduce the pressure and temperature conditions under which graphite transitions to the diamond phase, and are usually used as solvent catalysts in the form of alloys. With the help of solvent catalysts, the synthesis conditions can be reduced to 1300 °C–1500 °C and 5–6 GPa, while the direct phase transition of hexagonal graphite to cubic diamond requires a static pressure of 13 GPa and a temperature of about 2500 °C–3000 °C. The mechanism of diamond single-crystal synthesis by the catalytic method is not very clear. However, the generally accepted theory is that carbon re-precipitates on the molten catalyst surface and continues to grow on the diamond seed. Diamond crystals with dimensions of up to millimeters or even centimeters can be produced in this way. At present, the high-temperature and high-pressure catalytic method is the main process used to manufacture industrial diamond.

Recently, Walker/Kawai-type apparatuses were widely used to synthesize nano-diamond, which has greater hardness than single-crystal diamonds. Compared with the cubic high-pressure apparatus, this kind of press adopts a two-stage pressurization mode, in which eight tungsten carbide two-stage anvils are driven by six high-hardness first-stage steel anvils to produce a high pressure of 30 GPa. By changing the material of the secondary anvil, for example by using a polycrystalline diamond secondary anvil, an ultrahigh pressure of up to 50 GPa can be generated.

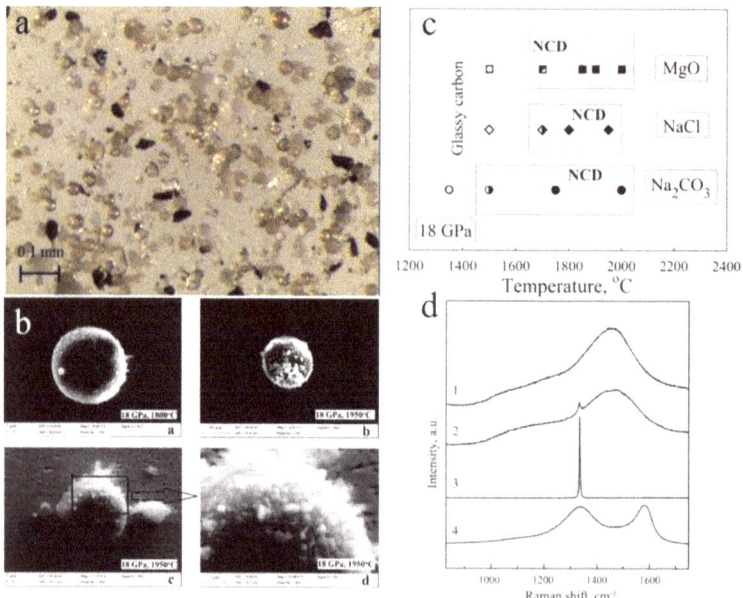

Figure 7.9. (a) Optical image of transparent nanocrystalline diamond (NCD) spheres synthesized from glassy carbon. (b) SEM images of NCD microspheres synthesized at 18 GPa at different temperatures in a NaCl pressure medium. (c) A schematic summarizing HPHT experiments on the direct transformation of glassy carbon into nanodiamond conducted in three different pressure media at 18 GPa. (d) Raman spectra of the synthesis products obtained in various HPHT experiments using different pressure and temperature parameters: 1—nanocrystalline diamond, 2—nanodiamond with a small amount of a polycrystalline component due to the fine-grained surface of the NCD spheres, 3—microcrystals of diamond on the surfaces of NCD spheres, 4—glassy carbon [59], reproduced with permission © Elsevier.

The secondary anvil is a chamfered cube. An octahedral void is formed in the middle of the eight chamfered carbide cubes, and an octahedral sintered by a mixture of the pressure-transmitting medium magnesium oxide and chromium oxide fills the void. In order to produce higher pressure, the cross section of the tungsten carbide anvil is usually several millimeters, which greatly limits the size of the sample cavity. Images of the equipment and the internal assembly of the octahedron are shown in figure 7.8. Graphite or rhenium is usually used as a heat generator. The temperature of the system is measured by a thermocouple. Onion carbon nanoparticles and glassy carbon spheres were used to synthesize nanocrystalline diamond under conditions of 18–25 GPa and 1850 °C–2000 °C [40, 57]. A schematic summarizing HPHT experiments on the direct transformation of amorphous carbon into nanocrystalline diamond conducted in three different pressure media is shown in figure 7.9.

7.3.2 CVD method

As mentioned above, the high-pressure, high-temperature method changes the temperature and pressure conditions to help the phase transition from various solid carbon sources to diamond to occur, either with or without a catalyst.

The CVD method is a way of obtaining diamond from gas-phase carbon source [60–62]. CVD refers to the process of introducing steam containing the reactant constituting the product elements and other gases required for the reaction into a reaction chamber to produce thin films or single crystals by chemical reaction on the surface of a substrate or seed crystal. When the CVD method was first used, the deposition efficiency was very low, about 20 nm per hour. With the introduction of hydrogen atoms into the gas-phase reactants, the deposition efficiency has been greatly improved. Nowadays, the CVD process activates a small number of hydrocarbons, usually methane, in a mixture of hydrogen and H_2 to form appropriate hydrogen atoms and C_xH_y radicals, which are transmitted to the surface of the diamond or the substrate through diffusion and convection and participate in the surface chemical reaction to grow diamond single crystals or films [63].

The most common gas excitation methods are hot filament, microwave plasma-assisted CVD, DC plasma, and DC arc-jet CVD. The hot filament method is the simplest way to activate a mixture of hydrogen and methane. A tungsten filament is set above the substrate and heated to 700 °C–1000 °C. A gaseous mixture of hydrocarbon and hydrogen is introduced at a predetermined position. The filament is then heated to about 2000 °C. The total gas pressure is 10–100 Torr, and the reaction lasts for 3 h. The microwave plasma-assisted CVD method is one of the CVD methods most widely used to synthesize diamond because of its simplicity, flexibility, and other advantages. The microwave plasma-assisted CVD reactor achieves a high growth rate and has great endurance; it can work for several hundred hours on thick films or single-crystal diamonds. As shown, the reactor consists of a cylindrical resonant cavity applicator that utilizes internal tuning which selects the electromagnetic mode excitation, positions and shapes the microwave discharge, and also matches the microwave power to the plasma loaded applicator. The DC plasma CVD and DC arc-jet CVD methods can effectively increase the crystal growth rate to as much as 900 mm h^{-1} [54, 64] (figure 7.10).

7.3.3 Shock wave

A schematic view of the configuration used for *in situ* x-ray diffraction measurements in shock-compressed graphite is shown in figure 7.11. Using a two-stage light gas gun or a powder gun, a polycarbonate or [100] oriented lithium fluoride (LiF) impactor was launched (from the right) onto a 2 mm thick graphite sample, as shown in figure 7.11(a). Upon impact, a leftward-traveling shock wave(s) with a peak longitudinal stress of between 9 and 61 GPa propagated into the graphite sample along the average *c*-axis, while a rightward-traveling shock wave traveled into the impactor material [65]. Four XRD frames (153.4 ns apart) were obtained during each impact experiment to characterize the crystal structures of either the compressed graphite or the high-pressure phases, depending on the peak stress achieved. During each frame, the x-rays probed multiple materials [19, 66].

The two-stage light gas gun configuration is schematically shown in figure 7.11(b). When the gunpowder is ignited, it drives the piston through the pump tube,

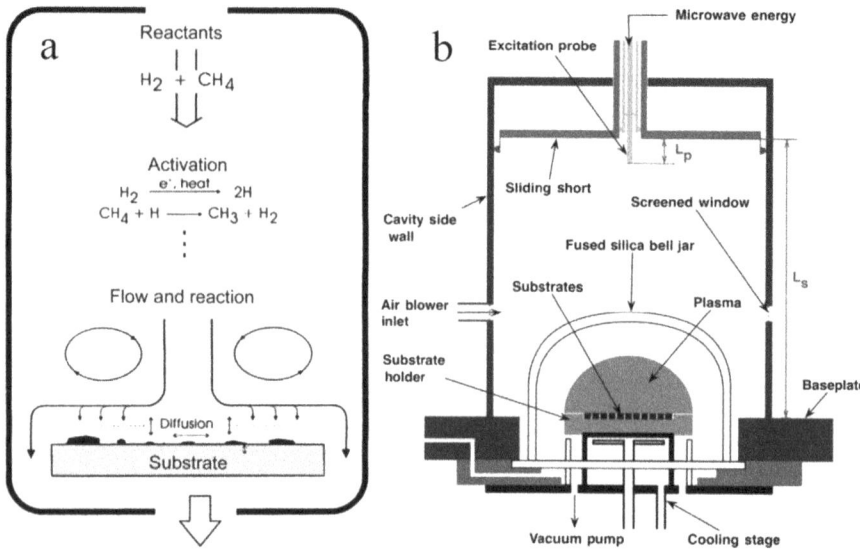

Figure 7.10. (a) A schematic showing the diamond CVD process: the flow of precursors (reactants) into the reactor, the thermal or plasma activation of the reactants, the gas-phase reactions and transport of species to the substrate, and the surface reactions that form diamond [62], reproduced with permission © Springer Nature. (b) Cross-sectional view of a microwave CVD reactor operated at 915 MHz to deposit diamond on multiple substrates [61], reproduced with permission © AIP Publishing.

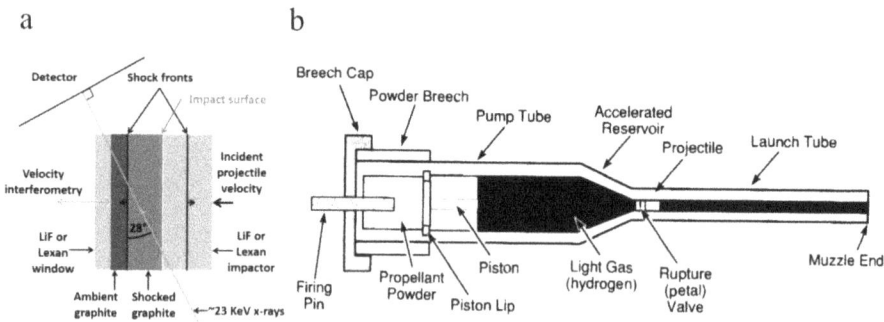

Figure 7.11. (a) Experimental configuration used for the shock wave *in situ* plate-impact XRD experiments. (b) A typical two-stage light gas gun [66, 67]. Source: [66, 67], reproduced with permission © American Physical Society and © Springer 1993, with permission of Springer.

compressing the hydrogen gas ahead of the piston. As the hydrogen gas pressure reaches the limit of the burst valve, the valve breaks, allowing the hydrogen to begin accelerating the smaller projectile in the launch tube. At about the same time, the front of the piston enters the conical section of the acceleration reservoir. As the front of the piston is squeezed down by the cone, it actually picks up speed and drives the entire hydrogen reservoir into the launch tube at high speed [67, 68].

7.3.4 Other methods

Here, we want to introduce two more synthesis methods. The first is the metallic reduction–pyrolysis–catalysis method. The main idea was inspired by the Wurtz reaction R1X + R2X + 2Na→ R1–R2 + 2NaX [69]. Accordingly, Yadong Li *et al* used the reaction between CCl_4 and Na at an appropriate temperature and pressure to obtain diamond [56].

$$CCl_4 \quad + \quad 4Na \frac{700\ °C}{Ni-Co} - \to C(diamond) \quad + 4NaCl$$

The process was referred to as reduction–pyrolysis–catalysis. In this reaction, CCl_4 was used as the carbon source and Na as the reductant and flux under the action of a Ni–Co alloy catalyst to synthesize diamond at 700 °C. A moderate amount of CCl_4 and an excess of metal Na and several pieces of Ni–Co alloy were loaded into a stainless steel hydrothermal reactor. The hydrothermal synthesis reactor was maintained at 700 °C for 48 h and then cooled down to room temperature. The product was a mixture of diamond and amorphous carbon. Although the reaction yield was only 2%, and the product contained a small amount of impurities, this method differs from the high-temperature and high-pressure method that uses a carbon phase transition; it is also different from the CVD method that deposits gaseous reactants. Instead, diamond is generated by replacing C in chloroform. It also provides some ideas and tips for further chemical synthesis methods. Another reduction reaction has also been reported between FeS and CO_2, which implies that there is another possibility for synthesizing diamond by a chemical method at normal pressure [55].

Another interesting method is to produce hexagonal or nanocrystalline cubic diamonds by applying large shear forces at low pressures [41, 70]. Compression and shear experiments were performed using a rotational apparatus containing two polycrystalline cubic boron nitride anvils. Pressure is generated by compressing the two anvils, and the shear stress can be generated by rotating one of the anvils. The success of this method indicates that a huge elastic shear deformation can significantly reduce the reaction pressure required.

7.4 The properties of diamonds

As a valuable material, diamond has many exciting properties that are applied in many fields. Due to its extraordinary optical properties, diamond is famed for its use in jewelry. Diamond also has exotic mechanical properties broadly used in cutting tools. With an electronic bandgap of up to 5.5 eV, diamond is considered to be a candidate for the fourth generation of semiconductors. Furthermore, the presence of nitrogen in the diamond vacancy can serve as a promising single-photon source and a significant platform for future quantum information. Moreover, its other novel properties, such as the superconductivity present in boron-doped diamonds, also make it attractive both for fundamental research and industrial applications.

7.4.1 The mechanical properties of diamonds

For a long time, natural diamond was considered to be the hardest material in the world, due to its strong chemical sp^3 hybridization bond. The Vickers hardness of natural diamond can reach up to 120 GPa [71], which is superior to the hardnesses of other superhard materials such as BN and WN. A goal that attracts researchers is to realize a material that is harder than natural diamond. On way to enhance the hardness of diamond is through the nanostructuring of diamond, which produces many nanograined or nanotwinned structures, according to the famous Hall–Petch effect which commonly appears in metals [72, 73]. In 2003, Irifune *et al* synthesized nanograined polycrystalline diamond from graphite at a temperature of 2300 °C–2500 °C and a pressure of 12–25 GPa [45]. The grain sizes of the synthesized diamonds ranged from 10 to 30 nm and their Knoop hardnesses reached up to 110–140 GPa. Other carbon precursors such as glassy carbon, amorphous carbon, and C_{60} were also successfully used to synthesize smaller diamonds (5–10 nm) at a relatively low temperature of about 1800 °C in 2007. However, the Knoop hardness of these diamonds dramatically dropped to 70–86 GPa due to the reverse Hall–Petch effect driven by grain-boundary sliding. In 2014, a research team at Yanshan University used onion carbon nanoparticles as a precursor and synthesized nanotwinned diamond with an average twin thickness about 5 nm under conditions of 1850 °C–2000 °C and 18–25 GPa [44]. A transmission electron microscopy (TEM) image and an HRTEM image of the synthesized nanotwinned diamond are shown in figures 7.12(a) and (b), which

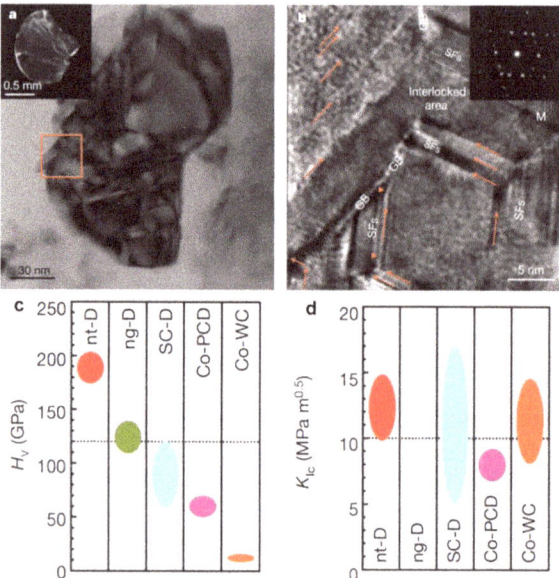

Figure 7.12. Synthesized nanotwinned diamond and its mechanical properties compared with other superhard materials. (a) TEM and (b) HRTEM images of a nanotwinned diamond sample synthesized at 2000 °C and 20 GPa. (c) Vickers hardnesses and (d) fracture toughnesses of different tool materials. Reproduced with permission from [44]; copyright 2014 Nature Publishing Group, a division of Macmillan Publishers Limited. All Rights Reserved.

clearly illustrate the nanotwinned grain boundaries. The typical mechanical properties (hardness and fracture toughness) of the nanotwinned diamond were significantly enhanced in comparison to other superhard materials such as natural diamond, single-crystal diamond, and cobalt-bonded tungsten carbide (Co–WC), as shown in figure 7.12(c) and (d). The enhancement in hardness was proposed to result from the nanotwinned microstructures via the Hall–Petch and quantum confinement effects, whereas the improvement of fracture toughness was attributed to the movement of dislocations along twin boundaries.

7.4.2 The optical properties of diamonds

As one of the most expensive jewels, their special optical properties are significant for diamonds. Ideal diamonds should be transparent, due to the very large intrinsic bandgap which can reach 5.5 eV. However, a number of diamonds, whether natural diamonds or synthesized diamonds, have colors due to the existence of strain, impurities, or defects [74]. Thus, these diamonds may absorb light in the ultraviolet (UV), visible, or even infrared spectral regions. For example, the brown color centers result from a relatively high N content (>100 ppm) for natural type Ia diamonds and extensive plastic deformation for type IIa diamonds [75, 76]. Single-crystal diamonds synthesized using the CVD method display various optical properties, ranging from transparency for the intrinsic bandgap to strong absorption of the whole visible spectrum, depending on the different synthesis conditions such as temperature, pressure, growth rate, or the synthesis process. Compared with natural brown diamonds, single-crystal CVD diamonds may have narrower x-ray rocking curves and higher fracture toughness [71]. With a lower density of dislocations than those of natural diamonds, these synthesized diamonds are considered to be optically homogeneous brown diamonds [77].

The process of annealing under HTHP conditions was developed to improve the optical properties of both natural diamonds and synthesized diamonds. Typically, the required temperatures range from 1800 °C to 2500 °C and the pressures reach more than 5 GPa, which help to restrict the graphitization of diamonds. The alteration of their optical properties is usually related to the degree of the defects, impurities, or strains [78] which act as the color centers of diamonds under HTHP conditions. For type Ia natural diamonds in which the nitrogen concentration is large, HTHP annealing helps the nitrogen aggregates to dissociate, and releases vacancies from the dislocations [79], both of which contribute to the change of absorption. Due to the removal of the strain of plastic deformation, HPHT-annealed type IIa natural diamonds also have a significant reduction in their visible absorption [77]. Since nitrogen and hydrogen are involved in the growth process, the CVD-grown brown diamonds contain various complex nitrogen or hydrogen defects, such as substitutional nitrogen N_s° and N_s^+ [80], nitrogen-vacancies (NV° and NV^-), vacancy hydrogen [81] and nitrogen-vacancy hydrogen [82]. Generally, because of the existence of hydrogen impurities and defects, the color centers responsible for the visible absorption of CVD-synthesized diamonds are less stable than those of brown natural diamonds under HTHP annealing.

To overcome the limitations of high-pressure synthesis, Meng *et al* developed low-pressure/high-temperature (LPHT) annealing in a hydrogen environment assisted by microwave plasma techniques to enhance the optical properties of CVD-synthesized diamonds [83]. Annealing temperatures of up to 2200 °C and pressures of less than than 300 Torr were used for the treatment of single-crystal CVD diamonds. After annealing and appropriate treatment, the brown diamonds became transparent, as shown in figure 7.13(a). There were significant decreases in the photoluminescence spectrum (figure 7.13(b)) as well as in the UV, visible (figure 7.13(c)), and infrared absorptions (figure 7.13(d)) arising from the changed defect structures, especially for hydrogen-related impurities in the CVD growth.

7.4.3 The electronic properties of diamonds

Due to the strong sp^3 hybridization interaction between neighboring carbon atoms in crystal as well as its wide electronic bandgap of 5.5 eV, diamond is highly appealing for a broad range of applications in electronics [84]. The breakdown field of diamond can reach 13 MV cm^{-1} [85], and the reported intrinsic carrier mobilities are 1945 and 2285 cm V^{-1}·s^{-1} for electrons and holes [86], respectively. The strong covalent bonds of diamond also facilitate heat dispersion via lattice vibrations, resulting in the extraordinary thermal conductivity of diamond. All the above characters ensure that diamond is the promising candidate for use as a wide-bandgap semiconductor, especially for high-frequency and high-power electronic devices [87, 88].

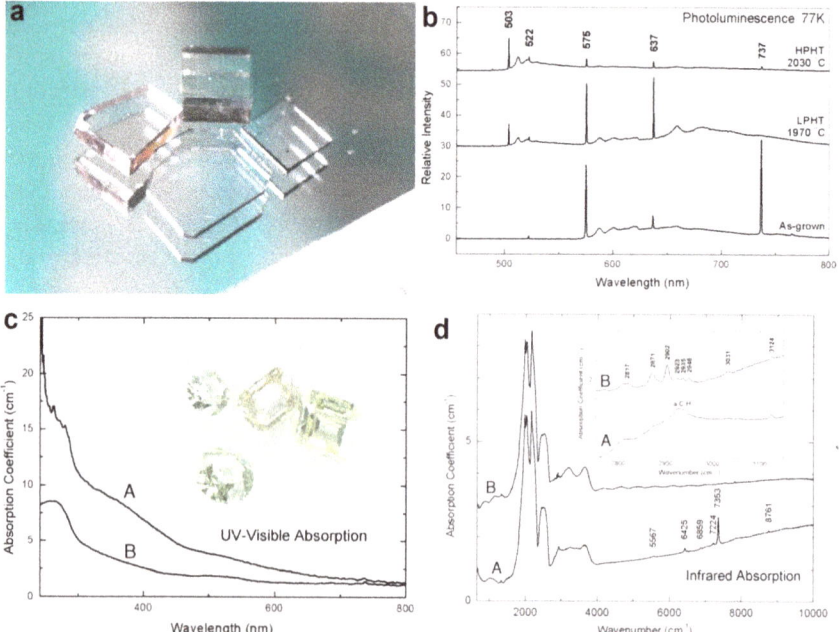

Figure 7.13. The optical properties of CVD-synthesized diamonds treated by LPHT annealing. (a) LPHT-treated transparent CVD diamonds. (b) Photoluminescence spectrum, (c) UV–visible absorption and (d) infrared adsorption of LPHT-treated CVD diamonds [84], reproduced with permission © Elsevier.

The electronic device applications of diamond require that it should be effectively doped to realize a high density of mobile carriers. In common with other wide-bandgap semiconductors such as GaAs, the objective is achieve shallow doping with a suitable activation energy for room-temperature applications; however, this is a big challenge at present. Although carbon is located in the same group as silicon in the periodic table, the small atomic radius and strong chemical bonding of carbon atoms in diamond make the doping of diamond much difficult than that of silicon. As is the case for other wide-bandgap semiconductors such as ZnO, diamond also displays an asymmetrical doping character such that p-type doping is much easier than n-type doping. Boron is considered to be the most promising impurity to function as a p-type acceptor in diamond [89], as shown in figure 7.14(a). However, the activation energy of boron doping reaches up to 0.37 eV [90], which still much higher than that of silicon (about 45 meV [91]). Depending on the quality of the diamond as well as the substrate, the doping density, and the method used for boron doping, the hole mobility in p-type diamond at room temperature varies dramatically, ranging from several to thousands of cm $V^{-1} \cdot s^{-1}$. For high-quality diamond with a light doping of boron, the hole mobility typically reaches up to 2200 cm $V^{-1} \cdot s^{-1}$ [92], which is larger than that of silicon or even GaAs. If the temperature and the dopant concentration are increased, the hole mobility drops significantly [93, 94], as shown in figure 7.14(b), due to the mechanisms of phonon scattering [95, 96] and neutral impurity scattering [97], respectively. On top of the above problems that have emerged with p-type doping, n-type doping of diamond is much difficult than p-type doping. Nitrogen has a similar atomic radius to that of carbon; thus, the doping of nitrogen in diamond avoids large lattice distortions, and several experiments have demonstrated the successful synthesis of nitrogen-doped diamond via CVD techniques [98, 99]. However, its activation energy is 1.7 eV [100] and the nitrogen impurity is a very deep donor, which limits its applications for most modern electronic devices [101]. The doping of sulfur [102] introduces large lattice distortions, which prevent the doped sulfur atoms from contributing mobile n-type carriers. Additionally, the electronic properties of sulfur-doped diamond become unstable when exposed to temperature variations, and it exhibits n-type conductivity at high temperatures but transforms to p-type conductivity at low temperatures [103]. Lithium has also been considered for use as a candidate shallow donor, as it has a small activation energy of 0.1 eV for interstitial lithium doped in diamond [104]. However, lithium also replaces a carbon site or occupies a vacancy position to act as a deep acceptor [105], which restricts the n-type conductivity of interstitial lithium. The fact of self-compensation limits the application of lithium as a promising n-type dopant for diamond [106]. At present, phosphorous still serves as the main donor; however, the activation energy of phosphorous-doped diamond can reach 0.6 eV [107], which limits the carrier concentration. For instance, a previous work reported a limited carrier concentration of 10^{11} cm^{-3} despite high donor concentrations of up to $6.8*10^{16}$ cm^{-3}, suggesting that most of the phosphorous impurities were not activated [108]. Moreover, increased temperature was able to activate the carrier concentration of phosphorous-doped diamond, yet it dramatically decreased carrier mobility [109], which also limits the use of

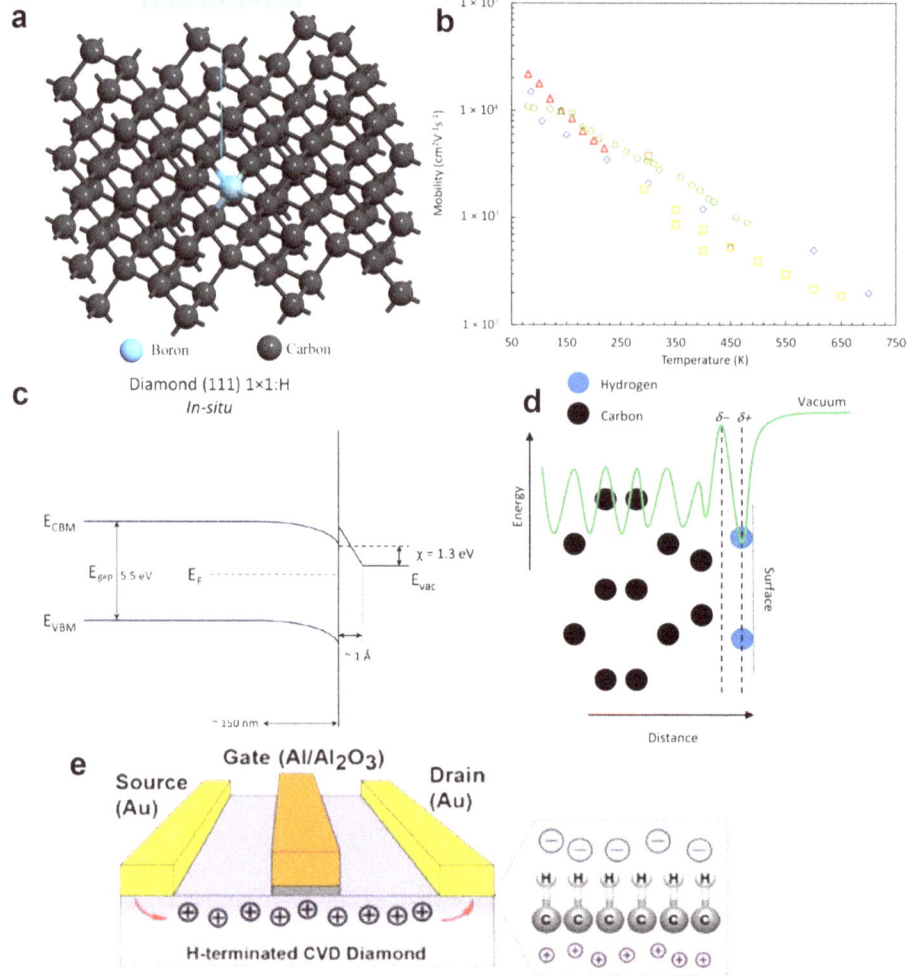

Figure 7.14. The electronic properties and potential applications of diamond. (a) and (b) Traditional impurity doping of diamond. (a) The atomic model of substitutional doping in boron-doped diamond. (b) Reported hole mobilities in boron-doped diamond with increasing temperature for several experimental groups marked as yellow squares, green circles, red triangles and blue diamonds [112]. (c)–(e) The strategy for surface transfer doping of diamond. (e) Illustration of band bending in hydrogen-terminated diamond in an ultrahigh vacuum. (d) Representation of the formation of a dipole at the (100) surface of hydrogen-terminated diamond, induced by the electronegativity difference between carbon and hydrogen. Sources: [112, 113], reproduced with permission © 2021, from Elsevier, and © 2014, The Materials Research Society, from Springer.

phosphorous-doped diamond in high-temperature electronic devices. Other methods, such as co-doping with two different dopants [110, 111], have also been proposed to improve the n-type conductivity of diamond. However, the possible occupied positions and the interactions between dopants are complex, and control of growth becomes difficult in experiments. Thus, the puzzle of n-type impurity doping of diamond is still the primary impediment for its use in electronics.

Another strategy that may potentially unlock the use of diamond-based electronic devices as typical wide-bandgap semiconductors uses the idea of surface transfer doping without the intentional introduction of a doped impurity [112]. Instead, surface transfer doping produces a thin semiconducting layer on the intrinsically insulating diamond surface. In the case of a hydrogen-terminated diamond surface, a dipole layer is formed because of the difference in electronegativity between the carbon and hydrogen atoms, which further results in band bending near the surface region, as shown in figures 7.14(c) and (d). By varying the construction of the diamond surface, termination species, and interfaces with different electron-accepting mediums, the conductivity of the diamond surface layer can effectively be tuned to produce n-type or p-type carriers. Figure 7.14(e) illustrates a transistor model based on the idea of surface transfer doping [113]. A comprehensive review provided details on this topic and described recent progress [112].

7.4.4 The superconductive properties of diamonds

Diamond is well known for wide-bandgap semiconductors; however, heavy doping [114] and the application of external stress [115] can achieve the goal of superconductivity in diamond. In 2004, Ekimov *et al* reported the discovery of superconductivity at a superconducting transition temperature T_c of about 4 K in boron-doped diamond as a bulk, type-II superconductor via electrical resistivity (figure 7.15(a)), magnetic susceptibility (figure 7.15(b)), specific heat, and field-dependent resistance measurements [114]. Their boron-doped diamond samples were synthesized using graphitic carbon and B_4C powder as the starting materials at a pressure of 8–9 GPa and a temperature of 2500–2800 K. The total B content of their diamond samples estimated using both nuclear magnetic resonance (NMR) and inductively coupled mass spectrometry approached $2.8 \pm 0.5\%$ and the carrier concentration determined by the zone-center phonon line was larger than 2×10^{21} cm^{-3}, which are similar to those of typical heavily boron-doped diamond films grown by CVD. The metallic states responsible for the superconductivity originate from the introduction of charge carriers into the intrinsic diamond bands instead of the impurity bands [3].

Another approach to the superconductivity of diamond is via strain engineering, as proposed by Liu *et al* in 2020 [115]; this approach does not require doping. As shown in figure 7.15(c), the increased strain in a deformed diamond lattice gradually closes the bandgap and then induces the accumulation of electronic states around the Fermi energy, which effects the transition from a semiconductor to a conductor. A further increase in the strain softens the deformed diamond lattice, generating a strong electron–phonon coupling and a sufficiently high electronic density of states at the Fermi level, which leads to superconductivity at an estimated critical transition temperature of 2.4–12.4 K for a Coulomb pseudopotential μ of 0.15–0.05. Figure 7.15(d) illustrates the evolution of phonon dispersion and the associated electron–phonon coupling with increasing strain in the deformed diamond. In particular, there is an obvious soft mode at the A point in q-space, whose vibration was also illustrated in figure 7.15(d). The softening of the deformed diamond lattice

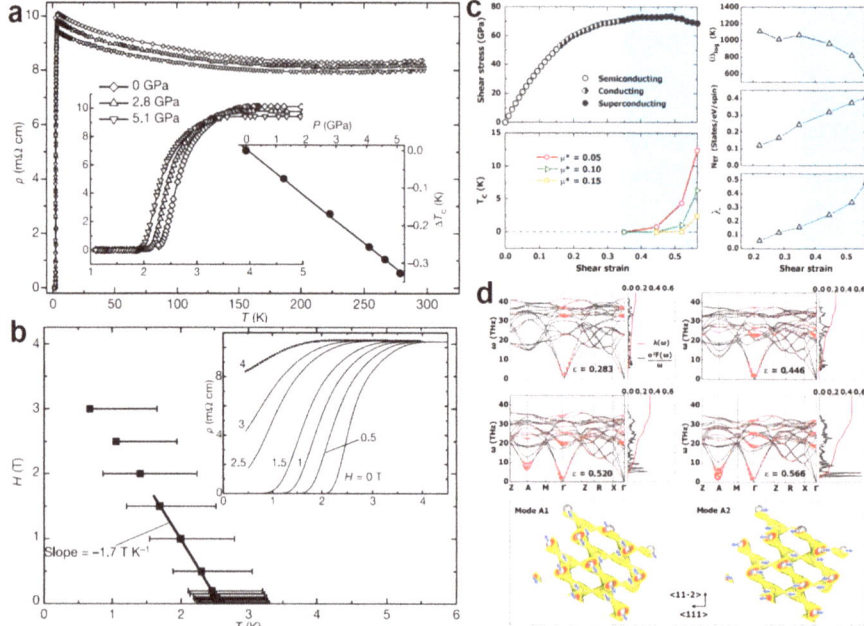

Figure 7.15. The superconductivity of diamond. (a) Temperature dependence of the electrical resistivity and (b) the upper critical field for boron-doped diamond. (c) The electronic evolution from semiconductivity and conductivity to superconductivity, estimated T_c and electron–phonon coupling parameter λ, and (d) phonon dispersion curves showing the strength of λ in deformed diamond with the variation of strain. (e) Illustration of the vibrational modes of the two lowest-frequency branches at the A point of the Brillouin zone. Source: [115], reproduced with permission © American Physical Society.

helps to generate strong electron–phonon coupling, which raises T_c. However, the strain required to induce superconductivity in diamond is large, which still is a challenge in experiments and needs further work.

7.4.5 The magnetic properties of diamonds

Magnetism in diamond has also received a great deal of research interest recently. Usually, the appearance of magnetic properties in diamond even or other forms of carbon is related to electronic instability, which can be induced by bonding defects such as a mixture of sp^2- and sp^3-coordinated carbon atoms [116]. The incorporation of other impurities such as hydrogen, boron, nitrogen, and phosphorous into carbon compounds also induces ferromagnetism. In 2005, Talapatra *et al* used nitrogen (^{15}N) and carbon (^{12}C) ion implantations to investigate magnetism in nanosized diamond crystals [117]. They observed a robust signature of ferromagnetism in both of the ion-implanted diamonds, as shown in figure 7.16. They further found that the magnetization was independent of the doped species at low doses, suggesting that it originates from bonding defects due to ion implantation in nanodiamonds. At high doses, however, ^{15}N implants have a larger saturated magnetization value than that

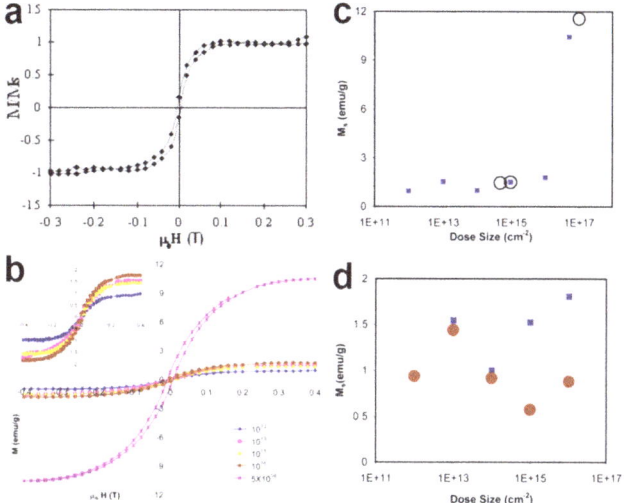

Figure 7.16. The magnetic properties of nanodiamond particles induced by ion implantation. (a) Ferromagnetic signals and (b) magnetic moments at saturation of ^{15}N-implanted nanodiamonds. (c) Magnetic moments at saturation as a function of dose size for ^{15}N implants and (d) a comparison with ^{12}C implants. Source: [118], reproduced with permission © American Physical Society.

of ^{12}C, which may result from the incorporation of nitrogen to form C–N bonds in nanodiamonds.

7.5 The applications of diamonds

Based on the above variously superior properties, diamond has many applications, both in fundamental science and industrial techniques. Colorless (transparent) and colorful diamonds are valuable and fashionable types of jewelry. As a wide-bandgap semiconductor with extraordinary thermal conductivity, diamond has very competitive uses in high-power electronic devices. The diamond anvil cell has been used as the basic item of equipment with which to produce static high pressure for the study of physics, materials science, chemistry, and geoscience [118]. The N_v center in diamond has also been an experimental platform for the study of quantum information [119] and single-photon emission [120].

7.5.1 Colorful diamonds as jewelry

In addition to colorless diamonds, colorful diamonds, such as yellow, blue, brown, red, green, pink, violet, and black diamonds are also valuable types of jewelry. The reason that diamond appears in so many colors is mainly due to doping, lattice defects, and radiation. Yellow is one of the most common diamond colors, especially for synthetic diamonds. Yellow diamond is mainly caused by N atoms. The less common type IIB diamond usually has a beautiful blue color. This is mainly due to the element boron. Black diamond is an amorphous diamond-like material with a polycrystalline porous structure and no cleavage plane. Its color is often caused by impurities such as the incomplete transformation of graphite. Natural single green diamond is rare, and its color is caused by the irradiation of diamond. Because particulate radiation only penetrates the surface of the diamond, the color is usually only a few microns deep. The formation mechanism of pink and red diamonds is different from those of the others and is possibly due to lattice distortion or heteroatom doping, for example, with small amounts of Fe, Na, or Cl [121].

7.5.2 The application of diamonds in electronics

Owing to their outstanding properties, such as a high breakdown field, high electron/ hole mobility, and the highest thermal conductivity of any known material, diamond has many potential applications in high-frequency and high-power electronics, detectors, electron emitters, and thermionic emitters. Profiting from the developed synthesis strategies, large area single-crystal diamond substrates (>1 cm^{-2}) with relatively low defect densities can be produced by the CVD method [122]. Moreover, the doping of diamond to obtain p-type and n-type carriers has also achieved some success. In spite of the deep dopant levels, the realized impurity concentrations of diamond can be up to 10^{20} cm^{-3}, which opens up the possibility of hopping conduction. An increase in temperature also helps to free the carriers and reduce the resistivity; thus, diamond-based devices favor high working temperatures.

7.5.3 The application of diamonds in fundamental research

Known as the hardest natural material in the world, diamond is widely used in the area of high-pressure research [123, 124]. Combined with synchrotron x-ray diffraction, neutron scattering, and Raman or infrared spectra, DAC techniques have been a typical method for research in physics, chemistry, materials science, and geoscience under static high pressure. Many exciting research achievements have been completed with the help of the DAC. For example, typical sodium metal is transformed into a transparent insulator at high pressures [125]. The critical transition temperatures for the superconductivity of sulfur hydride [126] and lanthanum hydride [127] are higher than 200 K, which approaches the dream of room-temperature superconductivity. Additionally, the N_v center in diamond provides a significant platform [128] with which to realize single-spin quantum sensors [119], quantum calculations [129–131], and single-photon emitters [120]. Figure 7.17 shows an atomic structural model of the N_v center in diamond and the scheme of an N_v quantum sensor used to measure nucleus–nucleus spin coupling.

7.6 Conclusions and outlook

In spite of the great success achieved in the recent decades, many problems related to the synthesis of diamond still remain to be solved. In addition to obtaining higher-quality large single crystals of diamond, the reduction of diamond preparation conditions by the chemical reaction method and the synthesis of different types of diamond, such as hexagonal and triclinic diamonds, at high temperatures and pressures will also attract more attention. In the field of the growth of high-quality jewelry-grade large single crystals, research and development into catalysts with

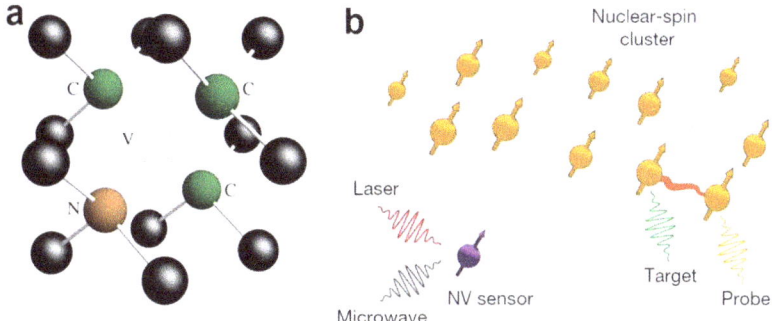

Figure 7.17. (a) Atomic model of the N_v center in diamond [128], reproduced with permission © Elsevier; (b) its application in a quantum sensor [130], reproduced with permission © Springer Nature.

higher efficiency and less residue may become popular. The chemical reaction has irreplaceable advantages in reducing the preparation conditions of diamond, but the yield and crystal quality still need to be improved. However, it is worrying that there are few studies and reports in this area, which requires more chemists to participate. Diamonds with other structures have always been powerful candidates for improving diamond hardness, especially the hexagonal diamonds; their denser hexagonal stacking mode leads to high hopes for ultrahigh hardness. Similarly, while diamond is getting harder and harder, we also hope to see the other excellent properties of diamond gain more extensive attention and application. In the future, it will be very attractive to explore the use of diamond combined other carbon materials such as graphene [132], carbon fullerene [133], and carbon nanotubes [134] to realize carbon-based insulators, semiconductors, and metals [135, 136] as the building blocks of electronic devices.

Acknowledgments

We are deeply thankful for the financial support of the National Natural Science Foundation of China (Grant No. 11704111) and the Fundamental Research Funds of the Central Universities (Grant No. 531107050916).

References

[1] Humble P and Hannink R H J 1978 *Nature* **273** 37–9
[2] Nianjun Y *et al* 2008 *Nat. Preced* 1–1
[3] Yokoya T, Nakamura T, Matsushita T, Muro T, Takano Y, Nagao M, Takenouchi T, Kawarada H and Oguchi T 2005 *Nature* **438** 647
[4] Sidorov V A, Ekimova E A, Rakhmanina A V, Mel'nik N N, Sadykov R A and Thompson J D 2006 *Sci. Technol. Adv. Mat.* **7** S2
[5] Ekimov E A *et al* 2004 *Nature* **428** 542–45
[6] Bannister F A and Lonsdale K 1943 *Nature* **151** 334–35
[7] Simon S F and Berman R 1955 *Zeitschrift für Elektrochemie, Ber. Bunsenges. Phys. Chem.* **59** 333–38
[8] Terrones H and Mackay A L 1991 *Nature* **352** 762

[9] Riley D P 1944 *Nature* **153** 587–88

[10] Bragg W L and Bragg W H 1913 *Proc. Royal Soc. Lond.* A **89** 277–91

[11] Ramachandran G N 945 *Nature* **156** 83–31

[12] Yuan Z J, Yao Y X, Zhou M and Bal Q S 2003 *CIRP Ann.* **52** 285

[13] Grimsditch M H, Anastassakis E and Cardona M 1978 *Phys. Rev.* B **18** 901

[14] Frondel C and Marvin U B 1967 *Nature* **214** 587

[15] Hanneman R E, Strong H M and Bundy F P 1967 *Science* **155** 995

[16] Ke F *et al* 2020 *Nano Lett.* **20** 5916

[17] Bundy F P and Kasper J S 1967 *J. Chem. Phys.* **46** 3437

[18] Chen H, Zhang W and Wang Z 2004 *J. Phys. Condens. Matter* **16** 741

[19] Kondo K and Hirai H 1991 *Science* **253** 772–74

[20] Abha Misra P K, Tyagi B S, Yadav P, Rai D S, Misra V, Pancholi and Samajdar I D 2006 *Appl. Phys. Lett.* **89** 071911

[21] Satpathy S and Salehpour M R 1990 *Phys. Rev.* B **41** 3048

[22] Sano T, Takahashi K, Sakata O, Okoshi M, Inoue N, Kobayashi K F and Hirose A 2009 *J. Phys. Conf. Ser.* **165** 012019

[23] Bagge-Hansen M, Stavrou E and Hammons J A *et al* 2020 *Phys. Rev.* B **102** 104116

[24] Yagi T, Utsumi W, Yamakata M, Kikegawa T and Shimomura O 1992 *Phys. Rev.* B **46** 6031

[25] Israde-Alcantara I *et al* 2012 *Proc. Natl. Acad. Sci.* **109** E738-E747

[26] Shiell T B, McCulloch D G, Bradby J E, Haberl B, Boehler R and McKenzie D R 2016 *Sci. Rep.* **6** 37232

[27] Gaspar Banfalvi 2012 *Open Chem.* **10** 1676

[28] Xiong G, Chakraborty P and Cao L *et al* 2018 *Carbon* **139** 85–93

[29] Le Guillou C, Rouzaud J N, Remusat L, Jambon A and Bourot-Denise M 2010 *Geochim. Cosmochim. Acta* **74** 4167

[30] Strong H M, Hanneman R E and Bundy F P 1967 *Science* **155** 995–7

[31] Ribeiro F J, Tangney P, Louie S G and Cohen M L 2005 *Phys. Rev.* B **72** 214109

[32] Nemeth P, Garvie L A, Aoki T, Dubrovinskaia N, Dubrovinsky L and Buseck P R 2014 *Nat. Commun.* **5** 5447

[33] Marvin U B and Frondel C 1967 *Nature* **214** 587–89

[34] Pan Z, Sun H, Zhang Y and Chen C 2009 *Phys. Rev. Lett.* **102** 055503

[35] Yagi T and Utsumi W 1991 *Proc. Japan Acad.* B **67B** 159–64

[36] Sidorov V A, Ekimov E A and Mel'Nik N N *et al* 2004 *J. Mater. Sci.* **39** 4957–60

[37] Ekimov E A, Sidorov V A, Rakhmanina A V, Mel'nik N N, Timofeev M A and Sadykov R A 2006 *Inorg. Mater.* **42** 1198

[38] Olivier E J, Neethling J H, Kroon R E, Naidoo S R, Allen C S, Sawada H, van Aken P A and Kirkland A I 2018 *Nat. Mater.* **17** 243

[39] Moore A E and Helmstaedt H 2019 *Nature* **570** E26

[40] Dubrovinskaia N, Dubrovinsky L, Langenhorst F, Jacobsen S and Liebske C 2005 *Diamond Relat. Mater* **14** 16

[41] Ma Y, Gao Y and An Q *et al* 2019 *Carbon* **146** 364–68

[42] Kulnitskiy B, Perezhogin I, Dubitsky G and Blank V 2013 *Acta Crystallogr.* B **9** 474

[43] Eisenhour D D, Daulton T L and Bernatowicz T J *et al* 1996 *Geochim. Cosmochim. Acta* **60** 4853–72

[44] Huang Q *et al* 2014 *Nature* **510** 250

[45] Irifune T, Kurio A, Sakamoto S, Inoue T and Sumiya H 2003 *Nature* **421** 599

[46] Natalia Dubrovinskaia, Dubrovinsky L, Solopova N A, Abakumov A, Turner S, Hanfland M, Bykova E, Bykov M, Prescher C and Prakapenka V B 2016 *Sci. Adv.* **2** e1600341

[47] Leonid Dubrovinsky, Dubrovinskaia N, Bykova E, Bykov M, Prakapenka V, Prescher C, Glazyrin K, Liermann H P, Hanfland M and Ekholm M 2015 *Nature* **525** 226

[48] Tanigaki K, Ogi H, Sumiya H, Kusakabe K, Nakamura N, Hirao M and Ledbetter H 2013 *Nat. Commun.* **4** 2343

[49] Dubrovinsky L, Dubrovinskaia N, Prakapenka V B and Abakumov A M 2012 *Nat. Commun.* **3** 1

[50] Zeng Z *et al* 2017 *Nat. Commun.* **8** 322

[51] Narayan J and Bhaumik A 2015 *J. Appl. Phys.* **118** 141

[52] Fujii Y, Maruyama M, Cuong N T and Okada S 2020 *Phys. Rev. Lett.* **125** 016001

[53] Naka S *et al* 1976 *Nature* **259** 38–9

[54] Sato Y, Matsumoto S and Tsutsumi M *et al* 1982 *J. Mater. Sci.* **17** 3106–12

[55] Melton C E, Langford R E and Giardini A A 1974 *Nature* **249** 647

[56] Qian Y, Li Y and Liao H *et al* 1998 *Science* **281** 246–47

[57] Han Q-G, Liu B, Hu M -h, Li Z -c, Jia X -P, Li M -Z, Ma H -A, Li S -S, Xiao H -Y and Li Y 2011 *Crystal Growth Design* **11** 1000

[58] Palyanov Y N, Kupriyanov I N, Khokhryakov A F and Ralchenko V G 2015 Crystal Growth of Diamond *Handbook of Crystal Growth(Second Edition)* (Boston, MA: Elsevier) ch17 pp 671–713

[59] Solopova N A, Dubrovinskaia N and Dubrovinsky L 2015 *J. Cryst. Growth* **412** 54

[60] Sevillano E 1998 Microwave-Plasma Deposition of Diamond *Low-Pressure Synthetic DiamondSpringer (Series in Materials Processing)* (Berlin: Springer) pp 11–39

[61] Asmussen J, Grotjohn T A, Schuelke T, Becker M F, Yaran M K, King D J, Wicklein S and Reinhard D K 2008 *Appl. Phys. Lett.* **93** 031502

[62] Butler J E and Windischmann H 1998 *MRS Bull.* **23** 22

[63] Amornkitbamrung V, Burinprakhone T and Jarernboon W 2009 *Surf. Coat. Technol.* **203** 1645

[64] Tallaire A, Achard J, Silva F, Brinza O and Gicquel A 2013 *C.R. Phys.* **14** 169

[65] Zhu Y Q, Sekine T, Kobayashi T, Takazawa E, Terrones M and Terrones H 1998 *Chem. Phys. Lett.* **287** 689

[66] Volz T J, Turneaure S J, Sharma S M and Gupta Y M 2020 *Physical Review* B **101** 224109

[67] Asay J R and Shahinpoor M 1993 High-Pressure Shock Compression of Solids *Shock Wave and High Pressure Phenomena* (New York, NY: Springer)

[68] Mrm Izawa, Flemming R L, Banerjee N R and McCausland P J A 2011 *Meteorit. Planet. Sci.* **46** 638

[69] Wurtz A 1855 *Ann. Chim. Phys.* **44** 275

[70] Shiell T B, Wong S and Cook B A *et al* 2019 *Carbon* **142** 475–81

[71] Yan C -S, Mao H -k, Li W, Qian J, Zhao Y and Hemley R J 2004 *Physica Status Solidi (a)* **201** R25

[72] Hall E O 1951 *Proc. Phys. Soc.* B 64B 747

[73] Petch N J 1953 *J. Iron Steel Inst.* **174** 25

[74] Fritsch E 1998 *The Nature of Diamonds* p 23

[75] Mao H K and Hemley R J 1991 *Nature* **351** 721

[76] Hounsome L S, Jones R, Martineau P M, Fisher D, Shaw M J, Briddon P R and Öberg S 2006 *Phys. Rev.* B **73** 125203

[77] Mora A E, Steeds J W, Butler J E, Yan C-S, Mao H K and Hemley R J 2005 *Physica Status Solidi (a)* **202** R69

[78] Collins A T, Kanda H and Kitawaki H 2000 *Diam. Relat. Mater.* **9** 113

[79] Collins A T, Connor A, Ly C-H, Shareef A and Spear P M 2005 *J. Appl. Phys.* **97** 083517

[80] Charles S J, Butler J E, Feygelson B N, Newton M E, Carroll D L, Steeds J W, Darwish H, Yan C-S, Mao H K and Hemley R J 2004 *Physica Status Solidi (a)* **201** 2473

[81] Glover C, Newton M E, Martineau P, Twitchen D J and Baker J M 2003 *Phys. Rev. Lett.* **90** 185507

[82] Glover C, Newton M E, Martineau P M, Quinn S and Twitchen D J 2004 *Phys. Rev. Lett.* **92** 135502

[83] Meng Y -F, Yan C -S, Lai J, Krasnicki S, Shu H, Yu T, Liang Q, Mao H -K and Hemley R J 2008 *Proc. Natl. Acad. Sci. U.S.A* **105** 17620

[84] Wort C J H and Balmer R S 2008 *Mater. Today* **11** 22

[85] Umezawa H 2018 *Mater. Sci. Semicond. Process.* **78** 147

[86] Gabrysch M, Majdi S, Twitchen D J and Isberg J 2011 *J. Appl. Phys.* **109** 063719

[87] Geis M W, Wade T C, Wuorio C H, Fedynyshyn T H, Duncan B, Plaut M E, Varghese J O, Warnock S M, Vitale S A and Hollis M A 2018 *Physica Status Solidi (a)* **215** 1800681

[88] Kasu M, Ueda K, Yamauchi Y, Tallaire A and Makimoto T 2007 *Diam. Relat. Mater.* **16** 1010

[89] Chrenko R M 1973 *Phys. Rev.* B **7** 4560

[90] Tsao J Y *et al* 2018 *Adv. Electron. Mater.* **4** 1600501

[91] Morin F J and Maita J P 1954 *Phys. Rev.* **96** 28

[92] Nesladek M, Bogdan A, Deferme W, Tranchant N and Bergonzo P 2008 *Diam. Relat. Mater.* **17** 1235

[93] Isberg J, Hammersberg J, Johansson E, Wikström T, Twitchen D J, Whitehead A J, Coe S E and Scarsbrook G A 2002 *Science* **297** 1670

[94] Isberg J, Lindblom A, Tajani A and Twitchen D 2005 *Physica Status Solidi (a)* **202** 2194

[95] Pernot J, Tavares C, Gheeraert E, Bustarret E, Katagiri M and Koizumi S 2006 *Appl. Phys. Lett.* **89** 122111

[96] Pernot J, Volpe P N, Omnès F, Muret P, Mortet V, Haenen K and Teraji T 2010 *Phys. Rev.* B **81** 205203

[97] Peterson R, Malakoutian M, Xu X, Chapin C, Chowdhury S and Senesky D G 2020 *Phys. Rev.* B **102** 075303

[98] Baranauskas V, Li B B, Peterlevitz A, Tosin M C and Durrant S F 1999 *J. Appl. Phys.* **85** 7455

[99] Müller-Sebert W, Wörner E, Fuchs F, Wild C and Koidl P 1996 *Appl. Phys. Lett.* **68** 759

[100] Farrer R G 1969 *Solid State Commun.* **7** 685

[101] Kalish R, Uzan-Saguy C, Philosoph B, Richter V, Lagrange J P, Gheeraert E, Deneuville A and Collins A T 1997 *Diam. Relat. Mater.* **6** 516

[102] Saada D, Adler J and Kalish R 2000 *Appl. Phys. Lett.* **77** 878

[103] Nakazawa K, Tachiki M, Kawarada H, Kawamura A, Horiuchi K and Ishikura T 2003 *Appl. Phys. Lett.* **82** 2074

[104] Kajihara S A, Antonelli A, Bernholc J and Car R 1991 *Phys. Rev. Lett.* **66** 2010

[105] Lombardi E B and Mainwood A 2007 *Physica* B **401-2** 57

[106] Goss J P and Briddon P R 2007 *Phys. Rev.* B **75** 075202

[107] Kalish R 1999 *Carbon* **37** 781

[108] Koizumi S, Teraji T and Kanda H 2000 *Diam. Relat. Mater.* **9** 935

[109] Koizumi S, Umezawa H, Pernot J and Suzuki M 2018 *Power Electronics Device Applications of Diamond Semiconductors* (Sawston: Woodhead Publishing)

[110] Hu X J, Li R B, Shen H S, Dai Y B and He X C 2004 *Carbon* **42** 1501

[111] Sergio Conejeros M, Zamir Othman A, Croot J N, Hart K M, O'Donnell P W, May and Allan N L 2021 *Carbon* **171** 857

[112] Crawford K G, Maini I, Macdonald D A and Moran D A J 2021 *Prog. Surf. Sci.* **96** 100613

[113] Pakes C I, Garrido J A and Kawarada H 2014 *MRS Bull.* **39** 542

[114] Ekimov E A, Sidorov V A, Bauer E D, Mel'nik N N, Curro N J, Thompson J D and Stishov S M 2004 *Nature* **428** 542

[115] Liu C, Song X Q, Li Q, Ma Y M and Chen C F 2020 *Phys. Rev. Lett.* **124** 147001

[116] Makarova T L 2004 *Semiconductors* **38** 615

[117] Talapatra S, Ganesan P G, Kim T, Vajtai R, Huang M, Shima M, Ramanath G, Srivastava D, Deevi S C and Ajayan P M 2005 *Phys. Rev. Lett.* **95** 097201

[118] Mao H-K, Chen X-J, Ding Y, Li B and Wang L 2018 *Rev. Mod. Phys.* **90** 015007

[119] Barry J F, Schloss J M, Bauch E, Turner M J, Hart C A, Pham L M and Walsworth R L 2020 *Rev. Mod. Phys.* **92** 015004

[120] Aharonovich I, Castelletto S, Simpson D A, Su C H, Greentree A D and Prawer S 2011 *Rep. Prog. Phys.* **74** 076501

[121] Wang W, Doering P, Tower J, Ren L and Eaton-Magaña S J 2010 *Geems & Gemology* **46** 4

[122] Schreck M, Asmussen J, Shikata S, Arnault J-C and Fujimori N 2014 *MRS Bull.* **39** 504

[123] Sheng-Yi Xie L, Wang F, Liu X-B, Li L, Bai V B, Prakapenka Z, Cai H -k and Mao S 2018 *J. Phys. Chem. Lett.* **9** 2388–2393

[124] Tang R, Li Y, Xie S, Li N, Chen J, Gao C, Zhu P and Wang X 2016 *Sci. Rep.* **6** 38566

[125] Yanming Ma, Eremets M, Oganov A R, Xie Y, Trojan I, Medvedev S, Lyakhov A O, Valle M and Prakapenka V 2009 *Nature* **458** 182

[126] Drozdov A P, Eremets M I, Troyan I A, Ksenofontov V and Shylin S I 2015 *Nature* **525** 73

[127] Drozdov A P *et al* 2019 *Nature* **569** 528

[128] Doherty M W, Manson N B, Delaney P, Jelezko F, Wrachtrup J and Hollenberg L C L 2013 *Phys. Rep.* **528** 1

[129] Waldherr G *et al* 2014 *Nature* **506** 204

[130] Abobeih M H, Randall J, Bradley C E, Bartling H P, Bakker M A, Degen M J, Markham M, Twitchen D J and Taminiau T H 2019 *Nature* **576** 411

[131] Humphreys P C, Kalb N, Morits J P J, Schouten R N, Vermeulen R F L, Twitchen D J, Markham M and Hanson R 2018 *Nature* **558** 268

[132] Novoselov K S, Geim A K, Morozov S V, Jiang D, Zhang Y, Dubonos S V, Grigorieva I V and Firsov A A 2004 *Science* **306** 666

[133] Kroto H W, Heath J R, O'Brien S C, Curl R F and Smalley R E 1985 *Nature* **318** 162

[134] Baughman R H *et al* 1999 *Science* **284** 1340

[135] Rizzo D J, Veber G, Jiang J, McCurdy R, Cao T, Bronner C, Chen T, Louie S G, Fischer F R and Crommie M F 2020 *Science* **369** 1597

[136] Xie S Y and Li X B 2021 *Nano-Micro Lett.* **13** 53

IOP Publishing

Functional Carbon Materials

Jianmin Ma and Jiantie Xu

Chapter 8

Activated carbon

Daxiong Wu, Wen Ma, Qinghe Yu and Jianmin Ma

Activated carbon (AC) is a low-cost, high-quality kind of porous carbon material, and it has a wide range of applications in various fields. This article briefly introduces the development history and the classification of ACs and summarizes the physical chemistry, the micro–nano structural properties of various ACs, and their basic synthesis methods. In addition, the applications of AC materials in supercapacitors, alkali-metal batteries, catalysts, field-effect transistor sensors, biomedicine, environmental governance, etc. are introduced. Finally, perspectives on AC for future research are discussed. We expect that the reader will obtain knowledge and information from this chapter.

8.1 History

AC is a kind of porous carbon material which has been widely applied worldwide in water treatment, seawater desalination, wastewater treatment, air purification, green gas capture, catalysts, and energy due to its unique physical and chemical properties [1–4]. AC is a black porous solid carbon made from wood, coal, lignite, coconut shell, and other carbon-rich raw materials. AC is a type of carbon material that has a high surface area, large porosity, a micro-/meso-/macroporous structure, and high surface reactivity; it is composed of up to 90% carbon [5, 6]. In addition, its carbon structure is rich in functional groups, such as carboxyls, carbonyls, phenols, lactones, and quinones, which in turn contain a large number of chemical elements such as oxygen, hydrogen, sulfur, and nitrogen. Therefore, the unique physical and chemical properties of AC depend on its functional groups, which make it a versatile material with a wide range of applications in many fields [7].

The existence of AC in human history can be traced back a long way, and its origin cannot be accurately determined. The predecessor of AC is charcoal. The earliest record dates back to 3750 BC. At that time, the Egyptians and Sumerians used charcoal to reduce the content of copper, zinc, and tin ore to make bronze, and it was also used as a smokeless fuel [8]. In 1550 BC, AC was used for medical

doi:10.1088/978-0-7503-4972-7ch8

purposes in Egypt. In 460–359 BC, the Greek physician Hippocrates used it to treat epilepsy in sheep. The Compendium of Materia Medica (Compendium of Materia Medica) by Li Shizhen (1518–1593) records that activated carbon was used to treat diseases [9]. It was not until 1773 that Scheele first discovered the special adsorption properties of charcoal for the treatment of gases. In 1786, he discovered the decolorization of solutions and first systematically illustrated the adsorption capacity of charcoal in the liquid phase [10, 11]. By 1872, the chemical industry had begun to use gas masks with carbon filters to prevent the inhalation of mercury vapor [9]. Later, with the promotion of gas masks, AC history entered its second stage. The AC market continues to expand; AC's adsorption and its catalytic properties are used in many industries such as refining, recycling, and synthesis. Following the development of these applications, AC plants have opened in the United States and elsewhere. AC as a man-made material was invented by the Swedish chemist Raphael von Ostreijko from 1900 to 1901. He obtained English and German patents in 1900 and 1901 for the manufacture of AC by the carbonization of metal chloride from plant-derived materials or by the reaction of carbon dioxide or water vapor with carbonized material [12]. First used in industrial production in a factory near Vienna in 1911, the product was powdered AC and its trade name was Epomit. In the same year, Norit was listed in the Netherlands. Carboraffin was sold in Czechoslovakia in 1912. Later, a factory in Ossin, Austria, used zinc chloride to chemically activate wood chips for industrial production, and Bayer's dye factory also adopted this method [13]. With the rapid development of modern society, environmental protection is increasingly taken seriously; progressively stricter government decrees promoted the rapid growth of the production and uses of AC in water purification, gas cleaning, air quality control, energy storage/conversion, and valuable chemicals and increased the amount of recycling. In the latter part of the 20th century, the environmental protection industry has become a large AC user [14]. In recent years, the demand for AC has been increasing: up until 2015, the demand for AC was as high as 12 804 million tons [15]. AC materials have attracted more and more researchers' attention in recent years. Figure 8.1 shows the numbers of papers published on the subjects of AC, graphite, graphene, carbon nanotubes, fullerenes, carbon fiber, and carbon black from 1995 to 2016. The number of published manuscripts related to AC is comparable to those for carbon fiber and graphite [15].

8.2 Classification of activated carbon

8.2.1 Classification by shape

AC is divided into five types according to shape, which are classified as follows: (1) powder AC: Its particle size is less than 0.18 mm, and this type of AC is the majority. (2) granular AC: its particle size is larger than 0.18 mm (about 80 mesh); (the sizes of amorphous granular AC particles are expressed in terms of the upper and lower limits of particle size); (3) cylindrical AC: its particle size is expressed in terms of the cross-sectional diameter of the cylindrical particles; (4) spherical AC: its particle size is expressed in terms of the diameter of the spherical particles; (5) honeycomb AC: its

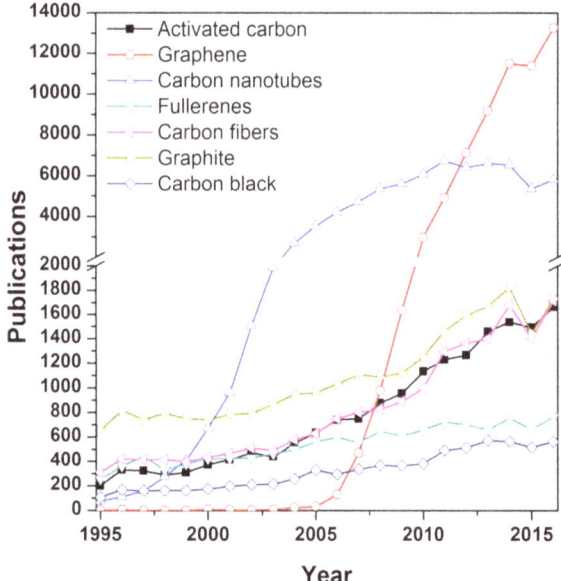

Figure 8.1. Number of articles published on carbon materials from 1995 to 2016 [15], reproduced with permission © Elsevier.

morphology is similar to a honeycomb and it is commonly marketed at a specified size of 100*100*100 mm. In addition, there are other special AC shapes. AC fiber has a diameter of 8–10 pm; its raw materials are viscose silk and polypropylene fiber. Compared to typical AC, AC fiber has a larger adsorption capacity, adsorption speed, and flexibility; it can be processed into various fabric shapes.

8.2.2 Classification by raw material

AC can be divided into six categories according to the raw material used, as follows: (1) wood AC, which is made of wood chips, charcoal, and so on; (2) shell AC is made of coconut shell, walnut shell, apricot stone, and other materials; (3) coal AC—its raw materials are lignite, peat, bituminous coal, anthracite, and so on; (4) petroleum AC, made from materials such as asphalt and other raw materials made of pitch-based spherical AC; (5) regenerated carbon, which uses waste carbon as a raw material via reactivation treatments of regenerated carbon (regenerated carbon is an important commodity and is available in large quantities); (6) organic waste and agricultural by-products containing carbon, such as rice husks, straw, cotton husks, coffee bean stalks, oil palm husks, bagasse, pulp waste, synthetic resins, can be converted into AC.

8.2.3 Classification by preparation method

AC can be prepared by a chemical method; for example, using zinc chloride as the activator for a mixed solution of raw material and zinc chloride heated to about 700 °C in the absence of air.

AC can be prepared by a physical method: for example, after the carbonization of the raw material, water vapor is used as the activator at a reaction temperature is about 1000 °C.

Impregnated AC: for example, AC is combined with iodine, silver, calcium or another inorganic impregnating agent; alternatively, special AC can be dipped in pyridine, ketone, a tertiary amine, or another organic impregnating agent.

Coated AC: for example, AC is coated with a biocompatible polymer to give it a smooth, permeable thin layer that does not clog the pores of the AC. This may be used, for example, to filter blood.

8.2.4 Classification by pore volume

AC is characterized by porosity, and its pore structure is formed through the synergistic effect of pore formation, pore expansion, pore combination, and pore collapse (figure 8.2) [1]. AC can be divided into large-pore and fine-pore AC according to the pore volume; large-pore AC refers to large pores, medium pores, or micropores in a large volume of AC and fine-pore AC refers to a large volume of micropore AC.

8.3 Fabrication of activated carbon

Depending on its shape and size, AC can be classified as granular carbon or powdered carbon; granular carbon is divided into shaped carbon (mainly columnar carbon, with a small amount of spherical carbon) and amorphous carbon (broken carbon). In general, AC preparation methods can be divided into three categories: [16, 17] the chemical activation method, the physical activation method, and the physicochemical activation method.

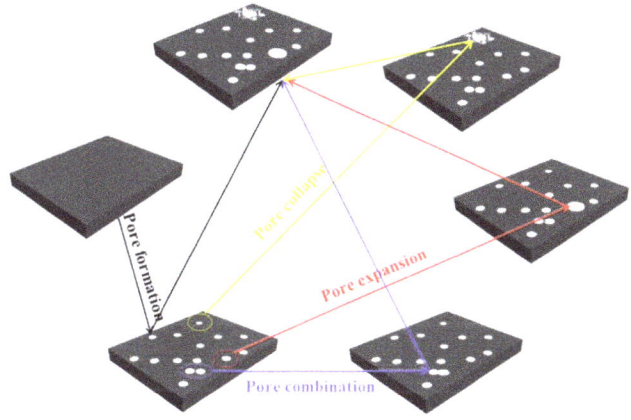

Figure 8.2. Pathway used to form a porous structure by chemical activation [1].

8.3.1 Chemical activation method

Acids (such as H_3PO_4), salts (such as $ZnCl_2$) (figure 8.3(a)), alkalis (such as KOH) (figure 8.3(b)), and other chemical agents are added to raw coal [18] and heated in an inert atmosphere. This method of carbonization and activation is called the chemical activation method [19–23]. The effects of chemical agents generally include infiltration, swelling, dissolution, oxidation, and dehydration. The principle of chemical activation is that the raw material is heated after dipping in chemical agent solution. Due to the dehydration caused by the chemical agent, the H and O elements in the raw material are released in the form of water vapor, forming a pore structure. Different chemical agents and activation temperatures, lead to different structure and performances of the resulting AC products. In one study, the activation temperature of H_3PO_4 was lower (400 °C–500 °C), and AC with a rich mesoporous structure was obtained. An AC with a specific surface area of more than 3000 $m^2\ g^{-1}$ can be obtained by activating coal with K_2CO_3 at 800 °C [24].

To date, the KOH alkaline fusion method (figure 8.4) has been the most effective method for improving the specific surface area of AC and reducing the ash content [24]. The alkali reacts with the silicon and aluminum compounds in the raw materials (such as kaolinite, quartz, etc.) to form soluble K_2SiO_3 or $KAlO_2$, which is washed away in the post-treatment, leaving a carbon skeleton with low ash

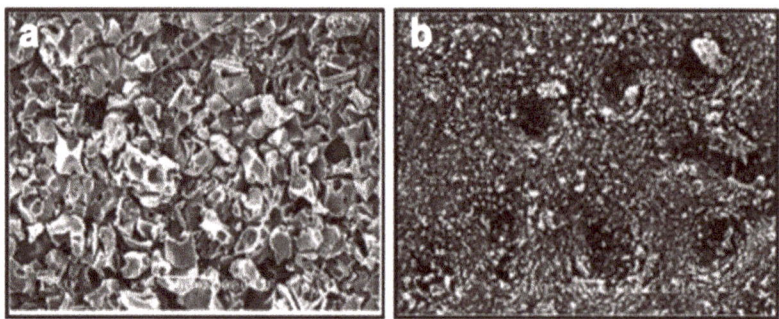

Figure 8.3. AC prepared using (a) $ZnCl_2$ and (b) KOH [18], reproduced with permission © Elsevier.

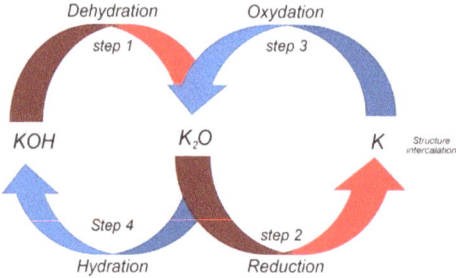

Figure 8.4. Mechanisms of KOH activation [24].

content. Alkali catalyzes the reaction of the carbon in coal to form the porous structure of AC. Baba *et al* believed that during infiltration by the KOH solution, potassium-bearing species could be embedded into the carbon structure of raw coal, expanding the carbon atomic layers. When the temperature rises, the KOH dehydrates and produces the metal oxide K_2O. When the temperature rises further, the K_2O reacts with the carbon in an inert atmosphere and is reduced to K, while the carbon escapes in the form of CO or CO_2 to form the pore structure of AC. When activated with CO_2, the embedded K_xO_y may be oxidized to K_xO_y +1 by CO_2, and then react with carbon to form K_xO_y and CO gases. K_xO_y reacts with CO_2 in the atmosphere and regenerates into K_xO_y +1, forming a catalytic cycle, and finally forming the pore structure of AC [24].

Chemical activation requires the raw materials to have high oxygen and hydrogen contents (no less than 25% and 5% as mass fractions, respectively) and is therefore suitable for wood raw materials (which have mass fractions of 43% for oxygen and 6% for hydrogen). However, except for a few young lignites (whose oxygen mass fractions reach 20%) and a few lignites and bituminous coals with medium and low degrees of metamorphism (whose hydrogen mass fractions are around 4.5%), few coals have sufficient oxygen and hydrogen contents to reach both of the above thresholds. In addition, although the pore structure of AC prepared by the chemical activation method is more developed, the activators are mainly corrosive substances, which corrode equipment and pollute the environment. Therefore, in industrial practice, the chemical method not commonly used to prepare coal-based AC.

8.3.2 Physical activation method

Physical activation is usually divided into two steps: [25] carbonization and activation. First, the coal is pyrolyzed at a low temperature to remove volatile substances from the coal, reduce the non-carbon elements (i.e. carbonization), and obtain carbonized material. The corresponding pore structure is then obtained by thermal destruction (i.e. activation) with water vapor or CO_2 at a high temperature (up to 1000 °C) and ablation of part of the carbon. In the process of physical activation, the gas activator has three kinds of activation effect on the carbonized material: (1) a reaction that removes tar and other amorphous carbon, so that the original closed pores become open and smooth; (2) the gas activator diffuses into the initial pore for the reaction, enlarging the pores (and sometimes causing wall burning, loss of adjacent pores, and pore consolidation); (3) the gas activator selectively activates the surface of the raw carbon and generates new pores. At the same activation temperature, the reaction speed of steam activation is higher than that of CO_2, and the specific surface area of AC produced using steam is higher than that of AC produced using CO_2. At the same loss rate, the AC obtained by lignite steam activation has a higher adsorption capacity than that obtained by CO_2 activation, its pore size distribution is wider, and its pore size is larger. The preparation of microporous AC by the physical activation method is relatively mature, but its disadvantages are that the activation time is long and is it hard to control the distribution of micropore sizes.

8.3.3 Physicochemical activation method

The physicochemical activation method is a combination of the chemical activation method and the gas activation method [26] and is also known as the catalytic activation method. First of all, certain amounts of chemical agents (additives, catalysts) are added to the raw coal, which is processed and formed. Carbonization and gas activation then produces AC with a special structure and performance. The main objective of mixing the raw materials with chemical agents is to use the chemical agents for different functions; the carbonization and activation of the AC additives controls the coking process; the effect of the residues is to change the phase of the carbonization activator and the carbon matrix reflexes to alter the composition, pore structure, surface chemistry and direction control of AC, thus laying the foundation for various kinds of application performance [16].

By changing the type and quantity of the additives, AC with well-developed pores and a reasonable pore size distribution can be prepared by the chemical physical activation method, which can improve the adsorption capacity. In particular, AC with well-developed pores can be prepared, and the adsorption capacity of AC for macromolecules in the liquid phase can be significantly improved. In addition, this method can be used to add special functional groups to the surface of AC materials; with the help of the chemical properties provided by the functional groups, their adsorption capacitites for specific pollutants can be improved, and AC adsorption materials can be used for chemical adsorption.

8.4 Applications of activated carbon

AC-based materials have large specific surface areas, porous structures and surface chemistries. Figure 8.5 shows the most common applications of AC from 1995 to 2016 according to the Scopus database. It is worth noting that the most studied area is the adsorption of heavy metal ions. In this chapter, we will focus on the applications of AC in supercapacitors, alkali-metal batteries, catalysts, field-effect transistor sensors, biomedicine, environmental governance, and so on.

8.4.1 Supercapacitors

The wide pore size distribution of AC plays an important role in improving its electrochemical performance during the charge and discharge of supercapacitors. Macropores ($\geqslant 50$ nm in size) are used as ion buffers for mesopores and micropores, while mesopores (2–50 nm in size), as known as transition pores, provide effective ion diffusion, and micropores (<2 nm in size) are used to store charges. In the case of graded porous carbon, there is an interconnection between these different pore structures, which helps the diffusion of electrolyte ions in the channel. Biomass-derived AC has an interconnected, multichannel, and porous structure, which is necessary for the design of better supercapacitors. Shijiao and his colleagues synthesized a hierarchical porous structure from corn shell and obtained a high specific capacitance of 356 F g^{-1} at 1 A g^{-1} [27]. Vijayan *et al* developed a thin cobalt film on porous carbon derived from biological waste to produce a layered

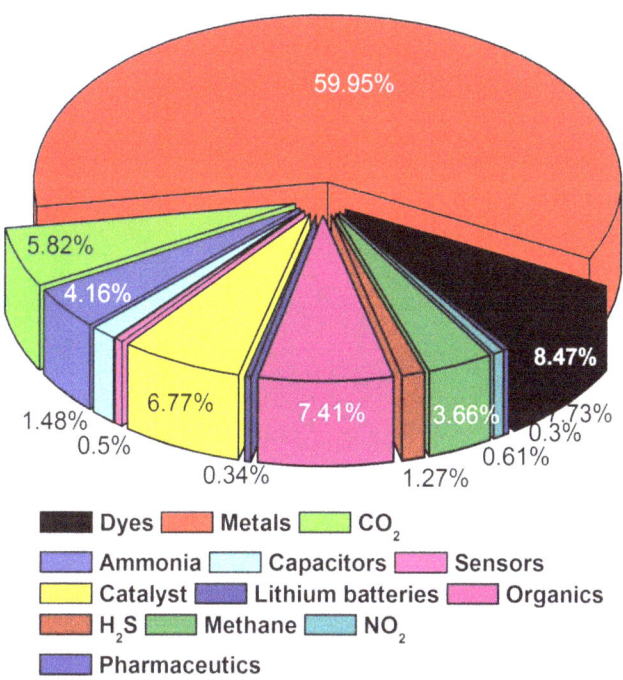

Figure 8.5. The percentages of the most common applications of AC from 1995 to 2016.

structure that provided excellent energy and power density [28]. The specific capacitance of electrode materials is also affected by surface wettability, that is, surface hydrophobicity/hydrophilicity, which is related to the surface functional groups. Therefore, the surfaces of electrode materials are enriched in oxygen-containing and nitrogen-containing functional groups by modification technology to obtain appropriate wettability. In addition, doping AC with heteroatoms such as boron, nitrogen, and sulfur produces more active centers and increases the hydrophilicity of the carbon surface, which is naturally present in some biological waste materials. Obviously, due to the hierarchical porous structure and good wettability of ACs, they are considered to be the best electrode materials for supercapacitors.

The raw materials of ACs are generally derived from metal–organic frameworks (MOFs) and biomass. Figure 8.6 shows a tree diagram of various carbon matrix composites and AC extracted from several biological wastes. Among them, MOFs have controllable morphology, variable pore size, and very high surface area with a 3D structure [30], which can produce ACs with large pore sizes. MOFs are highly crystalline materials obtained from the combination of a metal-containing anode and an organic connector, and can be used in various applications. Many researchers have reported that porous carbon prepared from MOFs has a high specific capacitance and a long cycle life [30, 31]. Osman *et al* successfully prepared activated porous carbon with a surface area of nearly 2315 m^2 g^{-1} from a Zn-based MOF [31]. It showed a specific capacitance of 325 F g^{-1} at a current density of 1 A g^{-1} and excellent capacitance retention after 150 000 cycles. Nevertheless, the AC derived

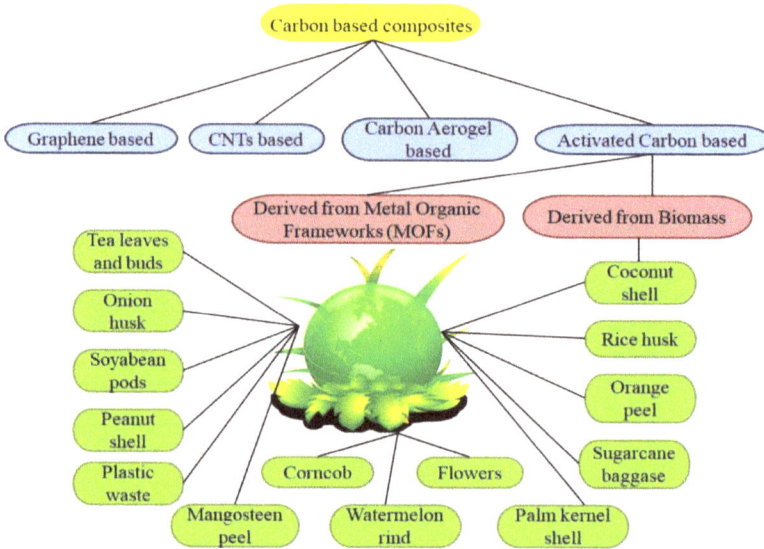

Figure 8.6. Various carbon-based composites and biowastes. Source: [29], reproduced with permission © Elsevier.

from MOFs is expensive. In this regard, it is necessary to produce low-cost carbon materials from biological waste, because biological waste is plentiful and environmentally friendly. It is an important source of carbon and helps to reduce dependence on fossil fuels. Many kinds of raw material, including corncob [32], tea waste [33], bagasse [34], rice husk [35], peanut shell [36], onion skin [37], olive pit [38], pine cone [39], watermelon skin [40], almond shell [41], coconut shell [42], banana skin [43], neem and ashoka leaf [44], walnut shell [45], and areca fiber [46] have already been used as precursors and described in the literature. Table 8.1 compares the performances of the biowaste-derived ACs studies reported in the literature to date. The selection of the biowaste precursor and the method used to activate it determine the surface area, the pore size distribution, the various functional groups on the surface, and the overall properties of electrode materials.

8.4.2 Rechargeable alkali-metal batteries

In the field of electrochemical energy storage and conversion, AC materials are widely used in lithium-ion batteries, metal–air batteries and hydrogen–oxygen fuel cells (figure 8.7) because of their excellent conductivity, large specific surface area, good chemical stability, and low cost. Some conductive agents, such as Vulcan XC-72 and super P, are made from AC [63, 64]. Conductive agents improve the electron transmission between electrode materials, so as to improve the electrochemical performance of batteries. In recent years, functional AC matrix composites for lithium secondary batteries have attracted great interest because they can overcome many disadvantages, such as low conductivity, large volume expansion, and dissolution in the electrolyte. The result of this combination is a battery with high

Table 8.1. Comparison of various biowaste-derived ACs as electrode materials.

Source of biowaste-derived AC	Activation method	Surface area of AC ($m^2 \, g^{-1}$)	Specific capacitance ($F \, g^{-1}$)	Electrolyte	Cycle stability	Reference
Coconut shell	Steam	1532	228 at 5 mV s^{-1}	6 M KOH	93% after 3000 cycles	[42]
Onion husk	K$_2$CO$_3$	2571	188 at 1 A g^{-1}	1 M TEABF$_4$/AC	92.5% after 2000 cycles	[37]
Peanut shell	ZnCl$_2$	1549	340 at 1 A g^{-1}	1 M H$_2$SO$_4$	95.3% after 10 000 cycles	[47]
Palm kernel shell	KOH		210 at 0.5 A g^{-1}	1 M KOH	95%–97% after 1000 cycles	[48]
Watermelon rind	KOH	2277	333.4 at 1 A g^{-1}	6 M KOH	96.8% after 10 000 cycles	[49]
Tea leaves	KOH	2841	330 at 1 A g^{-1}	2 M KOH	92% after 2000 cycles	[50]
Tea waste buds	KOH	1610	332 at 1 A g^{-1}	6 M KOH	97.8% after 100 000 cycles	[51]
Sugarcane bagasse	KOH	1939.9	298 at 1 A g^{-1}	1 M H$_2$SO$_4$	94.5% after 5000 cycles	[52]
Withered rose flowers	KOH/KNO$_3$	1980	350 at 1 A g^{-1}	6 M KOH	96.5% after 15 000 cycles	[53]
Albizia flowers	KOH	2757.6	406 at 0.5 A g^{-1}	6 M KOH	97% after 5000 cycles	[54]
Jujube fruit	NaOH	1135	460 at 1 A g^{-1}	6 M KOH	92.2% after 130 000 cycles	[55]
American poplar fruit waste	KOH	942	423 at 1 A g^{-1}	6 M KOH	97% after 200 000 cycles	[56]
Corncob	KOH	800	390 at 0.5 A g^{-1}	1 M H$_2$SO$_4$	94% after 5000 cycles	[57]
Plastic waste (polyethylene terephthalate)	KOH	2326	169 at 0.2 A g^{-1}	6 M KOH	90.6% after 5000 cycles	[58]
Rice husk	KOH	3145	367 at 5 mV s^{-1}	6 M KOH	≈100% after 30 000 cycles	[59]
Soybean pods	NaOH	2612	352.6 at 0.5 A g^{-1}	1 M Na$_2$SO$_4$	94.2% after 50 000 cycles	[60]
Orange peel	KOH	2160	460 at 1 A g^{-1}	1 M H$_2$SO$_4$	98% after 10 000 cycles	[61]
Mangosteen peels	NaOH	2623	357 at 1 A g^{-1}	6 M KOH	94.5% after 130 000 cycles	[62]

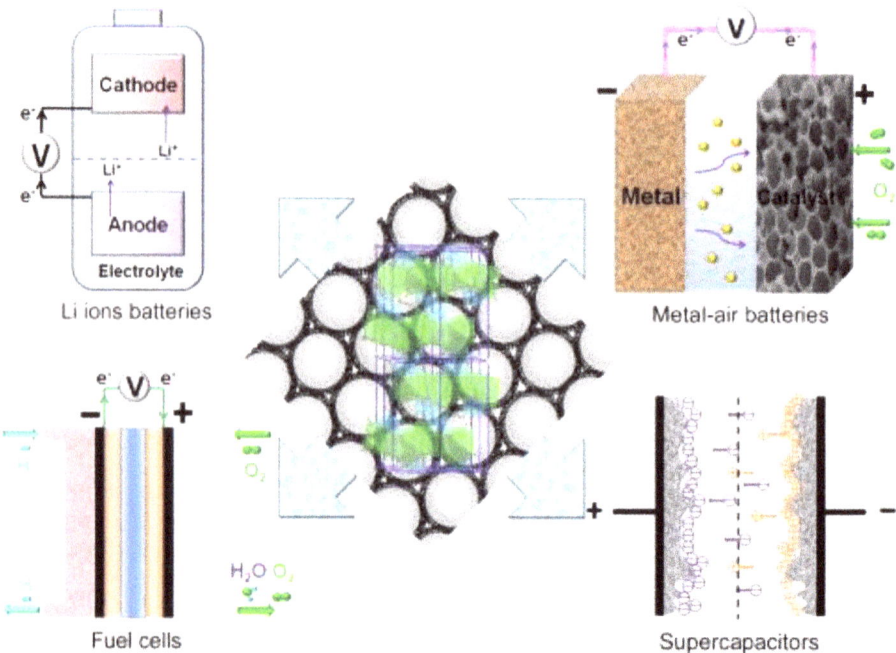

Figure 8.7. Typical applications of porous carbon in electrochemical energy conversion and storage (a lithium-ion battery, a metal–air battery, a fuel cell and a supercapacitor). Source: [65], reproduced with permission © Elsevier.

energy density. The composite electrodes have excellent rate performance, long cycle stability, and high catalytic performance.

Zhang *et al* reported a CoO/CMK-3 nanocomposite prepared by the impregnation method using $CO(NO_3)_2 \cdot 6H_2O$ as a cobalt source [66]. Compared to CMK-3, the composite had better specific capacity, coulombic efficiency, and cycle performance. Lee's team has successfully synthesized an Fe_3O_4-nanocrystal-impregnated mesoporous carbon foam (CF) composite material using $Fe(NO_3)_3$ as the raw material [67]. The results showed that the capacity of the composite was still more than 780 mAh·g^{-1} after 50 cycles. The CF widely dispersed the Fe_3O_4 nanocrystals, preventing their agglomeration and reducing capacity attenuation during charge/discharge. Shen *et al* designed a nanocasting method for the synthesis of a $Li_4Ti_5O_{12}$/CMK-3 composite, in which ordered mesoporous CMK-3 was used as a stable nanostructure matrix and $Li_4Ti_5O_{12}$ was used as the active material [68]. CMK-3 was not only used as a conductive substrate to provide fast electron transport, but also as a reaction vessel for the $Li_4Ti_5O_{12}$ precursor to prevent the $Li_4Ti_5O_{12}$ particles from agglomerating during heat treatment. In addition, various nanochannels provided a large contact area between the electrolyte and the electrode, ensuring that the electrolyte could easily be injected into the mesopores. As shown in figure 8.8, the high capacity of the $Li_4Ti_5O_{12}$/CMK-3 nanocomposites was 92.6 mAh g^{-1} at 40 °C, and their capacity loss was only 5.6% after 1000 cycles at 20 °C.

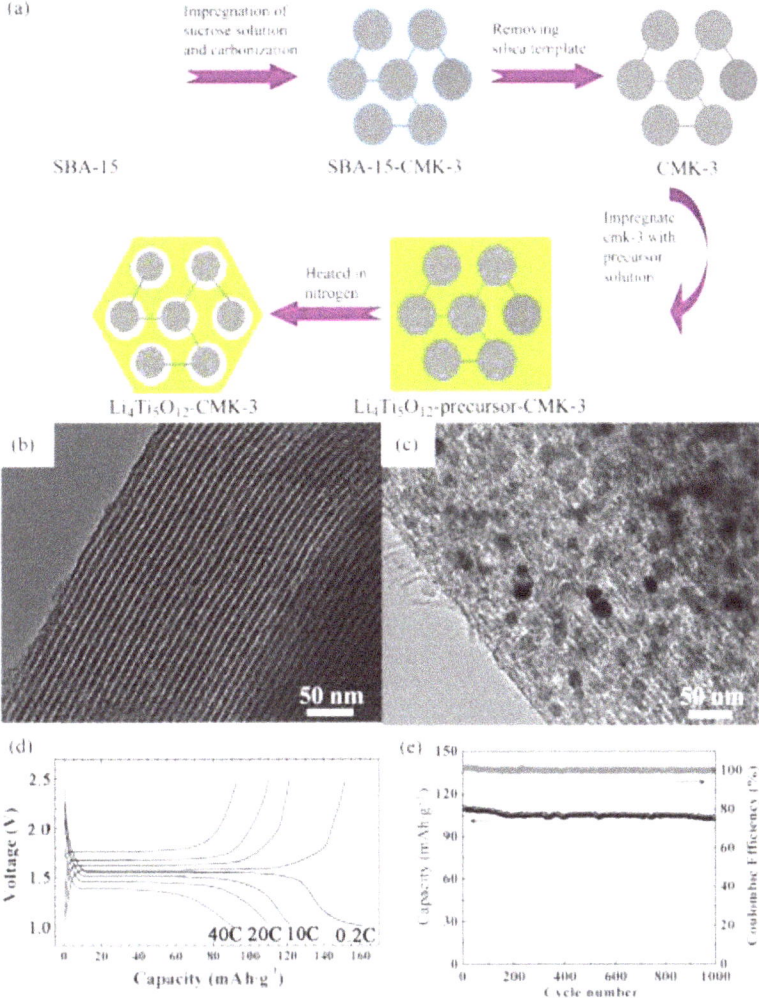

Figure 8.8. (a) Schematic diagram of the synthesis of the mesoporous $Li_4Ti_5O_{12}$/CMK-3 nanocomposite. (b) TEM image of the CMK-3 carbon template. (c) TEM image of the mesoporous $Li_4Ti_5O_{12}$/CMK-3 nanocomposite. (d) Constant-current charge–discharge curves of the mesoporous $Li_4Ti_5O_{12}$/CMK-3 nanocomposite electrode at different rates, and (E) the specific capacity and coulombic efficiency of the mesoporous $Li_4Ti_5O_{12}$/CMK-3 nanocomposite after 1000 cycles at 20 C. Source: [68], reproduced with permission © John Wiley and Sons.

8.4.3 Catalysts

Sustainable AC can be obtained by the pyrolysis/activation of biomass wastes from different sources. The carbon obtained in this way shows interesting properties, such as high specific surface area, conductivity, thermal and chemical stability, and porosity. These characteristics, in addition to customizable pore size distribution and the possibility of functionalization, have led to the increased use of AC in catalysis. The use of AC produced from biomass is a step toward the development of a more sustainable process and strengthens material recovery and reuse within the framework of the circular economy.

Porous carbon-based materials are used as general carriers for the preparation of heterogeneous catalysts [69–71]. For example, Mo nanoparticles have been incorporated into biochar to produce biochar nanostructured complexes that are active in the electrocatalytic hydrogen evolution reaction (HER). The Mo_2C nanoparticles were grown on biomass in the form of soybean. As a low-cost carbon carrier, its overpotential was 177 mV for a driving current density of 10 mA cm^{-2} [72]. In another study, biochar derived from sunflower seed shell biomass was modified with Mo_2C nanoparticles [73]. This nanoparticle-based catalyst required an overpotential of 60 mV to provide a current density of 10 mA cm^{-2}, as shown in figure 8.9. This catalyst had good molecular stability and a faradic efficiency of about 100%.

The oxygen reduction reaction (ORR) and the oxygen evolution reaction (OER) are the two most important reactions in energy storage and conversion systems, such as metal–air cells and fuel cells [75]. Wang and his colleagues reported a recent example of metal nanoparticles supported on carbon produced by biomass pyrolysis [76]. Co nanoparticles were dispersed in *Chlorella* biomass by impregnation. After pyrolysis at 900 °C, the biomass was placed in an Ar atmosphere for 1 h, producing carbon nanotube materials that contained Co nanoparticles in their structure

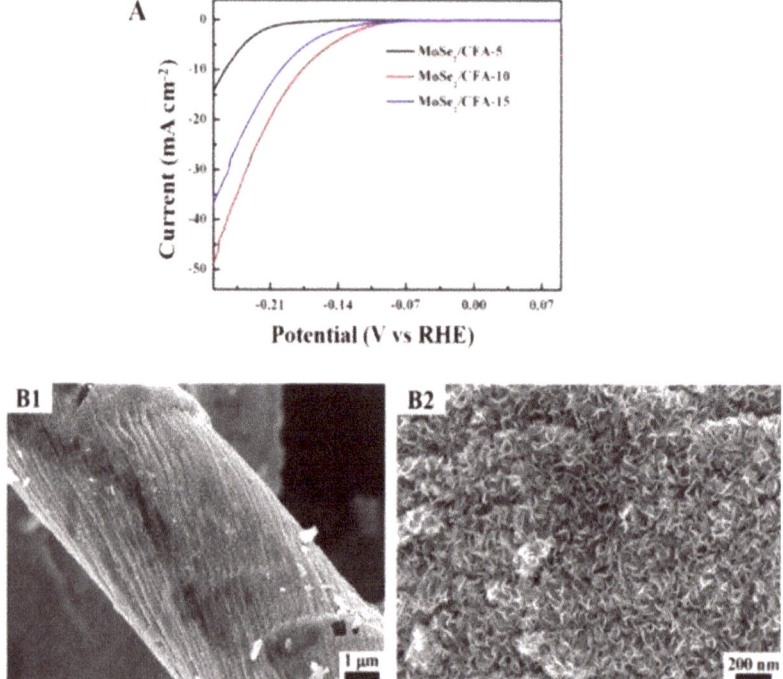

Figure 8.9. (a) Polarization curve of $MoSe_2$/carbon fiber aerogel (CFA) hybridized modified glassy carbon electrode (GCE) obtained using linear sweep voltammetry (LSV). (b) Field emission scanning electron microscopy (FESEM) images of the $MoSe_2$/CFA hybrids (B1, B2). Source: [74], reproduced with permission © American Chemical Society.

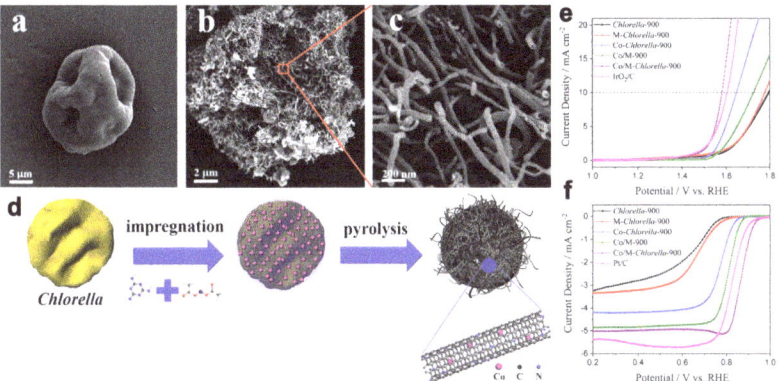

Figure 8.10. (a) SEM images of *Chlorella* and (b) and (c) SEM images of Co/M-Chloroella-900. (d) Schematic diagram of the preparation process of Co/M-Chlorella-900. (e) LSV curves of five *Chlorella* catalysts and 20 wt % commercially available Pt/C catalyst in 0.1 M KOH solution. (f) LSV curves of the *Chlorella* catalysts and a commercial IrO$_2$/C catalyst in 0.1 M KOH solution. Source: [76], reproduced with permission © American Chemical Society.

(figure 8.10). These nanoparticles can be used as bifunctional catalysts for the ORR and the OER. When the nanoparticles were used in the OER, the overpotential at 10 mA cm^{-2} was 23 mV lower than that recorded for the IrO$_2$/C reference catalyst (figure 8.10(e)). In the ORR, their half-wave potential was 40 mV higher than that of the reference catalyst based on 20% wt Pt/C (figure 8.10(f)).

8.4.4 Field-effect transistor sensors

The possibility of using cheap industrial waste to produce AC with special properties has always attracted broad attention worldwide. Waste resources such as carbon-containing agricultural and sideline products, low-quality coal, waste plastics, waste rubber, and petroleum by-products can be further efficiently utilized, and these waste materials can be used to prepare AC for a wide range of applications. AC has aroused great research interest because of its large specific surface area and abundant surface active groups [77, 78]. Due to its large specific surface area, high electric double-layer capacitance, and high electrostatic charge storage capacity, AC has become a potential gate electrode material. This material can realize a fast and highly reversible electrostatic process, therefore, its use as a quasi-reference electrode can greatly simplify device structure. Tang *et al* studied doped poly(3,4-ethylenedioxythiophene) (PEDOT) :poly(styrenesulfonate) (PSS) organic electrochemical transistors using AC gate electrodes with a high specific surface area [79]. Compared with a PEDOT:PSS grid with a similar geometric area, the AC grid achieved higher current modulation at a low voltage, and it was beneficial that there was no Faraday process at the interface between the AC electrode and the electrolyte. Through a cyclic voltammetry test, it was found that the AC used as the grid exhibited high double-layer capacitance. Therefore, the AC electrode was processed and integrated into the in-plane flexible device structure, providing a new reference for organic electrochemical transistors (OECTs) new electrode materials.

8.4.5 Biomedicine

AC materials have broad application prospects in the field of medicine and health; and medical materials made of AC materials, such as non-woven fabrics and other medical devices, have been widely used. Compared with traditional fibers, AC fiber has obvious advantages in terms of moisture absorption, adsorption, and biocompatibility. It has been used as a biological carrier in combination with copper, zinc, silver, and other nanoparticles to prepare new medical dressings with excellent antibacterial effects and good air permeability. The copper and zinc nanoparticles are loaded on an AC fiber substrate by chemical vapor deposition, and the carbon nanofibers are combined to prepare a composite material [80]. The asymmetric distribution of bimetals enables the composite material to effectively inhibit the growth of various bacteria and promote wound healing. Its inhibitory effect is more obvious than that of AC fiber cultured with Cu or Zn alone. The bimetallic composite material prepared as part of this research has broad application prospects in the medical and health field. In addition, when this AC fiber is used in medical accessories, it can maintain a long-term antibacterial effect; it has strong far-infrared/ radio activity that promotes skin collagen formation and growth factor secretion, and the fabric made has good mechanical properties that ensure the fabric is stable in use.

8.4.6 Environmental governance

In response to the gradual increase in the performance requirements of AC in worldwide markets, the pore structure of AC can be adjusted by process control and technical treatments, and its surface groups can be modified to improve its adsorption and catalytic performance. In addition to the fields of biomedicine and transistors, the important applications of AC also include the field of environmental protection, including wastewater treatment, gas purification, the adsorption of heavy metals and volatile organic compounds, and so on.

The enhancement of the characteristics of modern buildings in terms of heat preservation, noise reduction, and energy savings has also been accompanied by a gradual improvement in the airtightness of the building structure. The resulting problem of the elimination of harmful chemicals emitted by home building materials has become increasingly prominent. These harmful chemicals may cause asthma, allergies, immune diseases, and other hazards. The cleaning of the air in confined living spaces has attracted more attention. The fixed filter sheet is a filter material made by bonding and fixing AC to a foamed material that is used for ventilation; the content of AC in the fixed filter can reach 15%—80%. Sulfur pollution caused by the use of coal and petroleum is an important part of the prevention and control of air pollution. AC loaded with Co, Ni, and Mg compounds is used to remove sulfur dioxide gas in a new type of desulfurization technology. Metal ions are introduced into wood surfaces by methods such as ion exchange, and then activated by carbon.

AC treated with sodium carbonate solution has good selectivity for H_2S [81]. Gurwinder Singh *et al* prepared activated porous carbon spheres by carbonizing inexpensive D-glucose with non-corrosive potassium acetate, as shown in figure 8.11 [82]. The prepared activated porous carbon spheres effectively captured carbon dioxide [82]. The pyrolytic and non-pyrolytic lignocellulose residues produced in the food industry can be used as precursor materials. A chemical activation process that uses several activators to produce adsorbents is shown in figure 8.12 [83].

Water treatment problems such as the purification of pollution sources, the treatment of organic industrial wastewater and waste gas, and the removal of inorganic heavy metal ions can also be solved by modified AC. Ordinary AC fiber is modified to improve its adsorption performance [84]. Anwar *et al* oxidized AC to generate more adsorption sites and enhanced its adsorption performance for heavy

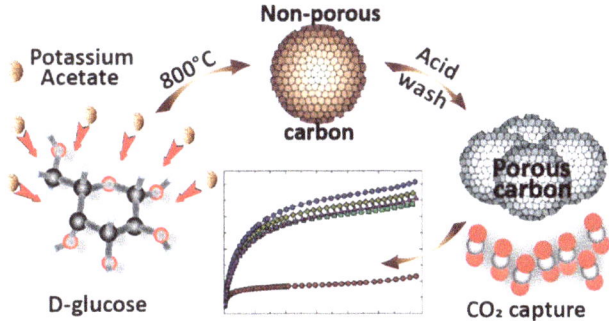

Figure 8.11. The use of carbonized D-glucose to produce porous carbon balls [82], reproduced with permission © Elsevier.

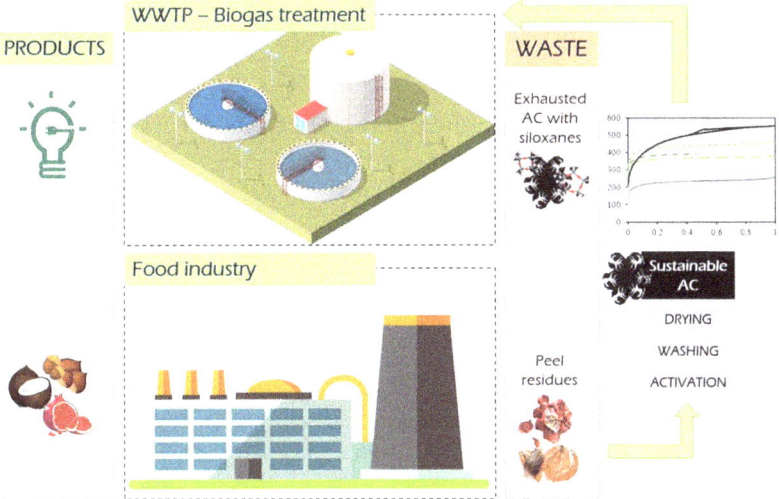

Figure 8.12. The use of lignocellulosic AC for efficient sewage purification [83], reproduced with permission © Elsevier.

metals such as Cr [85]. When AC fiber is used for waste disposal treatments, the number of fiber pores and their volume can affect the adsorption performance of the fiber for harmful gases such as formaldehyde. AC fiber mesh can be used instead of AC fiber cloth to treat high-volume organic waste gas, but it has the characteristics of low mechanical strength and large bed resistance, and thus requires special reinforcement.

8.5 Perspectives

AC is an the important research field in materials science. Due to its low preparation cost, simple synthesis process, high porosity, and the high specific surface area of its porous structure, it is a promising choice and is widely available for use in supercapacitors, alkali-metal batteries, catalysts, field-effect transistor sensors, biomedicine, environmental governance, etc. However, research into AC is still growing, and the future demand for AC products will increase year by year. The following points are proposed for future research:

(1) In future research, the preparation or mass production of AC should focus on reducing or even avoiding the potential pollution to the environment caused by chemical activation, the recovery of pyrolytic gas, and the disposal of waste generated in the process of chemical activation and paying attention to sustainable synthesis methods that have green or zero emissions.

(2) To date, few *in situ* techniques (XPS, FTIR, XRD, TEM, and SEM, etc.) or theoretical computational simulations (molecular dynamics models and density functional theory) have revealed the real-time interaction between the active carbon precursors and the chemical activators and the mechanism of pore formation. Therefore, an important research topic is the use of *in situ* characterization technology to understand the changes of pore structure and surface functional groups that occur under different activation conditions.

(3) With the widespread application of AC, the recovery of AC has begun to attract attention. If used AC cannot be recycled, not only will the treatment cost of each ton of wastewater increase, but such treatments will also cause secondary pollution of the environment. Therefore, the regeneration of AC is particularly important.

(4) it is usually considered that the application of AC has no safety problems, but in fact there is no absolute guarantee of its safety. The safety of AC applications should not be taken lightly; the nature of AC and the likelihood of any associated risks should be understood.

Although AC has achieved great success in wastewater treatment, adsorption/separation, fire prevention, catalysis, energy storage, and gas capture, with the rapid development of society, its use in different fields has to meet increasingly high requirements; therefore, greater effort is needed to update the related equipment to meet the needs of society, while ensuring its function and performance.

References

[1] Gao Y, Yue Q, Gao B and Li A 2020 Insight into activated carbon from different kinds of chemical activating agents: a review *Sci. Total Environ.* **746** 141094

[2] Le-Minh N, Sivret E C, Shammay A and Stuetz R M 2018 Factors affecting the adsorption of gaseous environmental odors by activated carbon: a critical review *Crit. Rev. Env. Sci. Tec.* **48** 341–75

[3] Yousefi M, Arami S M, Takallo H, Hosseini M, Radfard M, Soleimani H and Mohammadi A A 2019 Modification of pumice with HCl and NaOH enhancing its fluoride adsorption capacity: kinetic and isotherm studies *Hum. Ecol. Risk Assess.* **25** 1508–20

[4] Daud W M A W and Houshamnd A H 2010 Textural characteristics, surface chemistry and oxidation of activated carbon *J. Nat. Gas Chem.* **19** 267–79

[5] Morin-Crini N, Loiacono S, Placet V, Torri G, Bradu C and Kostić M *et al* 2019 Hemp-based adsorbents for sequestration of metals: a review *Environ. Chem. Lett.* **17** 393–408

[6] Jeirani Z, Niu C H and Soltan J 2017 Adsorption of emerging pollutants on activated carbon *Rev. Chem. Eng.* **33** 491–522

[7] Heidarinejad Z, Dehghani M H, Heidari M, Javedan G, Ali I and Sillanpää M 2020 Methods for preparation and activation of activated carbon: a review *Environ. Chem. Lett.* **18** 393–415

[8] Derbyshire F, Jagtoyen M and Thwaites M 1995 *Porosity in Carbons: Characterization and Applications* ed J W Patrick vol 1 (London: Edward Arnold)

[9] Inagaki M and Tascón J M D 2006 Pore formation and control in carbon materials *Interface Science and Technology* vol 7 (Amsterdam: Elsevier) ch 2 pp 49–105

[10] Wan Daud W M, Ali W S and Sulaiman M Z 2003 Effect of activation temperature on pore development in activated carbon produced from palm shell *J. Chem. Technol. Biot.* **78** 1–5

[11] Chen J (ed) 2016 *Activated Carbon Fiber and Textiles Woodhead Publishing Series in Textiles* (Sawston: Woodhead Publishing)

[12] Sontheimer H, Crittenden J C and Summers R S 1988 *Activated Carbon for Water Treatment* vol 90 (Karlsruhe: DVGW-Forschungsstelle)

[13] Dąbrowski A 2001 Adsorption – from theory to practice *Adv. Colloid Interface Sci.* **93** 135–224

[14] Mourão P A M, Laginhas C, Custódio F, Nabais J M V, Carrott P J M and Ribeiro Carrott M M L 2011 Influence of oxidation process on the adsorption capacity of activated carbons from lignocellulosic precursors *Fuel Process. Technol.* **92** 241–46

[15] González-García P 2018 Activated carbon from lignocellulosics precursors: a review of the synthesis methods, characterization techniques and applications *Renewable Sustainable Energy Rev.* **82** 1393–414

[16] Xie Q, Zhang X-l, Li L-T and JIN L 2005 Porosity adjustment of activated carbon: theory, approaches and practice *New Carbon Mater.* **2** 183–90

[17] Wei N, Zhao N-Q and Jia W J 2003 New progress in the fabrication and application of activated carbon *Mat. Sci. Eng* **5** 777–80

[18] Yahya M A, Al-Qodah Z and Zanariah Ngah C Z 2015 Agricultural bio-waste materials as potential sustainable precursors used for activated carbon production: a review *Renewable Sustainable Energy Rev.* **46** 218–35

[19] Gong G-z, Qiang X, Zheng Y-f, Ye S-f and Chen Y-f 2009 Regulation of pore size distribution in coal-based activated carbon *New Carbon Mater.* **24** 141–46

[20] Liu L, Liu Z, Yang J, Huang Z and Liu Z 2007 Effect of preparation conditions on the properties of a coal-derived activated carbon honeycomb monolith *Carbon* **45** 2836–42

[21] Kopac T and Toprak A J 2007 Preparation of activated carbons from Zonguldak region coals by physical and chemical activations for hydrogen sorption *Int. J. Hydrogen Energy* **32** 5005–14

[22] Nowicki P, Pietrzak R and Wachowska H 2008 Siberian anthracite as a precursor material for microporous activated carbons *Fuel* **87** 2037–40

[23] Lillo-Ródenas M, Cazorla-Amorós D, Linares-Solano A, Béguin F, Clinard C and Rouzaud J N 2004 HRTEM study of activated carbons prepared by alkali hydroxide activation of anthracite *Carbon* **42** 1305–10

[24] Hayashi J i, Uchibayashi M, Horikawa T, Muroyama K and Gomes V G 2002 Synthesizing activated carbons from resins by chemical activation with K_2CO_3 *Carbon* **40** 2747–52

[25] Jibril B Y, Al-Maamari R S, Hegde G, Al-Mandhary N and Houache O 2007 Effects of feedstock pre-drying on carbonization of KOH-mixed bituminous coal in preparation of activated carbon *J. Anal. Appl. Pyrol.* **80** 277–82

[26] Pastor-Villegas J and Durán-Valle C J 2002 Pore structure of activated carbons prepared by carbon dioxide and steam activation at different temperatures from extracted rockrose *Carbon* **40** 397–402

[27] Song S J, Ma F W, Wu G, Ma D, Geng W D and Wan J F 2015 Facile self-templating large scale preparation of biomass-derived 3D hierarchical porous carbon for advanced super-capacitors *J. Mater. Chem.* A **3** 18154–62

[28] Vijayan B L, Misnon I I, Karuppaiah C, Kumar G M A, Yang S Y, Yang C C and Jose R 2021 Thin metal film on porous carbon as a medium for electrochemical energy storage *J. Power Sources* **489** 229522

[29] Saini S, Chand P and Joshi A 2021 Biomass derived carbon for supercapacitor applications: review *J Energy Storage* **39** 102646

[30] Sun Y Z, Guo S C, Li W, Pan J Q, Fernandez C, Senthil R A and Sun X L 2018 A green and template-free synthesis process of superior carbon material with ellipsoidal structure as enhanced material for supercapacitors *J. Power Sources* **405** 80–8

[31] Osman S, Senthil R A, Pan J Q and Li W 2018 Highly activated porous carbon with 3D microspherical structure and hierarchical pores as greatly enhanced cathode material for highperformance supercapacitors *J. Power Sources* **391** 162–69

[32] Wang D B, Geng Z, Li B and Zhang C M 2015 High performance electrode materials for electric double-layer capacitors based on biomass-derived activated carbons *Electrochim. Acta* **173** 377–84

[33] Song X Y, Ma X L, Li Y, Ding L and Jiang R Y 2019 Tea waste derived microporous active carbon with enhanced double-layer supercapacitor behaviors *Appl. Surf. Sci.* **487** 189–97

[34] Rufford T E, Hulicova-Jurcakova D, Khosla K, Zhu Z H and Lu G Q 2010 Microstructure and electrochemical double-layer capacitance of carbon electrodes prepared by zinc chloride activation of sugar cane bagasse *J. Power Sources* **195** 912–18

[35] Wu M B, Li L Y, Liu J, Li Y, Ai P P, Wu W T and Zheng J T 2015 Template-free preparation of mesoporous carbon from rice husks for use in supercapacitors *New Carbon Mater.* **30** 471–75

[36] Purkait T, Singh G, Singh M, Kumar D and Dey R S 2017 Large area few-layer graphene with scalable preparation from waste biomass for high-performance supercapacitor *Sci. Rep.* **7** 15239

[37] Wang D W, Liu S J, Fang G L, Geng G H and Ma J F 2016 From trash to treasure: direct transformation of onion husks into three-dimensional interconnected porous carbon frameworks for high-performance supercapacitors in organic electrolyte *Electrochim. Acta* **216** 405–11

[38] Redondo E, Carretero-González J, Goikolea E, Ségalini J and Mysyk R 2015 Effect of pore texture on performance of activated carbon supercapacitor electrodes derived from olive pits *Electrochim. Acta* **160** 178–84

[39] Karthikeyan K, Amaresh S, Lee S N, Sun X L, Aravindan V, Lee Y G and Lee Y S 2014 Construction of high-energy-density supercapacitors from pine-cone-derived high-surface-area carbons *ChemSusChem* **7** 1435–42

[40] Lin X Q, Yang N, Lü Q F and Liu R 2019 Self-nitrogen-doped porous biocarbon from watermelon rind: a high-performance supercapacitor electrode and its improved electro-chemical performance using redox additive electrolyte *Energy Technol.* **7** 1800628

[41] Marcilla A, García-García S, Asensio M and Conesa J A 2000 Influence of thermal treatment regime on the density and reactivity of activated carbons from almond shells *Carbon* **38** 429–40

[42] Mi J, Wang X R, Fan R J, Qu W H and Li W C 2012 Coconut-shell-based porous carbons with a tunable micro/mesopore ratio for high-performance supercapacitors *Energy Fuel* **26** 5321–29

[43] Lv Y K, Gan L H, Liu M X, Xiong W, Xu Z J, Zhu D Z and Wright D S 2012 A self-template synthesis of hierarchical porous carbon foams based on banana peel for super-capacitor electrodes *J. Power Sources* **209** 152–57

[44] Biswal M, Banerjee A, Deo M and Ogale S 2013 From dead leaves to high energy density supercapacitors *Energy Environ. Sci.* **6** 1249–59

[45] Xu X Y, Gao J P, Tian Q, Zhai X G and Liu Y 2017 Walnut shell derived porous carbon for a symmetric all-solid-state supercapacitor *Appl. Surf. Sci.* **411** 170–76

[46] Natalia M, Sudhakar Y N and Selvakumar M 2013 Activated carbon derived from natural sources and electrochemical capacitance of double layer capacitor *Indian. J. Chem. Techn.* **20** 392–99

[47] Xiao Z A, Chen W W, Liu K, Cui P and Zhan D 2018 Porous biomass carbon derived from peanut shells as electrode materials with enhanced electrochemical performance for super-capacitors *Int. J. Electrochem Sc.* **13** 5370–81

[48] Misnon I I, Zain N K M, Abd Aziz R, Vidyadharan B and Jose R 2015 Electrochemical properties of carbon from oil palm kernel shell for high performance supercapacitors *Electrochim. Acta* **174** 78–86

[49] Mo R J, Zhao Y, Wu M, Xiao H M, Kuga S, Huang Y, Li J P and Fu S Y 2016 Activated carbon from nitrogen rich watermelon rind for high-performance supercapacitors *RSC Adv.* **6** 59333–42

[50] Peng C, Yan X B, Wang R T, Lang J W, Ou Y J and Xue Q J 2013 Promising activated carbons derived from waste tea-leaves and their application in high performance super-capacitors electrodes *Electrochim. Acta* **87** 401–8

[51] Khan A, Senthil R A, Pan J Q, Osman S, Sun Y Z and Shu X 2020 A new biomass derived rod-like porous carbon from tea-waste as inexpensive and sustainable energy material for advanced supercapacitor application *Electrochim. Acta* **335** 135588

[52] Wang B, Wang Y H, Peng Y Y, Wang X, Wang J and Zhao J B 2018 3-Dimensional interconnected framework of N-doped porous carbon based on sugarcane bagasse for application in supercapacitors and lithium ion batteries *J. Power Sources* **390** 186–96

[53] Khan A, Senthil R A, Pan J Q, Sun Y Z and Liu X G 2020 Hierarchically porous biomass carbon derived from natural withered rose flowers as high-performance material for advanced supercapacitors *Batteries Supercaps* **3** 731–37

[54] Wu F M, Gao J P, Zhai X G, Xie M H, Sun Y, Kang H Y, Tian Q and Qiu H X 2019 Hierarchical porous carbon microrods derived from albizia flowers for high performance supercapacitors *Carbon* **147** 242–51

[55] Yang V, Senthil R A, Pan J Q, Kumar T R, Sun Y Z and Liu X G 2020 Hierarchical porous carbon derived from jujube fruits as sustainable and ultrahigh capacitance material for advanced supercapacitors *J. Colloid Interf. Sci.* **579** 347–56

[56] Kumar T R, Senthil R A, Pan Z G, Pan J Q and Sun Y Z 2020 A tubular-like porous carbon derived from waste American poplar fruit as advanced electrode material for high-perform-ance supercapacitor *J. Energy Storage* **32** 101903

[57] Karnan M, Subramani K, Srividhya P K and Sathish M 2017 Electrochemical studies on corncob derived activated porous carbon for supercapacitors application in aqueous and non-aqueous electrolytes *Electrochim. Acta* **228** 586–96

[58] Wen Y, Kierzek K, Min J, Chen X, Gong J, Niu R, Wen X, Azadmanjiri J, Mijowska E and Tang T 2019 Porous carbon nanosheet with high surface area derived from waste poly (ethylene terephthalate) for supercapacitor applications *J. Appl. Polymer Sci.* **137** 48338

[59] Gao Y, Li L, Jin Y M, Wang Y, Yuan C J, Wei Y J, Chen G, Ge J J and Lu H Y 2015 Porous carbon made from rice husk as electrode material for electrochemical double layer capacitor *Appl. Energy* **153** 41–7

[60] Kong X D, Zhang Y, Zhang P, Song X L and Xu H M 2020 Synthesis of natural nitrogen-rich soybean pod carbon with ion channels for low cost and large areal capacitance supercapacitor *Appl. Surf. Sci.* **516** 146162

[61] Subramani K, Sudhan N, Karnan M and Sathish M 2017 Orange peel derived activated carbon for fabrication of high-energy and high-rate supercapacitors *Chemistryselect* **2** 11384–92

[62] Yang V, Senthil R A, Pan J Q, Khan A, Osman S, Wang L R, Jiang W C and Sun Y Z 2019 Highly ordered hierarchical porous carbon derived from biomass waste mangosteen peel as superior cathode material for high performance supercapacitor *J. Electroanal. Chem.* **855** 113616

[63] Zheng Z, Wang Y, Zhang A, Zhang T, Cheng F, Tao Z and Chen J 2012 Porous Li_2FeSiO_4/C nanocomposite as the cathode material of lithium-ion batteries *J. Power Sources* **198** 229–35

[64] Yoo H, Jo M, Jin B S, Kim H S and Cho J 2011 Flexible morphology design of 3D-macroporous $LiMnPO_4$ cathode materials for Li secondary batteries: ball to flake *Adv. Energy Mater* **1** 347–51

[65] Zhang K, Hu Z and Chen J 2013 Functional porous carbon-based composite electrode materials for lithium secondary batteries *J Energy Chem.* **22** 214–25

[66] Zhang H J, Tao H H, Jiang Y, Jiao Z, Wu M H and Zhao B 2010 Ordered CoO/CMK-3 nanocomposites as the anode materials for lithium-ion batteries *J. Power Sources* **195** 2950–55

[67] Yoon T, Chae C, Sun Y K, Zhao X, Kung H H and Lee J K 2011 Bottom-up *in situ* formation of Fe_3O_4 nanocrystals in a porous carbon foam for lithium-ion battery anodes *J. Mater. Chem.* **21** 17325–30

[68] Shen L F, Zhang X G, Uchaker E, Yuan C Z and Cao G Z 2012 $Li_4Ti_5O_{12}$ nanoparticles embedded in a mesoporous carbon matrix as a superior anode material for high rate lithium ion *Batteries. Adv. Energy Mater.* **2** 691–98

[69] Rodríguez-reinoso F 1998 The role of carbon materials in heterogeneous catalysis *Carbon* **36** 159–75

[70] Liu W J, Jiang H and Yu H Q 2015 Development of biochar-based functional materials: toward a sustainable platform carbon material *Chem. Rev.* **115** 12251–85

[71] Umeyama T and Imahori H 2013 Photofunctional hybrid nanocarbon materials *J. Phys. Chem.* C **117** 3195–209

[72] Chen W F, Iyer S, Iyer S, Sasaki K, Wang C H, Zhu Y M, Muckerman J T and Fujita E 2013 Biomass-derived electrocatalytic composites for hydrogen evolution *Energy Environ. Sci.* **6** 1818–26

[73] An K L, Xu X X and Liu X X 2018 Mo_2C-based electrocatalyst with biomass-derived sulfur and nitrogen co-doped carbon as a matrix for hydrogen evolution and organic pollutant removal *ACS Sustain. Chem. Eng.* **6** 1446–55

[74] Zhang Y F, Zuo L Z, Zhang L S, Huang Y P, Lu H Y, Fan W and Liu T X 2016 Cotton wool derived carbon fiber aerogel supported few-layered $MoSe_2$ nanosheets as efficient electrocatalysts for hydrogen evolution *ACS Appl. Mater. Inter.* **8** 7077–85

[75] Zhu Y G, Wang X Z, Jia C K, Yang J and Wang Q 2016 Redox-mediated ORR and OER reactions: redox flow lithium oxygen batteries enabled with a pair of soluble redox catalysts *ACS Catal.* **6** 6191–97

[76] Wang G H, Deng Y J, Yu J N, Zheng L, Du L, Song H Y and Liao S J 2017 From *Chlorella* to nestlike framework constructed with doped carbon nanotubes: a biomass-derived, high-performance, bifunctional oxygen reduction/evolution catalyst *ACS Appl. Mater. Inter.* **9** 32168–78

[77] Ahmed M B, Johir M A H, Zhou J L, Ngo H H, Nghiem L D, Richardson C, Moni M A and Bryant M R 2019 Activated carbon preparation from biomass feedstock: clean production and carbon dioxide adsorption *J. Clean. Prod.* **225** 405–13

[78] Liu G, Li C, Stewart B A, Liu L, Zhang M, Yang M and Lin K 2020 Enhanced thermal activation of peroxymonosulfate by activated carbon for efficient removal of perfluorooctanoic acid *Chem. Eng. J.* **399** 125722

[79] Tang H, Kumar P, Zhang S, Yi Z, Crescenzo G D, Santato C, Soavi F and Cicoira F 2015 Conducting polymer transistors making use of activated carbon gate electrodes *ACS Appl. Mater. Inter.* **7** 969–73

[80] Ashfaq M, Verma N and Khan S 2016 Copper/zinc bimetal nanoparticles-dispersed carbon nanofibers: a novel potential antibiotic material *Mater. Sci. Eng.* C **59** 938–47

[81] Karatepe N, Orbak İ, Yavuz R and Özyuğuran A 2008 Sulfur dioxide adsorption by activated carbons having different textural and chemical properties *Fuel* **87** 3207–15

[82] Singh G, Ismail I S, Bilen C, Shanbhag D, Sathish C I, Ramadass K and Vinu A 2019 A facile synthesis of activated porous carbon spheres from D-glucose using a non-corrosive activating agent for efficient carbon dioxide capture *Appl. Energy* **255** 113831

[83] Santos-Clotas E, Cabrera-Codony A, Ruiz B, Fuente E and Martín M J 2019 Sewage biogas efficient purification by means of lignocellulosic waste-based activated carbons *Bioresource Technol.* **275** 207–15

[84] Wang H, Xu J, Liu X and Sheng L J 2021 Preparation of straw activated carbon and its application in wastewater treatment: a review *J. Clean. Prod.* **283** 124671

[85] Anwar J, Shafique U, Salman M, Dar A, Anwar S and uz-Zaman W 2010 Removal of Pb (II) and Cd (II) from water by adsorption on peels of banana *Bioresource Technol.* **101** 1752–55

Chapter 9

Carbon aerogels

Wu Yang, Wang Yang, Yuling Chen and Xinwen Peng

Carbon aerogels are fascinating three-dimensional porous materials with a unique class of physicochemical properties, including high specific surface area and electrical conductivity, environmental compatibility, and chemical stability. These properties enable carbon aerogels to be used in many applications, such as energy storage, catalysis, sorbents, insulators, and desalination. In this chapter, we aim to provide a brief overview of the history and fundamental concepts of carbon aerogels and to present a comprehensive summary of recent developments in their application.

9.1 Introduction

Aerogels are the lightest solid materials; they are interconnected by colloidal particles and filled with gas; they possess abundant macropores and mesopores with a porosity of >95% in a nanoscale network whose pores have an average size of less than 100 nm. This unique structure endows aerogel with some excellent properties, such as low thermal conductivity, low density, high specific surface area, and porosity [1–3]. The first reported aerogel was SiO_2 aerogel, which was synthesized by Kistler in the 1930s (figure 9.1) [4]. In 1968, Teichner proposed the second generation of aerogels using safer, manageable precursors [5]. Although new synthesis routes were continually reported in the first few decades, the composition of aerogels was limited to metal oxides. At the end of the 1980s, Pekala synthesized a new class of organic aerogel through the sol–gel polymerization of resorcinol and formaldehyde precursors [6]. Since then, various types of aerogel have been reported. To date, aerogels have been divided into oxide aerogels (i.e. SiO_2 aerogel [7] and Al_2O_3 aerogel [8]), carbide aerogels (i.e. TiC aerogel [9] and SiC aerogel [10]), carbonitride aerogels [11], and carbon aerogels (i.e. carbon nanotube aerogel [12], graphene aerogel [13], and biomass aerogel [14]).

Carbon aerogels (CAs) are porous carbon materials with a 3D porous network structure formed of carbon nanomaterials, which have received considerable attention since they were first synthesized. The extremely high porosity of aerogels confers

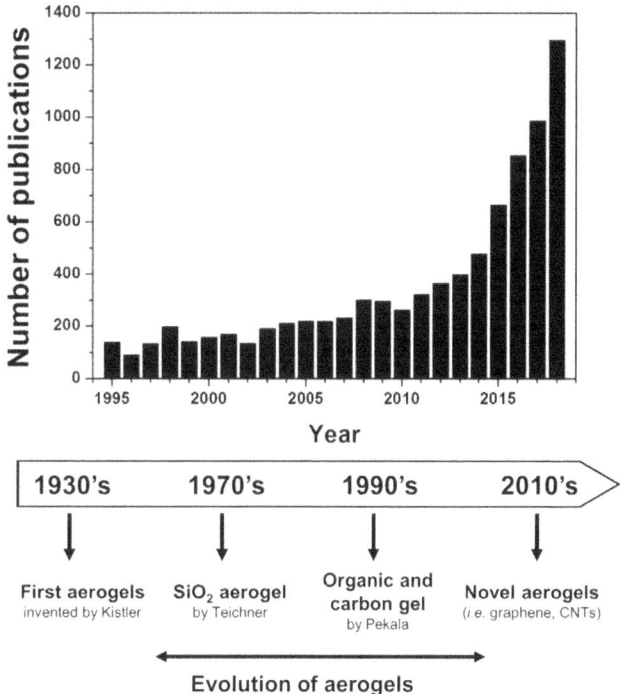

Figure 9.1. The number of published papers on aerogels along with the history of aerogel evolution after their invention [3], reproduced with permission © Elsevier.

various fascinating properties, such as ultralow density, large specific surface area, abundant pore structure, high electrical conductivity, environmental compatibility, and good thermal and chemical robustness [15]. Based on the above unique characteristics, CAs have been widely fabricated and used in various applications, such as energy storage, catalysis, sorbents, insulators, and desalination (figure 9.2) [16].

Notably, CAs have become a favored porous carbon material which can meet most high-performance application demands. Moreover, the particular properties of CAs are motivating research into their synthesis and application. To the best of our knowledge, only a few reviews have been related to the fabrication and application of carbon aerogels. This chapter gives a brief description of the history of CAs and presents a comprehensive overview of the preparation of carbon aerogels as well as their respective applications. Therefore, the purpose of this chapter is to provide comprehensive knowledge about CAs and to expound their engineering applications.

9.2 Preparation of carbon aerogels

CAs were first prepared by polymerizing resorcinol and formaldehyde under alkaline conditions, followed by supercritical drying and carbonization [17]. Since then, more and more monomers have been used for the preparation of CAs, and CAs have been produced with variations on essentially the same recipe. Generally, there are three main stages in the preparation of most CAs [18]: (1) gelation, (2)

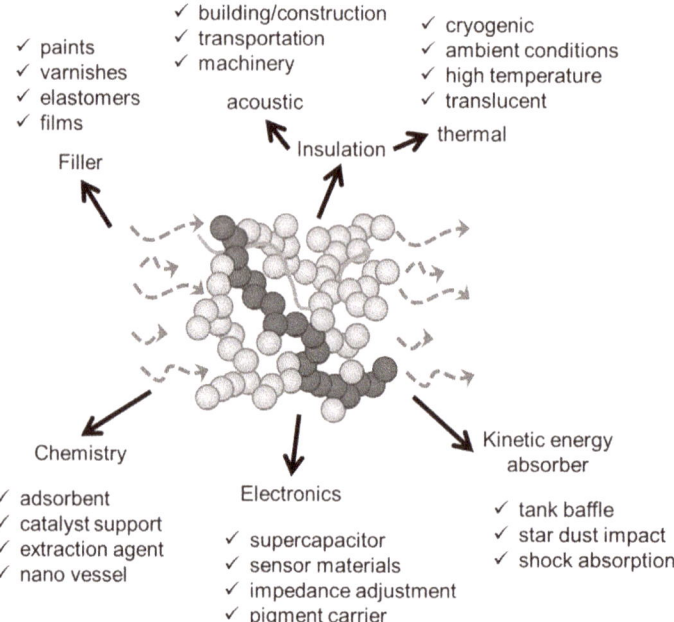

Figure 9.2. Application fields of carbon aerogel products [16], reproduced with permission © Elsevier.

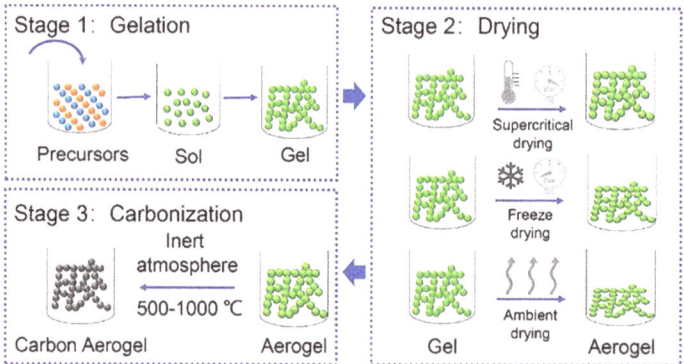

Figure 9.3. General preparation processes for CAs [18], reproduced with permission © John Wiley and Sons.

drying, (3) carbonization (figure 9.3), and each stage has a certain effect on the final properties.

9.2.1 Gelation

At the gelation stage, the hydrogel is formed via the polymerization and crosslinking of precursor molecules. However, polymerization reactions differ for the different precursors. Taking the example of resorcinol–formaldehyde aerogel, this stage involves the three chemical reactions shown in figure 9.4: (1) the formation of hydroxymethyl derivatives ($-CH_2OH$) derived from addition reactions between

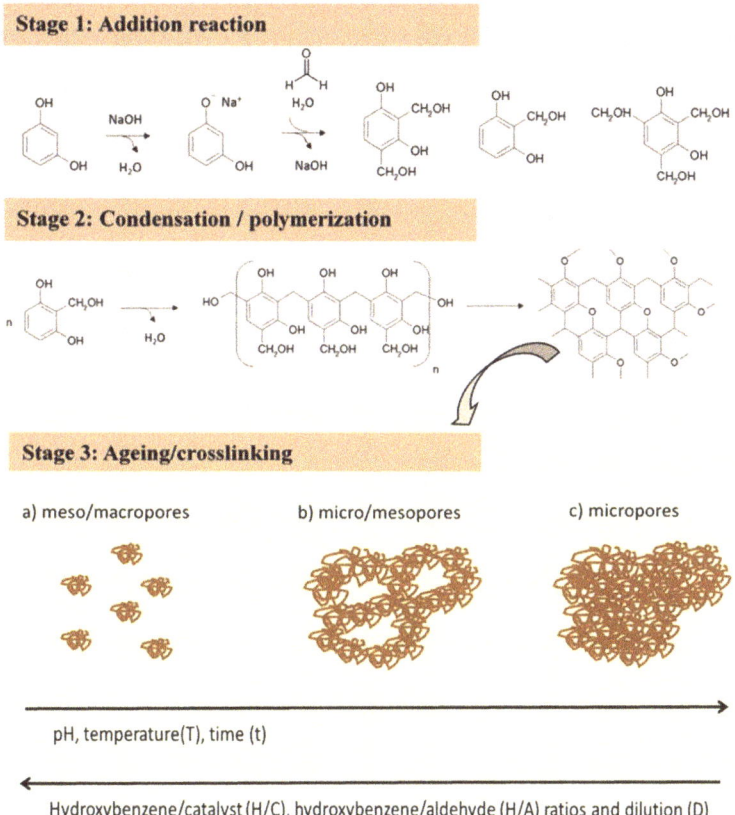

Figure 9.4. Gelation stage for the preparation of CAs [19], reproduced with permission © Elsevier.

formaldehyde and resorcinol; (2) the condensation/polymerization of hydroxymethyl resorcinol to form linear polymeric chains by the formation of methylene ($-CH_2$) and ether–methylene ($-CH_2OCH_2$) bridges; (3) the formation of the 3D hydrogel via crosslinking and agglomeration [16]. At the beginning of the addition reactions, alkali catalysts (Na_2CO_3, K_2CO_3, $Ca(OH)_2$, etc.) are used to catalyze a small portion of resorcinol to form active sites for the growth of the monomer particles [19]. Under alkaline conditions, the resorcinol anion can easily be formed, which is capable of a strong nucleophilic addition capability, and easily reacts with formaldehyde to form hydroxymethyl resorcinol. In addition to the alkaline catalysts, dilute acidic catalyst solutions (such as HCl) have recently been used to catalyze the electrophilic addition of formaldehyde to resorcinol, by increasing the electrophilicity of formaldehyde [20]. Organic carbon aerogels based on the polymerization of resorcinol and formaldehyde have traditionally employed various catalysts [21], which are also applied to other organic aerogels derived from melamine formaldehyde [22], phenol formaldehyde [23], cresol formaldehyde [24], phenol–furfural [25], and some polymers (i.e. poly-amide [26] and polyurethanes [27]). Furthermore, the preparation of templated carbon aerogel is assisted by some templates such as polymers, inorganic salts, and ceramic nanoparticles during the polymerization process [16]. Therefore, these templated carbon aerogels exhibit an ordered porous structure, narrow pore size distributions, and mechanical flexibility [28–30].

Among the other precursors, graphene, carbon nanotubes (CNTs), and biomass have recently been the most commonly used precursors, and have a simpler gelation stage than those of polymers [31]. Graphene sheets can be reassembled by the interaction force, leading to a 3D porous structure. To date, numerous methods have been devoted to fabricating graphene aerogels, such as the sol–gel, template, spacer support, self-support, and substrate methods [32–35]. The most commonly used method is gelation, in which graphene sheets are physically or chemically connected to each other by crosslinking agents to form a space network [36]. CNT aerogels are electrically conductive aerogels, in which CNTs can directly crosslink with each other through van der Waals interactions, forming electrically percolating networks [37]. However, these CNT aerogels have a few serious drawbacks, including poor mechanical instability and limited elasticity. It has been suggested that in order to overcome these challenges, surfactants and polymers should be added, since they can enhance the surface activity of CNTs or reinforce the network [38, 39]. Biomass is an attractive carbon aerogel precursor due to its low cost, abundance, and nontoxicity [40]. Cellulose-rich biomass is usually chosen as the precursor for the preparation of aerogels. The 3D interconnected framework of hydrated biomass contains a large void volume with large amounts of water (>80 wt%). After freeze-drying, the porous framework of a 3D aerogel can be obtained via water sublimation. However, a few biomasses with low water contents usually need to be homogenized to form a suspension, following by gelation and freeze-drying [41].

9.2.2 Drying

The second stage is drying, which is the critical step in the preparation of CAs. In this stage, the solvent in the hydrogel can be removed, leaving only the solid framework. The primary purpose of the drying process is to retain the initial pore structures, which have a significant effect on the final structure and properties of CAs. Therefore, drying hydrogels while minimizing the damage to the network caused by the capillary force is critical in order to obtain CAs with good textural properties. Nowadays, there are three typical drying methods for CAs (figure 9.3): (1) supercritical drying; (2) freeze-drying; and (3) ambient drying. However, the structural collapse and dimensional shrinkage of hydrogel might happen in the ambient drying process, due to the capillary tension at solid–liquid–vapor interface. Therefore, the most commonly used drying methods are supercritical drying and freeze-drying [42]. Generally, the dried gel products are classified as aerogels (supercritical drying), cryogels (freeze-drying), and xerogels (ambient drying), which are collectively known as aerogels [43].

9.2.2.1 Supercritical drying
Supercritical drying is the most efficient drying method for the preparation of CAs [44]. Theoretically, when the pressure and temperature of a vessel exceeds the critical

point of the solvent, the supercritical drying process eliminates the liquid surface tension at the solid–liquid–gas interface, and the shrinkage and collapse of the pore structure can be avoided during the drying procedure [45], as is evident in figure 9.5. Specifically, there is no interface between the liquid phase and the vapor phase in the supercritical state, but a uniform fluid is formed between the liquid phase and the vapor phase, which is gradually discharged from the gel due to the lack of a liquid–gas interface with no capillary effect, so that it does not cause contraction and structural damage. Therefore, supercritical drying endows aerogels with higher porosity and pore volume than the other drying methods [46]. However, incomplete solvent replacement can cause the slight shrinkage of gels, leading to a smaller pore size, more pore shrinkage, and a lower precursor concentration [47]. Moreover, the type of solvent is closely related to the drying effect, and CO_2 is thus widely used because of its lower critical pressure and temperature [48]. Barim *et al* [49] reported CAs derived using resorcinol and formaldehyde via the supercritical drying method, in which the acetone in the hydrogel was removed by supercritical extraction using supercritical CO_2 at 138 bar and 323 K. However, the working temperature and pressure were relatively high, and the process operations were complicated, time-consuming, and high risk; they were also inapplicable to continuous large-scale production. Therefore, the preparation of CAs by non-supercritical drying techniques has become a hot research topic.

9.2.2.2 Freeze-drying

Freeze-drying, which is an eco-friendly drying process, is mainly used to prepare CAs [50]. Traditionally, the hydrogel molecules are synthesized by the sol–gel process and then put into a vacuum vessel which is heated under a certain vacuum. The solid in the hydrogel network skeleton is directly sublimated, and the solvent in the hydrogel is frozen and removed by sublimation at low pressure without the formation of a gas–liquid interface, obtaining high-performance CAs. The

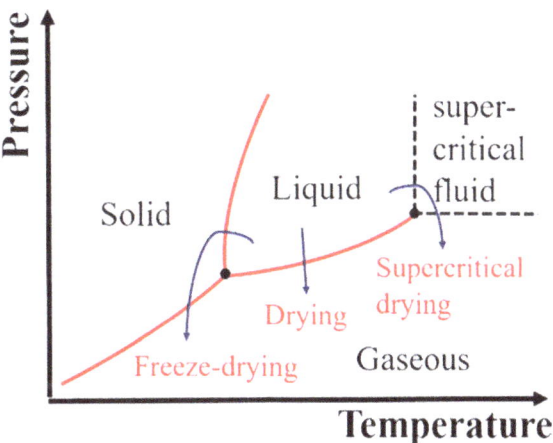

Figure 9.5. Phase diagrams of the removal of the liquid phase from the gel [45], reproduced with permission © Springer Nature.

properties of the CAs thus prepared are affected by many factors, such as pre-drying treatments, the freezing rate, and the precursor concentration. The freezing rate has an impact on the pore structure through changing the shape and size of ice crystals, and small ice crystals can be formed in the frozen samples at a rapid freezing rate, resulting in a small pores and a high specific surface area. A low precursor concentration is beneficial to the growth of ice crystals, which affect the porous structures of CAs. The structural shrinkage and collapse can be also reduced by using a solvent with a low expansion coefficient and a high sublimation pressure; and the most common practical solvent is water. When the pores are small, the water in the micropores is difficult to freeze, resulting in the cracking and fragmentation of CAs. The reason for this is believed to be that the tension formed by thermal shock destroys the pore structure of the gel. Therefore, freeze-drying is not suitable for the preparation of CAs with small pore sizes. As a substitute for supercritical drying, freeze-drying is mainly employed to prepare aerogels with large pore sizes due to its lower cost and simpler operation [51].

9.2.2.3 Ambient drying

Ambient drying is a safe, low-cost, and simple technique, which is a strong candidate for the large-scale production of aerogels in industrial settings [52]. Fischer *et al* [53] were the first to prepare a resorcinol–formaldehyde aerogel using ambient drying. Notably, ambient drying is simpler, quicker, and cheaper than supercritical drying and freeze-drying. However, ambient drying increases the capillary tension at the solid–liquid–vapor interface, which leads to structural collapse and dimensional shrinkage. There are two strategies that effectively prevent the destruction of the aerogel and retain good pore structure and mechanical stability under ambient drying conditions: the use of low surface tension to reduce the capillary force and enhancing the strength of the framework structure to resist capillary tension. Solvent replacement [54] and surfactant addition [55] are two methods used to reduce the solvent surface tension. When the strength of the framework structure is large enough to withstand capillary tension, collapse and shrinkage of the pore structure can be avoided.

In conclusion, small particles of CA can be prepared by the ambient drying method, but it is still difficult to prepare large blocks of CA in this way. CAs with good properties can be prepared by the impressive supercritical drying and freeze-drying technologies, but those technologies are limited by cost and time.

9.2.3 Carbonization

The last stage in the preparation of CAs is the carbonization of the dried gel. The carbonization process usually is performed by heating the sample under an inert atmosphere of N_2 or Ar at a high temperature between 600 °C and 2600 °C. With the temperature increase, the structure of the carbon skeleton stabilizes, and it reaches its stablest state at 800 °C [45]. In the carbonization process, hydrogen and oxygen functional groups decompose into gases and escape from the aerogel, generating a porous and carbonaceous 3D network [56]. This stage of great importance to polymer carbon aerogels and biomass carbon aerogels, but it is not necessary for graphene aerogels and CNT aerogels due to their completely graphitized structure. The macropores in the aerogels are decreased by structural shrinkage, while the mesopores and micropores are increased, especially the micropores. Previous results show that the micropore properties of CAs are mainly influenced by carbonization process, while the mesopore and macropore properties are mainly determined by the initial preparation parameters and the drying process [42]. Furthermore, the specific surface area is significantly increased by the carbonization treatment, especially at lower carbonization temperatures.

The carbonization temperature is one of the most crucial parameters that affect the physical properties of CAs. Higher carbonization temperatures lead to lower specific surface areas and pore volumes. When the carbonization temperature is low, the macropore volume decreases and the mesopore volume increases. The specific surface area of CAs does not change significantly at less than 600 °C. In the carbonization temperature range from 700 °C to 900 °C, the pore volume decreases with increasing temperature. However, the pore volume does not change significantly when the temperature is above 900 °C. Najeh *et al* [57] reported the influences of carbonization process on the electrical properties of CAs, and the conductivity of CAs reached its maximum value at 800 °C.

In conclusion, the synthesis conditions, such as the preparation of the sol mixture, gelation, drying, and carbonization can determine the surface properties (i.e. the specific surface area, pore volume, and pore size) of CAs. By controlling these synthesis conditions, various kinds of CA with different properties can be designed and prepared.

9.3 Applications of carbon aerogels

Over the past few decades, CAs have attracted increasing interest for industrial applications. Their high specific surface area/porosity, good electrical conductivity, and low density have made them promising candidates for several novel fields, such as energy storage, catalysis, sorbents, insulators, and desalination.

9.3.1 Energy storage

In order to meet the demand for the rapid development of electronic devices and applications, various advanced energy materials with high capacity and low cost have been investigated in recent years [58]. CAs with a 3D porous network are regarded as promising electrode materials for electrochemical energy storage systems, due to their rapid electron/ion transport, remarkable structural controllability, physicochemical stability, and excellent electrical conductivity. However, the different properties of CAs can cause conflicting electrochemical effects, which complicate their usage [59]. For example, the electrolytic wettability of CAs is necessary for electrochemical reactions. Repulsion between the electrode and the electrolyte enables rapid mass transfer [60]. More importantly, an appropriate pore size distribution is the key factor for good electrochemical performance in energy storage applications. In general, micropores (<2 nm) provide abundant adsorption sites for ions, while mesopores (2−50 nm) facilitate the rapid diffusion of electrolytic ions [61]. In addition, the specific capacitance of CAs is better than that of a conventional capacitor, which stems from the designability of their structural properties, their high electronic conductivity, and their machinability [62]. As we all know, capacitance and specific surface area, or pore volume, have a proportional relationship. Large specific surface areas increase specific capacitance, and large pores mean access to more electrolyte ions, while specific surface area and pore volume have a competitive relationship [63]. In conclusion, CAs possess several unparalleled advantages for electrochemical energy storage [64]: (1) the tremendous electrochemically effective surface area provides abundant active sites for ion storage; (2) the 3D interconnected porous structure offers numerous transport channels for electrolyte ions and shortens the transfer distance; (3) the high electron conductivity promotes the transportation of charges.

Shabangoli *et al* [65] reported a flexible and metal-free supercapacitor with a thionine-functionalized 3D graphene aerogel (Th–GA) that was fabricated via a convenient one-step hydrothermal approach. The functionalization of the 3D graphene aerogel with thionine resulted in a synergistic enhancement of the capacitance compared to that of a pure thionine electrode. The assembled Th–GA electrode exhibited a high specific capacitance of 512 F g^{-1} at 1 A g^{-1}, an ultrahigh specific energy density of 32.6 Wh kg^{-1}, and a specific power density of 12.8 kW kg^{-1} in a symmetric supercapacitor (figure 9.6(a)–(d)). Chen *et al* [66] prepared an N-self-doped carbon nanofiber aerogel (NCNF) with a highly porous, 3D interconnected network via the *in situ* growth of a zeolitic imidazolate framework on bacterial cellulose in an aqueous system (figure 9.6(e)). The highly interconnected 3D conductive network in the NCNF allowed efficient electron transport. The optimized aerogel electrode delivered remarkable capacitances of 224 F g^{-1} at 0.5 F g^{-1} and 612 mF cm^{-2} at 1.37 mA cm^{-2}, and an excellent energy density of 31.0 Wh kg^{-1} at a power density of 250 W kg^{-1}.

As the anode materials of lithium-ion batteries, CAs, with their 3D interconnected porous structure, can not only provide a stable network for lithium-ion transport but also offer a rapid electron transport pathway, thus achieving excellent cycle stability and rate performance [67]. For example, Shan *et al* [68] reported 3D reduced graphene oxide aerogels (GAs) with controlled surface defects which were produced by adjusting the hydrothermal reaction time. When used as anode materials for lithium-ion batteries, the 3D GA electrodes delivered a high reversible capacity of 1430 mAh·g^{-1} at 100 mA·g^{-1} and superior cycling stability. Lu *et al* [69] proposed intrinsic defect-rich hierarchically porous carbon architectures (DHPCs) as a host of sulfur (figure 9.7). The synergetic effect between the defects and the hierarchical structures promote a faster redox reaction, thus suppressing the shuttle effect. Lithium–sulfur batteries fabricated with DHPCs as the sulfur host offered a high specific capacity of 1182 mAh·g^{-1} at 0.5 C, excellent rate performance, and outstanding long-term stability.

Zhao *et al* [70] demonstrated the use of an oxygen and nitrogen co-doped holey graphene (ON/HG) aerogel as a high-performance additive-free sodium-ion battery anode (figure 9.8). In its preparation process, a polyurethane (PU) sponge served as both the sacrificing template and the nitrogen source. An elastic porous graphene monolith

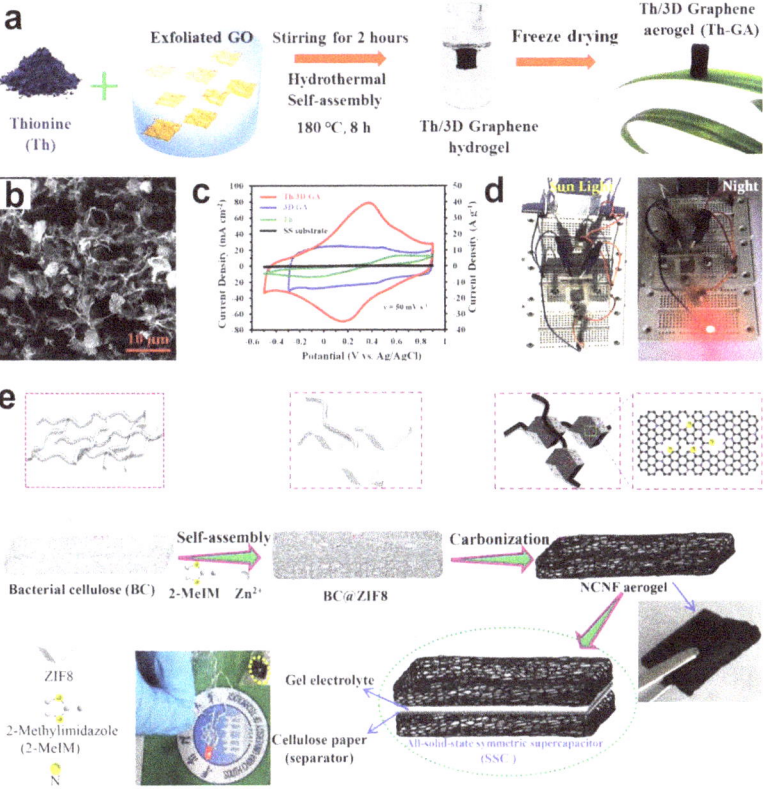

Figure 9.6. (a) Schematic illustration of the preparation of the lightweight thionine-functionalized 3D graphene aerogel (Th–GA). (b) SEM images of the Th–GA. (c) Cyclic voltammetry (CV) curves of the stainless steel (SS) substrate, pure thionine (Th), the graphene aerogel (GA), and thionine-functionalized GA (Th–GA) at 50 mV s⁻¹ in a 1 M H_2SO_4 electrolyte. (d) Demonstration of the practical applicability of the Th–GA//Th–GA device. Source for (a–d): [65], reproduced with permission © John Wiley and Sons. (e) Preparation of an all-solid-state symmetric supercapacitor based on freestanding N-self-doped carbon nanofiber (NCNF) aerogels [66], reproduced with permission © Elsevier.

with optimized crystallinity and abundant micropores on its mechanically strong 3D network was acquired by selective annealing at 450 °C in air. As an additive-free anode for sodium-ion batteries, the optimized ON/HG demonstrated a high specific capacity of 446 mAh·g⁻¹ at 0.1 A·g⁻¹, an excellent rate capacity of 189 mAh·g⁻¹ at 10 A·g⁻¹, and a long cycle life (81% capacity retention after 2000 cycles at 5 A g⁻¹).

CAs have a robust 3D network with numerous micro/mesopores and excellent electrical conductivity, which also can act as a promising anode material for potassium-ion batteries. Liu *et al* [71] prepared a 3D GA through the hydrothermal method, and a high specific capacity of 267 mAh·g⁻¹ at a rate of C/3 was obtained. Benefiting from the open porous structure of 3D gas, a relatively high diffusion coefficient of K-ions was observed. Recently, Lv *et al* [72] synthesized a 3D carbon nanofiber aerogel co-doped with sulfur and nitrogen (S/N-CNFAs) via the carbonization of sustainable seaweed

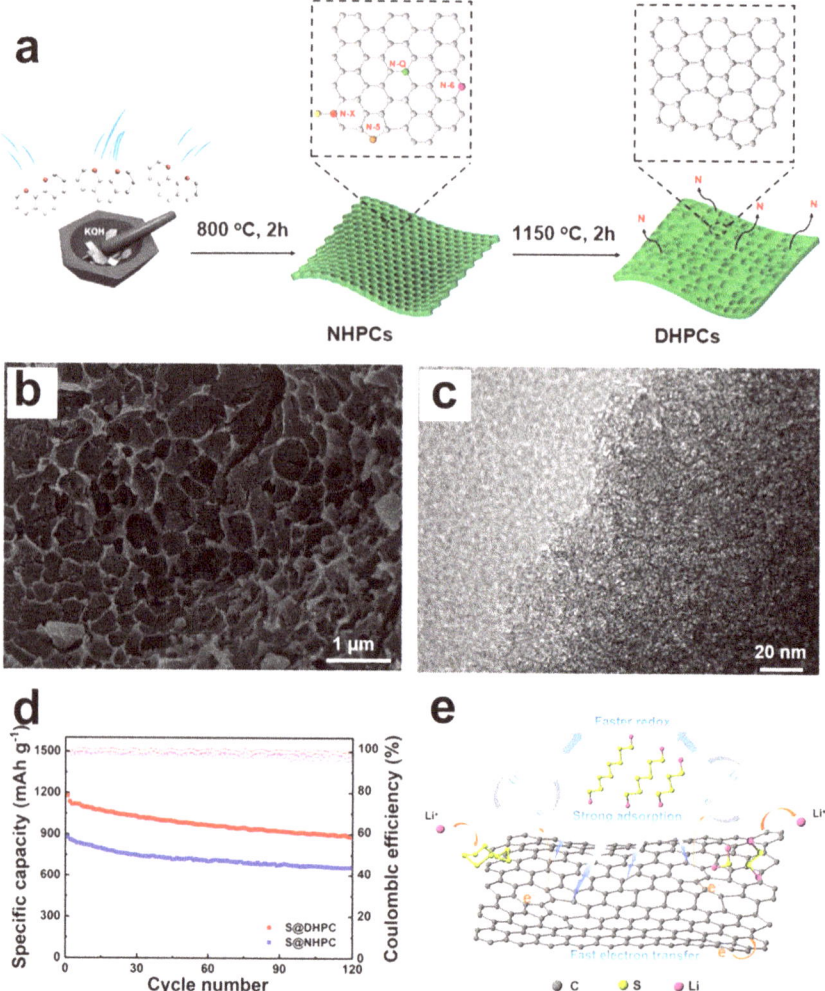

Figure 9.7. (a) Schematic illustration of the synthesis process of DHPCs. (b) SEM and (c) TEM images of DHPCs. (d) Cycling performances of S@NHPC and S@DHPC at 0.5 C. (e) Schematic illustration of the immobilization and conversion of sulfur species on the surface of DHPC [69], reproduced with permission © American Chemical Society.

(Fe-alginate) aerogels (figure 9.9). Benefiting from the optimized electronic structure provided by S/N co-doping, an interlayer spacing (0.41 nm) enhanced by S doping and a 3D interconnected porous structure were obtained. As a result, an S/N-CNFAs electrode exhibited a high specific capacity of 356 mAh·g^{-1} at 0.1 A·g^{-1} and an excellent cycling stability of 168 mAh·g^{-1} at 2 A·g^{-1} over 1000 cycles.

9.3.2 Catalysis

CAs can be designed to have a large surface area, controlled micro/mesoporosity, and narrow mesopore size distribution. These extraordinary properties, particularly the

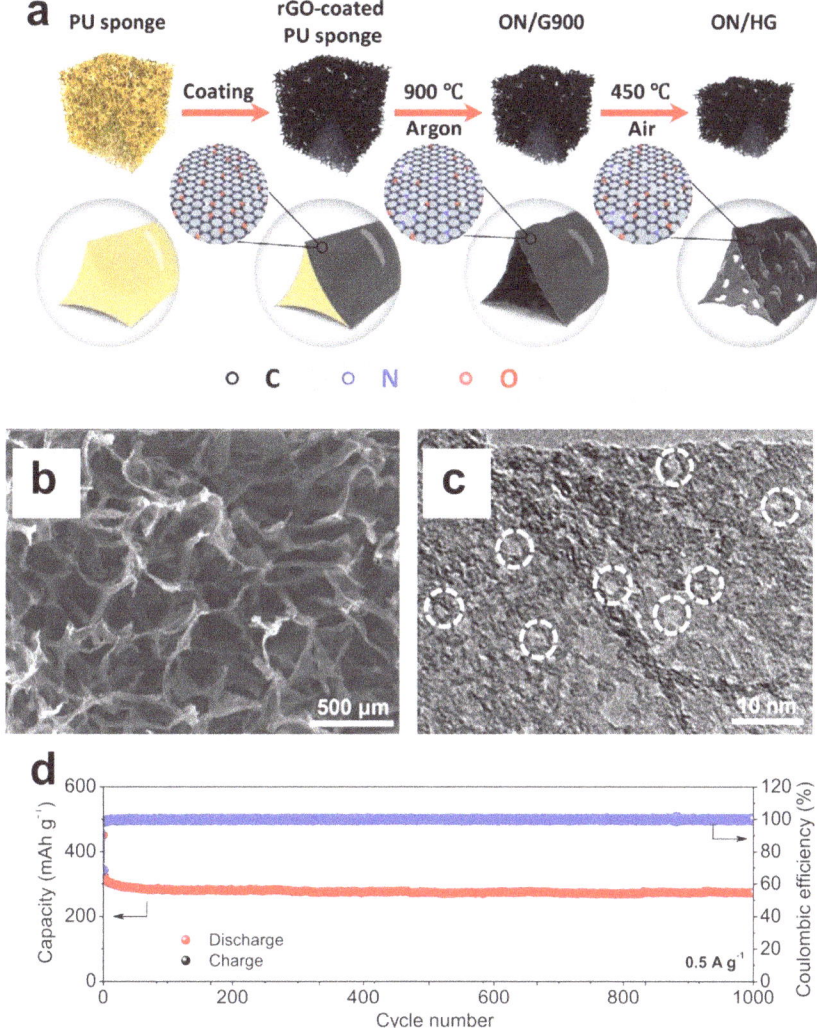

Figure 9.8. (a) Schematic illustration of the preparation procedure used for ON/HG. (b) SEM and (c) TEM images of ON/HG. (d) Cycling performance of ON/HG at 0.5 A·g^{-1} [70], reproduced with permission © John Wiley and Sons.

electrical conductivity, make CAs promising materials for application in catalysis. Furthermore, CAs have another advantage, namely, that they can be prepared in the form of monoliths, beads, powders, or thin films. Therefore, CAs exhibit attractive catalytic performance for chemical degradation, water splitting, and air purification. In this field, CAs are mainly exploited as attractive catalysts or catalytic supports.

9.3.2.1 Catalysts

The 3D interconnected network and high specific surface area of CAs can provide abundant active sites, making them efficient catalysts [73–78]. Compared with other

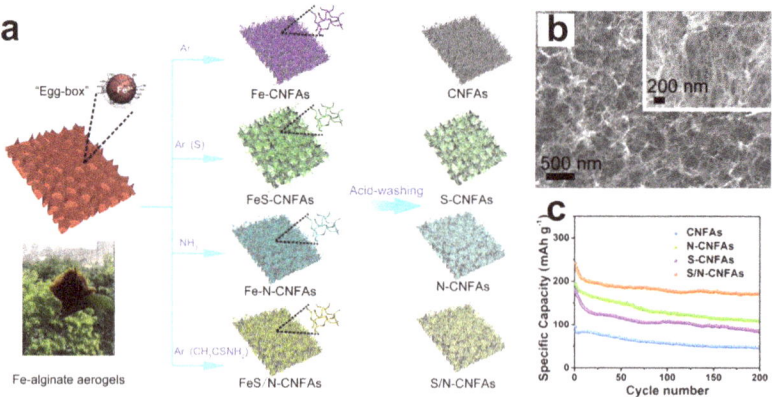

Figure 9.9. (a) Schematic illustration of the preparation procedure used for CNFAs, S-CNFAs, N-CNFAs, and S/N-CNFAs. (b) SEM image of S/N-CNFAs. (c) Cycling performance of CNFAs, S-CNFAs, N-CNFAs, and S/N-CNFAs electrodes at 0.5 A·g^{-1} [72], reproduced with permission © John Wiley and Sons.

catalysts, CAs exhibit an excellent adsorption capacity, which can markedly improve their catalytic performance. Therefore, CAs are widely used as photocatalysts or electrocatalysts for organic pollutant oxidation and air purification. For example, Tang *et al* [79] prepared a compressible g-C$_3$N$_4$/graphene oxide (GO) aerogel as photocatalyst for the degradation of dyes (methyl orange (MO) and methylene blue (MB)) and the reduction of bromates. The degradation rate of MO and MB (20 mg L^{-1}) reached 90% in 40 min, and the conversion rate of bromate (250 μg L^{-1}) was more than 80% in 60 min of visible-light illumination. Hu *et al* [80] demonstrated a graphitic carbon nitride (g-C$_3$N$_4$) aerogel modified by a perylene imide (PI) and graphene oxide as a visible-light photocatalyst for nitric oxide (NO) removal (figure 9.10). This composite aerogel (denoted as PICNGA) showed excellent activity in NO removal due to the strong light absorption, good planarity, and rapid charge transportation of graphene oxide. Therefore, an aerogel containing thiophene exhibited excellent photocatalytic activity for NO purification and had a removal ratio of up to 66%. Density functional theory calculations verified the photocatalytic effect of these aerogels, and the stability and recyclability of these aerogel photocatalysts increase their potential commercial value for the photocatalytic degradation of NO.

9.3.2.2 Catalytic supports

CAs possess outstanding structural and textural properties, which can be flexibly tailored on a molecular level by the sol–gel technique [18]. To design and tailor their chemistry for different catalytic reactions, various metallic compounds or metals have been dispersed in the CA matrix [81–85]. As a result, CAs are widely regarded as a superior support materials for catalysts. Their large-scale porous network that includes mesopores and macropores facilitates the loading of active materials without pore blocking.

Zou *et al* [86] reported a 3D nitrogen-doped graphene aerogel coupled with cobalt nitride nanoparticles (CoN$_x$/NGA) as an efficient trifunctional electrocatalyst (figure 9.11). The CoN$_x$/NGA nanohybrid, which had a hierarchical porous structure,

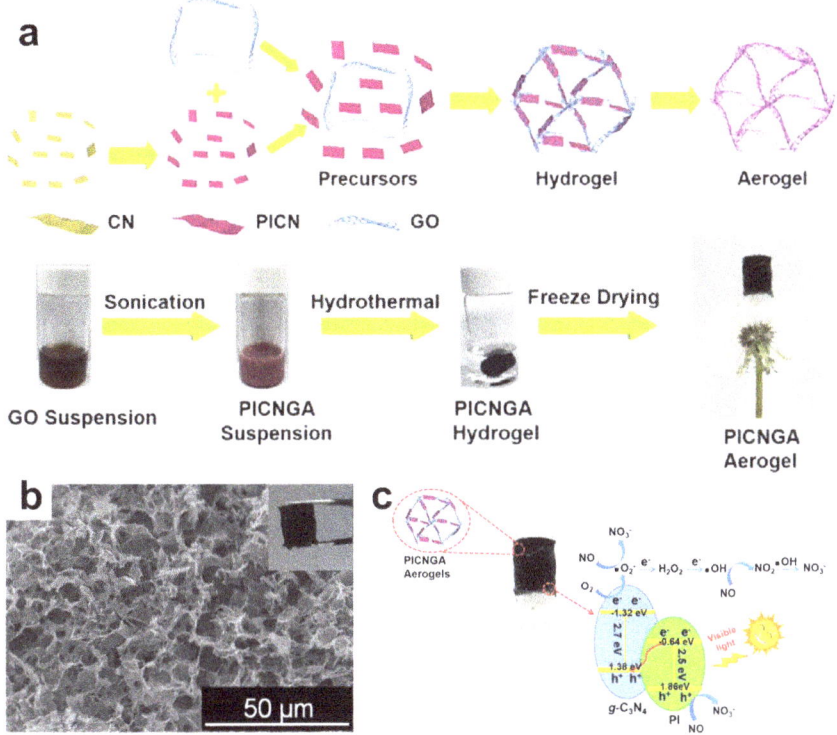

Figure 9.10. (a) Schematic illustration of the synthesis of the PICNGA aerogel. (b) SEM image of PICNGA. (c) Photocatalytic mechanism of NO removal under visible-light irradiation by PICNGA aerogels [80], reproduced with permission © John Wiley and Sons.

provided abundant dual active CoN_x and N_xC sites, thus exhibiting superior activities in the oxygen reduction reaction (ORR), the oxygen evolution reaction (OER), and the hydrogen evolution reaction (HER). Moreover, the highly open 3D hierarchical architecture of the CoN_x/NGA was beneficial in shortening the pathway for mass transportation. In general, overall water splitting (the OER combined with the HER) suffers from a high overpotential (η) and low energy efficiency [87]. During the ORR, the CoN_x/NGA exhibited an onset potential of 0.93 V and a half-wave potential ($E_{1/2}$) of 0.83 V, which are close to those of the noble metal benchmark Pt/C. To achieve the benchmark current density of 10 mA cm^{-2} for the OER, the CoNx/NGA delivered a low overpotential ($\eta 10$) of 295 mV. Analogously, CoN_x/NGA exhibited an impressive HER performance with a low overpotential of 198 mV. Moreover, the CoN_x/NGA exhibited good catalytic activity with 89% retention after 10 h operation, whereas a Pt/C electrode retained less than 67%. CoN_x/NGA-based zinc–air batteries connected in series were assembled to drive our electrolyzer, realizing self-driven electrochemical water splitting.

In the case of CO_2 reduction, CA-supported catalysts are frequently used for CO oxidation. Qu *et al* [88] demonstrated a metal–organic framework (MOF, HKUST-1) that decorated the surface of a 3D Ru/graphene aerogel hybrid (Ru/GA-HK) to perform catalytic carbon monoxide (CO) oxidation (figure 9.12). MOFs are

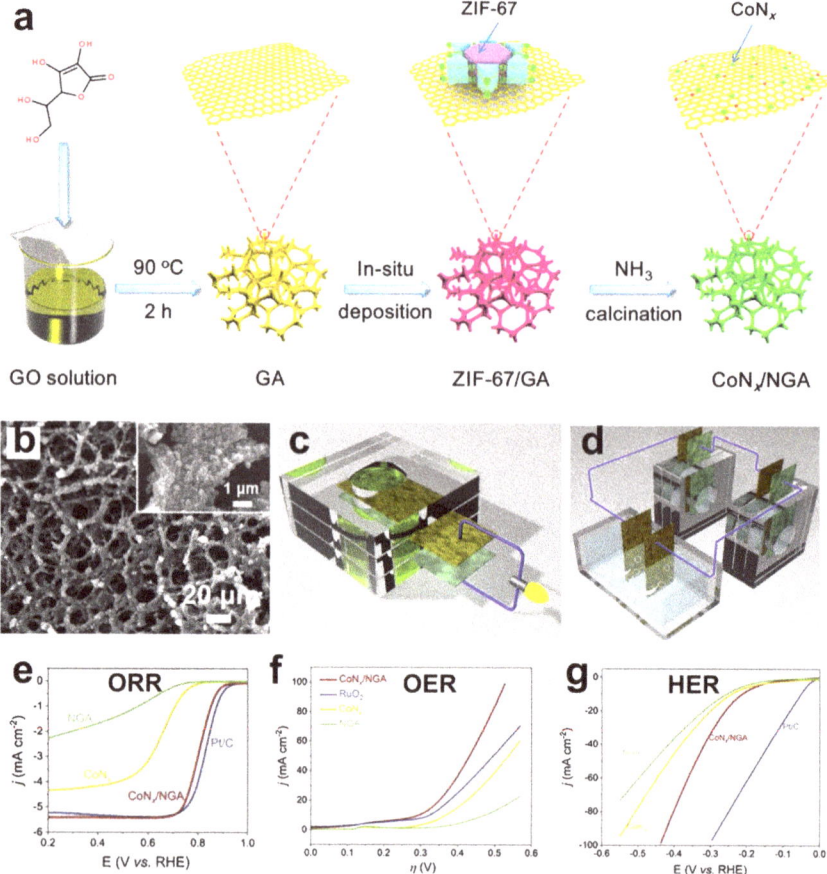

Figure 9.11. (a) Synthesis procedure used for the preparation of CoN$_x$/NGA. (b) SEM images of CoN$_x$/NGA. (c) Schematic illustration of a rechargeable Zn–air battery. (d) Schematic of the structural characterization of the self-driven overall water splitting device. Linear sweep voltammetry (LSV) curves of CoN$_x$, NGA, and CoN$_x$/NGA for the ORR (e), the OER (f), the HER (g). Source: [86], reproduced with permission © Elsevier.

potential gas adsorption materials for CO oxidation, due to their large surface area, excellent adsorption ability, and high chemical stability [89]. However, the kinetics of MOF-based catalytic CO oxidation are slow, due to the poor diffusion of these catalysts [90]. The open macroporous structure of the Ru/GA provided pathways for the access and diffusion of the reactant and product molecules. Thus, CO was simultaneously adsorbed and oxidized, leading to an enhanced reaction rate, and the CO conversion rate of the Ru/GA/HK catalyst was 100% at room temperature.

Wang *et al* [91] reported the production of 3D carbon quantum dot/graphene aerogel (CQD/GA) composites by a facile hydrothermal method for use as a photocatalyst. 3D GA provided suitable support for the immobilization of CQDs, which made the liquid-phase reaction system reusable. The CQD/GA catalyst prepared by a facile hydrothermal method was regenerated by a simple hydrothermal treatment that replenished the CQDs on deactivated catalyst. The CQD/GA

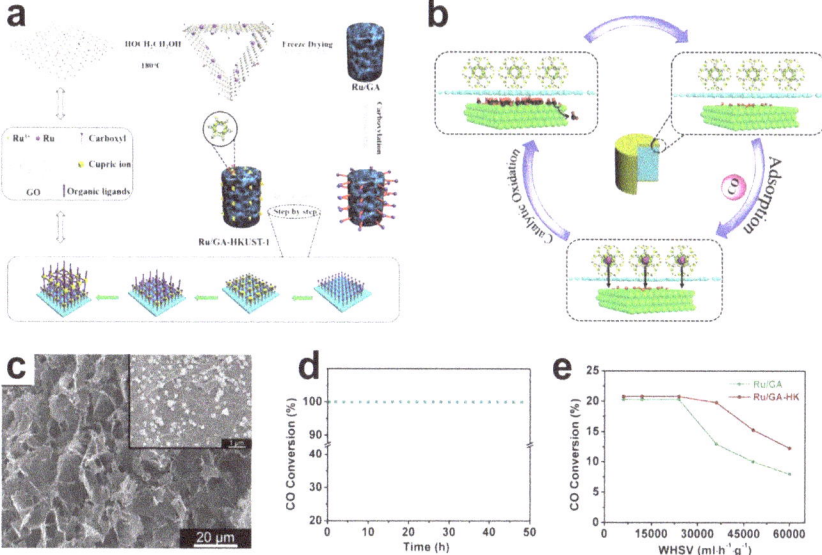

Figure 9.12. (a) Schematic illustration of the preparation of Ru/GA-HK. (b) The mechanism of the CO removal reaction. (c) SEM image of Ru/GA-HK. (d) The durability of Ru/GA-HK during CO oxidation. (e) The CO conversions of Ru/GA and Ru/GA-HK achieved by increasing the weight hourly space velocities [88], reproduced with permission © John Wiley and Sons.

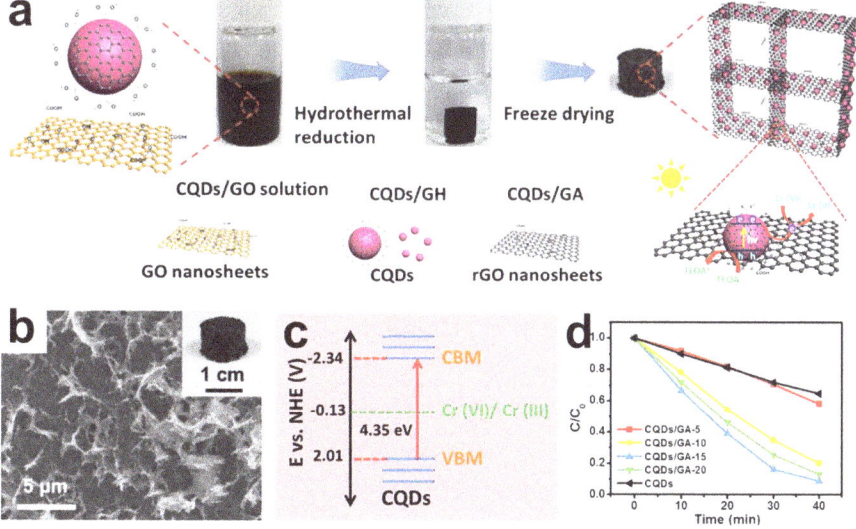

Figure 9.13. (a) Synthesis procedure of the 3D CQD/GA composite. (b) SEM images of CQD/GA. (c) Band structure diagram of the CQDs. (d) Photocatalytic performance of Cr(VI) reduction in an aqueous solution during irradiation with UV–Vis light. Source: [91], reproduced with permission © Elsevier.

catalyst reduced 91% of a hexavalent chromium (Cr(VI)) sample within 40 min by a photocatalytic reaction. The photocatalytic activity of the CQD/GA composite was fully restored after four successive regeneration cycles (figure 9.13).

9.3.3 Gas storage and separation

CAs have great potential for gas storage and separation, because their narrow and uniform pore size distribution is suitable for the storage and separation of gas molecules [92–94]. Their homogeneous micropores provide high affinity for the physisorption of gas molecules due to their small kinetic diameter and Lennard-Jones potential. In addition, metal atoms or heteroatoms can be incorporated into CAs, enhancing the affinity between the adsorbent and the gas molecules, and thus increasing the hydrogen uptake. Functionalization and the introduction of heteroatoms also improve CO_2 capture by increasing the affinity of the adsorbent surface for CO_2 molecules [95]. Consequently, CAs have been regarded as promising adsorbents for gas storage due to the ease of controlling their porous structures and molecular-level design [96–98].

Li *et al* [99] developed a facile process for the synthesis of a nitrogen-doped carbon aerogel (NCA) by a straightforward pyrolysis method that used a Schiff-base porous organic polymer (POP) aerogel as a precursor. The NCAs thus obtained exhibited good textural properties, such as a large specific surface area (up to 2356 $m^2\,g^{-1}$), high pore volume (1.12 $cm^3\,g^{-1}$), high bulk porosity (70%), low bulk density (5 $mg\,cm^{-3}$), and suitable nitrogen-doping amounts (2–5 wt%). The micro-pore volume and nitrogen content determined both the uptake of CO_2 and its CO_2/N_2 selectivity. As a result, remarkable CO_2 uptake capacities (6.1 $mmol\,g^{-1}$ at 273 K and 1 bar, 33.1 $mmol\,g^{-1}$ at 323 K and 30 bar) and high ideal adsorption solution theory (IAST) selectivity (47.8) at ambient pressure were achieved (figure 9.14).

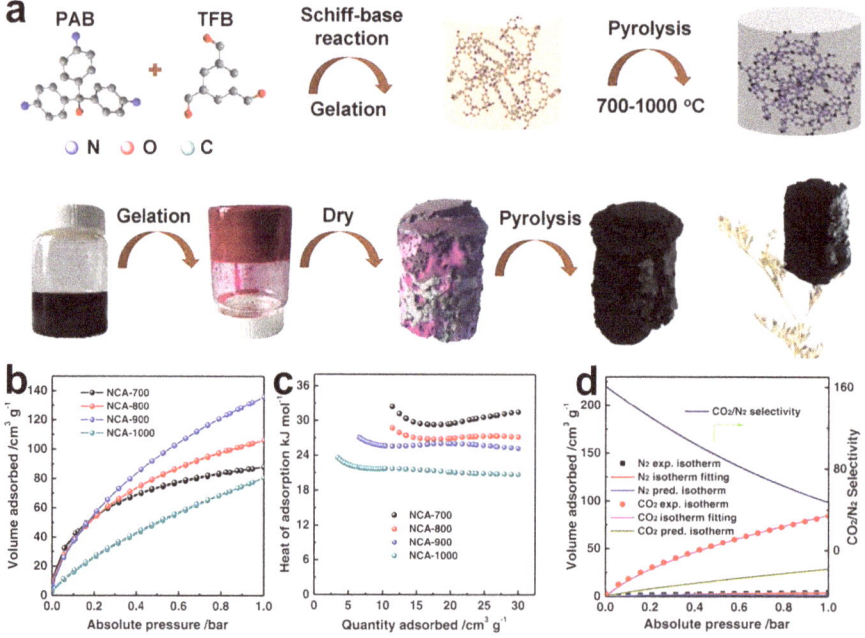

Figure 9.14. (a) Synthesis of NCAs. (b) CO_2 adsorption isotherms at 273 K. (c) Heat versus CO_2 adsorption curves of NCAs. (d) CO_2/N_2 selectivity of NCA-900 based on a typical fuel gas composition of 15% CO_2/85% N_2 (v/v). Source: [99], reproduced with permission © John Wiley and Sons.

Oh *et al* [100] reported a carbon-nitride-functionalized porous reduced graphene oxide aerogel (carbon nitride aerogel (CNA)), which functioned as a highly selective and regenerative adsorbent for CO_2. The templated growth of carbon nitride on porous reduced GO offers readily accessible CO_2 capture sites and sufficient pore volume to store gas. A strong interaction between gas molecules and adsorbents leads to poor reproducibility, thus it is imperative to balance the adsorption capacity for CO_2 capture against the release of gas molecules [101]. The strong dipole interaction of electron-rich nitrogen in a nonplanar microporous carbon nitride geometry delivers the specific and reversible adsorption of CO_2 under ambient conditions. Therefore, CNA exhibits a large CO_2 adsorption capacity of 0.43 mmol g^{-1}; it exhibits high CO_2 selectivity that is determined by a simple pressure swing; and it retains regenerability, desorbing 98% of its CO_2 (figure 9.15).

Liu *et al* [102] reported a 3D glucose/graphene-based aerogel (G/GA) produced using the hydrothermal reduction and CO_2 activation method for the adsorption of various target gases (figure 9.16). Glucose was used as the binder of a graphene oxide

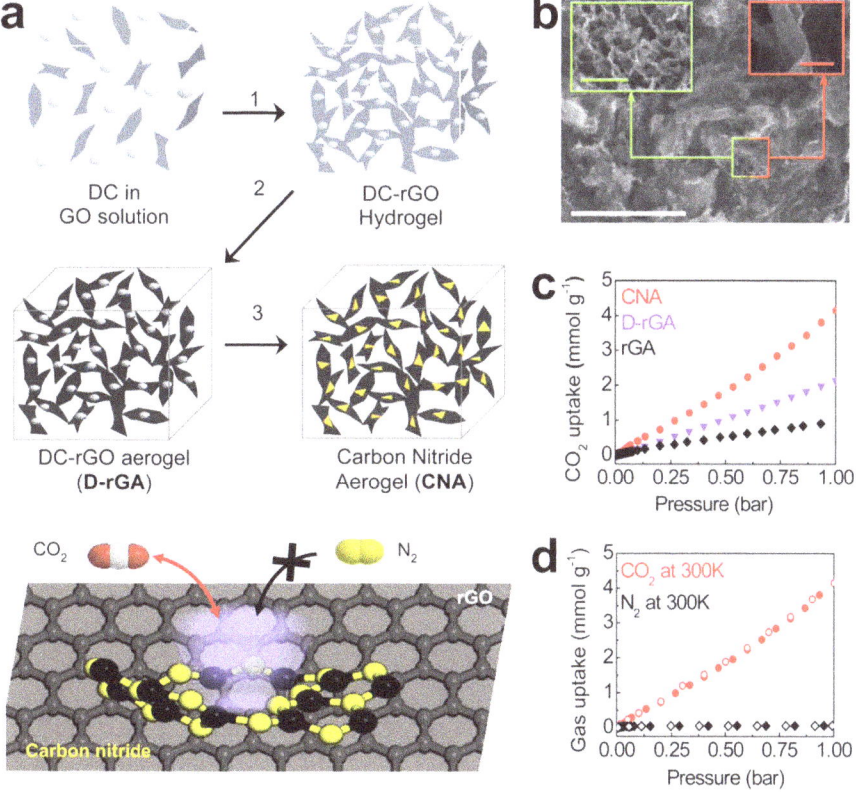

Figure 9.15. (a) Synthesis procedure used for the carbon nitride aerogel (CNA). (b) SEM image of CNA. (c) CO_2 adsorption isotherms of CNA, dicyandiamide-functionalized reduced graphene oxide aerogel (D-rGA), and pristine reduced graphene oxide aerogel (rGA). (d) CO_2 and N_2 adsorption isotherms of CNA, showing high selectivity for CO_2. Source: [100]; copyright (2015), reproduced with permission © American Chemical Society.

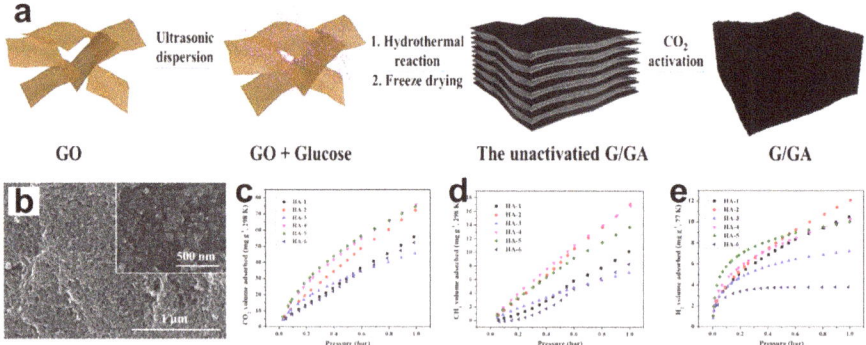

Figure 9.16. (a) Schematic illustration of the preparation of 3D G/GA. (b) SEM image of G/Gas. (c–e) Isotherms of CO_2, CH_4, and H_2 adsorption by G/GA. Source: [102]; copyright (2019), Multidisciplinary Digital Publishing Institute.

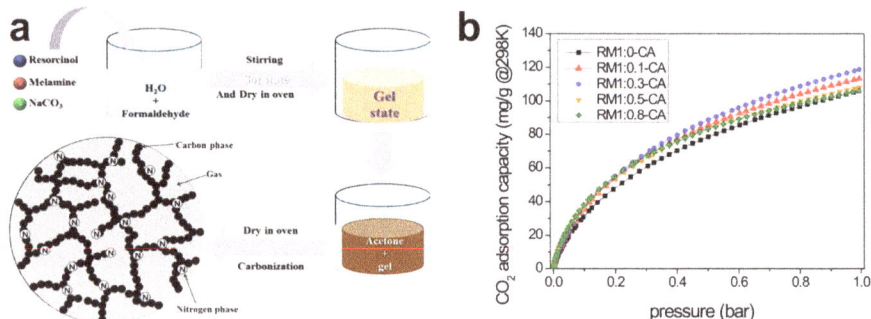

Figure 9.17. (a) Schematic diagram of the preparation of N-doped CAs. (b) CO_2 adsorption isotherms of the N-doped CAs [103], reproduced with permission © MDPI.

matrix to form a hierarchical morphology. The G/GA exhibited a large surface area (763 m^2 g^{-1}), a narrow mesopore size distribution, and hierarchical macroporous and mesoporous structures. As a result, G/GA displayed a promising adsorption performance for the removal of CO_2 (76.5 mg g^{-1} at 298 K), CH_4 (16.8 mg g^{-1} at 298 K), and H_2 (12.1 mg g^{-1} at 77 K) at 1 bar.

Jeon *et al* [103] demonstrated the effect of the nitrogen content of N-doped CAs for CO_2 capture using resorcinol and melamine as the carbon and nitrogen precursors (figure 9.17).

To evaluate the influence of the nitrogen moieties, N-doped CAs with varying resorcinol/melamine (R/M) ratios were prepared. The melamine-containing N-doped CAs had high nitrogen contents (5.54 wt%) and exhibited a high CO_2 capture capacity of 118.77 mg g^{-1} at R/M = 1:0.3. The results confirmed that the CO_2 adsorption capacity was strongly affected by the nitrogen moieties.

9.3.4 Water treatment

There is an urgent demand for feasible pollution remediation strategies to prevent deterioration of the global environment. To preserve the quality of aquatic ecosystems, the control and removal of hazardous organic compounds (oils, organic solvents, dyes, etc.) and heavy metal ions (Cr, Pb, Hg, Cu, Cd, Co, etc) are mandatory [104–106]. Their characteristics of high specific surface area, high porosity, and adjustable surface chemistry endow CAs with excellent adsorption capacity, which are widely used as attractive adsorbents for the separation of pollutants from water [107].

9.3.4.1 Oil/water separation

CAs are commonly used as selective adsorbents for oil/water separation due to their unique structural properties [108–113]. Wang *et al* [112] prepared N-rich CA (UFC foam) using commercially available poly(melamine formaldehyde) foams as the precursors. The UFC foam thus prepared possessed a highly efficient oil/water separation capability with a maximum sorption capacity of up to 158 times its own weight. Meanwhile, the UFC foam also exhibited superior recyclability due to its fire resistance and compressibility properties (figure 9.18). Chen *et al* [113] reported the production of a new CA by the carbonization of melamine foam, which exhibited a high specific surface area (268 m^2 g^{-1}), high porosity (over 99.6%), low density (5 mg cm^{-3}), and excellent absorptive properties for oil and organic solvents (148 to 411 times its own weight).

Graphene aerogels, fabricated using reduced graphene oxide (rGO) via self-assembly, have proven their capacity as superabsorbents for oil/water separation [114–117]. For example, Xu *et al* [116] developed a graphene aerogel using the hydrothermal chemical reduction and freeze-drying processes. When used as

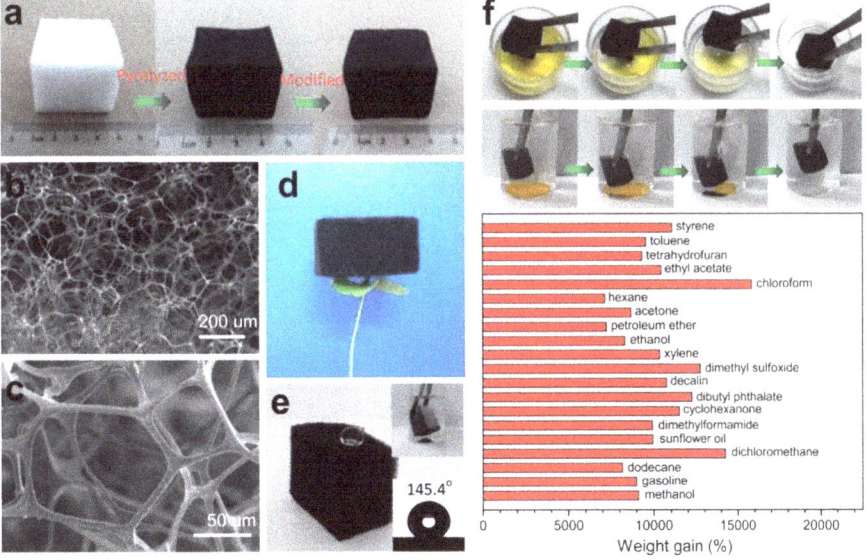

Figure 9.18. (a) Fabrication of the UFC foams. (b and c) SEM images of the UFC foams at different magnifications. (d) A 15 cm^3 UFC foam positioned on a piece of *Marsilea quadrifolia* grass. (e) Hydrophobicity of UFC foam. (f) Adsorption of organic liquids by UFC foams. Source: [112], reproduced with permission © Royal Society of Chemistry.

adsorbent for selective oil/water separation, the adsorption capacity was more than 100 times its own weight, which can be attributed to its excellent superhydrophobicity and superoleophilicity. Eom *et al* [117] reported an acrylamide (OctA)-functionalized MOF hybridized with rGO aerogel (OctA/rGA) for selective oil/water separation. Organic dye was clearly separated by the OctA membrane after mixture separation, and the OctA/rGA composites retained up to 85% of their relative adsorption capacities after ten cycles in a recyclability test (figure 9.19). CNT aerogels are also familiar to people because of their role in environmental clean-ups, especially for oil/water separation [118–120]. For example, Gui *et al* [118] prepared a magnetic CNT aerogel that exhibited a high mass sorption capacity for diesel oil of 56 g g^{-1}. Moreover, CNTs were added to the graphene aerogel to improve its mechanical strength and specific surface area.

In summary, polymer-, graphene-, and CNT-based aerogels demonstrate a fascinating capacity for selective oil/water separation, but they are not suitable for industrial application due to their high cost and complex preparation process. Biomass materials that can be converted into hydrophobic CAs are definitely a good choice.

9.3.4.2 Removal of heavy metal ions

Heavy metal ions are among the most common pollutants in the aquatic environment. Heavy metals, which have high toxicity and carcinogenicity, can remain in the environment for a long time due to their nonbiodegradability [121]. Consequently, it is urgently necessary to remove heavy metal ions from contaminated environments. Physical adsorption plays a dominant role in the removal of heavy metal ions and is

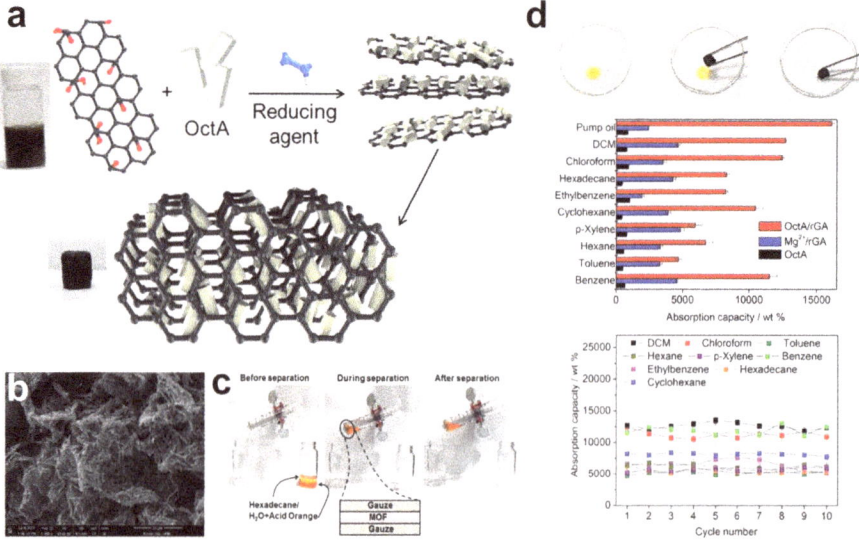

Figure 9.19. (a) Schematic illustration of the self-assembly process of rGO and OctA proposed for the synthesis of OctA/rGA. (b) SEM image of OctA/rGA. (c) Separation of oil and water by an OctA membrane. (d) Absorption performance of OctA/rGA. Source: [117], reproduced with permission © Royal Society of Chemistry.

significantly influenced by surface groups of adsorbents. In contrast to oil/water separation, adsorbents with hydrophilic groups containing N, O, S, and P exhibit high adsorption capacities for heavy metal ions [18].

CAs are frequently used in heavy metal ion adsorption due to their high specific surface area, large porosity, and controllable surface functional groups [122–126]. Meena *et al* [122] studied the adsorption of CAs for seven heavy metals (Pb, Cd, Hg, Cu, Zn, Ni, and Mn), and investigated the influences on the adsorption capacity such as pH, temperature, adsorbent dose, concentration, and contact time. The results indicated that the pH was the most influential parameter, and that the highest/lowest adsorption capacities were obtained for Cd (II) (400.8 mg g^{-1}) and Pb (II) (0.70 mg g^{-1}), respectively. The disparity between these can be attributed to the different chemical affinities and ion-exchange capacities of different heavy metal ions for the sorbent functional groups.

Over the past few years, tremendous efforts have been devoted to improving the adsorption performance of CAs by increasing the porosity and specific surface area, optimizing the surface composition, and adjusting the exposed active sites. Chen *et al* [124] designed a 3D macrostructured aerogel (3D GTs) through the self-assembly of 2D GO and 1D CNTs for high-performance pollutant removal (figure 9.20). The adsorption capacity of 3D GTs (Cd^{2+}, 235 mg g^{-1}) was higher than those of other adsorbents, which can be attributed to the synergistic effects of GO and CNTs in the microenvironment, nanosubstrate, and active sites. Kong *et al* [125] reported a N, S co-doped graphene-based aerogel for the adsorption of Cd^{2+} and organic dyes. The presence of the dyes improved the adsorption capacity of Cd^{2+}, whereas the dye capacities were not influenced by the presence of Cd^{2+}. Meanwhile, the results demonstrated that the surface diffusion was the vital factor in controlling the rate of adsorption.

Recently, a large number of biomass CAs were applied for heavy metal ion removal. Chen *et al* [126] developed cotton-derived porous carbon oxide (CDPCO) aerogels using alkaline etching methods for the removal of organic pollutants and heavy metal ions (figure 9.21). The adsorption capacity of cotton aerogels for heavy metal ions (Co(II) 71.4 mg g^{-1}, Cd(II) 40.2 mg g^{-1}, Pb(II) 111.1 mg g^{-1}, and Sr(II) 33.3 mg g^{-1}) were much higher than those of raw cottons. These results demonstrated that the adsorption capacity for heavy metal ions was strongly influenced by the functional groups on the aerogel surface.

9.3.5 Insulators

CAs also have the potential to be used for thermal insulation due to the controllability of their thermal conductivity. CAs possess small pore sizes of <100 nm and thus exhibit low thermal conductivity because of the Knudsen effect. The Knudsen effect is observed when gas molecules are confined within pores that have diameters smaller than the mean free path of the gas molecules (70 nm) at 1 bar [127–129]. However, CAs are brittle and fragile due to their low density [130]. Therefore, it is imperative to strengthen CAs for thermal insulation applications. To overcome the above issues, many attempts to strengthen the gel skeletons of CAs have been explored45. Hemberger *et al* [127] investigated the thermal conductivity of carbon aerogels with variable pore sizes (from 70 nm up to 11 000 nm). The relationship between the thermal conductivity and the

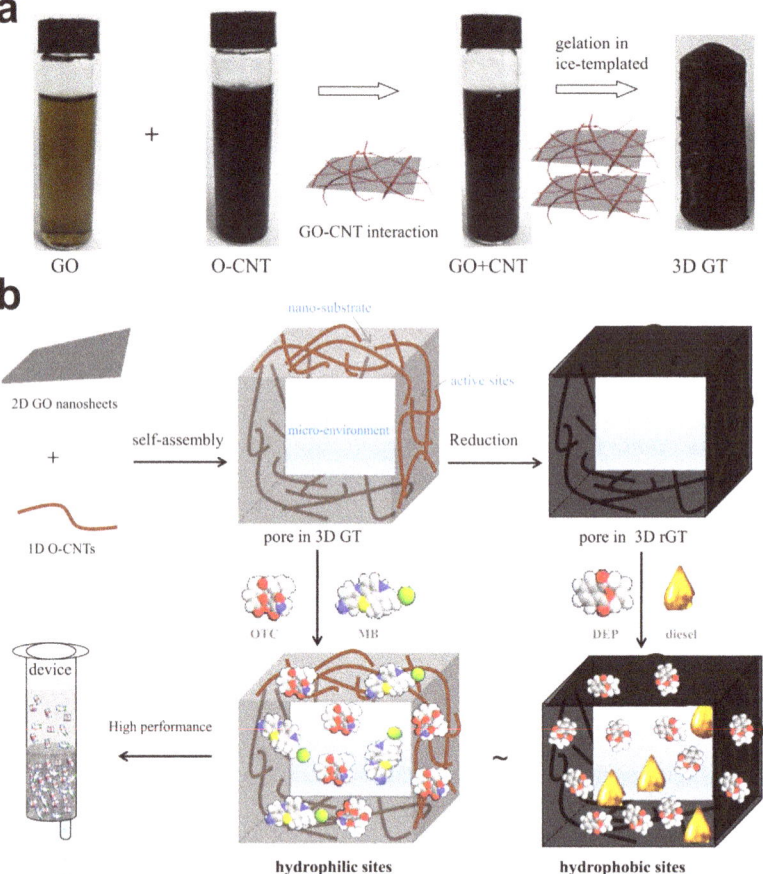

Figure 9.20. (a) A schematic of the fabrication of the 3D GT aerogel. (b) The synergistic effect of GO and CNTs on the 3D GTs' structure and adsorption performance. Source: [124], reproduced with permission © Elsevier.

pore size is displayed in figure 9.22. The actual measured thermal conductivity is about twice as high as the theoretical calculation results, which attributed to that the measured value is included in the coupling of thermal conductivity and the actual cytoskeletal particle structures of CAs. However, the theoretical results also verified that the trend of the thermal conductivity increased with the pore size.

Jia *et al* [131] reported the production of lightweight but mechanically strong carbon aerogel monoliths (CAMs) with low thermal conductivity and low density through the sol–gel polymerization of linear phenolic resin and hexamethylenetetramine (HMTA). This sol–gel polymerization reaction resulted in a rigid polymer chain at the molecular level and a robust framework at the macroscopic level, thus avoiding high capillary pressure and maintaining pore structure. Moreover, the CAMs thus prepared exhibited a hierarchical structure including macropores in the interstitial voids and micropores within the individual particles. The CAMs possessed mechanical strengths from 9 to 50 bar and thermal conductivities of 0.032–0.069 W m^{-1} K^{-1} (figure 9.23).

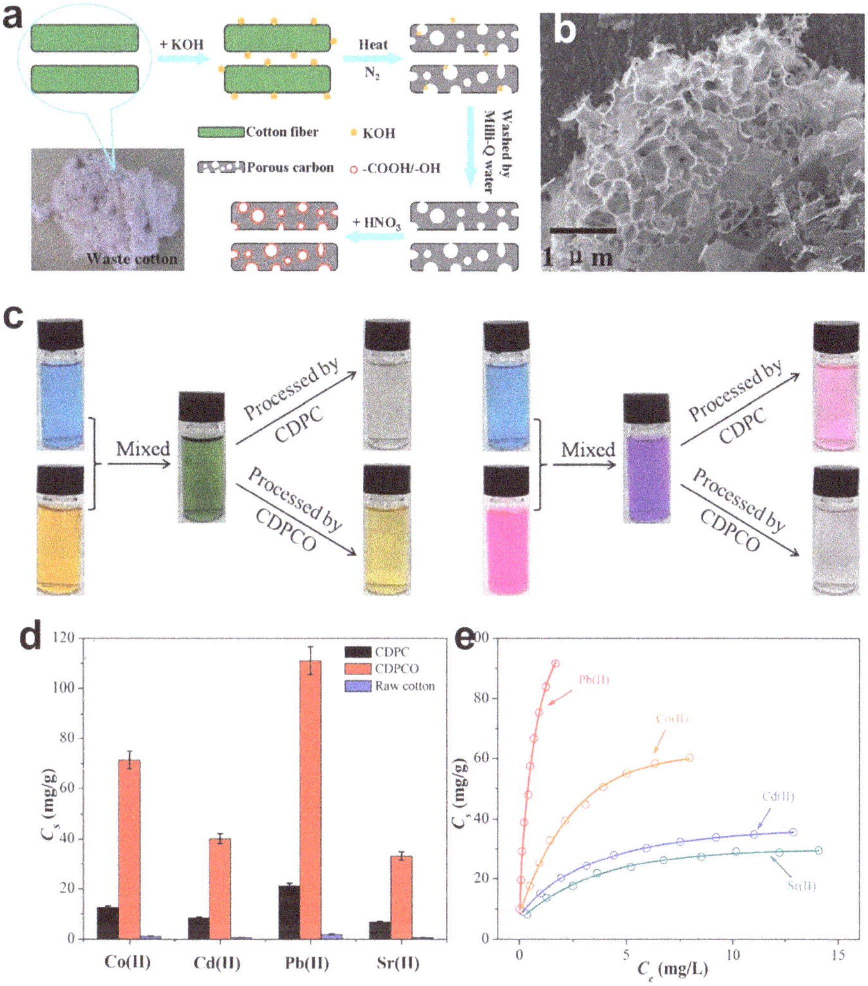

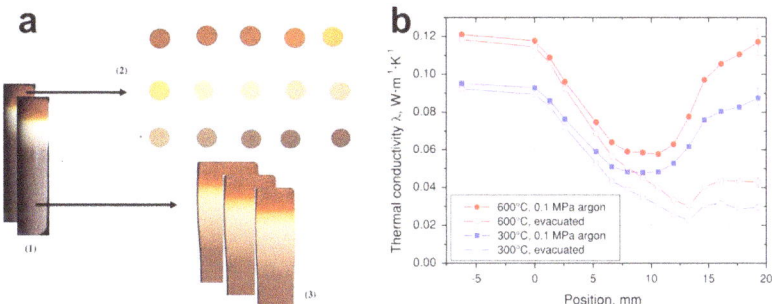

Figure 9.21. (a) Synthesis mechanism of cotton-derived porous carbon (CDPC) and CDPCO. (b) SEM image of CDPC. (c) Selective adsorption of MB from MB/MO and MB/RhB mixed solutions using CDPC and CDPCO. The adsorption of heavy metal ions using CDPCO (d) and the corresponding isotherms (e). Source: [126], reproduced with permission © Royal Society of Chemistry.

Figure 9.22. (a) Structural gradients of cylindrical RF aerogels. (b) Thermal conductivities of the graded carbon aerogel specimens at 300 °C and 600 °C . Source: [127], reproduced with permission © Springer Nature.

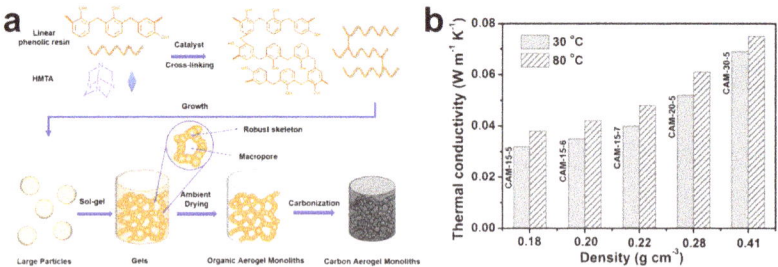

Figure 9.23. (a) Schematic illustration of the preparation of CAMs. (b) Thermal conductivities of the CAMs. Source: [131], reproduced with permission © Elsevier.

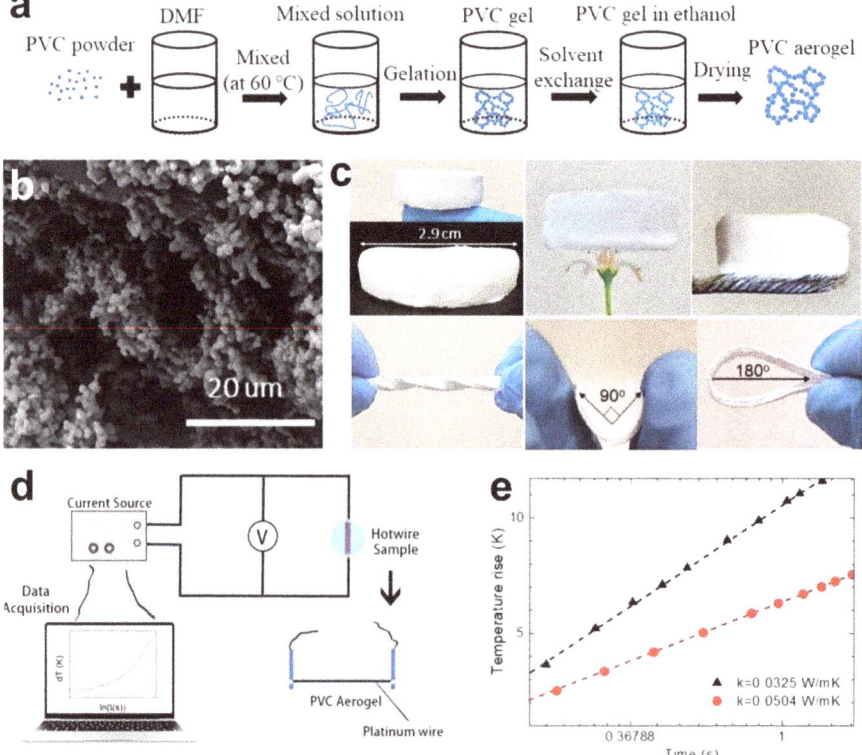

Figure 9.24. (a) Schematic illustrating the ambient synthesis process of PVC aerogels. (b) SEM image of PVC aerogel. (c) Photographs of twisting, the bending test, and an ultralight and flexible PVC aerogel balanced on a flower and a foxtail. (d) Schematic of hot-wire thermal conductivity measurement. (e) Typical experimental data (symbols) fitted to the transient heat conduction model (dashed lines). Source: [132], reproduced with permission © John Wiley and Sons.

Li *et al* [132] reported a facile ambient processing approach used to synthesize poly (vinyl chloride)-based aerogel (PVC aerogel), which had a flexible and low thermal conductivity (28 mW m^{-1} K^{-1}) without supercritical drying (figure 9.24). Thermal transport in the PVC aerogel depended on the contribution from air conduction. The heat transfer mechanism in the PVC aerogel was also investigated to maximize its insulative performance. The results indicated that smaller pore sizes and higher porosities increased the potential of the aerogel as a thermal insulator.

Men *et al* [133] demonstrated polysaccharide biomass-derived carbonaceous porous foams (PIL/cotton) produced using an ionic liquid (IL) or a poly(ionic liquid) (PIL) at the comparatively low temperature of 400 °C (figure 9.25). The IL

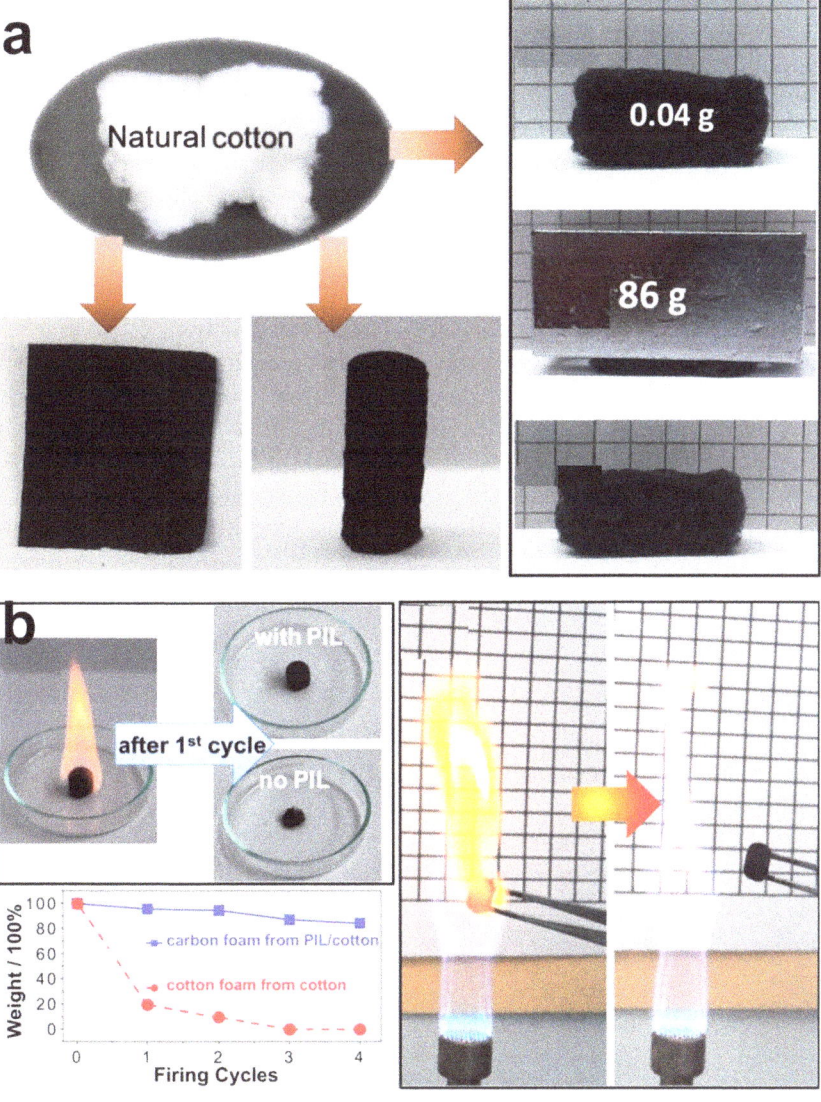

Figure 9.25. (a) Photographs of natural cotton, a carbonaceous thin film derived from a PIL/cotton composite film, a carbonaceous monolith derived from a PIL/cotton composite, and a carbonaceous cuboid cut from a carbonized PIL/cotton composite. (b) First flame retardance test and mass loss graphs of PIL/cotton foams. Source: [133], reproduced with permission © Royal Society of Chemistry.

and PIL acted as efficient carbonization/activation agents, promoting conversion and pore generation in the biomaterial. Its flame resistance was tested by burning samples which had been soaked in 500 wt% ethanol in air. The results demonstrated that the PIL/cotton lost only 4.1 wt%, while a natural cotton lost 80.3 wt%. The mass of the PIL/cotton sample remained at 85 wt% after four cycles of flame testing.

9.4 Conclusions and outlook

CAs have attracted great interest over the last few decades because of their fascinating characteristics, including their low density, abundant pore structure, high specific surface area, high electrical conductivity, good chemical stability, environmental compatibility, and controllable textural and structural features. This chapter summarized recent progress in the CA field for a wide range of applications, including energy storage, catalysis, gas storage, pollutant separation, and thermal insulation. In addition, CAs have great potential to be used for other applications, such as protective materials on nanometals, materials to shield against electromagnetic interference, and solar steam generators. Although CAs possess so many superiorities, there are numerous challenges for their practical application: (1) complicated synthesis procedures and high costs of production; (2) poor mechanical strength due to structural shrinkage and collapse caused by ambient-pressure drying; (3) structural instability; (4) limitation of selectivity in some applications. Therefore, it is imperative to develop advanced CAs to realize their practical application.

The future trends in the field of CAs may focus on the following five aspects: (1) decrease the costs of CA production by simplifying the production processes or employing low-cost precursors; (2) decrease the capillary tension caused by ambient-pressure drying; (3) increase their mechanical strength using new techniques; (4) regulate their microstructures to improve their selectivity; (5) expand the application fields.

Acknowledgments

We are grateful for the financial support of the Guangdong Basic and Applied Basic Research Foundation (Grant No. 2020A1515110705), the China Postdoctoral Science Foundation (Grant Nos. 2020M682711, 2020M682710, 2019M652882, and 2019T120725), the National Natural Science Foundation of China (Grant No. 31971614), the Guangzhou Science and Technology Funds (Grant No. 201904010078), and the State Key Laboratory of Pulp and Paper Engineering (Grant No. 2020C03).

References

[1] Pierre A C and Pajonk G M 2002 *Chem. Rev.* **102** 4243
[2] Hüsing N and Schubert U 1998 *Angew. Chem. Int. Ed.* **37** 22
[3] Maleki H 2016 *Chem. Eng. J.* **300** 98
[4] Kistler S S 1931 *Nature* **127** 741
[5] Teichner S J, Nicolaon G A, Vicarini M A and Gardes G E E 1976 *Adv. Colloid Interface Sci.* **5** 245

[6] Pekala R W 1989 *J. Mater. Sci.* **24** 3221

[7] Wang X-D, Sun D, Duan Y-Y and Hu Z-J 2013 *J. Non-Cryst. Solids* **375** 31

[8] Zu G, Shen J, Zou L, Wang W, Lian Y, Zhang Z and Du A 2013 *Chem. Mater.* **25** 4757

[9] Biedunkiewicz A, Figiel P, Krawczyk M and Gabriel-Polrolniczak U 2013 *J. Therm. Anal. Calorim.* **113** 253

[10] Chen K, Bao Z, Du A, Zhu X, Shen J, Wu G, Zhang Z and Zhou B 2012 *J. Sol–Gel Sci. Technol.* **62** 294

[11] Liu X, Li S, Mi R, Mei J, Liu L-M, Cao L, Lau W-M and Liu H 2015 *Appl. Energy* **153** 32

[12] Zestos A G and Venton B J 2017 *ECS Trans.* **80** 1497

[13] Arbizzani C, Righi S, Soavi F and Mastragostino M 2011 *Int. J. Hydrogen Energy* **36** 5038

[14] Aaltonen O and Jauhiainen O 2009 *Carbohydr. Polym* **75** 125

[15] Zhao W, Zhang H, Liu J, Xu L, Wu H, Zou M, Wang Q, He X, Li Y and Cao A 2018 *Small* **14** 1802394

[16] Lee J-H and Park S-J 2020 *Carbon* **163** 1

[17] Pekala R W, Alviso C T and LeMay J D 1990 *J. Non-Cryst. Solids* **125** 67

[18] Gan G, Li X, Fan S, Wang L, Qin M, Yin Z and Chen G 2019 *Eur. J. Inorg. Chem.* **2019** 3126

[19] Enterría M and Figueiredo J L 2016 *Carbon* **108** 79

[20] Mulik S, Sotiriou-Leventis C and Leventis N 2007 *Chem. Mater.* **19** 6138

[21] Zhang H, Feng J, Li L, Jiang Y and Feng J 2019 *RSC Adv.* **9** 5967

[22] Rasines G, Lavela P, Macías C, Zafra M C, Tirado J L, Parra J B and Ania C O 2015 *Carbon* **83** 262

[23] Seraji M M, Kianersi S, Hosseini S H, Davarpanah J and Elahi S 2018 *J. Non-Cryst. Solids* **491** 89

[24] Zhu Y, Hu H, Li W-C and Zhang X 2006 *J. Power Sources* **162** 738

[25] Wu D and Fu R 2006 *Micropor. Mesopor. Mater* **96** 115

[26] Williams J C, Nguyen B N, McCorkle L, Scheiman D, Griffin J S, Steiner S A and Meador M A B 2017 *ACS Appl. Mater. Interfaces* **9** 1801

[27] Ma Q, Cheng H, Fane A G, Wang R and Zhang H 2016 *Small* **12** 2186

[28] Chen Y, Zhang Z, Lai Y, Shi X, Li J, Chen X, Zhang K and Li J 2017 *J. Power Sources* **359** 529

[29] Salihovic M, Hüsing N, Bernardi J, Presser V and Elsaesser M S 2018 *RSC Adv.* **8** 27326

[30] He T *et al* 2019 *J. Mater. Chem.* A **7** 20840

[31] Wu X-L, Wen T, Guo H-L, Yang S, Wang X and Xu A-W 2013 *ACS Nano* **7** 3589

[32] Qiu B, Xing M and Zhang J 2014 *J. Am. Chem. Soc.* **136** 5852

[33] Fang Q and Chen B 2014 *J. Mater. Chem.* A **2** 8941

[34] Jahan M, Bao Q and Loh K P 2012 *J. Am. Chem. Soc.* **134** 6707

[35] Yin S, Zhang Y, Kong J, Zou C, Li C M, Lu X, Ma J, Boey F Y C and Chen X 2011 *ACS Nano* **5** 3831

[36] Huang X, Sun B, Su D, Zhao D and Wang G 2014 *J. Mater. Chem.* A **2** 7973

[37] Bryning M B, Milkie D E, Islam M F, Hough L A, Kikkawa J M and Yodh A G 2007 *Adv. Mater.* **19** 661

[38] Zeng S, Chen H, Wang H, Tong X, Chen M, Di J and Li Q 2017 *Small* **13** 1700518

[39] Liu T *et al* 2018 *Nanoscale* **10** 4194

[40] Han Y, Zhang X, Wu X and Lu C 2015 *ACS Sustain. Chem. Eng.* **3** 1853

[41] Bi H, Huang X, Wu X, Cao X, Tan C, Yin Z, Lu X, Sun L and Zhang H 2014 *Small* **10** 3544

[42] Job N, Théry A, Pirard R, Marien J, Kocon L, Rouzaud J-N, Béguin F and Pirard J-P 2005 *Carbon* **43** 2481

[43] White R J, Brun N, Budarin V L, Clark J H and Titirici M-M 2014 *ChemSusChem* **7** 670

[44] Maleki H, Durães L and Portugal A 2014 *J. Non-Cryst. Solids* **385** 55

[45] Hu L, He R, Lei H and Fang D 2019 *Int. J. Thermophys.* **40** 39

[46] Moreno-Castilla C and Maldonado-Hódar F J 2005 *Carbon* **43** 455

[47] Yang J, Li S, Yan L, Liu J and Wang F 2010 *Micropor. Mesopor. Mater.* **133** 134

[48] Schwan M and Ratke L 2013 *J. Mater. Chem.* A **1** 13462

[49] Barım Ş B, Bayrakçeken A, Bozbağ S E, Zhang L, Kızılel R, Aindow M and Erkey C 2017 *Micropor. Mesopor. Mater.* **245** 94

[50] Zuo L, Zhang Y, Zhang L, Miao Y-E, Fan W and Liu T 2015 *Materials* **8** 6806

[51] Kalinin S V, Kheifets L I, Mamchik A I, Knot'ko A G and Vertigel A A 1999 *J. Sol–Gel Sci. Technol.* **15** 31

[52] Maleki H, Durães L and Portugal A 2015 *J. Phys. Chem.* C **119** 7689

[53] Fischer U, Saliger R, Bock V, Petricevic R and Fricke J 1997 *J. Porous Mater.* **4** 281

[54] Kraiwattanawong K, Tamon H and Praserthdam P 2011 *Micropor. Mesopor. Mater.* **138** 8

[55] Lee K T and Oh S M 2002 *Chem. Commun.* 2722

[56] Hanzawa Y, Hatori H, Yoshizawa N and Yamada Y 2002 *Carbon* **40** 575

[57] Najeh I, Ben Mansour N, Mbarki M, Houas A, Nogier J P and El Mir L 2009 *Solid State Sci.* **11** 1747

[58] Armand M and Tarascon J M 2008 *Nature* **451** 652

[59] Canal-Rodríguez M, Ramírez-Montoya L A, Villanueva S F, Flores-López S L, Angel Menéndez J, Arenillas A and Montes-Morán M A 2019 *Carbon* **152** 704

[60] Canal-Rodríguez M, Menéndez J A, Montes-Morán M A, Martín-Gullón I, Parra J B and Arenillas A 2019 *Electrochim. Acta* **295** 693

[61] Kakunuri M and Sharma C S 2018 *J. Mater. Res.* **33** 1074

[62] Zhang Y *et al* 2019 *J. Mater. Chem.* A **7** 19668

[63] Lu C, Huang Y H, Hong J S, Wu Y J, Li J and Cheng J P 2018 *J. Colloid Interface Sci.* **524** 209

[64] Gao X-r, Xing Z, Li Z-j, Dong X-y, Ju Z-c and Guo C-l 2020 *New Carbon Mater.* **35** 486

[65] Shabangoli Y, Rahmanifar M S, El-Kady M F, Noori A, Mousavi M F and Kaner R B 2018 *Adv. Energy Mater.* **8** 1802869

[66] Chen H, Liu T, Mou J, Zhang W, Jiang Z, Liu J, Huang J and Liu M 2019 *Nano Energy* **63** 103836

[67] Wang S, Wang R, Zhao Q, Ren L, Wen J, Chang J, Fang X, Hu N and Xu C 2019 *J. Colloid Interface Sci.* **544** 37

[68] Shan H, Xiong D, Li X, Sun Y, Yan B, Li D, Lawes S, Cui Y and Sun X 2016 *Appl. Surf. Sci.* **364** 651

[69] Guan L *et al* 2020 *ACS Nano* **14** 6222

[70] Zhao J, Zhang Y-Z, Chen J, Zhang W, Yuan D, Chua R, Alshareef H N and Ma Y 2020 *Adv. Energy Mater.* **10** 2000099

[71] Liu L, Lin Z, Chane-Ching J-Y, Shao H, Taberna P-L and Simon P 2019 *Energy Storage Mater.* **19** 306

[72] Lv C, Xu W, Liu H, Zhang L, Chen S, Yang X, Xu X and Yang D 2019 *Small* **15** 1900816

[73] Long D, Chen Q, Qiao W, Zhan L, Liang X and Ling L 2009 *Chem. Commun.* **2009** 3898

[74] Wu Z-S, Yang S, Sun Y, Parvez K, Feng X and Müllen K 2012 *J. Am. Chem. Soc.* **134** 9082

[75] Fu G, Yan X, Chen Y, Xu L, Sun D, Lee J-M and Tang Y 2018 *Adv. Mater.* **30** 1704609

[76] Job N, Marie J, Lambert S, Berthon-Fabry S and Achard P 2008 *Energy Convers. Manage.* **49** 2461

[77] Cai J, Liu W and Li Z 2015 *Appl. Surf. Sci.* **358** 146

[78] Hardjono Y, Sun H, Tian H, Buckley C E and Wang S 2011 *Chem. Eng. J.* **174** 376

[79] Tang L, Jia C-t, Xue Y-c, Li L, Wang A-q, Xu G, Liu N and Wu M-h 2017 *Appl. Catalysis* B **219** 241

[80] Hu J, Chen D, Li N, Xu Q, Li H, He J and Lu J 2018 *Small* **14** 1800416

[81] Thirumalraj B, Rajkumar C, Chen S-M, Veerakumar P, Perumal P and Liu S-B 2018 *Sensors Actuators* B **257** 48

[82] Liu Q, Shen J, Yang X, Zhang T and Tang H 2018 *Appl. Catalysis* B **232** 562

[83] Hu P, Long M, Bai X, Wang C, Cai C, Fu J, Zhou B and Zhou Y 2017 *J. Hazard. Mater.* **332** 195

[84] Hu E, Wu X, Shang S, Tao X-M, Jiang S-X and Gan L 2016 *J. Clean. Prod.* **112** 4710

[85] Tian H, Wu J, Zhang W, Yang S, Li F, Qi Y, Zhou R, Qi X, Zhao L and Wang X 2017 *Chem. Eng. J.* **313** 1051

[86] Zou H, Li G, Duan L, Kou Z and Wang J 2019 *Appl. Catalysis* B **259** 118100

[87] Reddy D A, Choi J, Lee S, Ma R and Kim T K 2015 *RSC Adv.* **5** 18342

[88] Qu J, Chen D, Li N, Xu Q, Li H, He J and Lu J 2018 *Small* **14** 1800343

[89] Ji W *et al* 2017 *ACS Appl. Mater. Interfaces* **9** 15394

[90] Huang B, Kobayashi H, Yamamoto T, Matsumura S, Nishida Y, Sato K, Nagaoka K, Kawaguchi S, Kubota Y and Kitagawa H 2017 *J. Am. Chem. Soc.* **139** 4643

[91] Wang R, Lu K-Q, Zhang F, Tang Z-R and Xu Y-J 2018 *Appl. Catalysis* B **233** 11

[92] Heo Y-J and Park S-J 2018 *Green Chem.* **20** 5224

[93] Heo Y-J, Yeon S-H and Park S-J 2019 *Carbon* **143** 288

[94] Lee J-H, Heo Y-J and Park S-J 2018 *Int. J. Hydrogen Energy* **43** 22377

[95] Wang W, Motuzas J, Zhao X S and Diniz da Costa J C 2019 *ACS Appl. Mater. Interfaces* **11** 30391

[96] Alhwaige A A, Ishida H and Qutubuddin S 2016 *ACS Sustain. Chem. Eng.* **4** 1286

[97] Alhwaige A A, Agag T, Ishida H and Qutubuddin S 2013 *RSC Adv.* **3** 16011

[98] Moon C-W, Kim Y, Im S-S and Park S-j 2014 *Bull. Korean Chem. Soc.* **35** 57

[99] Li H, Li J, Thomas A and Liao Y 2019 *Adv. Funct. Mater.* **29** 1904785

[100] Oh Y, Le V-D, Maiti U N, Hwang J O, Park W J, Lim J, Lee K E, Bae Y-S, Kim Y-H and Kim S O 2015 *ACS Nano* **9** 9148

[101] Kong Y, Jiang G, Fan M, Shen X, Cui S and Russell A G 2014 *Chem. Commun.* **50** 12158

[102] Liu K-K, Jin B and Meng L-Y 2019 *Polymers* **11** 40

[103] Jeon D-H, Min B-G, Oh Jong G, Nah C and Park S-J 2015 *Carbon Lett.* **16** 57

[104] Wei G, Miao Y-E, Zhang C, Yang Z, Liu Z, Tjiu W W and Liu T 2013 *ACS Appl. Mater. Interfaces* **5** 7584

[105] Mi X, Huang G, Xie W, Wang W, Liu Y and Gao J 2012 *Carbon* **50** 4856

[106] Dietrich D, Licht C, Nuhnen A, Höfert S-P, De Laporte L and Janiak C 2019 *ACS Appl. Mater. Interfaces* **11** 19654

[107] Li K, Zhou M, Liang L, Jiang L and Wang W 2019 *J. Colloid Interface Sci.* **546** 333

[108] Zhan W, Yu S, Gao L, Wang F, Fu X, Sui G and Yang X 2018 *ACS Appl. Mater. Interfaces* **10** 1093

[109] Zhao H, Wang Q, Chen Y, Tian Q and Zhao G 2017 *Carbon* **124** 111

[110] Li L, Hu T, Sun H, Zhang J and Wang A 2017 *ACS Appl. Mater. Interfaces* **9** 18001
[111] Yang Y, Tong Z, Ngai T and Wang C 2014 *ACS Appl. Mater. Interfaces* **6** 6351
[112] Yang Y, Deng Y, Tong Z and Wang C 2014 *J. Mater. Chem.* A **2** 9994
[113] Chen S, He G, Hu H, Jin S, Zhou Y, He Y, He S, Zhao F and Hou H 2013 *Energy Environ. Sci.* **6** 2435
[114] Chen C, Li F, Zhang Y, Wang B, Fan Y, Wang X and Sun R 2018 *Chem. Eng. J.* **350** 173
[115] Liu B, Ren X, Chen L, Ma X, Chen Q, Sun Q, Zhang L, Si P and Ci L 2019 *J. Hazard. Mater.* **373** 705
[116] Xu L, Xiao G, Chen C, Li R, Mai Y, Sun G and Yan D 2015 *J. Mater. Chem.* A **3** 7498
[117] Eom S, Kang D W, Kang M, Choe J H, Kim H, Kim D W and Hong C S 2019 *Chem. Sci.* **10** 2663
[118] Gui X, Zeng Z, Lin Z, Gan Q, Xiang R, Zhu Y, Cao A and Tang Z 2013 *ACS Appl. Mater. Interfaces* **5** 5845
[119] Ma C-B, Du B and Wang E 2017 *Adv. Funct. Mater.* **27** 1604423
[120] Lv P, Tan X-W, Yu K-H, Zheng R-L, Zheng J-J and Wei W 2016 *Carbon* **99** 222
[121] Kawata K, Yokoo H, Shimazaki R and Okabe S 2007 *Environ. Sci. Technol.* **41** 3769
[122] Meena A K, Mishra G K, Rai P K, Rajagopal C and Nagar P N 2005 *J. Hazard. Mater.* **122** 161
[123] Zhang Y, Zhang D, Zhou L, Zhao Y, Chen J, Chen Z and Wang F 2018 *Chem. Eng. J.* **336** 690
[124] Shen Y, Zhu X, Zhu L and Chen B 2017 *Chem. Eng. J.* **314** 336
[125] Hou P, Xing G, Tian L, Zhang G, Wang H, Yu C, Li Y and Wu Z 2019 *Sep. Purif. Technol.* **213** 524
[126] Chen H, Wang X, Li J and Wang X 2015 *J. Mater. Chem.* A **3** 6073
[127] Hemberger F, Weis S, Reichenauer G and Ebert H-P 2009 *Int. J. Thermophys.* **30** 1357
[128] Feng J, Feng J, Jiang Y and Zhang C 2011 *Mater. Lett.* **65** 3454
[129] Wiener M, Reichenauer G, Hemberger F and Ebert H P 1826 *Int. J. Thermophys.* **2006** 27
[130] Lee Y J, Jung J C, Yi J, Baeck S-H, Yoon J R and Song I K 2010 *Curr. Appl Phys.* **10** 682
[131] Jia X, Dai B, Zhu Z, Wang J, Qiao W, Long D and Ling L 2016 *Carbon* **108** 551
[132] Li M, Qin Z, Cui Y, Yang C, Deng C, Wang Y, Kang J S, Xia H and Hu Y 2019 *Adv. Mater. Interfaces* **6** 1900314
[133] Men Y, Siebenbürger M, Qiu X, Antonietti M and Yuan J 2013 *J. Mater. Chem.* A **1** 11887

Lightning Source UK Ltd.
Milton Keynes UK
UKHW050651200123
415674UK00003B/34

9 780750 349703